Sustainable Construction: Use of Incinerator Ash

Proceedings of the International Symposium organised by the Concrete Technology Unit, University of Dundee and held at the University of Dundee, UK on 20 - 21 March 2000

Edited by

Ravindra K. Dhir
Director, Concrete Technology Unit
University of Dundee

Tom D. Dyer
Research/Teaching Fellow, Concrete Technology Unit
University of Dundee

Kevin A. Paine
Research/Teaching Fellow, Concrete Technology Unit
University of Dundee

Thomas Telford

Published by Thomas Telford Publishing, Thomas Telford Ltd, 1 Heron Quay, London E14 4JD.

URL: http://www.t-telford.co.uk

Distributors for Thomas Telford books are
USA: ASCE Press, 1801 Alexander Bell Drive, Reston, VA 20191-4400, USA
Japan: Maruzen Co. Ltd, Book Department, 3–10 Nihonbashi 2-chome, Chuo-ku, Tokyo 103
Australia: DA Books and Journals, 648 Whitehorse Road, Mitcham 3132, Victoria

First published 2000

Also available from Thomas Telford Books
- *Use of Recycled Concrete Aggregate*
- *Performance of Multi-component Binder Concrete in Coastal and Highway Environments*

A catalogue record for this book is available from the British Library

ISBN: 0 7277 2861 X

Printed and bound in Great Britain by Bookcraft (Bath)

PREFACE

That the sustainable use of materials is an absolute requirement for mankind is now accepted worldwide. Moreover, the opportunities for developing sustainable use of materials are becoming increasingly advantageous, and the recovery of maximum value from waste materials is a key concern in this process. One of the most effective approaches to value recovery is through the generation of heat and power from waste incineration. Whilst the benefits of this approach are significant, disposal of the ash is the main difficulty in justifying incineration as a wholly sustainable solution.

The construction industry can offer a number of potential applications for using incinerator ash. Such opportunities offer means of maximising the value of these wastes and allowing them to be regarded as a useful resource.

This International Workshop is the first of its kind held in the UK and is contributed to by some of the World's leading experts in the field of incinerator ash processing and recycling. The proceedings contain forty-two papers presented under four sections: (i) Generating, Processing and Handling, (ii) Use of Incinerator Bottom Ash in Fill and Highway Works, (iii) Use of Incinerator Bottom Ash in Concrete Construction, (iv) Use of Incinerator Fly Ash in Construction. Thus, they comprehensively summarise the state-of the art in this field.

The Workshop Opening Addresses were given by Mr G. Atkins OBE, Chairman of the Energy from Waste Association, and Mr W.K. Townend OBE, President of the Institute of Waste Management. The Opening Keynote paper was presented by Professor J. Péra, Institut National des Sciences Appliquées de Lyon, France. The Closing Keynote paper was given by Professor J.H. Tay, Nanyang University, Singapore.

The Editors would like to thank Dr S.B. Desai OBE, Department of the Environment, Transport and the Regions, UK; Dr ir. C. Zevenbergen, Vermeer Infrastructure Development B.V., The Netherlands; Professor R.C. Amme, University of Denver, USA; and Mr O. Hjelmar, VKI, Denmark for chairing the Workshop sessions.

The Workshop was sponsored by Aggregate Industries (UK) Limited; Ballast Pheonix; Castle Cement Limited; Energy from Waste Association; Feniks Recycling, The Netherlands; Greater Manchester Waste Limited; and Onyx Hanson Limited. All sponsors are gratefully acknowledged for their invaluable support for the Workshop.

The Workshop is the second in a series of events intended by the Concrete Technology Unit to promote sustainable concrete construction. Others in the series include the first International Symposium on "Use of Recycled Concrete Aggregate" and the performance of multi-component binder concrete in coastal and highway environments, to be held at a later date.

These proceedings have been prepared from the camera-ready manuscripts provided by the authors and therefore editing has been restricted to minor changes, where it was considered absolutely necessary.

Ravindra K Dhir
Thomas D Dyer
Kevin A Paine

Dundee
January 2000

CONTENTS

OPENING KEYNOTE PAPER

INCINERATION OF WASTES AND THE ENVIRONMENT

J Péra

J Ambroise

National Institute of Applied Sciences

France

ABSTRACT. Incineration is one of the best methods of reducing the volume and hazard of organic hazardous wastes. Through incineration, more than 90 percent of the volume of the original waste is typically reduced. Matter is neither created nor destroyed in an incinerator. Individual constituents are converted to more stable forms that are less hazardous to human health and the environment. In addition, the stable forms are more easily managed in an environmentally sound manner. This paper presents the impact of coal or waste incineration on the environment. The emissions are kept low through expensive flue-gas cleaning systems that remove harmful components, but such systems can generate other types of wastes which are difficult to manage. Emerging technologies such as gasification, pyrolysis and thermolysis, which reduce the amount of residual waste and the flue gas volume can represent good alternatives to incineration.

Keywords: Cement, Coal, Combustion, Desulfurization, Dioxin, Emission, Gasification, Heavy metals, Incineration, Pyrolysis, Thermolysis, Waste.

Professor Jean Péra, is Director of the Inorganic Materials Unit, National Institute of Applied Sciences, Lyon, France. He specialises in the development of new binders for the concrete industry. His interests also include the use of different wastes in cement and concrete. He has published widely and is member of different scientific societies such as ACI, RILEM, Materials Research Society and American Ceramic Society. He serves as expert for the Fifth (EC) Framework Programme - "Promoting competitive and sustainable growth".

Jean Ambroise, is Assistant Professor in the Inorganic Materials Unit, National Institute of Applied Sciences, Lyon, France. His main research interests include the design and the production of innovative building materials as well as the management of industrial wastes.

INTRODUCTION

Environmental issues are of major importance in most combustion applications because low rank fuels and essentially all waste materials contain compounds which, when improperly oxidized, and not collected, are air pollutants.

Combustion of most coals produces both NO_x and SO_x emissions which can ultimately lead to acid rain; hazardous emissions may also be produced by the incomplete destruction of originally toxic material during hazardous waste incineration or by the formation of new toxic compounds during the incineration of municipal or hospital wastes. Fortunately, cost effective NO_x control can be achieved by the application of combustion modification technologies (e.g., burner modifications, staged combustion, or reburning) but emission levels below 250 ppm are difficult to achieve with pulverized coal firing. SO_2 emissions can be controlled with either wet or dry scrubbing systems; however, both are relatively expensive, require a large initial capital investment, and may be unsuitable for retrofit applications with space and/or lifetime constraints. Increasingly stringent incinerator performance standards, and the need to handle larger quantities of different types of waste streams (due to the elimination of landfill disposal sites) are demanding the development of new, efficient waste incineration technology.

ENVIRONMENTAL AND HEALTH IMPACTS OF ENERGY PRODUCTION.

Impacts of energy production on the environment and human health depend on the form of the energy source concerned. If they are solid fuels (mainly coal), oil, gas, nuclear or renewable energies (such as hydropower, solar, biomass, wind and wave or geothermal energies), the impacts differ markedly (Table I). They also involve the stage of the energy cycle concerned, and the technologies and quantities used.

Major environmental consequences of energy production based on fossil fuels include [1]:

- carbon dioxide build-up in the atmosphere, possibly triggering global warming;
- acid precipitation, with far-reaching impacts on vegetation, soil and water as well as on corrosion of buildings and materials;
- effects on land and water quality as the result of the extraction and processing of coal, in terms of trace metals (mercury, lead, cadmium, etc.) released to the terrestrial and aquatic environments;
- environmental effects of solid waste disposal and,
- health problems arising from the combustion of oil and coal, releasing car dioxide, sulfur dioxide, nitrogen oxides and other constituents.

CLEAN COAL TECHNOLOGIES

Clean Coal Technologies (CCTs) are defined [2] as "technologies designed to enhance both the efficiency and the environmental acceptability of coal extraction, preparation and use". These technologies reduce emissions and waste, and increase the amount of energy gained from each tonne of coal.

Table 1 Survey of Environmental and Health Effects of Energy Production (Source: Adapted from OECD, State of the Environment, 1991)

ENERGY SOURCES	ENVIRONMENTAL AND HEALTH IMPACTS				
	Air	Land and soils	Water	Biota	Humans
Coal	Emissions of SO_x, NO_x, and particulates	Land subsidence; solid waste and ash disposal	Acid mine drainage; Storage heaps pollution	Exploitation of natural habitats	Visual intrusions; transport noise; occupational risks
Petroleum products	Emissions of H_2S, SO_x, NO_x, CO, CO_2, HC and NH_3	Land use for facilities and pipes	Oil spills; water availability	Natural habitat disturbances	Blow-outs; explosions; odour
Gas	Emissions of H_2S, methane and trace metals	Land use for facilities and pipes	Liquid residual disposal	Pipeline impacts on wildlife	Blow-outs; leaks; spills; explosions
Electricity generation - Fossil fuels	Long-range transports of SO_x, NO_x and CO_2	Solid waste; ash and slag disposal	Water availability; thermal releases	Secondary effects on biota	Visual impacts; noise
- Nuclear	Radioactive dust and gaseous effluents emissions	Land subsidence; radioactive waste	Mine drainage; thermal releases; radionuclide emissions	Secondary effects on biota	Visual impacts; radiation and occupational risks
- Hydropower		Landslide risks; land irreversibly flooded	Hydrologic cycle and water quality disturbances	Ecosystem changes; fish migration influence	Visual impacts; dam rupture risks
- Renewable energy resources biomass, geothermal, wind and solar origin	Combustion emissions	Land use demand	Water pollution through biomass conversion and geothermal exploration	Ecosystem disruption	Visual impacts; noise; occupational risks

Stack Gas Treatment Applied to Gaseous Emissions from Pulverized Fuel Combustion (PFC)

PFC is the most widely used method for burning coal for power generation. In PFC, coal is milled to a powder and blown into the boiler with air. As a powder, the coal has a large surface area and is easily combusted in burners. This provides the heat which is used to produce superheated steam to drive turbines and hence generate electricity. At present, nearly all of the world's coal-fired electricity is produced using PFC systems.

Emissions from PFC can be reduced by post-combustion CCTs. Electrostatic precipitators and/or fabric filters can remove more than 99 % of fly ash from flue gases. Flue gas desulfurization (FGD) methods can remove 90-97% of SO_x from flue gases. Emissions of NO_x are reduced by using low-NO_x burners, which modify emissions by up to 40 % and reburning techniques. Together these modify the combustion process to reduce NO_x emissions by up to 70 %, and are being widely adopted as they can be installed into existing plants. Selective catalytic NO_x reduction, a post-combustion technique, can achieve reductions of 80-90 %.

Industry has continuously striven to increase efficiencies of such conventional plants; for example, the average thermal efficiency of US power plants has increased from 5 % in 1990 to around 35 % currently. New conventional PF power plants achieve above 40 % efficiency. This results in corresponding reduction in CO_2 emissions as less fuel is used per unit of electricity generated.

Fluidized Bed Combustion (FBC)

Fluidized bed combustion is a method of burning coal in a bed of heated particles suspended in a gas flow. At sufficient flow rates, the bed acts as a fluid resulting in rapid mixing of the particles. Coal is added to the bed and the continuous mixing encourages complete combustion and a lower temperature than that of PFC. The advantages of fluidized beds are they produce less NO_x in the outlet gas, because of lower combustion temperatures, and they produce less SO_x when limestone is continuously added with the coal. They can also use a wider range of fuels than PFC. Atmospheric-pressure fluidized beds are commercially available now as two types, bubbling-bed (known as Atmospheric Fluidized Bed Combustion - AFBCs) and circulating-bed (CFBCs). The efficiency of most fluidized beds used for power generation is similar to that of conventional plant. However, use of this technology has been stimulated by its better environmental performance. Pressurised fluidized beds, which can achieve efficiencies of 45 %, are at advanced stages of demonstration.

Gasification and IGCC

An alternative to coal combustion is coal gasification. When coal is brought into contact with steam and oxygen, thermochemical reactions produce a fuel gas, largely carbon monoxide and hydrogen, which when combusted can be used to power gas turbines. Integrated Coal Gasification Combined Cycle (IGCC) power generating systems are presently being developed and demonstrated. These systems give increased efficiencies by using waste heat from the product gas to produce steam to drive a steam turbine, in addition to a gas turbine. Existing commercial systems achieve efficiencies close to 45 %. IGCC systems additionally produce less solid waste and lower emissions of SO_x, NO_x, and CO_2. Over 99 % of the sulfur present in coal can be recovered for sale as chemically pure sulfur.

Hybrid Systems

Hybrid combined cycles are also under development. These combine the best features of both gasification and combustion technologies, using coal in a two-stage process. The first stage gasifies the majority of the coal and runs as gas turbine, the second combusts the residual "char" to produce steam. Efficiencies greater than 50 % are possible.
In addition to these CCTs, a development which can apply to all of the generating systems is the co-firing with coal of biomass or wastes. Benefits can include reductions in CO_2, NO_x, and SO_x emissions relative to coal-only fired plants.

RESIDUES PRODUCED BY CLEAN COAL TECHNOLOGIES

In the recent past, legislation to control air pollution has been the focus of regulatory activity associated with using coal. Some measures to control atmospheric pollution, for example the installation of flue gas desulfurization systems, have resulted in an increase in residue production. An integrated, holistic approach to pollution control requires that reductions in emissions do not result in an increase in pollution from solid wastes. Desulfurization wastes may pose an environmental problem in the near future, as they form in large amounts, their quality may vary and possibilities for their utilization are limited [3].
More than 2,000 FGD plants are already in operation throughout the world [4]. The wet limestone (or lime)/gypsum is now a mainstay FGD process in the world as shown in Figure 1.

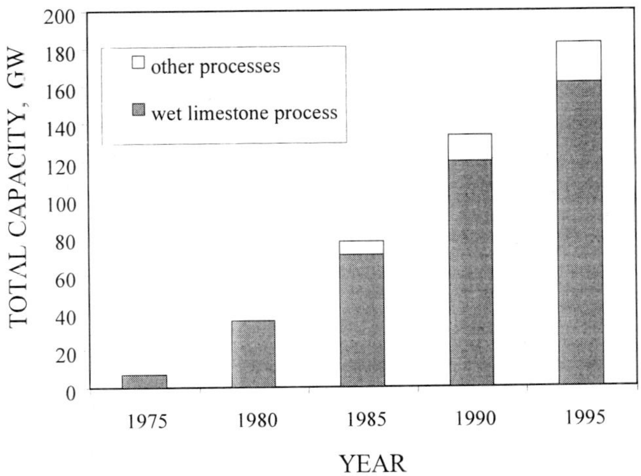

Figure 1 Trend in FGD Unit Installation Throughout the World [5]

The process uses limestone as an absorbent, and limestone is abundantly available all over the world, easily handleable and cheap [6]. Utilization of limestone has popularized the process. Recovery of by-product gypsum is considered another reason for the spread. However, it is difficult to imagine that the by-product gypsum is commercially valuable, because gypsum is not generally in demand except in countries such as Japan [3]. It is reported that the annual amount of the by-product gypsum produced by the limestone FGD process will be 5 million tonnes greater than the total demand of 22 million tonnes in the USA in the near future [3].

Limestone ($CaCO_3$) reacts with the flue gas in the slurry scrubber to produce $CaSO_3$. Because most flue gases contain oxygen, the solids will also contain variable quantities of $CaSO_4$, depending on the ratio of oxygen to SO_2. Generally in the USA and Eastern Europe, the $CaSO_3/CaSO_4$ or $CaSO_4$ solids are disposed of in ponds or landfills [7]. Conversion of the $CaSO_3/CaSO_4$ sludge to gypsum ($CaSO_4.2H_2O$) by air oxidation has also been used to make the sludge more easily dewatered and suitable for landfill.

The chemistry of the desulfurization process is complex and can be represented as a three-stage process comprising [3]:

Absorption : $SO_2 + H_2O \rightarrow H_2SO_3$

Neutralization : $H_2SO_3 + CaCO_3 \rightarrow CaSO_3 + CO_2 + H_2O$

Oxidation : $CaSO_3 + \frac{1}{2}O_2 + 2H_2O \rightarrow CaSO_4.2H_2O$

The above-mentioned reactions are summarized as follows:

$$SO_2 + 2H_2O + CaCO_3 + \frac{1}{2}O_2 \rightarrow CaSO_4.2H_2O + CO_2.$$

As is seen in the above chemical equation, the limestone, oxygen and water react with sulfur dioxide to produce gypsum and carbon dioxide. It is estimated that 11.2 million tonnes of CO_2 gas is annually discharged into the atmosphere from the wet limestone-gypsum FGD plants throughout the world, and it is clear that the CO_2 emission increases in proportion to the number of such FGD plants.

Solid waste is discharged from the wet limestone-gypsum FGD process, and this is the most significant problem. A typical 1,000 MW power plant with high-sulfur coal would produce 760,000 tonnes per year of wet $CaSO_3/CaSO_4$ sludge plus an additional 340,000 tonnes per year of dry fly ash, and this material typically contains 50 % water and is thixotropic, so it is not suitable for landfill [7]. Gypsum slurry from the scrubber contains dissolved solids and potentially hazardous elements (Pb^{2+}, As^{3+}, Cr^{6+}, Cd^{2+}), ... from the fly ash [3]. Potential problems may arise with the disposal of some residues due to the high contents of CaO and $CaSO_3$ or $CaSO_4$ in these residues and the alkalinity of any leachate [4].

From this example, it is pointed out that processes that reduce a pollutant but emit another pollutant should be avoided. It is important to introduce the concept of total environmental quality into the selection philosophy of FGD [3].

Fluidized bed combustion (FCB) is one of the direct coal-burning technologies and enables desulfurization to occur in the furnace. FCB has recently received increased attention. The waste solid is left as an environmental problem in a FBC plant. Coal ash does not fuse at the relatively low combustion temperature of around 850°C used in most FCB plants. BFBC and CFBC residues from hard coal contain up to 50 % illite, together with quartz, hematite, and magnetite. Additional crystalline phases appear in the residues due to limestone or dolomite added as sorbent: lime (CaO), anhydrite ($CaSO_4$) and calcium carbonate ($CaCO_3$) [8]. A significant quantity of carbonaceous material may be present in some FBC residues (up to 30 % unburned carbon). Such types of residues are less easy to manage than glassy fly ash produced by PFC.

INCINERATION OF WASTES

Incineration can be defined as the thermal processing of solid waste by chemical oxidation with stoichiometric or excess amounts of air. The types of wastes range from solid household rubbish and infectious hospital waste through to toxic solid and liquid chemical wastes. End products include hot incineration gases, composed primarily of nitrogen, carbon dioxide, water vapor (fuel gas) and non-combustible residue (ash). Energy can be recovered by heat exchange from the hot incineration gases. Burning of solid waste can fulfil up to four distinctive objectives:

1. Volume reduction: volume reduction is, on average 90 %; weight reduction is about 70%.

2. Stabilization of waste: incinerator output (ash) is more inert than incinerator input (due to oxidation of organic compound).

3. Recovery of energy from waste: energy recovered is used to generate steam for use in on-site electricity generation or export to local factories.

4. Sanitarization of waste: it is of primary importance in incineration of clinical waste. Incineration of Municipal Solid Waste (MSW) will ensure destruction of pathogens prior to final disposal or landfill.

History of Incineration

The history of incineration is summarized in Table 2.

Types of Incinerators

Municipal solid waste (MSW) incinerators

Municipal solid waste (MSW) incinerators burn what is mostly considered to be non-hazardous waste. MSW generally consists of paper, garden wastes, food wastes, metals, rubber, textiles, wood, plastics and glass. However, it is becoming increasingly hazardous with the rise in use of disposable inherently toxic products such as consumer batteries and PVC plastics. Municipal waste incinerators are also known as Mass-Burn Incinerators (Fig. 2), Waste-to-Energy plants or Resource Recovery facilities, since some of the heat produced is used to generate steam or electricity.

Table 2 History of Incineration (Source: Internet)

DATE	COUNTRY	COMMENTS
1870	Paddington, UK	First incinerator for steam (energy) production using waste. It did not meet energy targets and was closed. Local communities opposed smell and smoke.
1876-1908	Europe	Over 210 incineration plants constructed across Europe including England, Denmark, Sweden, Belgium, Switzerland, Czech Republic and Poland. Most of these incinerators were closed because it was more profitable to burn gas, oil and coal.
1885-1908	USA	Construction of 180 MSW incinerators. By 1909, 102 of them were closed due to design flaws.
1883	Budapest, Hungary	First waste sorting and recycling plant in Europe
1900	Vienna, Austria Stuttgart, Germany Paris, France	First pyrolysis plants (anaerobic decomposition) built to produce methane for steam generation. Projects failed to meet energy production and waste reduction targets.
1920-1950	Worldwide	Introduction of new incinerator designs, raised combustion temperatures, dispersed emissions further through taller chimneys. Construction of 700 incinerators in USA, many hundreds in Europe.
1950's	Worlwide	Waste becomes increasingly toxic requiring more treatment. Cheaper options of dumping waste at land and sea gain popularity over incineration.
1970's	Europe	Resurgence of incineration building due to higher volume and increasingly combustible make-up of waste. The combustible content of MSW increased with rise in use of packaging and disposable products.
1970-1979	USA	Introduction of Clean Air Act (1970). All but 64 of 364 incinerators operating in 1964 closed.

Types of Incinerators

Municipal solid waste (MSW) incinerators

Municipal solid waste (MSW) incinerators burn what is mostly considered to be non-hazardous waste. MSW generally consists of paper, garden wastes, food wastes, metals, rubber, textiles, wood, plastics and glass. However, it is becoming increasingly hazardous with the rise in use of disposable inherently toxic products such as consumer batteries and

PVC plastics. Municipal waste incinerators are also known as Mass-Burn Incinerators (Fig. 2), Waste-to-Energy plants or Resource Recovery facilities, since some of the heat produced is used to generate steam or electricity.

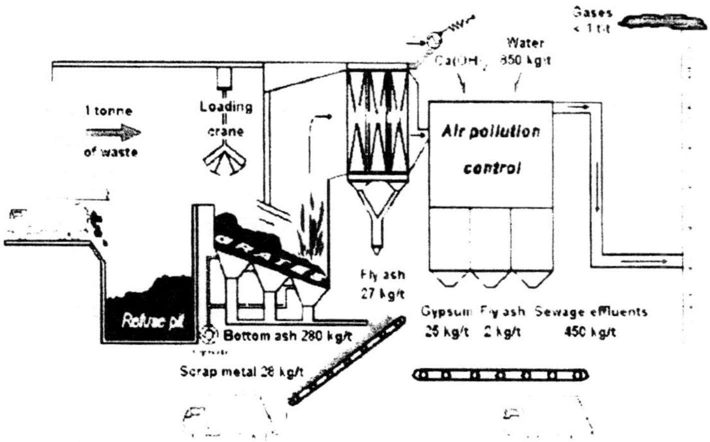

Figure 2 Typical Mass Burn Incinerator [9]

Hazardous waste (HZW) incinerators

Hazardous waste (HZW) incinerators burn toxic industrial waste. These include sewage sludge, used oils, resins, plastics, pharmaceuticals, paints, dyes, solvents, pesticides and polychlorinated biphenyls (PCBs). Most HZW incinerators use rotary kiln technology fired at very high temperatures. Due to the hazardous nature of the by-products of incineration additional 'pollution control' devices are essential.

Hospital waste (HW) incinerators

Infectious waste from hospitals and clinics is classified as hospital waste (HW). It includes used medical instruments, chemotherapy agents, anatomical parts and dressings. Since HW incinerators burn a high percentage of PVC plastic and are often equipped with no or little pollution control equipment, these incinerators can be a large source of dioxins.

Studies have shown that, in fact, only about 10 % of HW streams actually contain infectious material such as disposable syringes, gloves and blood bags. The remainder is made up of a mixture of HZW and MSW as below:

- Toxic chemicals from medications, disinfectants and cleaning agents;
- Metals and PVC plastics;
- Paper, cardboard and textile (often chlorine-bleached) products;
- Glass;
- Organic food and packaging waste.

To avoid infection, HW incinerators require special supervision and equipment. The post-combustion ashes, dusts and filters have to be stored in special landfills as they are highly toxic. Pollutants resulting from incineration, including waste water and gases, are directly or indirectly released into the environment.

Environmental Impacts of Waste Incineration

Contrary to some claims, incinerators do not break waste down into water and carbon dioxide. The waste is simply transformed from primary waste to secondary wastes-ashes and gases. Given the mixed and toxic composition of common waste streams, residual materials of incineration include heavy metals and organochlorine compounds, contained in the bottom ash, fly ash, gases and products of incomplete combustion (PICs). Ever under ideal conditions incinerators emit toxic substances.

Toxic Emissions [10]

Hydrogen chloride

Most of the chlorine in the waste stream is converted into hydrogen chloride, a strong acid gas which at high temperatures will attack most metals it meets. Most of the hydrogen chloride can be removed with alkaline scrubbing devices before the flue gases leave the stack, but not necessarily before this acid gas has damaged some of the materials from which the incinerator is built. Furnace linings, ductwork and boiler tubes need frequent and costly attention.

Nitric oxide

At the high temperatures of combustion the nitrogen and oxygen in the air combine to form nitric oxide (NO). Because this gas is neutral, it cannot be removed by scrubbers using alkaline chemicals, such as lime. Systems involving the injection of ammonia or urea can convert some of the nitric oxide back into nitrogen, but these high-energy reagents are expensive (they are normally used as fertilizers) and the removal of the nitric oxide is only about 60 % effective [10]. Any nitric oxide that is not removed is later converted by sunlight into nitrogen dioxide (NO_2) which contributes to photochemical smog and acid rain.

Toxic metals

At the temperatures of combustion many of the toxic metals such as lead, cadmium, arsenic, mercury and chromium are liberated from otherwise fairly stable matrices like plastics. Furthermore, they are liberated in the form of tiny particles or gases, which if they escape from the stack, vastly increase the potential surface area of contact between themselves and the environment. They also penetrate deep in human lungs, where they are rapidly exchanged with the bloodstream. The traditional method of removing metals from emissions is via particulate control devices such as electrostatic precipitators or baghouses (fabric filters). The former, while being very robust, are less efficient at removing the tiniest particles of concern. The latter are more efficient but suffer from breakage and blockage and need careful maintenance.

A particularly problematic metal is mercury [11]. A gram of mercury can contaminate a 10-hectare lake. Incineration releases mercury in two gaseous forms: elementary mercury and mercuric chloride. Mercury combines with chlorine from the waste stream to form a higly volatile species of mercuric chloride. This form of mercury trends to bioconcentrate and bioaccumulate in aquatic food chains at a rapid rate, through a process called methylation. Mercuric chloride is soluble in water and can be washed out of the atmosphere onto land by heavy rains. Elemental mercury remains in the atmosphere for longer periods. Mercury released in gaseous form is difficult to contain and control.

A three year study by the Swedish National Energy Administration and EPA (Internet source) estimated that waste incineration is responsible for 55 % of annual mercury emissions. Scientific data on mercury points to incinerators as the second largest source of mercury emissions, surpassed only by coal burning plants. Many modern incinerators now employ activated carbon to absorb the mercury. Activated carbon improves mercury removal efficiencies to the high 90 % range [12].

Dioxins and furans

Dioxins $C_{12}H_4Cl_4O_2$ are a group of compounds formed from halogenated hydrocarbons in the presence of oxygen and at relatively high temperatures. Kees Olie et al [13], in The Netherlands, identified the chemical 2, 3, 7, 8 - Tetra Chlorinated Dibenzo para-Dioxin (TCDD) in the emissions from trash incinerators. They, and subsequent workers, also found many other members of the dioxin family (there are 75 poly chorinated dibenzo para dioxins, or PCDDs) and members of the furan family (there are 135 poly chlorinated dibenzo furans, or PCDFs) in these emissions. The major response to this discovery from consultants representing the incinerator industry was to claim that as long as the incinerator was operated at a high temperature all the dioxins and furans would be destroyed [14]. However, these claims were subsequently found to be based on fraudulent manipulation of the data [15].

In 1985, the reason why high temperatures alone could not solve the dioxin problem was revealed at the International Symposium on Dioxin held in Bayreuth, Germany. Two groups showed that dioxins could be reformed after the flue gases left the combustion chamber [16, 17]. It is now well established that if the flue gases from an incinerator are passed through air pollution control devices operating at temperatures in the range 200-400°C, more than a hundred fold increase in dioxin and furan formation can take place [18]. A strategy that would essentially minimize post combustion formation of dioxin would require the quenching of the flue gases immediately after they emerge from the combustion chamber. However, this strategy conflicts with the aim of generating electricity, because this requires the flue gases to go through boilers to generate steam to drive turbines, thus delaying the moment when flue gas quenching occurs.

Without the immediate quenching system, the ash collected in the scrubbing devices will be contaminated with dioxins and furans.

While some persons have argued that modern incinerators are net destroyers of dioxins and furans [19] this argument does not hold if more appropriate dioxin levels in the incoming waste are assumed and if the dioxins in the fly ash and the bottom ash are included [20]. A hundred times more dioxin may leave the facility on the fly ash, than from the air emissions. However, until recently, regulatory agencies, particularly the US EPA, have turned a blind eye to the dioxins and furans left on the fly ash, even though in some cases the combined ash (a combination of bottom ash and fly ash) is being used as daily cover in some US landfills. In stark contrast, in Japan, as a result of growing concern about the dioxin problem there, the government announced in 1997 that they were limiting the total dioxine emissions (i.e. emissons plus fly ash plus bottom ash) to 5 micrograms of dioxin International Toxic Equivalents (I-TEQ) per tonne of trash burned. According to presentations made at Dioxin'97 in Indianapolis, this will almost certainly require the fly ash from Japanese incinerators to be vitrified, which will still further escalate the costs of incineration [21, 22].

Any dioxin released from an incinerator is readily captured by grazing animals and fish [23, 24]. In 1986, 16 dairy farmers downwind of a huge incinerator in Rotterdam, were told not to sell their milk, because it contained three times higher dioxin levels than anywhere else in The Netherlands. This situation continued until 1995 by which time the incinerator has be retrofitted. A 2,000 tonnes per day trash incinerator built in Norfolk, Virginia, USA, in 1988, was found in 1994, to be putting out more dioxin (approximately 2,000 grams of toxic equivalents per year) than the combined emissions from all the traffic, incinerators, industry and all other sources in Sweden, Germany and The Netherlands added together [25].

The dioxin problem can only be solved by phasing out the use of chlorinated plastics and the industrial use of chlorine [10].

Ash disposal

There are two kinds of ash generated by an incinerator: the bottom ash which falls through the grate system in the furnace (about 90 % of the ash), and the fly ash, which is the very fine material which is collected in the boilers, the heat exchangers and the air pollution control devices. As far as toxic metals are concerned, it is a chemical truism to state that the better the air pollution control the more toxic the fly ash becomes.

In some countries (Germany, France), the fly ash is assumed to be a highly toxic material and is automatically sent to hazardous waste containment facilities. In Japan, current regulations will probably force the vitrification of the fly ash. However, in other countries, the toxicity of the fly ash is obscured by three things:

a) the mixing of the fly ash with the bottom ash before testing and disposal;

b) not testing for the absolute levels of toxics like metals and dioxins in the ash, but rather only looking at what dissolves out the ash during a leaching test;

c) the interference of the lime present in the ash with some of these leaching tests [26].

Further, in some countries the ash is being handled and disposed of in a highly unsatisfactory fashion from an environmental point of view. For example, in The Netherlands, as of 1994, 35 % of the fly ash was going into asphalt [27]. In the USA, combined ash has gone directly to municipal landfills and mixed with trash containing organic material. In many instances it is used for landfill cover. Elsewhere, the fly ash has been used to make concrete, with no warning on the label product that it contains toxic metals or dioxins.

Products of Incomplete Combustion (PICs)

PCIs are formed during routine incinerator operations as well as during upsets and malfunctions, efficiency depends on the type of waste and on the temperature, length of incineration time and amount of oxygen in the furnace. Some PICs can be more toxic and more difficult to destroy than the original wastes. PICs are emitted from incinerators in the waste water, in ash residues and as emissions to air. The buring of chlorinated plastic, such as PVC, in MSW causes organochlorine emissions. An average sized incinerator produce 265 to 2,230 tonnes of PIC's a year, depending on waste and operating conditions.

ENVIRONMENTAL CONCERNS RAISED BY COMBUSTION OF WASTES IN CEMENT KILNS

Annually, the energy equivalent of about 25 million tonnes of coal is required by CEMBUREAU members to service the demand for cement in Europe [28]. This is a significant use of non-renewable primary fossil fuel, and therefore the industry is committed to seeking out more energy efficient ways of producing cement as well as alternative, more sustainable energy sources.

The use of alternative fuels is a well established technology in most of the European and US cement industry. In 1995, about 10 % of the thermal energy consumption in the European cement industry originated from alternative fuels. This is equivalent to 2.5 million tonnes of coal. The proportion is gradually increasing and figures above 50 % are already achieved in certain regions.

Waste materials which the cement industry has used as alternative fuels include used tyres, rubber, paper waste, waste oils, waste wood, paper sludge, sewage sludge, plastics and spent solvents.

The use of waste as alternative fuels in the cement industry has numerous environmental benefits such as:

- reduction of the use of non renewable fossil fuels such as coal,
- contribution towards a lowering of emissions such as greenhouse gases,
- maximization of the recovery of energy from waste: all the energy is used directly in the kiln for clinker production,
- maximization of the recovery of the non-combustible part of the waste and elimination of the need for disposal of slag or ash, as the inorganic part substitutes raw material in the cement.

Normal operation of cement kilns provides combustion conditions which are more than adequate for the destruction of even the most difficult to destroy organic substances. This is primarily due to the very high temperatures of the kiln (2,000°C in the combustion gas from the main burners and 1,100°C in the gas from the burners in the precalciner). The gas residence time at high temperature in the rotary kiln is of the order of 5 to 10 seconds and in the precalciner more than 3 seconds.

Because a cement kiln is a large manufacturing unit operating as a continuous process and with a high heat capacity and thermal inertia, a significant change in kiln temperature is not possible. The kiln can readily switch back to conventional fuel. The cement kiln therefore offers an intrinsically safe thermal environment for the use of alternative fuels.

Directive 94/97/EC regulates at EU level the use of hazardous waste as an alternative fuel in cement kilns. It stipules an emission limit of 0.1 nanogram of dioxins per cubic metre of gas emitted. Cement kilns can meet this emission limit: the values reported by CEMBUREAU are in the range of 0.001 to 0.5 ng/Nm^3.

Extensive studies investigating the behavior of metals in cement kilns have shown that the vast majority are retained in the clinker; i.e., antimony, arsenic, barium, beryllium, cadmium, chromium, copper, lead, nickel, selenium, vanadium and zinc. Extremely volatile metals such as mercury and thallium are not incorporated into the clinker to the same degree and these metals are, therefore, carefully controlled in the alternative fuel.

Cement production operations generate large quantities of cement dust (CKD). Generally, dry process kilns emit more CKD than wet process kilns. Cement kiln dust can be either disposed or some of it may be recycled into the cement making process.

Around half of the CKD collected from air pollution control equipment can be recycled back into the waste feed. However, in waste burning kilns, cycling of volatile metals, such as lead, may pose certain problems as a result of the recycling of CKD. CKD that is not recycled is generally stored or disposed on-site in piles unless it is used for construction, agriculture, transportation or other product purposes.

The US EPA investigated and evaluated risks to human health and the environment both from existing on-site management practices and a variety of off-site uses, and finally released a Draft Report to Congress on CKD (RTC) in December 1993 [29].

The RTC documents numerous instances of groundwater and surface water contamination from CKD disposal practices, many of them at hazardous waste burning kilns. Common management practices mentioned in the report, such as storage in exposed, unlined piles, disposal in water-filled abandoned quarries and landfills, caused contamination of nearly air, surface water, and groundwater.

The most prevalent metals in CKD were cadmium, chromium, arsenic, lead, and selenium. Data from the Report indicate that CKD from kilns burning hazardous waste contains on average 9 times more lead, 5 times more cadmium, and 7 times more chromium than kilns which do not burn hazardous waste. The Report concluded that the risks associated with CKD are "generally low, however, there is a potential under certain circumstances for CKD to pose a danger to human health and environment, and it may do so in the future".

ADVANCED THERMAL CONVERSION THECHNOLOGIES FOR ENERGY FROM SOLID WASTE: GASIFICATION/PYROLYSIS/THERMOLYSIS

Gasification is a high-temperature process that is optimized to produce a fuel gas with a minimum of liquids and solids. Gasification, which is more proven than pyrolysis, consists of heating the feed material in a vessel with or without the addition of oxygen. Water may or may not be added. Decomposition reactions take place and a mixture of hydrogen and carbon monoxide (CO) are the predominant gas products, along with water, methane, and CO_2. Pyrolysis is a medium to high temperature (500-1000°C) process for converting solid feedstocks into a mixture of solid, liquid, and gaseous products. Pyrolysis to maximize production of liquid fuels and chemical feedstocks directly from a feedstock requires careful reaction control and fast heating and cooling rates the liquids that do form from breaking down to gases.

Thermolysis is a vacuum pyrolysis of organics present in the waste ($P = 100$ mbar - $\theta = 500°C$). The reaction products include a gas phase that supplies the energy needed for thermolysis and superheating steam for power generation, and a carbon-rich solid phase that retains most of metals, chlorine and sulfur, by the action of basic materials that are always present in wastes. The residue is then washed to yield a solid fuel resembling hard-ash coal.

A French magazine, "Environnement et Technique" has published a 5-page article in its October'96 issue n° 160 about the inauguration by Siemens AG of a 100,000 t/y integrated thermolysis unit at Fürth, Germany. PCDDs/Fs concentration in flue gases is lower than 0.006 ng TEQ/m^3, which is 16 times lower than the limit of 0.1 ng TEQ/m^3, required by the European Directive n° 94/67 applicable to MSW incinertors.

CONCLUSION

Destroying waste through incineration has been practised for many years. Today, waste disposal by incineration is common in many developed countries. A major benefit of incineration is the large reduction in the volume of waste (up to 90 %) required to be landfilled. A more recent trend is the combination of incineration and energy recovery in so-called waste-to-energy (WtE) plants and most incinerators built today incorporate some form of energy recovery.

As environmental concerns have heightened over the last 30 years, waste incinerators have been required to install sophisticated equipment for cleaning exhaust gases. Depending on the regulations of the country concerned, this gas cleaning equipment can be large and expensive.

However, the efficiency of modern WtE plants is limited by restrictions imposed on the tube metal temperature of the superheater to prevent excessive high temperature corrosion by hydrogen chloride (HCl) present in the flue gas, which means that the overhall plant efficiency is low.

Despite advances in material technology have allowed the steam temperature to be increased so that the efficiency of WtE is typically 25 %, new technologies - i.e., gasification, pyrolysis, thermolysis - still have several advantages:

- the amount of residual solid waste is limited: thermolysis yields about 330 kg of semi-coke, containing 30-40 percent of inerts, per tonne of MSW; this coke can be used in thermal electric plants, in blast furnaces, in cement kilns as substitution fuel. The amount of ash is less than 2 %;

- the gas generated in such technologies is only about 15 to 30 % of the volume of the flue gas from incineration and thus the gas cleaning equipment required is significantly smaller and cheaper;

- air pollution is widely lower than requirements of any present and future European Directives applicable to MSW incinerators.

REFERENCES

1. RYDING, S.O.. Environmental Management Handbook. IOS Press, Amsterdam, The Netherlands, 1992, 777 p.
2. WORLD COAL INSTITUTE - Coal - Power for Progress. 3rd Edition, January 1999, WCI, London (UK), 29 p.
3. KIKUCHI, R.. Secondary Pollution Associated with Flue Gas Desulfurization. In : Proceedings of the Global Symposium on Recycling, Waste Treatment and Clean Technology, REWAS'99, San Sebastian, Spain, September 5-9, 1999, Published by TMS and INASMET. Edited by I.Gaballah, J. Hager and R. Solozabal, 1999, Vol. III, pp. 2045-20053.
4. CLARKE, L.B.. Management of FGD Residues. IEA CR/62, IEA Coal Research, London, UK, 1993, 82 p.
5. KLINGSPOR, J., TOREKUD, A. and AHMAN, S.. Low-Cost FGD System for Emerging Market. ABB Review, 1998, 1, pp. 30-40.
6. SATO, T.. Fixaton of CO_2 with Lime. Inorganic Material, 1996, 3, pp. 67-80.
7. ROCHELLE, G Flue Gas Desulfurization. In: Coal Processing and Pollution Control, Gulf Publishing Company, Houston, USA, Edited by T. Edgar, 1983.
8. CRE-COAL RESEARCH ESTABLISHMENT. Disposal and Utilisation of Ash Residues. Final report - ESSC Project 722-ED/803; Cheltenham, UK, 1996, 91 p.
9. REIMANN, D.O., and HAMMERLI, H. Verbrennungstechniken Bedarf - Entwicklung - Berechnung - Optimierung. Abfallwirschafts Journal, 1992, N° 8, p. 612
10. CONNETT, P. Municipal Waste Incineration: A Poor Solution for the Twenty First Century. Paper presented at the 4th Annual International Management Conference Waste-To-Energy, Amsterdam, The Netherlands, November 24-25, 1998.
11. CONNETT, E., and CONNETT, P.. Mercury in Massachussetts: An Evaluation of Sources, Emissions, Impacts and Controls. Waste Not # 363, 1996.
12 GUEST, T.L., and KNIZEK, O.. Mercury Control at Burnaby's Municipal Waste Incinerator. In: Proceedings of the 84 th Annual AWMA Meeting, Vancouver, Canada, Paper 91-103.30, 1991.
13. OLI, K., VERMEULEN, P.L., and HUTZINGER, O.. Chloridibenzo-p-Dioxins and Chlorodibenzofurans are Trace Components of Fly Ash and Flue Gases of Some Municipal Incinerators in the Netherlands. Chemosphere, 1977, 6, p. 455.
14. HASSELRIIS, F.. Relationship Between Combustion Conditions and Emission of Trace Pollutants. Paper presented at the New-York State Air Pollution Association, May 2, 1984.

15. COMMONER, B., Mc NAMARA, M., SHAPIRO, K., and WEBSTER, T.. The Origins of Chlorinated Dioxins and Dibenzofurans Emitted by Incinerators that Burn Unseparated Municipal Solid Waste and an Assessment of Methods for Controlling Them. Center for the Biology of Natural Systems, Queens College, Flushing, USA, Dec. 1, 1984.

16. OZVACIC, V.. A Review of Stack Sampling Methodology for PCDDs, and PCDFs. Chemosphere, 1986, 15, p. 1173.

17. VOGG, H., and STIEGLITZ, L.. Thermal Behavior of PCDD/PCDF in Fly Ash from Municipal Incinerators. Chemosphere, 1986, 15, p. 1373.

18. US EPA. Municipal Waste Combustors-Background Information for Proposed Standards: Post Combustion Technology Performance". EPA-450/3-89-27C, August 1989.

19. INTERNATIONAL ASH WORKING GROUP (IAWG). Municipal Solid Waste Incinerator Residues. Studies in Environmental Science 67, Elsevier (publ.), Amsterdam, The Netherlands, 1997, 974 p.

20. ENDS Report. Incinerators Remain Net Dioxin Sources, Says ETSU, Oct. 1997.

21. ABE, S., KANABAYASHI, F., KIMURA, T., and KOKADO, M.. Decomposition of Dioxins and Related Compounds in MSW Ash Melting. Organohalogen Compounds, 1997, 31, p. 348.

22. SAKAI, S., HIRAOKA, M., ISHIDA, M., SHIJI, R., NIE, P., and NAKAMURA, N.. A Study of Total PCDDs/Fs Release to Environment from MSWI. Organohalogen Compounds, 1997, 31, p. 376.

23. CONNETT, P., and WEBSTER, T. An Estimation of the Relative Human Exposure to 2, 3, 7, 8-TCDD Emissions via Inhalation and Ingestion of Cow's Milk. Chemosphere, 1987, 16, p. 2079.

24. Mc LACHLAN, M.S.. Accumulation of PCDD/F in an Agricultural Food Chain. Organohalogen Compounds, 1995, 26, 105.

25. US EPA. Compilation of MWC Dioxin Data. Office of Air Quality Planning and Standards, July 27, 1995.

26. CONNETT, E., and CONNETT, P.. The Great Incinerator Ash Scan. Waste Not ≠ 316-319, March 1995.

27. BREMMER, H.J., TROOST, L.M., KUIPERS, G., DE KONING, J, and SEIN, A.A.. Emissions of Dioxins in The Netherlands. National Institute of Public Health and Environmental Protection (RIVM), Bilthoven, The Netherlands, Report 770501018, February 1994.

28. CEMBUREAU. Alternatives Fuels in Cement Manufacture. Ciments, Bétons, Plâtres, Chaux. 1999, N° 838/3, pp. 179-183.

29. US EPA. Office of Solid Waste. Report to Congress on Cement Kiln Dust. December 1993, pp. 3-57

PART ONE:

GENERATION, PROCESSING AND HANDLING OF INCINERATOR ASH

FACTORS THAT CONTROL THE QUALITY OF MUNICIPALS SOLID WASTE INCINERATION RESIDUES

H Belevi

Swiss Federal Institute for Environmental Science and Technology (EAWAG)

Switzerland

ABSTRACT. This paper presents the factors determining element behavior in municipal solid waste incinerators. The method is primarily based on material flow analyses in full-scale incinerators and laboratory experiments with synthetic samples. Processes in the furnace are mainly focused on.

The following master variables explain the element transfer behavior to the raw gas in the furnace:

- Occurrence and distribution of the elements in the input waste,
- temperature, redox conditions, chlorine content, content of the other reaction partners than oxygen and chlorine in the furnace bed as well as
- residence time and mixing conditions in the furnace bed.

The hypotheses on the influence of these factors on the element behavior were validated by the thermal treatment of bottom ash samples in the laboratory.

The main conclusions are:

- The transfer behavior of major, minor and trace elements from the municipal solid waste to the incinerators products can be explained by the method applied.
- The knowledge obtained in the present study can be used to achieve "tailor made" product qualities by designing new incinerators.

Keywords: Material flow, Bottom ash, ESP dust, Municipal solid waste, Incinerator, Transfer coefficient, Heavy metal, Behavior, Composition

Dr H Belevi is a senior researcher at the Swiss Federal Institute for Environmental Science and Technology (EAWAG) and lecturer in the Swiss Federal Institute of Technology Zurich (ETH). His research interests include many aspects of the resource and waste management, especially investigation of factors that control the quality of products in waste treatment systems by combining the material flow analyses with process understanding.

INTRODUCTION

Less than 15% of municipal solid waste (MSW) generated in the world is incinerated. Although about 30% of MSW are incinerated in the industrialized nations, this number increases rapidly because of the new strategies and guidelines. These guidelines for a sustainable resource management require that municipal solid waste incinerators (MSWI) produce well designed products. These "tailor made" products will either be used as secondary raw materials (to save the resources) or can be landfilled without leaving to the next generations polluted aquatic and terrestrial ecosystems (to protect the resources). For this purpose, incinerators should separate the compounds from MSW, which are harmful to the environment, and concentrate them in the gas cleaning residues. Thus bottom ash could be used as a construction material or can be landfilled over long-term periods without an adverse environmental impact, and gas cleaning residues would be attractive for resource recovery processes. However, “tailor made” product qualities can be only achieved, if the factors that control the element transfers from MSW to incinerator products are understood.

In this paper, a method is presented to investigate the factors determining element behavior in incinerators. It is based on material flow analyses in full-scale incinerators and laboratory experiments with synthetic samples.

The following specific questions are asked:

- How do major, minor and trace elements behave in a municipal sold waste incinerator?
- What are the factors determining the element behavior?
- Is it possible to use the knowledge obtained by the field and laboratory experiments to achieve "tailor made" product qualities?

EXPERIMENTAL DETAILS

Full Scale Experiments, Sampling, Sample Preparation and Analytical Techniques

The field experiments were carried out in the MSWI of St. Gallen, Switzerland. The grate type furnace is designed to incinerate 5.2 tons of waste per hour [1]. The incinerator was fed with two different waste types; on the first day with waste from residential source (household waste) and on the second day with waste from commercial and residential source (municipal solid waste). The average furnace room temperature is maintained between 820°C and 880°C. Material flow analyses were performed for both types of the waste input. The quantities of all goods entering and leaving the incinerator are measured during 8 to 10 hours.

The waste input was weighted by a calibrated crane balance, but not analyzed. The sampling of bottom ash took 10 hours and was carried out in one hour sampling periods (totally 10 sampling per day). The boiler ash sampling was conducted before it was mixed with gas cleaning residues in one hour periods over 10 hours. The gas cleaning residues were sampled over 9 hours in 30 minutes sampling periods. The chemicals and water consumption were registered continuously during 10 hours. Samples of these materials were also taken. The exhaust gas mass flow and analysis were carried out according to the suggestions set by the Swiss Agency for the Environment, Forests and Landscape [2].

All solid samples were dried at 105°C for 24 h. About 500 ml of each aqueous sample was acidified with HNO_3 to a pH of about 2. The residue and all other samples were kept at 4°C.

Bottom ash samples were ground for 4 h in a steel ball mill and sieved through a 0.5 mm screen. The amount of oversize material (sieving residues) was between 4 wt% and 11 wt%. The whole oversize material of 8 bottom ash samples is chosen randomly to assess the concentrations in sieving residues. The particle size of sieving residues does not allow digesting them with conventional methods. A procedure with reasonable expenses is developed to determine the concentrations with an acceptable precision and accuracy [1]. The material passed the screen was used as fine ground laboratory samples for element analyses. The same procedure was used for boiler ash samples. Gas cleaning residue samples were ground in a hammer mill. The resulting powdered boiler ash and gas cleaning dust completely passed the 0.5 mm screen and were used for element analysis.

According as element to be analyzed, different methods were used for sample dissolution and concentration determination. Sample dissolution was carried out in aqua regia, in a mixture of hydrofluoric acid, hydrochloric acid and nitric acid, in hot sodium hydroxide, according to Wurzschmitt [3] and in deionized water. Concentrations were measured by atomic absorption spectrometry, inductively coupled plasma atomic emission spectrometry, ionchromatography, potentiometry using ion selective electrodes, flow injection analysis, polarography, X-ray fluorescence spectrometry, CNS-Analyzer and coulometry.

Calculations of Mass and Element Flows As Well As Transfer Coefficients

Element flows are simply calculated by multiplication of mass flows with element concentrations. Element concentrations in the waste input were not measured. They were calculated by performing mass balances for each element.

Transfer coefficients are defined by the following equation:

$$k_{i,j} = \frac{A_{i,j}}{A_{i,W}}$$

where

$k_{i,j}$ is the transfer coefficient of the total mass or of the element i to the output j.

$A_{i,j}$ is the total mass flow or the element flow of the element i through the output j.

$A_{i,W}$ is the total mass flow or the element flow of the element i through the waste input.

Transfer coefficients describe the distribution of the input mass among the various products. For example, a high transfer coefficient of an element to the bottom ash indicates that this element is mainly transferred to the bottom ash.

Laboratory Experiments

The apparatus for the evaporation experiments in the laboratory consists of a furnace with a gas collection device, a condensation flask in an ice water bath, three gas washing bottles, a rotameter and a suction pump. After the air feed rate is set, the furnace is heated up to the desired temperature. 10 g of the sample is put in a ceramic crucible with 7 cm diameter. The crucible is then placed in the furnace. At the end of the residence time, the crucible is taken out and cooled. After cooling, the sample is pressed into a pellet. The concentration determination in the pellet is performed with a x-ray fluorescence spectrometer. The experiments are carried out with synthetic mixtures as well as with bottom ash samples.

RESULTS AND DISCUSSION

Daily Average Concentrations in the Municipal Solid Waste

Figure 1 shows the element concentrations in the municipal solid waste. As expected, organics are (besides water) substances with the highest concentration. The sum of the elements Al, Ca, Fe, K, Mg, Na and Si content is about 11 wt%. Non-metals P, S, Br, Cl and F make up less than 1 wt% and minor and trace elements less than 1 wt%. As, Co, Hg, Li and Mo show the lowest concentrations among measured elements (≤10 mg/kg).

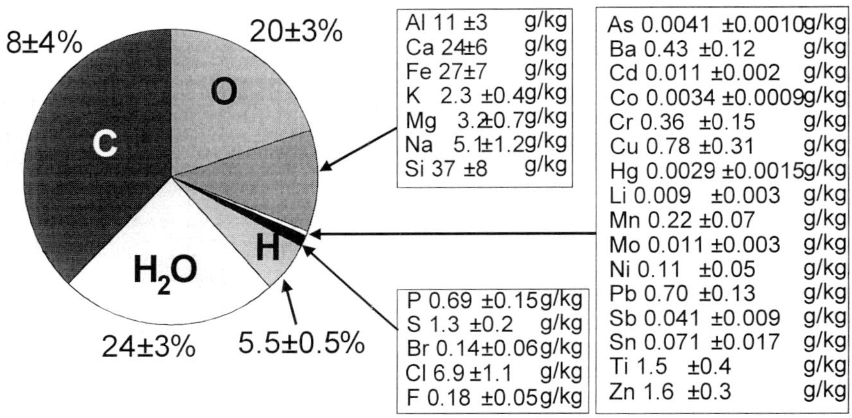

Figure 1 Daily average concentrations in the municipal solid waste
% corresponds to wt-%

Daily Average Element Concentrations in the Incinerator Products

Table 1 shows the daily average element concentrations in the solid residues and the abundances in the Earth's crustal rocks. Al, Ca, Fe, K, Mg, Na, Si are the major elements with concentrations ≥1% in the bottom ash. The carbon concentration is also higher than 1%. The concentrations of major elements and P, F, Co, Hg, Li, Mn, Ti are of the same order as the abundances. The other elements are enriched in the bottom ash.

As mentioned in the experimental details, the concentrations in the several fractions of the bottom ash are separately determined. The mean concentrations of all elements except Fe, Co, Cr, Cu, Mn, Mo, Ni, Pb, Sn and Zn in the sieving residues are lower than or equal to those in the fine ground laboratory sample. About 15% of Pb, 20% of Co and Mn and 28% of Zn exist in the sieving residues and coarse goods. About 32% of Cr, 34% of Sn, 36% of Mo, 47% of Fe and Cu and 67% of Ni come from the sieving residues and coarse goods.

In the boiler ash, the elements Al, Ca, Fe,K, Mg, Na, Si C, S, Cl, Pb, Ti and Zn exist in concentrations ≥1%. In the gas cleaning residues, Al, Ca, Fe, K, Mg, Na, Si, C, S, Cl, Pb, and Zn exist in concentrations ≥1%. Except metals and metalloids as well as P, Co, Li, Mn and Ti all measured elements are enriched in the boiler ash and gas cleaning residues compared to the Earth's crustal rocks.

Table 1 Daily average element concentrations in the incinerator residues and abundance in the Earth's crustal rocks [4,5]. Element concentrations are average values of 8 to 10 hourly samples

ELEMENT	CONCENTRATION, g/kg dry mass			
	Bottom Ash	Boiler Ash	Gas cleaning residues	Abundance in the Earth's crustal rocks
Non-metals				
TC	19±1	28±8	18.5±1.6	0.2
TIC	7.0±0.8	4.2±1.0	0.9±0.2	n.d.
TOC	12.0±1.3	24±9	17.6±1.5	n.d.
N	0.38±0.09	0.36±0.09	0.38±0.07	0.02
P	3.2±0.3	5.1±0.6	2.8±0.3	1
S	4.3±0.4	62±15	30±3	0.3
Halogens				
Br	0.08±0.02	0.39±0.12	5.5±2.7	0.003
Cl	3.4±0.5	26±4	282±18	0.2
F	0.33±0.11	2.2±0.8	5.1±1.5	0.5
Metals and metalloids				
Al	53±7	68±10	33±4	80
Ca	117±14	145±16	175±18	50
Fe	142±18	28±3	9.6±0.8	50
K	8.0±0.8	31±4	29±3	15
Mg	16±1	14±1	7.1±0.4	30
Na	25±3	29±3	38±4	25
Si	188±12	110±9	40±6	270
Minor and trace elements				
As	0.015±0.003	0.026±0.004	0.040±0.005	0.002
Ba	2.0±0.5	2.7±0.5	1.8±0.3	0.3
Cd	0.006±0.001	0.21±0.03	0.41±0.05	0.0002
Co	0.016±0.003	0.022±0.006	0.015±0.002	0.03
Cr	1.8±0.7	1.5±0.5	0.74±0.23	0.1
Cu	4.0±1.4	0.98±0.17	1.09±0.17	0.05
Hg	<0.0002	0.0008±0.0006	0.09±0.03	0.00008
Li	0.045±0.011	0.052±0.011	0.034±0.009	0.015
Mn	1.1±0.3	1.1±0.2	0.55±0.08	1
Mo	0.050±0.014	0.065±0.014	0.058±0.013	0.001
Ni	0.57±0.25	0.43±0.20	0.12±0.04	0.07
Pb	1.6±0.3	21±5	14.4±1.6	0.01
Sb	0.026±0.005	0.76±0.09	1.55±0.36	0.0002
Sn	0.14±0.05	0.33±0.07	2.0±0.5	0.002
Ti	6.8±1.5	13.2±2.7	5.4±0.9	5
Zn	3.5±1.2	26±3	36±4	0.08

TC: Total carbon TIC: Total inorganic carbon TOC: Total organic carbon
n.d.: not determined

Daily Average Element Concentrations in the Exhaust Gas

In the exhaust gas, daily average carbon concentration is 8.4 vol%, oxygen concentration 10.9 vol%, the water content of the wet gas 19.3 vol%, sulfur concentration 19±6 mg/Nm^3dry and mercury concentration 0.14±0.13 mg/Nm^3dry. Dust concentrations are very low due to the very efficient gas cleaning system (4.8±3.2 mg/Nm^3dry). Therefore concentrations of the remaining elements, which exist mostly particle bound, are not measured, and are estimated by using the measured dust concentrations and literature values.

Transfer Coefficients

Figure 2 shows mass and element transfer coefficients. About 79 wt% of MSW mass are transferred to the exhaust gas. Each ton of household waste produces 190 kg bottom ash dry matter, 3.6 kg boiler ash and 19 kg gas cleaning residues.

About 99% of carbon is transferred to exhaust gas, 1.0 % to bottom ash, less than 1% to. boiler ash and gas cleaning residues. Phosphorus is mainly transferred to bottom ash. More than 50% of sulfur leave the incinerator by gas cleaning residues and boiler ash and about 13% by exhaust gas. This corresponds to a gas cleaning efficiency of about 80%. Bromine and chlorine mainly enter the raw gas and removed from the raw gas by gas cleaning devices. Fluorine is distributed between bottom ash and gas cleaning residues.

About 90% of aluminum, calcium, magnesium and sodium, about 67% of potassium and about 97% of silicon are transferred to bottom ash. Iron shows the highest transfer coefficient to bottom ash among the investigated elements. This is due to bulk metallic iron in the waste input. Barium, lithium and titanium are also mainly transferred to bottom ash (88% to 90%). About 76% of arsenic leave the incinerator by bottom ash.

More than 90% of metals chromium, copper, manganese and nickel, 88% of cobalt and 86% of molybdenum are transferred to bottom ash. About 90% of cadmium and almost about 100% of mercury is transferred to the raw gas and mostly concentrated in gas cleaning residues. While cadmium is removed efficiently from the raw gas, mercury removal is not very efficient probably because of the concentration fluctuations in the input. The high uncertainty of the transfer coefficients occurs because of the short timed mercury peaks [1]. About 44% of lead, 37% of tin and 43% of zinc are transferred to bottom ash. Antimony is more volatile; 88% of antimony enters the raw gas and is concentrated in gas cleaning residues.

Main indicators and their indications

The following variables are defined here as master variables determining the transfer of elements to the raw gas in the furnace:

- Occurrence and distribution of the element in the input waste such as existence in bulky or in easy entrainable inorganic or organic goods
- Physical and chemical conditions in the furnace bed such as temperature, redox conditions, chlorine content and content of other reaction partners than oxygen and chlorine
- Kinetics in the furnace bed such as residence time and mixing conditions

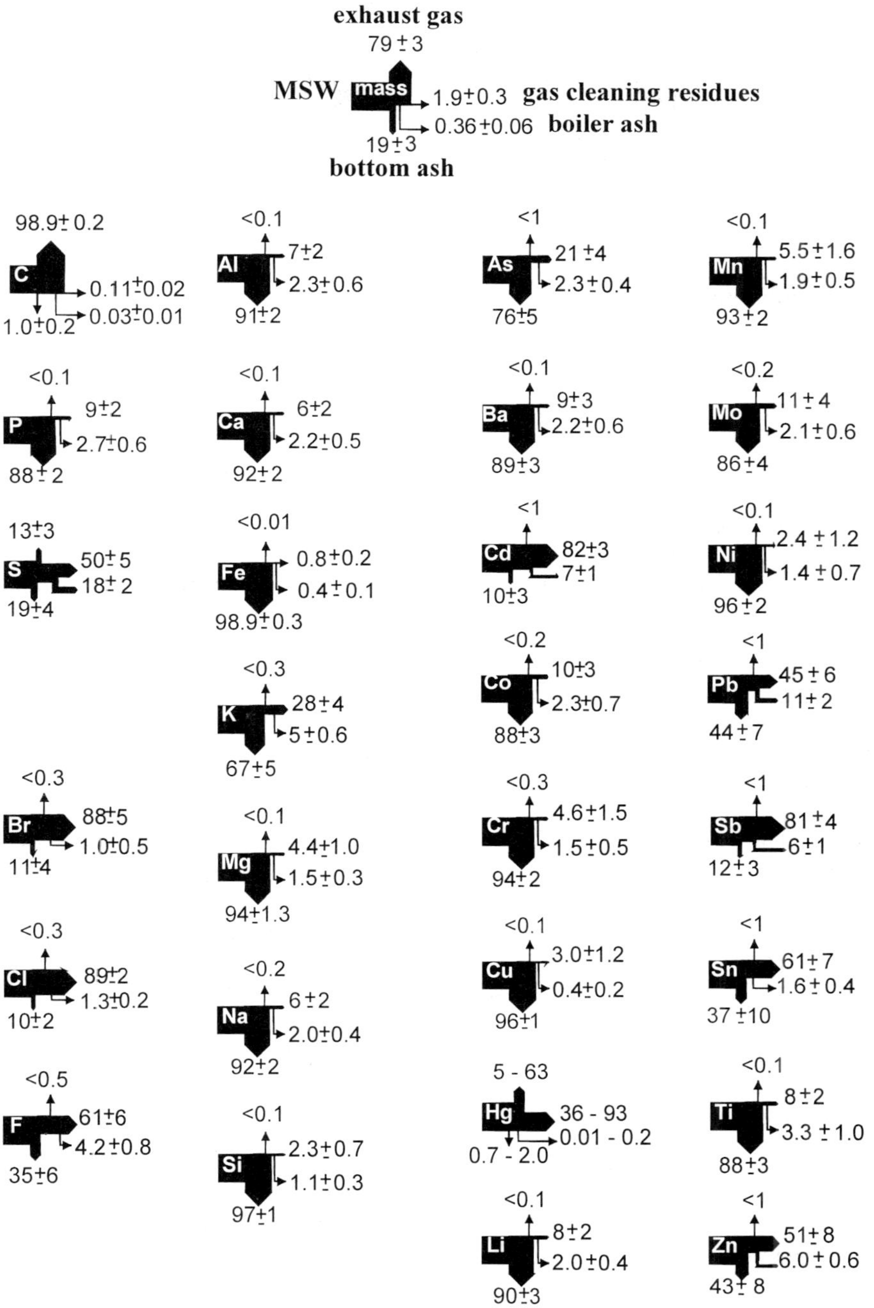

Figure 2 Mass and element transfer coefficients
All values are expressed in %

First information on the effect of these master variables on the element transfer behavior is obtained by the element transfer characteristics of the furnace bed and boiler as well as by the concentration dependence on particle size in bottom ash. Following calculations were performed with the results obtained by material flow analyses:

- T-tests to establish if there are differences between the household waste and the municipal solid waste incineration regarding the transfer coefficients to the raw gas $k_{i,RAW}$.
  Different $k_{i,RAW}$ values indicate high sensitivity to process conditions in the furnace.
- Ratio transfer coefficient into the gas cleaning devices to the transfer coefficient into the boiler ash $k_{i,CD} / k_{i,BRA}$ (CD: gas cleaning devices, BRA: boiler ash).
  The lower $k_{i,CD} / k_{i,BRA}$ values than about 4.5 indicates that the transfer of the element to the raw gas mostly occurs by entrainment in the furnace bed.
- T-tests to establish if there are differences between the household waste and the municipal solid waste incineration regarding the transfer coefficient ratios $k_{i,CD} / k_{i,BRA}$.
  Different $k_{i,CD} / k_{i,BRA}$ values for mixed municipal solid waste incineration and household waste incineration indicate that the transport of the element in the raw gas occurs in particle matrix.
- Comparison of concentrations in sieving residues with the concentrations in fine ground samples of bottom ash.
  Higher concentrations in the sieving residues than in the fine ground samples of bottom ash indicates that the element can also exist in bulky goods in the waste input.

Additional indicators such as acid soluble parts in bottom ash and in boiler ash were also used to explain the transfer behavior of some elements.

Table 2 shows the indications on the element transfer to the raw gas, transport in the raw gas, occurrence and distribution in the waste input and sensitivity to process conditions in the furnace bed. The hypotheses are first approaches and do not claim to be complete. The transfer of the elements of the group 1 and 2 to the raw gas occurs by entrainment and they are mainly transported in the raw gas by entrained particle matrix. While the elements of the group 1 mostly exist in bulky goods, the elements of the group 2 also exist in easy entrainable inorganic goods or in organic goods. The low transfer to the raw gas ($k_{i,RAW} \leq 0.15$) can be explained by the low dust generation in the furnace ($k_{mass,CD}/k_{mass,BMA} < 0.15$). Potassium (Group 3) is transferred to the raw gas by entrainment (perhaps also by evaporation) and is transported in the raw gas by fine particles. It mostly exists in easy entrainable inorganic or organic goods in MSW. Incineration of different types of waste could not significantly indicate if the transfer of the elements of the groups 1, 2 and 3 to the raw gas is sensitive to input variations.

The elements of the groups 4, 5 and 6 are transferred to the raw gas by evaporation and transported in the raw gas by fine particles and as gaseous species. The wide range of transfer coefficients of the group 4 can be explained to a certain extent by the occurrence of these elements in bulky goods in MSW. The evaporation rate is decreased by low surface to volume ratio of bulky goods. Therefore these elements are also sensitive to input variations (the mixed municipal solid waste incineration produces different $k_{i,RAW}$ values than the household waste incineration). Bromine and antimony (Group 5) mostly exist in easy entrainable inorganic or in organic goods. They are also sensitive to input variations. The elements of the group 6 also exist mostly in easy entrainable inorganic or in organic goods. But they do not show high sensitivity to input variations.

Table 2 Main indications on the element behavior

ENTR or EVAP: Transfer to the raw gas occurs mostly by entrainment or by evaporation

PART or FINE/GAS: Transport in the raw gas occurs either by entrained particle matrix or by fine particles and as gaseous species

BULK or NON-BULK or ALL: Occurrence in bulky or in easy entrainable inorganic and organic goods or in both

$k_{i,RAW}$ Element transfer coefficient to the raw gas during household and MSW incineration

ELEMENT	TRANSFER TO THE RAW GAS $k_{i,RAW}$	INDICATIONS			
		Main processes responsible for the transfer to the raw gas	Transport in the raw gas	Occurrence and distribution in MSW	Sensitivity to input variations
Group 1 Si, Fe, Co, Cr, Mn, Ni	<0.08[1)]	ENTR	PART	BULK	cannot be shown
Group 2 P, Al, Ca, Mg, Na, Ba, Li, Ti	0.08-0.15[2)]	ENTR	PART	ALL	cannot be shown
Group 3 K	0.33	ENTR and EVAP?	FINE/ GAS?[4)]	NON-BULK	cannot be shown
Group 4 F, Cu, Mo, Pb, Sn, Zn	0.03-0.65	EVAP	FINE/ GAS	BULK	SENSITIV
Group 5 Br, Sb	0.75-0.89	EVAP	FINE/ GAS	NON-BULK	SENSITIV
Group 6 C, S, Cl, As, Cd, Hg	0.24-1.00	EVAP[3)]	FINE/ GAS	NON-BULK	cannot be shown

1): $k_{Co,RAW}$= 0.08-0.10 1): $k_{Mg,RAW}$= 0.06-0.08

2): S is transferred to the raw gas by evaporation as well as by entrained particle matrix [1]

3): Acid solubility indicates that a high part of K in the raw gas is not bound in the silicate matrix [1]

These results indicate that the occurrence and distribution is an important master variable for the elements of the group 1, 2 and 3 with respect to the transfer to the raw gas. The parameter variations that influence the entrainment in the furnace such as furnace construction, primary air feed rate, mixing on the grate may influence the transfer coefficients of these elements. Evaporation is the main process responsible for the transfer of the elements of the groups 4, 5 and 6. Beside the occurrence and distribution, physical and chemical conditions as well as kinetics in the furnace influence the transfer to the raw gas. More information on the transfer mechanisms of these elements can be obtained by additional laboratory experiments.

The detailed results of the laboratory experiments with synthetic mixtures are given elsewhere [1]. Table 3 is a first attempt to simplify the effects of the temperature, redox conditions, chlorine content and other reaction partners than oxygen and chlorine on the element transfers to the raw gas in a furnace within the investigated parameter range.

Table 3 Effect of physical and chemical conditions on the element transfer to the raw gas in the laboratory experiments.

Residence time: 30 min air feed rate: 15 $Nm^3\ kg^{-1}\ h^{-1}$
chlorine addition: 2.9 mol/kg carbon addition: 10.7 mol/kg
Bottom ash addition: replacement of 500 g/kg quartz with bottom ash

ELEMENT	TEMPERATURE INCREASE BETWEEN 500°C AND 900°C	DECREASE OF OXIDATIVE CONDITIONS (CARBON ADDITION)	INCREASE OF CHLORINE CONTENT	INCREASE OF THE CONTENT OF OTHER REACTION PARTNERS THAN OXYGEN AND CHLORINE (BOTTOM ASH ADDITION)
Zn	Higher transfer	Lower transfer	Higher transfer	Lower transfer
Cu	Higher transfer	Lower transfer	Higher transfer	Low sensitivity
Cd	Higher transfer	Low sensitivity	Higher transfer[1)]	Lower transfer
Pb	Higher transfer	Low sensitivity	Higher transfer[1)]	Low sensitivity
Sn	Higher transfer	Higher transfer	Higher transfer	Lower transfer
Mo, Sb	Lower transfer	Lower transfer	Higher transfer	Lower transfer
As	Lower transfer	Higher transfer	Higher transfer	Lower transfer

1): Higher temperatures are more important than higher chlorine contents for the element transfer into the gaseous phase within the investigated parameter range.

High temperatures and high oxidative conditions favor the transfers of Cu and Zn to the raw gas. The transfers of Cd and Pb are higher by a temperature increase. But redox conditions do not play any significant role within the investigated range. More Sn is transferred to the raw gas by decreasing oxidative conditions and increasing temperatures. Mo and Sb do not need higher temperatures than 500°C for being transferred to the raw gas. A decrease of oxidative conditions causes lower transfers. Arsenic transfer to the raw gas is adversely affected by an increase of oxidative conditions and temperature. An increase of chlorine content favors all the metals mentioned above. Other reaction partners than oxygen and chlorine negatively affect the transfer of Zn, Cd, Sn, Mo, Sb and As. However, this effect strongly depends on the excess chlorine and temperature. For example, if excess chlorine is present, this effect is very low for Sb. If this is not the case, antimony transfer into the gaseous phase is prevented by the reaction with the other waste components.

Effect of kinetics on evaporation

The effect of residence time on the evaporation behavior of investigated elements in the laboratory experiments is given elsewhere [1]. Information is obtained on kinetics in the furnace bed by a combination of the results of the laboratory experiments with the element behavior in the incinerator. The elements Cd, Cu, Pb and Sn are chosen for this purpose.

Copper is transferred to the raw gas by evaporation (Table 2). Thermodynamic calculations suggest a transfer coefficient of higher than 0.9 to the raw gas [6,7]. The measured values are however lower than 0.05. High copper transfers to the raw gas need high temperatures and excess chlorine under oxidative conditions. Although these conditions exist in the furnace bed on average basis, they seem not to be present on the same location at the same time. The residence time of solid goods at high temperatures and under oxidative conditions with excess chlorine must be very low (seconds to minutes).

A decrease of oxidative conditions and an increase of chlorine availability favor tin transfer (Table 3). Although tin mostly exists in bulky goods and therefore the evaporation rate is decreased due to mixing controlled kinetics, the transfer coefficient to the raw gas is relatively high ($k_{Sn,RAW}$ =0.63, Figure 2). This suggests that not highly oxidative conditions together with high chlorine availability exist in the furnace bed. For the transfer of Pb and Cd, the temperature is the most important factor (Table 3). High Cd transfer to the raw gas ($k_{Cd,RAW}$ =0.90, Figure 2) suggests that residence times up to hours at temperatures >700°C exist in furnace bed. Higher transfer is probably prevented by the reaction of Cd with other reaction partners than chlorine and oxygen to non-volatile species. Lower transfer coefficients of Pb are probably caused by its occurrence in bulky goods.

Based on these indications, hypotheses are postulated on kinetics in the furnace bed. In the gasification/ pyrolysis zone of the furnace, the decomposition of organic materials starts, temperature and HCl concentration increase and oxygen concentration decreases. The residence time of solid goods must be in the range of minutes, since a high amount of tin is evaporated in this zone. In the combustion zone, the temperature is higher, oxygen concentration also increases, but HCl concentration is not as high as in the gasification/pyrolysis zone. Otherwise the transfer coefficient of copper to the raw gas would not be very low. In the burnout zone, the temperature is high (>700°C) and the conditions are oxidative. The total residence time of solid goods in the combustion and burnout zones reach up to hours, since high amounts of lead and cadmium are transferred to the raw gas.

Thermal treatment of bottom ash in the laboratory

In order to validate the hypotheses mentioned above, evaporation experiments were carried out in the laboratory with two bottom ash samples. Sample A is a bottom ash sample taken during the field measurements. Sample B is taken in a way that coarse materials (>2 cm) are excluded. It can be assumed that this sample represents a mechanically treated bottom ash by grinding and separation processes. Both bottom ash samples are thermally treated analogously to the synthetic mixtures. The temperature is between 800°C and 1000°C and the residence time is 2.5 hours. The air feed rate is 15 $Nm^3\ kg^{-1}\ h^{-1}$, and 14.5 mol chlorine is added per kg bottom ash (chlorine input to the incinerator is about 35 mol/kg bottom ash).

Table 4 shows the concentrations in these samples before and after the thermal treatment as well as in Earth's crustal rocks. High transfers into the gaseous phase are expected and measured for Cd, Cu, Pb and Zn. Higher residence times increase the transfers of Pb and Cu, but not Zn and Cd. They probably have previously reacted in the incinerator or reacts during these experiments to stable species as silicates or spinels (effect of other reaction partners than oxygen and chlorine). For cadmium, this effect is more significant, since only a small part (<10%) of Cd input is transferred to the bottom ash in the incinerator and therefore a high part may be present in bottom ash as stable species. About 50% of Mo and 20% to 30% of Sn is transferred into the gaseous phase. This is also to expect. Mo evaporation is low at high temperatures (800°C to 1000°C) and oxidizing atmosphere is not favorable for Sn volatilization. As expected, the remaining elements are not volatilized to a significant extent.

Table 4 Element concentrations in the bottom ash samples before and after the thermal treatment and a comparison with the concentrations in the Earth's crustal rocks [4,5]

ELEMENT	CONCENTRATION, mg/kg				
	Sample A		Sample B		Rocks
	before	after	before	after	
Cd	4.5	2.9	3.5	1.8	0.2
Co	26	26	28	28	30
Cr	1900	1900	1700	1700	100
Cu	3000	470	1400	90	50
Mn	1400	1400	900	900	1000
Mo	33	16	11.0	4.9	1
Ni	340	340	200	200	70
Pb	1900	65	1100	55	10
Sn	230	160	50	40	2
Zn	4500	1000	2600	430	80

These results indicate that an optimization of master variables might produce bottom ashes with the element concentrations of the same order as crustal rocks regarding the elements Cd, Co, Cu, Mo, Mn, Pb, Sn and Zn. Ni and Cr cannot be volatilized in incinerator. However, a part of Ni and Cr could be removed from bottom ash by mechanical separation.

CONCLUSIONS

Tailor made product qualities seem to be achieved by optimizing existing incinerators and by designing new plants. For this purpose, mechanical processing techniques (grinding and separation processes) and thermal treatments have to be combined.

REFERENCES

1 BELEVI H. Environmental Engineering of Municipal Solid Waste Incineration, vdf University Press, Zürich, 1998.

2 BUWAL Suggestions for emission measurements of air pollutants at stationary plants. Swiss Agency for the Environment, Forests and Landscape, Bern. 1987.

3 WURZSCHMITT B. Ein neues Schnellverfahren mit Alkaliperoxid in einer Universalbombe für Mikro- Halbmikro. und Makroeinwaagen. Microchimica Acta, Vol. 36/37, 1951, p. 369.

4 COX P.A. The elements, their origin, abundance and distribution. Oxford university press, London, 1989.

5 BACCINI P. (Ed), Deponierung fester Rückstände aus der Abfallwirtschaft. vdf university press, Zürich, 1993.

6 FERNANDEZ M.A., MARTINEZ L., SEGARRA M., GARCI J.C. AND ESPIELL F. Behavior of heavy metals in the combustion gases of urban waste incinerators. Environmental Science & Technology Vol 26, 1992, pp 1040-1047.

7 VERHULST D., BUEKENS A., SPENCER P.J. AND ERIKSSON G. Thermodynamic behavior of metal chlorides and sulfates under the conditions of incinerator furnaces. Environmental Science & Technology. Vol 30, 1996, pp 50-56.

GLASS AND GLASS-CERAMICS FROM INCINERATOR FLY ASH

A R Boccaccini
H Kern
Technical University of Ilmenau
Germany

J Ma. Rincón
M Romero
Institute Torroja
Spain

ABSTRACT. Powder technology and sintering were used to transform a vitrified municipal incinerator fly ash into glass-ceramic products. Highly crystalline and dense products were fabricated developing an optimised heat-treatment in order to influence the crystallisation and densification kinetics of powder compacts. The materials exhibited a microstructure composed mainly of elongated crystals of pyroxene type in a residual glassy matrix. The remaining porosity was low and limited to small isolated pores. The sintered material showed attractive mechanical properties, i.e. relatively high fracture strength and toughness and high thermal shock resistance. The products have a large application potential, especially as construction and building materials. Open questions concerning the chemical durability and toxicity of the products remain, however, and they must be fully answered before these glass-ceramics can be realistically considered for the envisaged applications.

Keyword: Incinerator fly ash, Glass-ceramics, Power technology, Mechanical properties.

Dr A R Boccaccini, is a research associate at the Department of Materials Technology, Technical University of Ilmenau. He has published and travelled widely. One of his research interests is the development of glass-ceramics and composite materials from silicate residues.

Prof Dr H Kern, is Professor and Director of the Institute of Materials Technology at the Technical University of Ilmenau. His current research interests include recycling technologies and composite materials with emphasis on mechanical properties and interfaces.

Dr J Ma Rincón, is Head of the Glass-ceramic Laboratory at the Institute Torroja, Madrid. He is an internationally recognised expert on the field of glass, glass-ceramics and recycling of silicate residues. He has lead numerous international projects related to these topics.

Dr. M. Romero, is a research associate at the Glass-ceramic Laboratory at the Institute Torroja. She has published and travelled widely. Her current interests include the processing and characterisation of glass-ceramics from silicate residues.

INTRODUCTION

The significance of municipal solid waste incineration is continuously increasing in countries where the population density is high and the availability of space for land-filling is limited. Although incineration reduces the volume of municipal waste by approximately 90%, it leaves considerable amounts of solid residues, such as bottom and boiler ashes, and filter fly ashes. Filter ashes, which are produced at a rate of 25-30 kg/ 1000 kg incinerated waste, are particularly problematic since they contain significant concentrations of heavy metals, as well as trace amounts of organic pollutants (e.g. polychlordibenzo-dioxins and -furanes) [1-3].

Due to increasingly stringent environmental regulations, these residues are regarded as hazardous in most countries [3]. Therefore, they must be deposited in special landfills equipped with careful control of the effluents. This is a costly and environmentally unsatisfactory solution.

Vitrification is one of the most promising technological options for the inertization of these residues, with the objective of render a product that can be reused or, at least, deposited in standard land-fill sites without any risk [4-9]. By melting the residues at temperatures above 1300°C, a relatively inert glass is produced: the high temperatures involved in the process lead to the complete destruction of the organic pollutant compounds, and heavy metals can be either incorporated in the glassy product or separated from the residue by evaporation or differential precipitation [4-9]. The inert vitreous product can, in principle, be utilised for urban furniture, landscaping decoration or for road construction. As a disadvantage, vitrification is an energy-intensive process involving relatively high costs. Therefore, its use can only be fully justified if high-quality products with optimised properties can be fabricated, which can thus compete with current materials, for example for building, architectural or insulation applications. One effective way to improve the properties of the vitrified products without major alterations to the process itself is the induction of a controlled crystallisation, i.e. by forming a glass-ceramic. In previous reports, the possibility of obtaining bulk glass-ceramic materials from vitrified municipal incinerator fly ash was demonstrated [10-15]. The mechanical properties and toxic potential of the glass-ceramics were also determined [12] and an extensive microstructural characterisation of the glass-ceramics was carried out [15].

Glass-ceramic articles can be produced also by sintering of glass powders. In the ideal case, glassy amorphous powders are densified at relatively low temperatures by exploiting a viscous flow sintering mechanism. After densification, the material must be subjected to a crystallisation heat-treatment to obtain the required glass-ceramic microstructure [16,17]. A second possibility is the fabrication of dense glass-ceramics by a sintering process in which both densification and crystallisation take place simultaneously. The technological significance of this process as well as the theoretical complexities of its kinetics have been discussed recently by Gutzow et al. [18].

Several advantages appeared when using powder technology and sintering to fabricate glass-ceramics. For example, the products can be manufactured using ordinary equipment available in a ceramic plant. Thus, the technology does not require major investment, being particularly suitable for production of articles of complicated shape.

Limited work has been published regarding the fabrication of glass-ceramics from powdered waste by sintering. These few works include glass-ceramics from metallurgical slag [19,20], from coal-fired power station fly ashes combined with waste glass [21,22] and from iron-rich jarosite zinc hydrometallurgical residues [23,24]. To our knowledge, however, municipal waste (MW) incinerator fly ashes have not yet been considered in this regard.

In this study, the possibility of fabricating glass-ceramic products by sintering of glass powders obtained from vitrified MW incinerator fly ashes is explored. The mechanical and technological properties of the new sintered materials (fracture strength, toughness, hardness, thermal shock resistance, brittleness, etc.) were measured and compared with those of similar products.

EXPERIMENTAL DETAILS

The chemical composition of the incinerator fly ash used is shown in Table 1. In addition to the oxides listed, traces of Cl, Sn, Sb, Cd, Ba, As, Sr, Zr, Pb and Mo were also present.

Table 1 Composition of the investigated incinerator fly ash [11]

OXIDE	WEIGHT (%)
Na_2O	3.5
MgO	2.4
Al_2O_3	17.5
SiO_2	38.0
P_2O_5	1.6
SO_3	0.2
K_2O	1.8
CaO	21.1
TiO_2	1.7
MnO	0.4
Fe_2O_3	8.0
PbO	0.3
ZnO	3.5

Batches of the as-received fly ash, placed in alumina crucibles, were heat-treated at 600°C for 2 h in air, followed by melting in an laboratory furnace at 1300°C for 2 h. No additives, fluxes or nucleating agents were added. The glass was quenched in air at room temperature. A shiny black-colour glass was thus obtained. The density of this glass was 2.80 g/cm^3, as reported elsewhere [11]. Glass powder suitable for sintering experiments was obtained by crushing the as-quenched product and by subsequent ball-milling to a mean particle size of 35 μm, with particle sizes varying between 2 and 90 μm. Cylindrical compacts (10 mm diameter, 10 mm height) were fabricated by uniaxial cold-pressing at a pressure of 250 MPa without using any binder. The green densities achieved were in the range 62-65 % of theoretical density. The sintering parameters were chosen according to previous results on nucleation and crystallisation of the bulk as-quenched product [10,11]. Using differential thermal analysis, it was shown that nucleation takes place at 880°C and that crystal growth occurs at temperatures > 950 °C [10]. Accordingly, samples were sintered following a 2-step schedule, as shown in Figure 1, where T_l was either 950 or 1000°C, and t_l was 1, 3, 6 or 10 hours. The heating and cooling rates were set at 10 °C/min. After sintering, the density of the samples was determined geometrically. X-ray diffraction (XRD) analysis (D 5000, Siemens, CuK_α radiation, Ni filtered) of the as-quenched glass and of sintered samples was conducted. Polished sections were prepared to a 1 μm finish for scanning electron microscopy (SEM) observations (CamScan 44, Cambridge Inst.) by standard grinding and polishing procedures using SiC paper and diamond paste. Samples were also prepared for transmission electron microscopy (TEM) examination (Philips CM 30, LaB_6-cathode, acceleration voltage of 300 kV). These specimens were prepared and thinned by using a diamond cutting saw, a diamond grinding pad (15 μm) and a dimple grinder with 2 μm diamond-paste. To reach electron transparency the specimens were ion beam thinned. Finally they were sputtered with a thin carbon layer to prevent electrostatic charging.

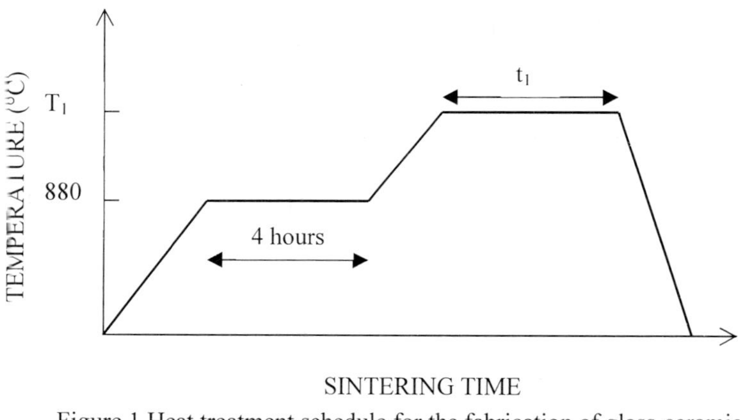

Figure 1 Heat treatment schedule for the fabrication of glass-ceramics from vitrified flyash by sintering

Large cylindrical samples (60 mm diameter, 5 mm height) were also fabricated by cold pressing and sintering. These samples were used to prepare specimens suitable for the determination of thermomechanical properties. The elastic constants were determined by means of the ultrasonic technique on cylindrical specimens with previously ground and polished parallel faces. Four-point flexural strength (40 mm outer span, 20 mm inner span) was measured at a cross-head speed of 1 mm/min using rectangular bars of 3x4 mm^2 cross section. Fifteen specimens were tested, and the data averaged. SEM was used to observe fracture surfaces. Short bars (25 mm length) were used to determine the thermal expansion coefficient between room temperature and 700 °C using a dilatometer (Netzsch 402 E). Vickers indentation was used to characterise the hardness Hv and indentation fracture toughness K_{Ic} of the sintered samples using indentation loads of 20 N. For calculation of K_{Ic}, the equation of Anstis et al. [25] was used. The relation between Hv and K_{Ic} yields the brittleness index B [26]. Thermal shock testing involved heating test cylinders in air to the desired temperature and equilibrating for 30 min. Then the samples were quenched in distilled water maintained at room temperature and examined for surface cracks using a stereo light microscope. The appearance of superficial cracks was used as the criterion to determine the critical temperature difference under thermal shock.

RESULTS AND DISCUSSION

Densification, microstructure and mineralogical composition

Table 2 summarises the results of the density of samples sintered under different conditions. Referring the quoted density values to the density of the equivalent bulk glass-ceramics, 2.89 g/cm^3 [11], it is seen that a high degree of densification by sintering can be achieved, especially at temperatures > 950 °C. No „barrelling“ or other shape distortion effects occurred during sintering, and the samples retained the cylindrical shape of the „green“ bodies after sintering. This indicates that the shrinkage was isotropic and that appropriate shape control of the products is possible. This is of practical importance with the aim to fabricate components of complex shapes and high dimensional tolerances for technical applications.

Table 2 Sintered density of glass-ceramics

T_l (°C), t_l (h)	Density (g/cm^3)
950, 1	2.45
950, 3	2.54
950, 6	2.60
950, 10	2.67
1000, 6	2.74
1000, 10	2.74

Figure 2 shows a SEM image of a polished surface of a sample sintered for 6 hours at 1000°C. These can be considered to be the optimised sintering parameters, since the sintered material is fairly dense with only small dispersed pores. Some areas of the sample, however, exhibited higher porosity as result of incomplete densification around large grains. This microstructure seems to indicate that the viscous flow assisted densification mechanism stops as crystallisation and crystal growth proceeds and individual crystallised grains encounter each other.

The deleterious effect of crystallisation during viscous flow densification has been observed by several investigators in different glass-ceramic systems [17, 27-30]. Rabinovich, for example, has indicated that glasses intended for convertion into a glass-ceramic via powder technology and sintering must have a moderate tendency to surface crystallisation [31]. This is because although glass particles would crystallise from their surface, the significant surface area of the powdered material will ensure volume fine crystallisation similar to that in bulk glass-ceramics with heterogeneous nucleation.

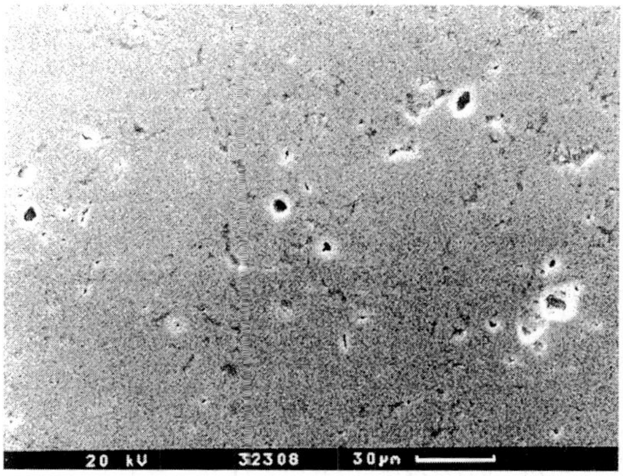

Figure 2 SEM micrograph of a polished section of a glass-ceramic sample sintered at: T_1= 1000 °C, t_1 =6 h (T_1, t_1, see Figure 1)

In general, the interaction between crystallisation and densification can be altered by changing the heating rate [27,28,30]. The general finding is that rapid heating provides better sintering (enhanced densification and retarded crystallisation), otherwise crystallisation arrests sintering due to the increase of the effective viscosity, thus jeopardising the viscous flow assisted densification mechanism [27,28,30]. Another possibility for controlling the relative rate of sintering and crystallisation of glass powders is to chemically treat the glass particle surfaces [17]. Thus, eliminating impurities as well as defects and surface flaws on vitreous particles by chemical etching, the tendency of glass powders to crystallise before sintering can be reduced. For the sintering conditions in the present experiments, the glass powders obtained from vitrified filter ashes seem to behave well in that the densification and crystallisation kinetics are such that fairly dense, highly crystalline products can be fabricated without the need to manipulate the glass particle surfaces by chemical etching, which would add an extra complexity to the process

Figure 3 is a TEM micrograph of a dense area of a sample sintered for 6 h at 1000 °C. The densified material shows a high degree of crystallisation, characterised by a network of interlocked elongated crystals (axial ratio: 3-5) between which smaller, equiaxed crystals of nanometric dimensions are situated. The crystallites are embedded in a residual glassy matrix. The result of XRD analyses of the same sample is shown in Figure 4, demonstrating the crystallisation achieved. The peaks of the XRD pattern could be associated with those of crystalline phases belonging to the pyroxene group, mainly augite, in agreement with the results on bulk glass-ceramics obtained from the same vitrified flyash [15]. Another interesting result here is that the crystalline pattern obtained was reproducible and did not depend on the batch of as-received material taken for the experiments. This indicates that the chemical composition of this particular incinerator fly ash was very homogeneous.

According to previous results [15] on the crystallisation of bulk samples, the elongated crystalline phase, as seen in Figure 3, corresponds to the pyroxene (augite) phase identified by XRD analysis (Figure 4). Although diopside is a pyroxene crystal of general composition $CaO.MgO.2SiO_2$, it is well-known that crystals belonging to the pyroxene group, i.e. augite, can be formed by substitutions of the type: $[CaO.TiO_2]_x$ $[MgO.FeO.ZnO]_{1-x}$ $[Al_2O_3.Fe_2O_3]_y.2SiO_2$. Thus, in the present glass-ceramics, the presence of Fe, Mg, Ca, Zn and Al suggests that the main crystalline phases are in fact pyroxenes of the type augite, similar to those formed in crystallised basalt glasses [32], and in model glasses formulated to simulate the composition of vitrified fly ash [33]. The formation of augite has also been reported during crystallisation of a waste material obtained by vitrification of electric arc furnace dust [34]. Moreover, pyroxene crystals have been also observed in crystallised glasses obtained from other industrial waste rich in Fe, such as fly-ash from thermal power stations [35] and goethite industrial waste [36,37].

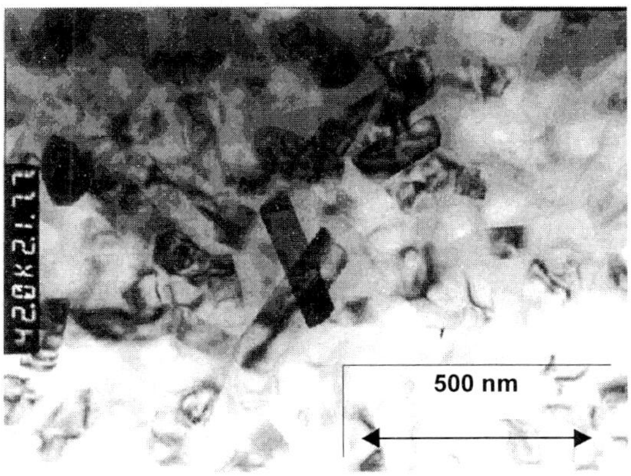

Figure 3 SEM micrograph of a dense area of a highly crystalline sample sintered for 6 h at 1000 °C showing the presence of elongated and equiaxed nanocrystals in a residual glassy matrix

The assessment of the crystallisation kinetics and the accurate identification of crystalline phases and of their chemical composition have been reported to be difficult tasks in this kind of glass-ceramics obtained from vitrified incinerator fly ash [4,9]. Microstructural observations and XRD data have been used in a separate study to determine the crystallisation mechanism active in the bulk materials [15]. It was suggested that crystallisation occurs by the epitaxial growth of the pyroxene crystals from an spinel phase, which being the phase that forms more quickly, would then provide heterogeneous nucleation sites for the further crystallisation progress. According to the morphology of the crystals in the present sintered glass-ceramics, it appears that the crystallisation process follows a similar mechanism as that in the bulk material. Nevertheless no spinel phase was detected by XRD analysis in the present material. Therefore, further studies regarding the crystallisation process of the present materials are needed. In this context, the occurrence of simultaneous densification and crystallisation under different heating rates is being investigated with the aim to elucidate the factors affecting the kinetics of both processes and their complex interaction during sintering.

Mechanical and technical properties

Table 3 shows a summary of the properties measured on a sample sintered using the two step schedule (880 °C/ 4 h, 1000 °C, 6 h). Data from the as-quenched vitrified fly-ash product, the bulk glass-ceramic [12], and from other glass-ceramic materials taken from the literature [14], are also shown in Table 3 for comparison purposes.

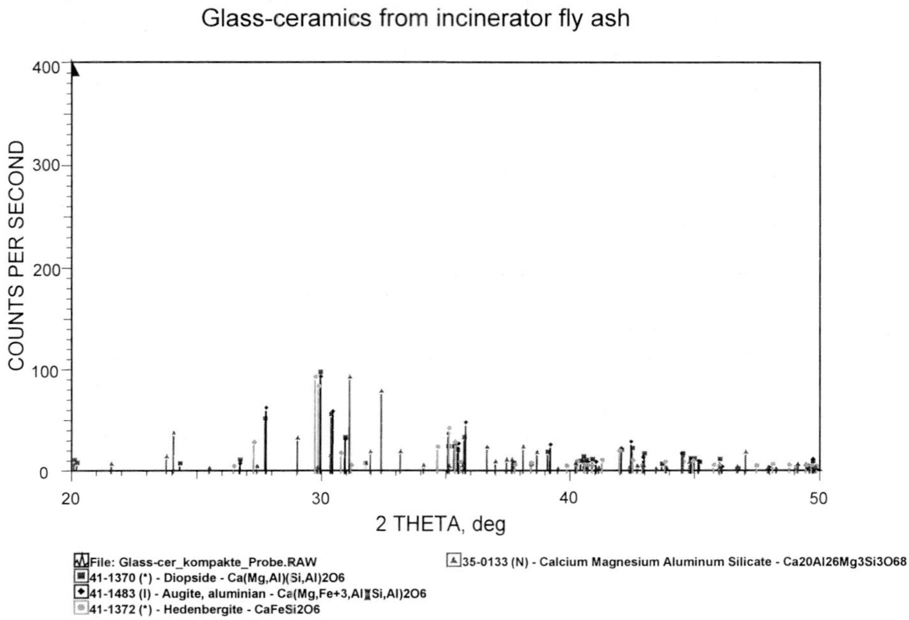

Figure 4 XRD pattern of a glass-ceramic sample sintered for 6 h at 1000 °C showing high level of crystallisation. The peaks correspond to crystalline phases of the type pyroxene, as explained in the text

A SEM image of the fracture surface of the sample sintered under the two-step schedule (880 °C/ 4 h, 1000 °C, 6 h) is shown in Figure 5. The fracture follows a tortuous path, as the crack front encounters the elongated crystallites. This fracture mode may contribute to the higher fracture toughness of this glass-ceramic in comparison to the amorphous glass, as measured by the indentation technique. The presence of the crystalline phase should act also as reinforcement, leading to the higher mechanical strength of the glass-ceramic over that of the amorphous glass. However, the presence of residual porosity in the sintered product may offset the improvement of fracture strength due to crystallisation, resulting in both products having similar fracture strength (see Table 3). The negative effect of porosity on fracture strength of brittle materials is well-documented in the literature [38, 39].

The combination of a relatively high fracture toughness (as compared with that of the vitrified amorphous product, K_{Ic} = 0.6 MPa $m^{1/2}$ [12]) and a relatively low hardness in the present sintered glass-ceramic, results in a very low brittle index (B = 2.5 $\mu m^{-1/2}$ in comparison with B = 9.1 $\mu m^{-1/2}$ for the amorphous glass [12]). As shown elsewhere [40], low brittleness indexes of glass-ceramics and other ceramic materials should lead to relatively high resistance to erosion and abrasion and to good machinability [26].

The improvement of the thermal shock resistance over that of the bulk glass, as indicated by the higher critical temperature difference measured, is also very remarkable. Possibly the presence of residual closed pores in the sintered material is contributing to this enhancement of thermal shock resistance. It has been theoretically and experimentally demonstrated that a low fraction of small, isolated pores enhances the critical temperature difference for crack initiation due to the combined effect that porosity has on the basic properties affecting thermal shock resistance, e. g. Young's modulus, thermal conductivity and fracture strength [41].

Table 3 Measured properties of the sintered glass-ceramic sample (880 °C/ 4h; 1000 °C/ 6h). For comparison, properties of the as-quenched glass, of the bulk glass-ceramic and of other glass-ceramic material are shown

PROPERTY	MATERIAL			
	Sintered glass-ceramic (this work)	As-quenched glass [12]	Bulk glass-ceramic [12]§	Bulk glass-ceramic [13]*
Density (g/cm^3)	2.74	2.80	2.89	2.95
Hardness (GPa)	3.8 ± 0.2	5.5 ± 0.2	7.9 ± 0.3	5.9
Young's modulus (GPa)	85	93	124	134
Poisson's ratio	0.24	0.28	0.26	-
Thermal expansion coefficient (10^{-6} °C^{-1})	6 (20-700°C)	5.9 (20-600°C)	6.5 (20-700°C)	9 (20-700°C)
Indentation fracture toughness (MPa $m^{1/2}$)	1.5 ± 0.1	0.6 ± 0.1	1.7 ± 0.1	1.5
Fracture strength (modulus of rupture, MPa)	88	90	240	-
Thermal shock resistance (water-quench test, ΔT_c, (°C)	~ 300	~ 150	~ 280	-
Brittleness Index ($\mu m^{-1/2}$)	2.5	9.1	4.6	3.9

§ Bulk glass-ceramic material obtained from the same vitrified flyash used in the present study [12]

* Bulk glass-ceramic material obtained from vitrified residues from a Spanish municipal incinerator [13]

CONCLUSIONS

Sintering and crystallisation was shown to be a convenient way for transforming vitrified municipal incinerator fly ashes in glass-ceramic products with attractive mechanical properties. The products have a large application potential. They could be, for example, candidate materials for applications in floors of industrial buildings and constructions, and for outside and inside facings of walls as tiles and pavements. Their relatively high fracture toughness and low brittleness should result in good machinability and high erosion resistance enabling applications also in specialty areas, for example for constructing diverse machine tool pieces. The chemical durability of the products remains to be studied. In particular, it should be investigated how the residual porosity on the products affects the leaching of heavy metals. Further studies of this matter are, thus, necessary, for example by conducting leaching tests of sintered glass-ceramic with different porosities under several conditions. The question regarding the chemical resistance of the sintered products must be fully answered before these glass-ceramics can be realistically considered for the envisaged applications.

Figure 5 SEM image of the fracture surface of a glass-ceramic sample (T_l= 1000 °C, t_l =6 h) showing the tortuous crack path during fracture as a result of crack deflection due to the presence of elongated crystallites in the glassy matrix

ACKNOWLEDGEMENTS

Part of these investigations were carried out with financial assistance of DAAD (Bonn, Germany) and Ministerio de Educación y Cultura (Madrid, Spain) (Project: Acciones Integradas). The experimental assistance of Dr. K. Pfeiffer and Dr. J. Schawohl (TU Ilmenau) and Dr. B. Schunck (RWTH Aachen) is gratefully appreciated.

RERERENCES

1. HJELMAR, O. Disposal strategies for municipal solid waste incineration residues, J. of Hazardous Mat. 47, 1996, pp. 345-358.
2. DERIE, R. A New way to stabilize fly ash from municipal incinerators, Waste Management 16, 1996, pp. 711-716.

3. WILLIAMS, P. T. Pollutants from incineration: an overview. In: Waste Incineration and the Environment, Issues in Environmental Science and Technology, Hester R.E. and Harrison R.M. (eds), Royal Scociety of Chemistry, p. 27-52 (1994).
4. GUTMAN, R. Thermal technologies to convert solid waste residuals into technical glass products, Glastech. Ber. Glass Sci. Technol. 69, 1996, pp. 285-299.
5. TRABER, D., MÄDER, U., EGGENBERGER, U, SIMON, F.-G., WIECKERT, C. Phase Chemistry Study of Products from the Vitrification Processes AshArc and Deglor, Glastech. Ber. Glass Sci. Technol. 72, 1999, pp. 91-98.
6. KRAUß, M. Vitrification of waste materials, Glastech. Ber. Glass Sci. Technol. 70, 1997, pp. 375-381.
7. De LABARRE, D. Vitrification des residus d'incineration, Verre 3, 1997 pp. 33-39.
8. BICKFORD, D.F. and SCHUMACHER R. Vitrification of hazardous and radioactive wastes. Ceram. Eng. Sci. Proc., 1995, pp. 1-10.
9. CHAPMAN, C. C. Evaluation of Vitrifying Municipal Incinerator Ash, Ceram. Trans. Vol. 23, 1991, pp. 223-231.
10. BOCCACCINI A.R., KOPF, M. and ONDRACEK, G. Neue Techniken der Abfallverwertung: Zu den Möglichkeiten einer Verwertung von Filterstauben aus der Abfallverbrennung durch werkstofftechnische Verfahren. Z. Angewandt. Umweltforschung, 7, 1994, pp. 357-367.
11. BOCCACCINI A.R., KOPF, M., and STUMPFE W. Glass-ceramics from filter dusts from waste incinerators, Ceram. Int., 21, 1995, pp. 231-235.
12. BOCCACCINI, A. R., PETITMERMET, M. and WINTERMANTEL, E., Glass-ceramics from municipal incinerator fly ash, Ceram. Bull. 76 [11], 1997, pp. 75-78.
13. ROMERO, M., RAWLINGS, R. D., RINCON, J. Ma., Development of a New Glass Ceramic by Means of Controlled Vitrification and Crystallisation of Inorganic Wastes from Urban Incineration, J. Europ. Ceram. Soc. 19, 1999, pp. 2049-2058.
14. ROMERO, M., RAWLINGS, R. D., RINCON, J. Ma., Crystal Nucleation and Growth in Glasses from Inorganic Wastes from Urban Incineration, J. Noncryst. Solids (1999) submitted.
15. RINCON, J. Ma., ROMERO, M., BOCCACCINI, A. R., Microstructural Characterisation of a Glass and a Glass-ceramic Obtained from Municipal Incinerator Fly Ash, J. Mat. Sci. 34, 1999, pp. 4413-4423.
16. GONZALEZ OLIVER, C. J. R., Sintering of Glass-Ceramics, in: Glass Ceramic Materials, Fundamentals and Applications, Ed. by T. Manfredini, G. C. Pellacani and J. Ma. Rincón, Mucchi Editore, Modena, 1997, pp. 93-100.
17. CLARK, T. J., and REED, J. S., Kinetic Processes Involved in the Sintering and Crystallisation of Glass Powders, J. Am. Ceram. Soc. 69, 1986, pp. 837-846.
18. GUTZOW, I., PASCOVA, R., KAMARANOV, A., SCHMELZER, J., The Kinetics of Surface Induced Sinter Crystallisation and the Formation of Glass-Ceramic Materials, J. Mat. Sci. 33, 1998, pp. 5265-5273.
19. KIM, H. S., RAWLINGS, R. D., ROGERS, P. S., Sintering and Crystallisation Phenomena in Silceram Glass, J. Mat. Sci. 24, 1989, pp. 1025-1037.
20. BERZINA, L., CIMDINS, R., ROZENSTRAHUA, I., BOSSERT, J. and KRAVTCHENKO, I., Glass-ceramics with Multibarrier Structure Obtained from Industrial Waste, Key Eng. Mat. 132-136, 1997, pp. 2228-2231.
21. BOCCACCINI, A. R., BÜCKER, M., BOSSERT, J., Glass and Glass-Ceramics from Coal Flyash and Waste Glass, Tile and Brick International 12, 1996, pp. 515-518.
22. BERZINA, L., CIMDINS, R , ROZENSTRAHUA, I., BOSSERT, J., BÜCKER, M. Environment Friendly Glass Ceramic From Metallurgical Waste, Br. Ceram. Proc. Nr. 60, Vol. 1, 1999, pp. 179-180.
23. RINCON, J. Ma., PELINO, M., ROMERO, M., Glass-Ceramics Obtained from Axial Pressing and Sintering of Vitrified High Iron Content Red Muds, in: Proc. 1st Nat. Congress

Valorisation and Recycling of Incustrial Wastes, Ed. by M. Pelino and G. C. Pellacani, Mucchi Editore, Modena , 1997, pp. 169-182.

24. KAMARANOV, A., PELINO, M., TAGLIERI, G., CANTALANI, C., Sintered Building Glass-Ceramics Based on Jarosite, in: Proc. XVIII International Congress on Glass, San Francisco (USA), July 5-10, 1998, The American Ceramic Society, Ohio, M. K. Choudhary, N. T. Huff and Ch. H. Drummond III (Eds.).
25. ANSTIS, G. R., CHANTIKUL, P., LAWN, B. R., MARSCHALL, D. B., A Critical Evaluation of Indentation Techniques for Measuring Fracture Toughness: Direct Measurements, J. Am. Ceram. Soc. 64, 1981, pp. 533-538.
26. BOCCACCINI, A. R., Machinability and brittleness of glass-ceramics, J. Materials Processing Technology 65, 1997, pp. 302 - 304.
27. BOCCACCINI, A. R., STUMPFE, W., TAPLIN, D. M. R., PONTON, C. B., Densification and Crystallization of Glass Powder Compacts during Constant Heating Rate Sintering, Mat. Sci. Eng. A219, 1996, pp. 26-31.
28. PANDA, P. C. and RAJ, R. Sintering and Crystallisation of Glass at Constant Heating Rates, J. Am. Ceram. Soc. 72, 1989, pp. 1564-66.
29. MONTEIRO, R. C. C., LIMA, M. M. R., Coalescence and Crystallisation in Powdered Glasses, in Third Euroceamics, Vol. 2, Ed. by P. Duran and J. F. Fernandez, Faenza Ed. Iberica, 1993, pp. 1181-1186.
30. ZIMMER, J., RAETHER, F., MÜLLER, G., In-Situ Investigations of Sintering and Crystallisation on Lithium Aluminosilicate Glass-Ceramics, Glastech. Ber. Glass Sci. Technol. 70, 1997, pp. 186-188.
31. RABINOVICH, E. M., Cordierite Glass-Ceramics Produced by Sintering, in: Advances in Ceramics Vol. 4, Ed. by J. H. Simmons, D. R. Uhlmann, G. H. Beall, The American Ceramic Society, Columbus, 1982, pp. 327-333.
32. CHICK, L. A., LOKKEN, R. O. and THOMAS, L. E., Basal glass-ceramics for the immobilization of transuranic nuclear waste, Ceram. Bull. 62, 1983, pp. 505-516.
33. KIPKA, A., LUCKSCHEITER, B. and LUTZE, W., Melting of fly ashes and product properties, Glastech. Ber. 66, 1993, pp. 215-220.
34. GAO, Z., DRUMMOND III, C. H., Thermal Analysis of Nucleation and Growth of Crystalline Phases in Vitrified Industrial Wastes, J. Am.Ceram. Soc. 82, 1999, 561-565.
35. BARBIERI, L., MANFREDINI, T., QUERALT, I., RINCON, J. Ma. and ROMERO, M., Vitrification of fly ash from thermal power stations, Glass Technol. 38, 1997, 165-170.
36. PELINO, M., CANTALANI, C. and RINCON, J. Ma., Preparation and properties of glass-ceramic materials obtained by recycling goethite industrial waste, J. Mat. Sci. 32, 1997, pp. 4655-4660.
37. ROMERO, M. and RINCON, J. Ma., Preparation and properties characterisation of high iron content glasses obtained from FeOOH industrial wastes, J. Europ. Ceram. Soc. 18, 1998, pp.153-160.
38. BOCCACCINI, A. R., Fabrication, Microstructural Characterisation and Mechanical Properties of Glass Compacts Containing Controlled Porosity of Spheroidal Shape, J. Porous Materials 6, 1999, pp. 369-379.
39. RICE, R. W., Porosity of Ceramics, Marcel Dekker, New York, 1998.
40. BOCCACCINI, A. R., The Relationship Between Wear Behaviour and Brittleness Index in Engineering Ceramics and Dispersion Reinforced Ceramic Composites, Interceram 48, 1999, pp. 176-187.
41. ARNOLD, M., BOCCACCINI, A. R , ONDRACEK, G., Theoretical and experimental considerations on the thermal shock resistance of sintered glasses and ceramics using modeled microstructure-property correlations, J. Mat. Sci. 3, 1996, pp. 463-469.

REUSE PERSPECTIVES OF INCINERATOR FLUE GAS RESIDUES FOLLOWING $NaHCO_3$ OR $Ca(OH)_2$ INJECTION

V Campanaro
L Liberti
G F Liberti
M Notarnicola
University of Bari
Italy

F Roethel
L Swanson
State University of New York
United States of America

ABSTRACT. $NaHCO_3$ has recently been proposed as $Ca(OH)_2$ substitute for neutralizing acid flue gases in MSW incineration plants, but the impact of such substitution on the resulting solid residues has not been fully assessed yet. To this aim, mixed residue (fly ash + powdered products from acid gas neutralization sections) following $NaHCO_3$ or $Ca(OH)_2$ injection in a recently upgraded 440 t_{msw}/d waste-to-energy facility has been characterized and processed by a cement-based stabilization/solidification (S/S) technique for possible reuse as construction material. Experimental results (workability, strength, porosity and leachability) show that the two neutralizing agents greatly influence the S/S technique as well as potential reuse of the final product.

Keywords: MSW, Incineration, Ash, $NaHCO_3$, $Ca(OH)_2$, Cement, Stabilization/Solidification.

Dr V Campanaro, is a PhD researcher at the Polytechnic University of Bari, Italy, specializing in solidification/stabilization techniques for hazardous waste treatment.

Professor L Liberti, is a full Professor of Environment Protection at the Polytechnic University of Bari. His present research interests are in the field of advanced treatment and disinfection of municipal wastewater as well as solid waste management.

Dr GF Liberti, is a research student at the Polytechnic University of Bari currently involved in incinerator flue gases advanced treatment and resulting residues management.

Dr M Notarnicola, is an Assistant Professor of Materials Science and Technology at the Polytechnic University of Bari. His research interests include inertization treatment and reuse of incineration residues in civil engineering.

Professor F Roethel, is a research Professor at the Waste Reduction and Management Institute, State University of New York at Stony Brook, USA.

Professor L Swanson, is a full Professor and Director of the Waste Reduction and Management Institute, State University of New York at Stony Brook, USA.

INTRODUCTION

Acceptance of municipal solid waste (MSW) incineration is resulting in the evolution of novel and enhanced techniques at waste-to-energy (WTE) plants. Innovation in plant configuration can improve performance, especially on atmospheric emissions and economics, but quite often the resulting modification of solid residues is not properly assessed particularly if these latter undergo specific treatment (e.g., Solidification/Stabilization) to ease disposal and even reuse.

A typical such innovation is the substitution of hydrated lime ($Ca(OH)_2$) with baking soda ($NaHCO_3$) as neutralizing agent of acidic flue gas, promoted in the last decade in particular by Solvay Ltd., Europe's largest producer of baking soda (product trade name "BICAR") and actually employed in several WTE plants in Northern Italy. Reason for this substitution is that soda allows faster kinetics and higher reaction efficiency, thus yielding smaller amount of finer solid residue (classified as "hazardous waste" by Italian law), whose safe disposal is a major cost and concern.

Solidification/Stabilization (S/S) techniques are largely used to decrease pollution impact of incineration residues, making acceptable their eventual reuse [1-2]. S/S techniques employ inorganic or organic reagents and, within the former, cement is widely used. Cement based S/S processes may be applied to solidification of dust, liquid and sludge, ensuring low cost, wet matter compatibility, basic environment with immediate solubility decrease of many potentially toxic anions and cations, with the strength of the products growing with time [3]. While other matrices (i.e., organic polymers) achieve purely physical confinement of wastes, cement provides a significant chemical immobilization potential [4].

By reference to Coriano's WTE plant (400t/d) treating MSW (municipal solid waste) and IHW (infectious hospital waste) where soda has recently substituted lime for neutralization of acidic flue gas, this paper describes influence of such substitution on characteristics of solid residues, further submitted in laboratory to inorganic binder (cement) based S/S treatment.

EXPERIMENTAL

Plant Description

WTE plant configuration is essential to understand the conditions of formation of solid residues. The plant investigated, located in Coriano (FO, Italy), owned and managed by CIA-AMIA, a public Consortium among seven municipalities (Cattolica, Gabicce, Misano, Rimini, Riccione, S.Giovanni in M. and Santarcangelo di R.), consists of three parallel lines with moving grate furnace, two constructed in 1976, the third added in 1992, for an overall capacity of approx. 440 MSW t/d. Following Italian regulations on waste incineration (D.C.I.27/7/84) and atmospheric emission control (D.P.R. n.203/88), flue gas treatment in Lines 1 and 2 (120 t/d each), initially carried out by an air heat exchanger and electrostatic precipitator, was upgraded in 1991 to the configuration shown in Figure 1.

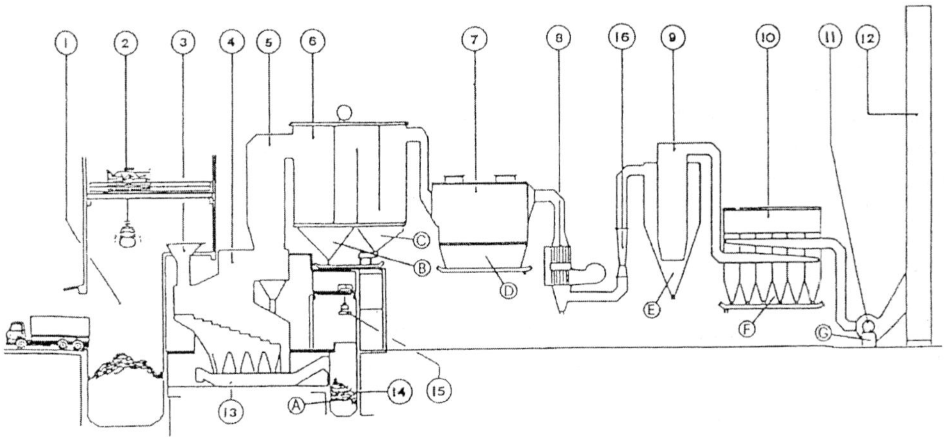

LEGEND

1 refuse pit
2 refuse bridge crane
3 charging hopper
4 furnace
5 post combustion chamber
6 steam boiler
7 electrostatic precipitator
8 air-flue gas heat exchanger
9 cyclonic reactor
10 baghouse
11 flue gas fan
12 stack
13 bottom ash chain conveyor
14 bottom ash pit
15 bottom ash bridge crane
16 Venturi tube

Figure 1 Coriano plant scheme

First, a postcombustion chamber was added in order to destroy organo-chlorinated compounds (gas temperature ≥950°C, contact time ≥2s , inlet gas speed ≥10m/s and oxigen concentration ≥6% by volume). An air heat exchanger with a PTFE tube bundle was also placed after the electrostatic precipitator to reduce gas temperature to approx.150°C. Then hydrated lime was injected into a Venturi tube followed by a cyclone reactor to neutralize halogenated acids (HX) and sulphur oxides (SO_x). Final dedusting and heavy metals removal was carried out in a new textile PTFE baghouse. Partially unexhausted lime collected by the cyclone and the baghouse was recirculated to increase reaction efficiency. Since March 1992, powdered activated carbon is injected in the same way as hydrated lime to control Hg emissions [5].

Line 3, constructed in 1992, is almost identical to Lines 1 and 2, only bigger (200 t/d). Due to continuous increase of MSW calorific value (1800 kcal/kg in 1975 compared to 2500 kcal/kg in 1992), in 1994 all three Lines were equipped with a steam boiler to achieve energy recovery in the form of steam and, since 1997, electricity. In 1991 the plant received a regional permit to incinerate also infectious hospital waste (IHW) in approx. 10% (IHW) to 90% (MSW) ratio. In 1995 hydrated lime was replaced by baking soda (BICAR) with no more reagent recycle.

Materials and Methods

Residue from final silo (FSR), where fly ash from dedusting devices and powders from acidic gas neutralization are collected, was investigated as representative of different fine solids wasted from the plant. Eight hourly samples (3 dm^3 each) were taken daily from 8 a.m. to 4 p.m. and then mixed to yield a representative sample. Sampling operations were carried out before and after the change from lime to soda (1995) as deacidificating agent, when plant operated in steady conditions.

FSR was characterized for dry (to avoid soluble salts elution) particle size analysis according to UNI 8520.

Loss on ignition (LOI) was further determined to account for uncombusted components, hydration water and carbonates present in the finer fractions [6]. To this aim a 30g sample oven dried to constant weight at 105°C was kept two hours at 550°C and then correspondingly weight loss determined.

FSR was also characterized for content of Cd, Cr, Cu, Pb, heavy metals with reportedly relevant influence on S/S treatment. A microwave extraction technique was used according to USEPA "Test Methods for Evaluating Solid Waste" SW 846 (0,5g sample in 10 ml concentrated HNO_3 for 10 minutes in a microwave oven) followed by atomic absorption spectrophotometry in flame and graphite furnace (AAS Mod. Spectra 800 by Varian).

Leaching with acetic acid agent (IRSA-CNR Metodi Analitici per i Fanghi) was performed on the residue for the metals indicated upon.

X-ray powder diffrattometry tests were carried out to evidence inorganic compounds on sample with size reduced to less than 300μm, using a Philips PW1710 apparatus (Cu anode tube Kα 1,54018, 30kV-55mA, 2Θ = 5-55°).

After FSR characterization, the cement based S/S technique was used to prepare mortars. As experienced by other Authors, hydraulic binder (CEM II A/L) based mortars, where cement was partially substituted by incineration residue, were prepared using duplicate FSR representative samples (wt. ratio cement to sand 1/3, water to cement ratio ½)[7-8-9]. In all cases siliceous sand according to UNI EN 196/1 was used. No more than 30% cement substitution with FSR was tested to avoid excessive worsening of specimen strength.

Sodium bicarbonate based FSR showed very poor mortars workability and requested addition of a commercial water reducer (MAPEI N200, 1,5% of binder weight) as well as increase of water to binder ratio from 0,5 to 0,65-0,70. Unsubstituted mortars were obviously prepared as control. Each mortar was submitted to flow test (UNI 7044). Time of set was also determined by Vicat device according to EN 196-3. Prismatic 40x40x160mm specimens were prepared and aged under water at 7, 28 and 60 days. Compression tests were then carried out on specimens previously broken by flexure.

Material microstructure was also assessed by SEM (Philips mod.XL20) observations and micrographs of fracture surfaces of broken samples.

RESULTS AND DISCUSSION

FSR Characterization

Granulometric curves in Figure 2 show that finer part is most abundant with both reagents (≥74% wt. passing 0,075 mesh).

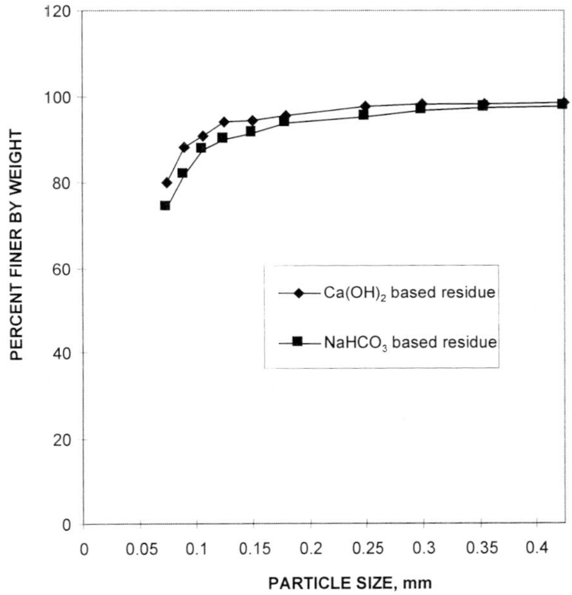

Figure 2 Granulometric curves of $Ca(OH)_2$ and $NaHCO_3$ based residues

LOI for $Ca(OH)_2$ (5,67%) was appreciably higher than for $NaHCO_3$ based residue (3,15%) pointing out sensible difference in corresponding formation conditions.

Heavy metals concentration in either lime or soda based FSR (Table 1) seems to indicate a slightly smaller capturing ability of soda, except for Pb that, accordingly, is also less abundant in leachate (Table 2).

Table 1 Heavy metals in final silo residues (mg/kg)

ELEMENT	$CA(OH)_2$ BASED	$NAHCO_3$ BASED
Cd	175	166
Cr^{VI}	230	191
Cu	736	633
Pb	5409	5648

Table 2 Heavy metals in CH_3COOH test leachate (mg/l)

ELEMENT	$CA(OH)_2$ BASED FSR	$NAHCO_3$ BASED FSR
Cd	0,20	0,37
Cr^{VI}	0,55	2,46
Cu	0,30	0,43
Pb	6,40	1,59

It should be noted that these values call for the costly disposal of both residues in the II-C type sanitary landfills, according to Italian regulation (DCI 27/7/84).

Diffrattometry evidenced presence of $CaCl_2$, $CaSO_3$, SiO_2, Al_2O_3, Fe_2O_3 and $Ca(OH)_2$, for lime added residues. In sodium bicarbonate added residues it was relieved presence of NaCl, Na_2SO_3. Na_2CO_3 and ferrous oxide.

Cement Based S/S Treatment

Mortars: rheological characteristics

Flow test evidenced sensible difference among mortars prepared (see Table 3). As already mentioned $NaHCO_3$ caused very poor workability that, despite of water reducer and higher w/c ratio, decreased so rapidly to make the mortar significantly stiffer in just few seconds.

Table 3 Mortars workability (flow test)

$Ca(OH)_2$ based	$NaHCO_3$ based	Control
31	20	70

As expected, time of set showed similar behaviour with Vicat needle not even able to penetrate soda mixes (see Table 4). Furthermore, attempts to evaluate water for "normal" paste were quite difficult as few milliliters of water added changed it from too liquid to too stiff.

Table 4 Time of set of mortars

	WATER FOR "NORMAL" PASTE (g)	INITIAL SETTING TIME	FINAL SETTING TIME
Control	123	2h 00'	3h 45'
$Ca(OH)_2$ based FSR paste	143	2h 15'	4h 05'
$NaHCO_3$ based FSR paste	97	immediate	immediate

Mortars: mechanical strength

Compression strength of cured specimens previously broken by flexure is reported in Figure 3. As expected, residue added mortars have strength significantly lower than control but $NaHCO_3$ based mortars show very poor behavior at all curing times [10].

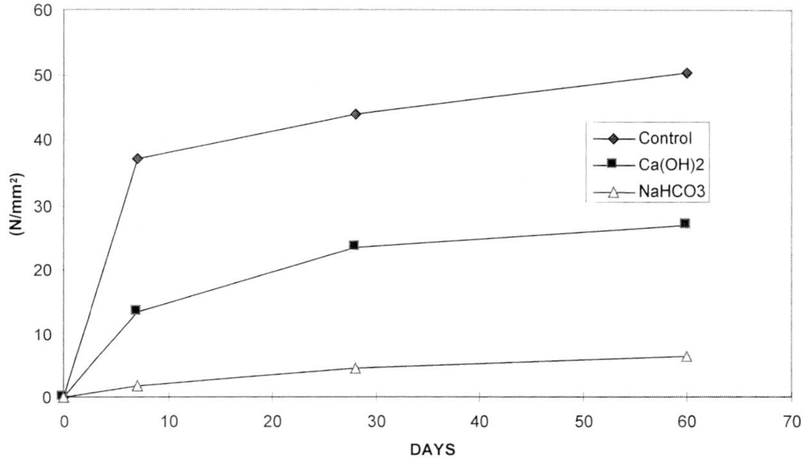

Figure 3 Compression strength of $Ca(OH)_2$ and $NaHCO_3$ based residue mortars

Neither the higher w/c nor heavy metals concentration can justify the performance breakdown of $NaHCO_3$ based FSR mortars compared with control and even $Ca(OH)_2$, calling probably for interaction between cement paste and residues added (e.g., unreacted $NaHCO_3$ and salts from neutralization reactions). Infact residues collected before acidic flue gas neutralization showed lower mortars resistance than lime but consistently higher than soda based residues.

These results suggest that the cement based S/S technique is only attractive towards lime based FSR mortars and that reuse perspectives in civil engineering construction of soda based residues should be confined to instances where poor workability and strength are required.

Mortars: microstructure

SEM micrographs (Figure 4a, b, c) of fracture surface of control, $Ca(OH)_2$ and $NaHCO_3$ based residue mortars tend to confirm that development of typical hydration products is inhibited or slower with soda based residues characterized by a less compact microstructure.

Mortars: leaching

It must be pointed out that in force Italian regulations do not permit yet reuse in civil engineering of waste incineration residue that, therefore, must be disposed of in sanitary landfills with increasing protection degree (II-B or even II-C type) related to its leaching behavior. As shown in Table 5, cement S/S treatment allows disposal of both FSR in II-B rather than in more expensive II-C sanitary landfills.

a) Control

b) $Ca(OH)_2$ based FSR

c) $NaHCO_3$ based FSR

Figure 4 SEM of fracture surface of control, $Ca(OH)_2$ and $NaHCO_3$ based residue mortars

Table 5 Leaching behavior of cement mortars (mg/l)

ELEMENT	$CA(OH)_2$ BASED FSR	$NAHCO_3$ BASED FSR	ITALIAN LIMIT FOR DISPOSAL IN II-B TYPE LANDFILL (*)
Cd	0.08	0.12	0.2
Cr^{VI}	0.25	0.38	2
Cu	0.03	0.05	1
Pb	0.16	0.26	2

(*) Italian regulation DCI 27/7/84

CONCLUSIONS

Acidic flue gas neutralization with $NaHCO_3$ instead of $Ca(OH)_2$ in MSW waste-to energy plants appreciably influences physical-chemical and mechanic characteristics of ash residues as well as the behavior of their cement Stabilization/Solidification derivates. In particular, poor workability of fresh mixes and rapid strength decay seem to deny reuse opportunities for cement mortars prepared with $NaHCO_3$ based residues.

ACKNOWLEDGEMENTS

The authors wish to thank CIA-AMIA, owner and manager of the incineration plant, for cooperation and assistance throughout the investigation.

REFERENCES

1. INTERNATIONAL ASH WORKING GROUP, "An International Perspective on Characterisation and Management of Residues from Municipal Solid Waste Incineration - Summary Report", Ed.Netherlands Energy Research Foundation, Petten the Netherlands, 1994

2. COBB, J T, MANGELSDORF, C P, BLACHERE J R. High-strength concrete containing MSW incinerator ash in Clean energy from waste and coal. American Chemical Society, 1993, pp264-275

3. ALI, M T, CHANG, W F. Strength properties of cement-stabilized Municipal Solid Waste incinerator ash masonry bricks", ACI Materials Journal, Vol.91, 1993, pp 256-263.

4. GLASSER, F P ."Chemistry of Cement-Solidified Waste Forms", in Chemistry and Microstructure of Solidified Waste Forms , Ed. By R.D.Spence, Lewis Publishers Boca Raton Fla USA, 1992, pp 2-14

5. LIBERTI, L, NOTARNICOLA, M, AMICARELLI, V, CAMPANARO, V, ROETHEL, F, SWANSON, L. Mercury removal with powdered activated carbon from flue gases at the Coriano municipal solid waste incineration plant. Waste Management and Research, Vol.16, 1998, pp183-189.

6. U.S. ENVIRONMENTAL PROTECTION AGENCY - RISK REDUCTION ENGINEERING LABORATORY EPA/600/R-93/167: Evaluation of solidification/stabilization treatment processes for municipal solid waste combustion residues, 1993.

7. HWA, T J, KIAT, C H. Use of ash derived from refuse incineration as a partial replacement of cement. Cement and Concrete Research., Vol.13, 1991, pp 171-175,.

8. HAMERNIK, J D AND FRANTZ G C. Physical and chemical properties of municipal solid waste fly ash. ACI Materials Journal, Vol.88, 1991, pp 294-301.

9. MONZO', J , PAYA', J , BORRACHERO, M V , AND PERIS-MORA. Mechanical behaviour of mortars containing sewage sludge ash (SSA) and Portland cements with different tricalcium aluminate content. Cement and concrete research, Vol.29, 1999, pp 87-94.

10. HAMERNIK, J D AND FRANTZ G C. Strength of concrete containing municipal solid waste fly ash. ACI Materials Journal, Vol.88, 1991, pp 508-517.

CHARACTERISATION OF Cr, Pb AND Zn IN MUNICIPAL WASTE INCINERATION ASHES BY MICRSCOPIC AND SPECTROSCOPIC METHODS

B Gagnepain
French Agency for Energy and Environment

R Flamand
D Gauthier
G Flamant
Institute of Material Science and Process Engineeirng, (IMP-CNRS)
France

ABSTRACT. A whole set of residues (bottom ashes, superheater ashes, economiser ashes and electrostatic precipitator dusts) coming from the same mass burn Municipal Solid Waste incinerator was studied by X-ray microscopy and spectroscopy techniques, plus a series of chemical extractions, in order to draw general trends on heavy metal chemical and physical forms in these residues. The study was mainly focused on Cr, Pb and Zn. X-ray microanalysis (XRM) showed that Pb and Cr are very often concentrated in little nodules, occluded in larger particles made of Ca, Mg, K silicates or aluminosilicates. Zn is widely dispersed in little or large particles associated with Si, Al, O, Ca, Mg, Fe, Na or K. X-ray photoelectron spectroscopy (XPS) proved that alkaline chlorides (and perhaps $ZnCl_2$) are enriched on ash surface. The comparison between the results from both XRM and sequential chemical extractions indicates clearly that for Pb, the possible chemical forms existing in the whole set of residues are the pure metal, its oxide, perhaps its sulfate and complex silicates; in economizer ashes, Pb chloride exists additively.

Keywords: Heavy metals, MSW incineration ashes, Characterization, SEM/XRM, XPS, Speciation

Dr B Gagnepain, is currently an engineer employed by the French Agency for Energy and Environment (ADEME). He obtained his PhD in December '98 on the subject of heavy metals speciation in solid residues from municipal waste incineration.

Mr R Flamand, is a senior engineer at the Institute of Material Science and Process Engineering (IMP), belonging to CNRS (Centre National de la Recherche Scientifique), located in Odeillo, France. He is the responsible of the Microanalysis Department.

Dr D Gauthier, is is a senior research engineer at the Institute of Material Science and Process Engineering (IMP-CNRS), Odeillo, France. His reseach interests include the fate of heavy metals and trace elements in incineration and combustion processes.

Dr G Flamant, is Head of the Group "High Temperature Reactors" at the Institute of Material Science and Process Engineering (IMP-CNRS), Odeillo, France. He is the author or co-author of more than 100 papers in international journals in the fields of energy and high temperature processes.

INTRODUCTION

Incineration appears to be a convenient strategy to solve the more and more worrying problem of household waste management. Indeed, municipal solid waste (MSW) incineration permits a great reduction of waste volume (about 70%) and energy recovery by power generation. Unfortunately, unlike organic matter which is destroyed during the combustion, metals and especially harmful heavy metals (HM) are only transformed and concentrated in the different produced residues (solid and gaseous emissions), which means that these residues may represent potential sources of pollution, mainly by leaching.

Most previous studies related to MSWI ashes are mainly concerned with the elemental composition [1-5]. But it is proved that the toxicity of these ultimate residues depends for a large part on the physical and chemical forms (so-called speciation) of toxic elements in MSWI ashes [6]. Moreover, the residue recycling capacity is strongly related to these properties; this is especially the case of bottom ashes, which represent about half the total volume of ultimate residues, and that engineers try to reuse either in concrete or in road construction. It is thus essential to improve our knowledge about HM speciation in ashes.

In this mind, Plüss and Ferrell [7] attempted to characterize electrostatic precipitator (ESP) dusts with scanning electronic microscopy-energy dispersive scanning (SEM/EDS) and tried to identify possible Pb chemical forms in these residues with X-ray quantitative microanalysis. With X-ray diffraction technique, Artelt et al [8] detected the presence of Zn chlorides and Pb sulphates in ESP dusts. Recently, several teams carried out more thorough speciation studies on bottom ashes [9], ESP dusts [10], mixtures of both these ashes [11] and air pollution control (APC) residues [12]. These studies were based on a comprehensive approach [9] using association of spectroscopy and microscopy techniques. By combining the informations about the ash surface environment provided by the different techniques, the mapping of HM geographic distribution, and the element or phase associations, the authors tentatively identified probable and possible chemical and physical forms of HM in ashes. In fact, a complete review of the applications of all physico-chemical techniques for characterizing ashes was given in the treatise on MSW incinerator residues characterization, disposal, treatment and utilisation [13] elaborated by the International Ash Working Group.

The purpose of our study was to develop a characterization study, on a wide scale, by analyzing all the residues generated by a MSW incinerator. X-ray microanalysis (XRM) using an electronic microprobe provided information about the ash morphology, the HM localization in ashes and the composition of ash particles containing HM. X-ray photoelectron spectroscopy (XPS) permitted to detect elements which are present only on the ash particle surface and non detected with the microprobe, and to check the possible surface enrichment in HM; we also tried to determine chemical bindings between HM and the other components of ashes by this technique. Finally, taking into account the results of chemical extractions series performed on these ashes, we intended to establish correlations between the different informations obtained, and we drew general trends about Pb speciation in incineration residues.

SAMPLING AND ANALYSIS METHODS

Samples preparation

Four kinds of residues coming from the same incinerator (a French mass burn MSW waste-to-energy facility, of capacity: 28 t/h) were analyzed: bottom ashes (BA), super heater ashes (SHA), economizer ashes (EA) and electrostatic precipitator dusts (ESPD). The corresponding temperatures of the sampling zones are respectively: 900-1000 °C (BA), 650 °C (SHA), 450 °C (EA), and 200 °C (ESPD).

The weight of initial samples ranged from 1 kg up to 5 kg. After air-drying, the ashes were homogeneized as much as possible in order to work on most possibly representative samples, then sieved to less than 4 mm, crushed, coned and quartered to prepare working samples of weight 20-30 g. Working samples were stored at ambient temperature and protected from light and moisture. Then, before any analysis, they were dried at 60°C for one hour.

X-ray microanalysis by microprobe

Ash particles were examined by using an electronic microprobe "CAMECA" Microbeam, controlled by a PC, operating in wavelength-dispersive scanning (WDS) mode with an accelerating voltage of 15 kV and intensity ranging from 10 nA (for major elements) to 15 nA (for trace elements). The microanalysis principle is as follows: the morphology is first visually observed with the scanning electronic microscope (SEM) technique, examining polished sections of ash particles embedded in epoxy resin. Then the selected ash particles are observed with four different X-ray WDS detectors: LIF for metals, PET for metals and major elements, TAP for major elements and PC2 for light elements, and their components are located.

X-ray microanalysis was divided in two parts. First, the qualitative work which can be defined as:

* Research of particles containing HM in the whole sample
* Acquisition of an X-ray spectrum to identify all the components of the selected ash particles (those containing HM)
* Localization and mapping of HM geographic distribution.

Then, the wavelength-dispersive semiquantitative approach, achieved with the standardless method [7], which gives more information about the composition of the selected ash particles.

X-ray photoelectron spectroscopy (XPS)

Samples-to-be-observed were obtained by compacting ash powders, then cut into thin lozenges (5 mm × 5 mm and 1 mm high). The surface analysis is done as follows: samples are sticked onto a tantalum tape and the chamber vacuum is around 10^{-7}-10^{-8} Torr. Samples are submitted to an Al Kα X-ray source, and scanned in a low energy resolution survey from 1450 $e^{-}V$ to 150 $e^{-}V$, with acquisition every 0.5 $e^{-}V$ in 100 ms. The surface sample composition is obtained after calculating the 2p peak area.

RESULTS AND DISCUSSION

Characterisation of ashes

First, HM contents of ash samples were determined by inductively coupled plasma - atomic emission spectrometry (ICP-AES), after total digestion of a 1g-sample in a Teflon flask, using $HCl/HNO_3/HF$ acid mixture as defined by Gagnepain [14]; the objective of determining the global HM content was to select the metals to be searched in ash particles with the electronic microprobe (the ponctual detection limit is rather high: 300 to 500 ppm depending on the metal). The results are summarized hereafter, in ppm or µg/g of solid:

As: 2.5-20 ppm
Cd: 5-205 ppm
Cr: 204-445 ppm
Cu: 440-930 ppm
Ni: 120-1050 ppm
Pb: 675-6200 ppm
Zn: 1360-20180 ppm

According to these concentration levels, only Zn, Pb, Cu, Ni, and scarcely Cr, are likely to be detected and scanned with microprobe. Therefore, our research was mainly focused on Pb, Cr and Zn, but we also payed attention to other metals if present. In addition, when metals are very localized and concentrated in these points, they should be detected even if their mean elemental concentration is under the detection limit.

Ash morphology

The great heterogeneity of MSW ashes is noticeable when examining their morphology. We observed several kinds of particles in our samples, extremely different. Figures 1, 2, 3 and 4 (secondary electron or SE images) give some examples of particles found respectively in bottom ashes (BA), superheater ashes (SHA), economizer ashes (EA) and electrostatic preçipitator dusts (ESPD). Generally, the samples of MSW residues which were analyzed contained the following types of particles:

* smooth siliceous particles, with uniform aspect or slightly cracked, slightly colored, spherical or ovoid , sized from several tens to several hundreds of µm; observed in all residues;
* particles with a slightly colored matrix, containing black dots, mainly made of CaO, of different shape (oval, spherical, or some times of very complicated shape) and size (from several tens to several hundreds of µm); essentially observed in BA;
* composite particles, presenting both smooth and rough zones; observed in all residues;
* very large particles (several hundreds of µm), made of a dark matrix and containing little occluded spherical nodules, either white or dark; observed in SHA, EA and ESPD (that-is-to-say in all residues but BA).

Similarly, Flüss and Ferrell [7] observed different morphologies in ESP dusts:

* slight gray, spherical to angular particles
* individual small particles with diameters between 1 to 10 µm
* large heterogenous glassy particles
* aggregates of small particles.

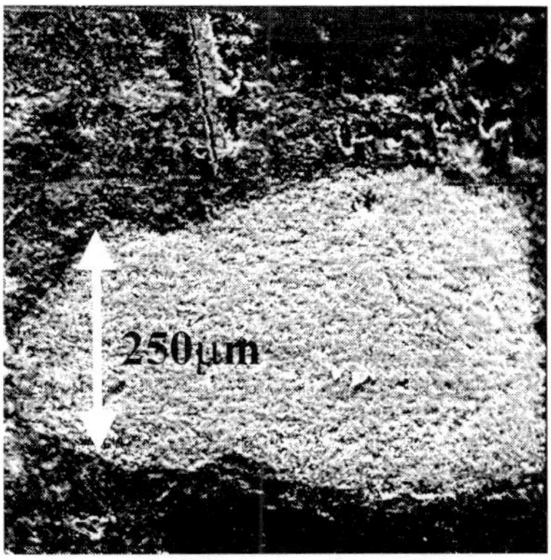

Figure 1 Bottom ash particle

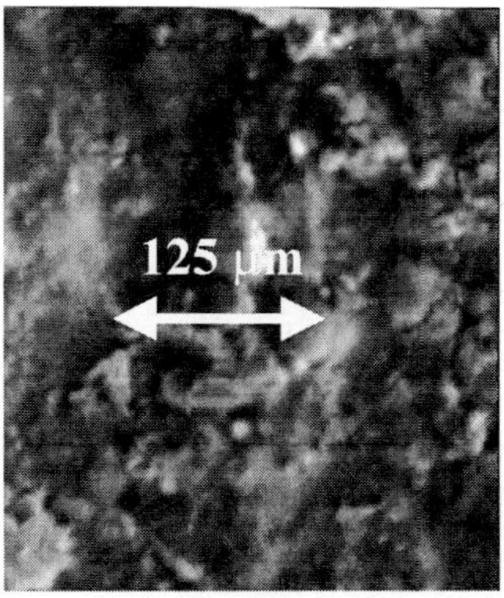

Figure 2 Superheater ash particle

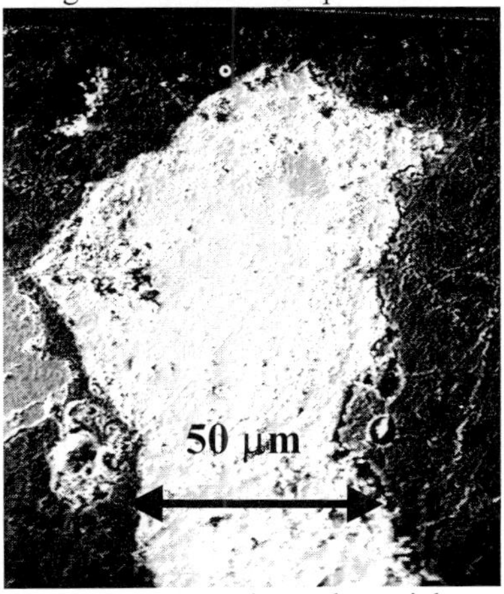

Figure 3 Economiser ash particle

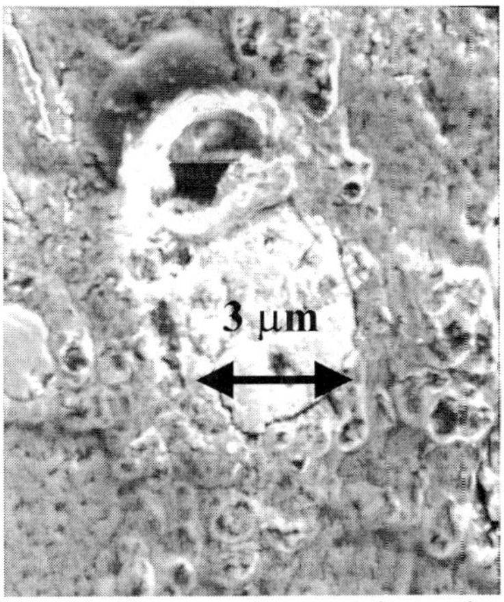

Figure 4 Electrostatic precipitator dusts

Our observations are closer to Plüss and Ferrell's results than to those of Eighmy et al's [10] whose residues seemed to be more homogeneous. The latter depicted ESP dusts as very fine-grained and agglomerated material constituted of 0.05-1 μm-sized polycristalline platelets accumulated in 0.1-10 μm aggregates, covering 5-100 μm in diameter spheres. Nevertheless, we must keep in mind that MSWI ashes can be slightly different from one incinerator to another, according to the types of waste burnt and technology employed.

Localisation of heavy metals and mapping of HM-containing particles

First, we noticed that Zn is widely distributed in the samples and thus easily detectable and localizable, whereas Pb is more scarcely detected due to its X-ray absorption properties and hyper localization. In order to tentatively establish correlations between the presence of several components in the particles, we did first a general mapping of the scanned particle components, then we drew the cartography of each particle including the respective geographic distributions of these components. Two examples of such cartographies, in the cases of a superheater ash particle and of an economizer ash particle, are given in Figures 5 and 6 respectively: the particle SE image, several X ray dot maps and the resulting cartography are displayed in each case.

In the analyzed SHA particle, XR Microanalysis reveals the presence of Al, Ca, O, Si, S, Mg, Cl and Pb which is very much segregated in this large area since it is located in one small nodule and in a lesser extent in a second one. Ca and Al exist in the whole scanned zone, but non uniformely and jointly in some cases, and Si is found mostly in regions where the latter do not exist. The Pb

–containing areas coincide partially with those containing Si, Ca and O, whereas they almost never coincide with areas containing Mg, Cl, or Al. Since Ca and S exist together most of the times, we may think they are as $CaSO_4$.

WDS semi-quantitative X Ray Microanalysis (see Table 1) pointed out that Pb is the main component of nodule 1, slightly more abundant than O, whereas Si, Ca and S are present as minor elements. By putting together all these results, we can deduce that Pb is most probably present in nodules as an oxide linked to SiO_2, CaO, or $CaSO_4$, perhaps as the free metal itself, and the nodules are included within a complex matrix composed of Si or Ca oxides, or even of some alumino-silicates.

SE Image

X ray dot map of Pb

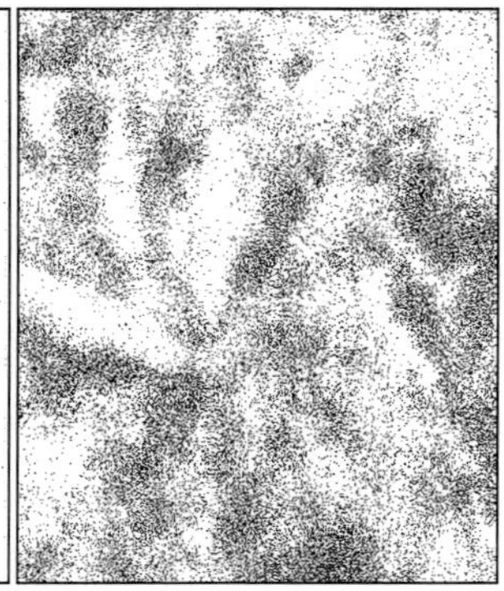

X ray dot map of Ca

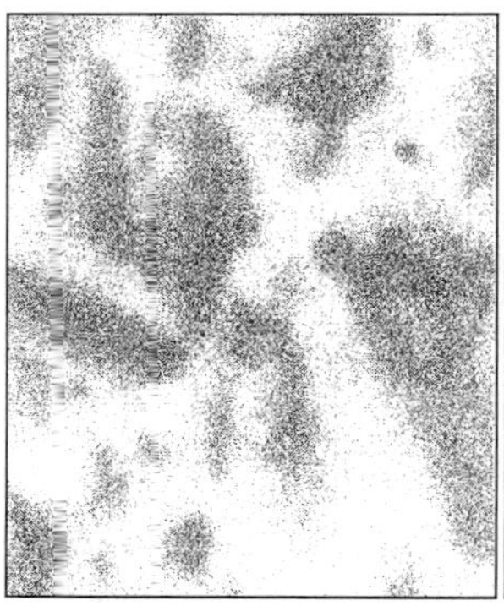

X ray dot map of Si

X ray dot map of Cl

X ray dot map of O

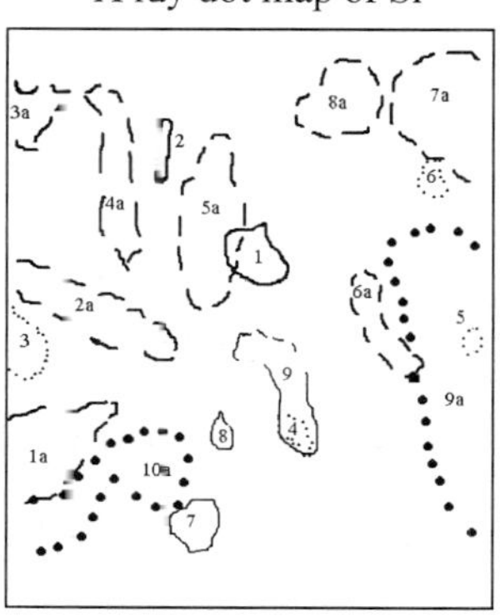

Figure 5 Example of a Pb-enriched superheater ash particle : secondary electron image, X ray dot maps of several components, and resulting cartography – 1,2: Pb areas; 3 to 6: Ca and Al-enriched zones; 1a to 8a: Ca-depleted zones; Si, O present in the whole particle, but mainly concentrated in zones 7 to 9.

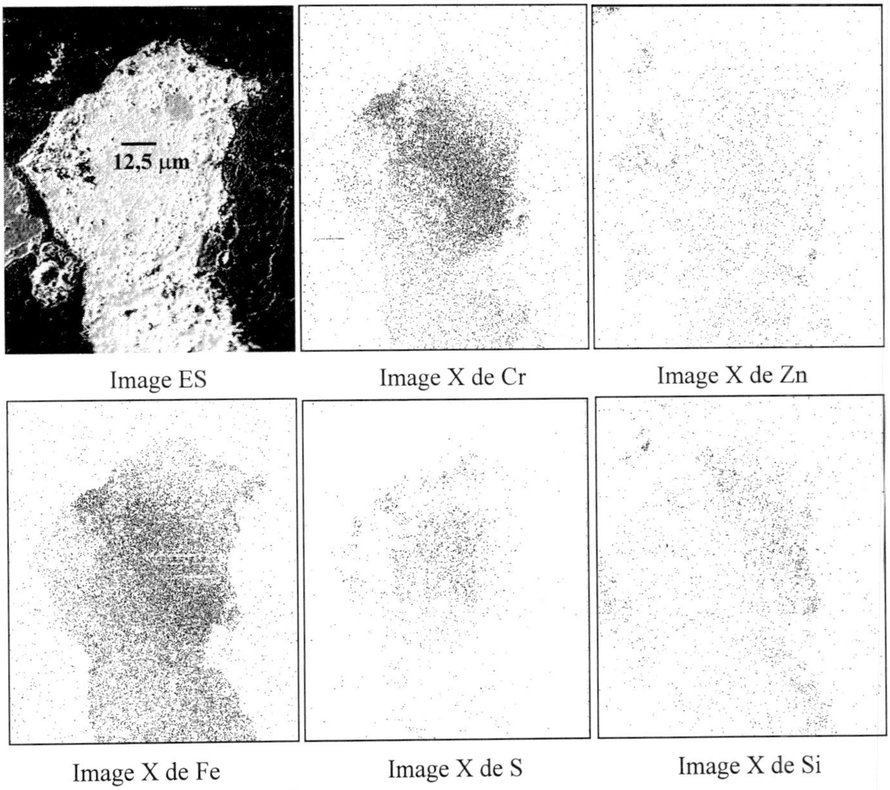

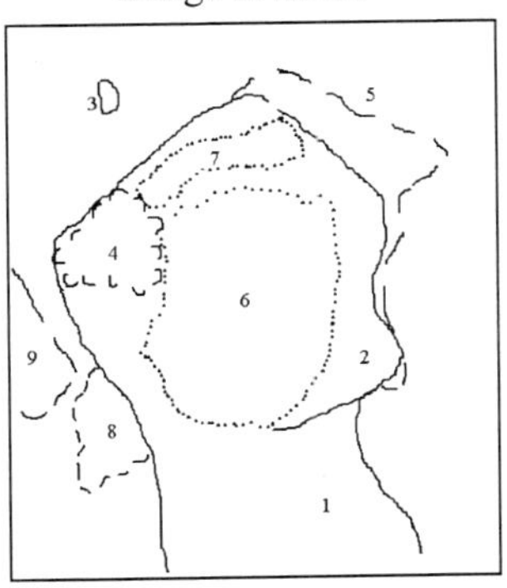

Figure 6 Example of a Cr-enriched economizer ash particle : secondary electron image, X ray dot maps of several components, and resulting cartography - 1, 2: Cr areas; 3: zone containing Zn, Si, Ca; 4: Ca area ; 5: zone containing Fe, Zn, O; 6, 7: S areas; 8, 9: zones containing Si, O.

Globally, the study showed that Pb (always) and Cr (often) are localized and very concentrated in little spherical nodules (2-10 µm size) occluded in larger particles (several hundreds of µm). Generally, Zn is widely dispersed in tens to hundreds µm-sized particles, where it is often associated with other metals such as Fe, Cu. This kind of distribution is also sometimes observed for Cr. In this configuration, Cr is nearly always associated with Fe, Zn and other elements such as Co, V: this means that the scanned particle might be a steel piece.

Table 1 WDS semiquantitative (standardless) X-ray microanalysis of a Pb-enriched superheater ash particle

ELEMENT	I.x/I.std	K. RATIO	CONCENT.	ATOM. CONC.
Pb	0.9898	0.9898	0.99303	0.4546
Si	0.0199	0.0199	0.02134	0.0721
Al	0.0005	0.0005	0.00064	0.0023
O	0.0830	0.0202	0.06455	0.3827
S	0.0711	0.0239	0.02180	0.0645
Ca	0.0267	0.0098	0.01012	0.0240
Total:		1.06399	1.11149	

XRM showed that in little nodules, Pb is generally the main constituent associated with O, sometimes with Ca. The matrix of larger particles is not uniform, but on the contrary it is made of calcium, magnesium, potassium silicates or aluminosilicates, silica or other oxides. Our results concerning Pb are consistent with literature: Plüss and Ferrell [7] found by SEM/XRM characterization of ESP dusts that Pb is probably as oxide or hydroxide; Wearing [15] pointed out a strong incompatibility between Pb and most crystalline phases; Kirby and Rimstidt [11], as well as Clozel and Legendre [16] found Pb enrichment in glass phases; and Eighmy et al [10] observed lead-containing minerals such as Pb_3SiO_5, $Pb_3O_2SO_4$, $Pb_3Sb_2O_7$, K_4PbO_4 in XPRD analyses carried out on ESP dusts. These mineral species may exist in our residues since we found several times Pb associated with Si and O.

Concerning Zn, the microanalysis showed that this metal is included in large particles whose matrix is constituted of a composite structure (aluminosilicates and silicates of calcium, sodium, magnesium, potassium or calcium, silicon, magnesium oxides). This is in good agreement with Wearing [15] who, in his study dealing with metallurgical slags, indicates that Zn is compatible with several crystalline phases such as melilite, magnetite, pyroxene.

Study of ash-surface environment (by XPS)

An example of XPS spectrum is given in Figure 7, although the results are not as fruitful than expected. Indeed, we succeeded in detecting only one metal, Zn, in SHA, EA and ESP dusts, and we did not find any in BA. Other HM may be present on the ash surface, but they are not concentrated enough in the scanning zone to be detected by this technique. Nevertheless, XPS analyses proved the presence of Ca, O, Na, Cl, S and sometimes K in SHA, EA and ESP dusts, but in BA, we observed only Ca, O and Na. Since the photoelectron peaks were not high enough in the obtained spectra to permit high resolution survey and good curve fittings, we can only emit some hypotheses concerning possible associations between the elements. Nevertheless, SHA, EA and ESP dusts particle surfaces are clearly enriched with chlorides (probably NaCl, KCl and perhaps Zn chlorides), which confirms other authors' observations [17]. In fact, in the first part of the study – microanalysis - , chlorides were not detected, and this is most probably because they were dissolved during the water polishing stage. In their study, Eighmy et al [10] managed to identify more precisely probable Pb and Zn components such as $PbSO_4$, $PbCl_2$, Pb, PbO, ZnO, Zn chlorides, $ZnBr_2$ located on the ash surface, but they were dealing with much higher concentrations in their samples than we did (Zn: 100,000 ppm; Pb: 30,000 ppm versus 20, 000 and 6000 respectively in our case), which permits to carry out high resolution surveys.

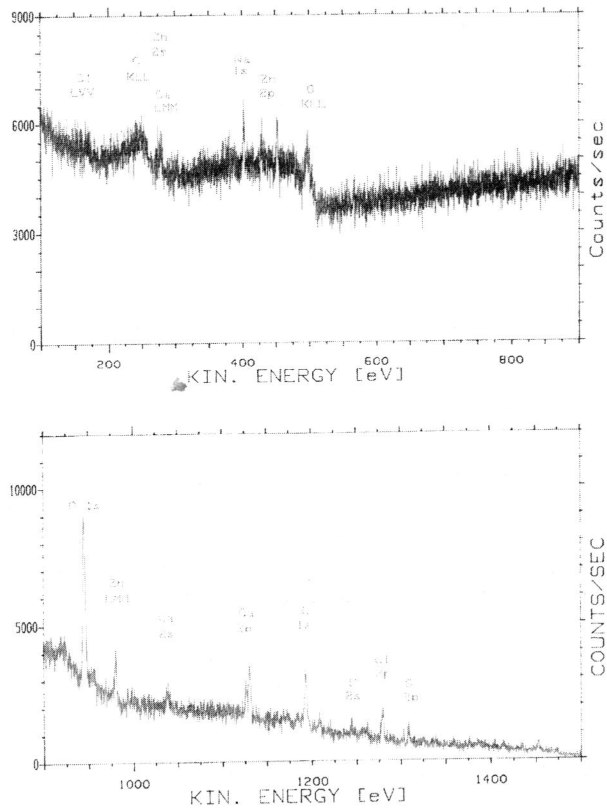

Figure 7 Example of an ESP dusts particle XPS spectrum

Study of HM speciation in ashes

In this last part, we focus on lead which is a particularly worrying heavy metal for MSWI manufacturers and operators.

Our microanalysis showed that in the whole set of MSWI ashes studied, Pb is always located in little nodules, where it is often the main component, associated with little amounts of Ca, Si, S or K. These nodules are generally occluded in large (tens to hundreds of μm) particles with Ca, Mg, potassium silicates, aluminosilicates, lime, or silica matrix.

In fact, we studied also the chemical and physical trapping of HM in our set of ashes by sequential chemical extractions series [14]. This 7-step technique, which is derived from mineralogical methods, gives a general view on the way metals are sequestered in residues and consequently some information about their mobility from ashes [11]. For Pb, our sequential extractions results showed that:

- in BA, Pb is mainly trapped in iron oxide phases and, in a minor part, in glasses;
- in SHA, Pb is partitioned nearly equally in crystalline iron oxide phases and glasses;
- in EA, Pb is mainly trapped in glasses, and it is in little amounts in crystalline iron oxide phases and water soluble phases;
- in ESP dusts, Pb is almost totally within glasses and only in minor part in crystalline iron oxide phases.

Table 2 indicates and compares the probable or possible Pb components identified by both microanalysis and chemical extraction series methods. The combination shows clearly that possible Pb chemical forms are, for all the studied residues, the pure metal its oxide, perhaps its sulphate and complex silicates, and in addition its chloride in the economizer ashes.

Table 2 Pb possible components identified by sequential extraction series and XRM
*: those components typed in italic are possible lead components, but they do not exist since Pb was never detected at these steps in our analyzed samples

EXTRACTION STEP	IDENTIFIED PHASE	CORRESPONDING COMPONENTS	RESULTING POSSIBLE PB COMPONENTS	
			Chemical extractions	XRM
1	water soluble	chlorides, nitrates	$PbCl_2$, *$Pb(NO_3)_2$**	-
2	Ca - exchangeable	surface-adsorbed metals		-
3	acid soluble	carbonates, hydroxydes	*$PbCO_3$**	-
4	organically bound	organometallics sulphides	*PbS**	-
5-6	crystalline and amorphous iron oxides	sulphates, pure metals, oxides	$PbSO_4$, PbO, Pb	PbO, Pb, $PbSO_4$
7	residual fraction	silicates, aluminosilicates, oxides, metals	PbO, Pb, Pb-silicates	Pb, PbO, Pb-silicates

ACKNOWEDGMENTS

This study was partially supported by ADEME (Agence de l'Environnement et de la Maîtrise de l'Energie) and GEC ALSTHOM Stein Industrie. XPS analyses were carried out with the help of Dr Berjoan.

REFERENCES

1. REIMANN, D.O., Waste Man. & Res. 1989, 7, pp 57-62.
2. CLAPP, T.L. ; MAGEE II, J.F. ; AHLERT, R.C. ; KOSSON D.S., Environ. Prog. 1988, 7(1), pp 22-30.

3. KORZUN, E.A. ; HECK H.H., J. Air Waste Manage. Assoc. 1990, 40(9), pp 1220-1226.
4. RIGO, H.G. ; CHANDLER, A.J., In National Waste Processing Conference ASME (16th biennal), 1994; pp 49-63.
5. VELHÖW, J. , Rückstände aus der Müllverbrennung; Martin Faulstich - EF Verlag für Energie und Umwelttechnik, Berlin, 1992; pp 161-191.
6. RIZEQ, R.G. ; CLARK, W.D. ; SEEKER, W.R. , In 11th Incineration Conference, Knoxville, 1991, pp 623-630.
7. PLÜSS, A. ; FERRELL, R.E.JR. , Hazard. Waste Haz. Matl. 1991, 8, pp 275-292.
8. ARTELT, H.M. ; GREINACHER, E. ; MEYER, G. , J. Prakt. Chem. 1993, 335, pp 225-261.
9. EIGHMY, T.T. ; EUSDEN, J.D.JR ; MARSELLA, K. ; HOGAN, J., DOMINGO, D., KRZANOSWKI, J.E. ; STÄMPFLI, D., Environmental aspects of construction with waste; Elsevier Science BV : Amsterdam, 1994 , pp 111-136.
10. EIGHMY, T.T. ; EUSDEN, J.D.JR ; KRZANOWSKI, J.E.; DOMINGO, D.S.; STÄMPFLI, D.; MARTIN, J.R.; ERICKSON, P.M., Environ. Sci. Technol., 1995, 29, pp 629-646.
11. KIRBY, C.S.; RIMSTIDT J.D., Environ. Sci. Technol. 1993, 27, pp 652-660.
12. STUART, B.J., KOSSON, D.S. , Combust. Sci. and Tech. 1994, 101, pp 527-548.
13. CHANDLER, A.J.; EIGHMY, T.T.; HARTLEN, J.; HJELMAR, O.; KOSSON, D.S.; SAWELL, S.E.; VAN DER SLOOT, H.A.; VEHLOW, J., "Municipal Solid Waste Incinerator Residues", Studies in Environmental Science 67, Elsevier Science BV : Amsterdam, 1997, pp 203-254.
14. GAGNEPAIN, B., " Spéciation des métaux lourds dans les résidus solides d'usines d'incinération d'ordures ménagères et contribution à l'interprétation des processus de vaporisation", PhD thesis, N° 295D , 1998, Université de Perpignan, France.
15. WEARING, E. , Mineral. Mag. 1983, 47, pp 335-345.
16. CLOZEL, B.; LEGENDRE, O., "Caractérisation minéralogique et cristallochimique de vitrifiats", in International Congress on Waste Solidification and Stabilisation Processes, Nancy, 1995.
17. FERNANDEZ, M.; MARTINEZ, L.; SEGARRA, M.; GARCIA, J. AND ESPIELL, F. , Environ. Sci. Technol. 1992, 26(5), pp 1040-1046.

PHYSICAL PROPERTY AND pH DEPENDENCY OF SOLUBILITY TEST ABOUT PLASMA MOLTEN SLAG FROM FLY ASH

Naoki Furuta

Toshihiro Hosoi

Masaki Ohata

Tsuginori Inaba

Chuo University

Japan

ABSTRACT. Hazardous wastes such as fly ash can be melted by plasma treatment, reducing its volume by 70% from the original. The molten slag from fly ash protects heavy metal pollution. However, if it will be buried in the ground it will be exposed in acid rain in the future. So we investigated solubility test of the molten slag by changing the pH values from 1 to 9. In order to quantify the watertightness characteristics of the slagged substances, static capacitance, etc. were measured again after 1 year. The molten slag has a good endurance even after eluting for a period of 1 year.

Keywords: Plasma treatment, Solubility test, pH dependency, Fly ash, Molten slag

Professor N Furuta, is a member of Plasma group of HTRC of Chuo University. His main area of interest is analytical spectroscopy with emphasis on the development and application of spectrochemical measurement system. His particular research interests include emission spectrometry and mass spectrometry using an inductively coupled plasma (ICP). One of his professional activities are Asia-Pacific Associate Editor of Journal of Analytical Atomic Spectrometry(The Royal Society of Chemistry).

Mr T Hosoi, is a student of Master course of Chuo university. He is conducting research related to the processing of hazardous wastes by a torch plasma.

Mr M Ohata, is a research student at Chuo University. He is currently completing his PhD on the subject of spectroscopic characterization and analytical application of a high power nitrogen microwave-induced plasma.

Professor T Inaba, is the project manager of Plasma group of High-Tech Research Center of Chuo University. His interests are in the area of plasma treatment of hazardous wastes, application of radiation power and so on. His activities are a past Member of the Board of IEE Japan, and Plasma Committee of the UIE, a Chairperson of High Current Committee, IEE Japan etc.

INTRODUCTION

Waste treatment is a very serious problem all over the world today. It is difficult to secure reclaimed land in Japan so requiring immediate solution. Hazardous wastes such as fly ash can be melted by plasma treatment and it makes possible to reduce its volume by 70% from the original volume.[1]

By stabilizing the arc plasma by an axial gas flow, it is possible to obtain ultrahigh temperatures in excess of 3,000 to 5,000°C and to generate extremely high thermal, radiant, and acoustic energies. A characteristic of arc plasma is that its output energy can be easily and precisely controlled by magnitude of the electric current and by ambient conditions.[2] In order to utilize the potential of arc plasmas to the maximum, we used basic experimental facilities for studies of arc plasma.

By plasma treatment, molten slag from fly ash protects the heavy metal pollution.[1][2] However, if molten slag from fly ash will be buried in the ground it will be exposed in acid rain in the future. Heavy metal elements may loose out of the slag again and flow into the soil. So that in this time we investigated solubility test of the plasma molten slag from fly ash by changing pH index from 1 to 9.

PRINCIPLE OF PLASMA TORCH OPERATIONS

Figure 1 shows a schematic setup of the experimental equipment used in this study. First, a 70mm-diameter carbon disc electrode is placed on a 60mm-diameter stainless steel rod is electrode inside a crucible filled with argon gas in a chamber. On the upper surface of the carbon electrode, a hole is shaped, which measures 15mm-diameter and 10mm-depth. The substance to be melted is placed in the hole. The reason for using argon gas is to eliminate the effect of chemical changes caused by interaction of the processed materials with the surrounding atmosphere.

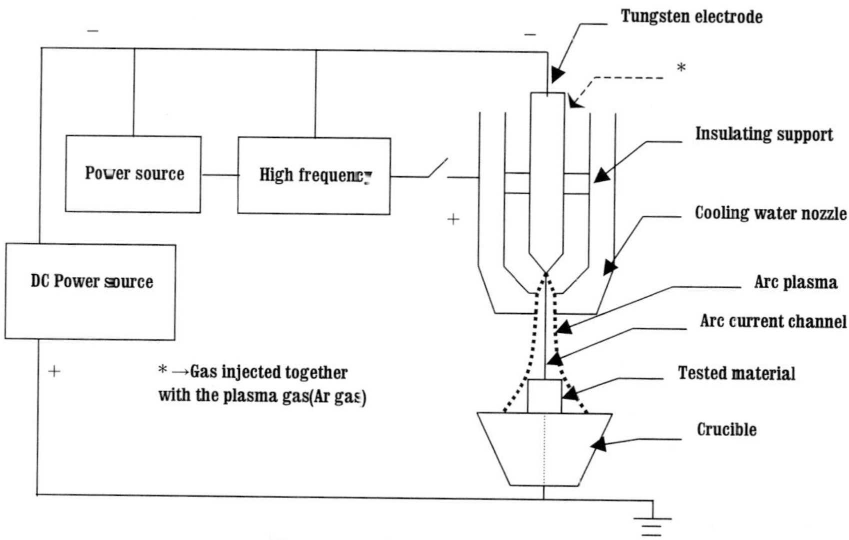

Figure 1 Principle of plasma torch

In order to generate the arc plasma used in the plasma torch, a pilot arc is started by a high-frequency discharge initiated between the electrode and the nozzle. Then, using the conductivity of the pilot arc, the main arc is started in the electrode-crucible gap. Once the main arc is started, the pilot arc is usually turned off. This main arc is ejected from the nozzle by the pressure of the arc gas and operating gas (Ar gas); however, in this case, the high-temperature plasma current is enhanced by efficient use of the thermal pinch produced by the nozzle and the operating gas. Tested specimens are slagged by high-temperature main arc. In order to prevent the electrode from being worn out by oxidation, argon gas was used as the plasma gas.

After slagging, the specimens were naturally cooled in the crucible.

pH DEPENDENCY OF SOLUBILITY TEST

Plasma Slagging of Incinerator Fly Ash

There are two types of ash produced by refuse incinerators: one is ash consisting of large particles that remain inside the incinerator, and the other is so-called fly ash consisting of fine particles collected by precipitation from the incinerator gas emissions. For the purposes of this study, we used fly ash from the Sanitation Center of Okegawa City.

Since the incinerator fly ash consists of very fine particles, it cannot be placed on the electrode as is, because it would be blown away by the plasma jet. To prevent this, we prepared a paste by mixing it with distilled water and forming square blocks having sides 1 cm long. It was confirmed that the treatment converts fly ash into a slag having a smooth vitrified surface. Figure 2 and Figure 3 shows two specimens.

Judging from appearances, the weight was reduced by about half and the volume by two-thirds of the original amount. Fly ash contains moisture and metals which have low boiling point. We believe that the weight decrease of 50% consists of fly ash deposited in the form of vitrified slag on other parts of the equipment, and fly ash evaporated and dispersed by the gas flow. Since the latter component still represents an environmental hazard so that we settled filters and a scrubber type washing precipitator before the exhaust.

Slagging condition was 240 A in electric current and about 55 V in arc voltage. Heating time was 180 seconds and gas flow rate was 20 litters per minute.

Figure 2 Appearance of original fly ash

Figure 3 Appearance of molten slag

About 8 samples were investigated on the solubility test by changing pH dependency from 1 to 9.

Solubility Test

Molten slag has various shapes and different surface area. So we used ground slag less than 0.1mm diameter. The original fly ash was also ground, and the diameter was less than 0.1mm just the same as the ground slag. Solubility test was examined in 6 hours stirring a solution. Solubility condition was 10 ml/g. It means that 1 g sample was immersed in 10 ml water following to the regulation of the Environment Agency of Japan. This result is shown in Table 1 and Figure. 4.

Analysis method was ICP-MS. Analysis elements were selected to be higher concentrated ones compared with soil components. We considered that these elements were man-made origin. If these elements were buried in the ground, it will become dangerous for ecology and environment.

The result shows that the ground slag prevents to be dissolved in water rather than the ground fly ash. Although this time the solubility test condition was severe by using the ground slag, still it had been keeping water proof tightness enough. In addition, with decreasing the pH values, more elements were dissolved in water. And it was common result for the original fly ash and the ground slag. The pH value of the acid rain is about 4.5 in Japan. Our result was safe enough at pH 4 compared with the drainage standard of 0.3 mg/l concerning to Cd and Pb by the Prime Minister's Office. The drainage standards in Japan are shown in Table 2. However, if the ground slag is exposed at pH 1, its safety won't be guaranteed because some results are over the drainage standard as masked in Table 1.

Table 1 pH dependency of solubility test result

	GROUND SLAG				GROUND FLY ASH			
	pH1	pH4	pH7	pH9	pH1	pH4	pH7	pH9
^{11}B	973	1430	899	415	5910	4240	2560	2130
^{53}Cr	5490	2.26	0.840	2.50	1090	19.7	31.2	12.7
^{63}Cu	3280	136	162	244	29000	1490	1520	1520
^{66}Zn	553	20.4	14.1	<2.36	76700	44.6	64.2	40.4
^{75}As	20.2	<0.204	15.4	<0.204	362	34.0	35.9	35.3
^{77}Se	26.3	2.20	1.26	<2.40	53.1	46.4	41.4	40.1
^{95}Mo	34.5	4.71	4.80	8.25	33.3	136	131	137
^{107}Ag	<0.073	<0.073	<0.073	<0.073	<0.073	<0.073	<0.073	<0.073
^{111}Cd	26.4	1.17	2.78	<0.085	549	4.85	4.69	<0.085
^{118}Sn	32.9	0.156	0.204	0.886	1030	3.14	3.53	9.09
^{121}Sb	83.8	40.3	52.7	63.9	775	211	216	217
^{202}Hg	<0.246	<0.246	<0.246	<0.246	<0.246	<0.246	<0.246	<0.246
^{208}Pb	232	4.57	<0.187	<0.187	10000	7.64	7.72	<0.187
^{209}Bi	<0.094	<0.094	<0.094	<0.094	6.71	6.71	<0.094	<0.094

[ng/ml]

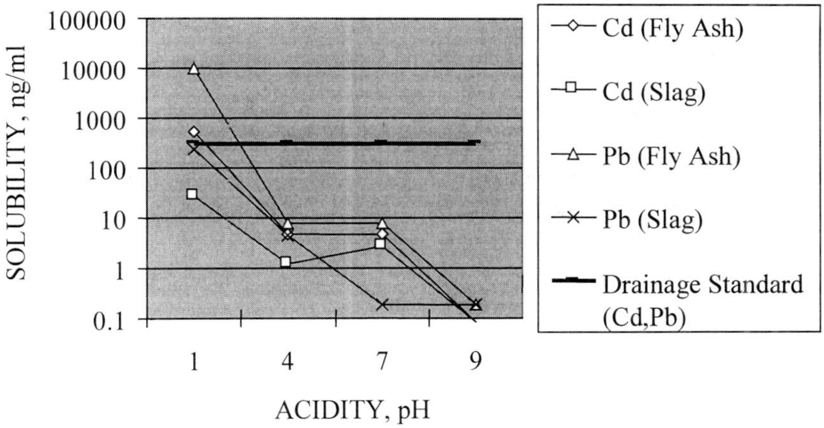

Figure 4 pH dependency of solubility test for Cd and Pb

Table 2 Drainage standards by the Prime Minister's Office in Japan

ITEM (HAZARDOUS MATERIAL)	DRAINAGE STANDARD
Cd	0.3
Pb	0.3
As	0.3
Hg	0.005
Se	0.3
Cr (VI)	1.5

[mg/l]

The purpose to prevent dissolving toxic elements from molten slag in the acid rain was satisfied. Nevertheless, in the more hard condition just like pH 1, it won't be safe. Next time by using without ground slag, it would be improved. In this case, molten slags which have similar size and shape have to be produced.

TEST OF ELECTRICAL PROPERTIES OF AQUEOUS SOLUTIONS

In order to quantify the watertightness characteristics of the slagged substances, we measured the following three parameters.

Static Capacitance

For this test, we fabricated a cubic acrylic tank with an internal side of 10 cm (wall thickness 2 mm) having square copper electrodes (side length 7 cm) attached to the outside surfaces (see Figure 5). The tank was filled with filtered aqueous solution and its static capacitance was measured with an LCZ-meter. Results are shown in the form of initial values in Table 3

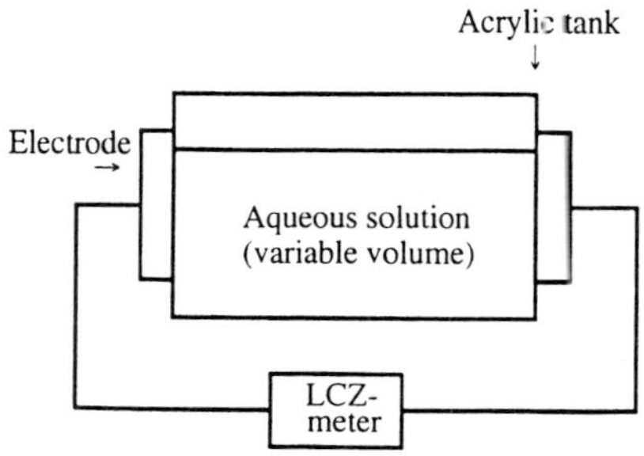

Figure 5 Circuit for measurements

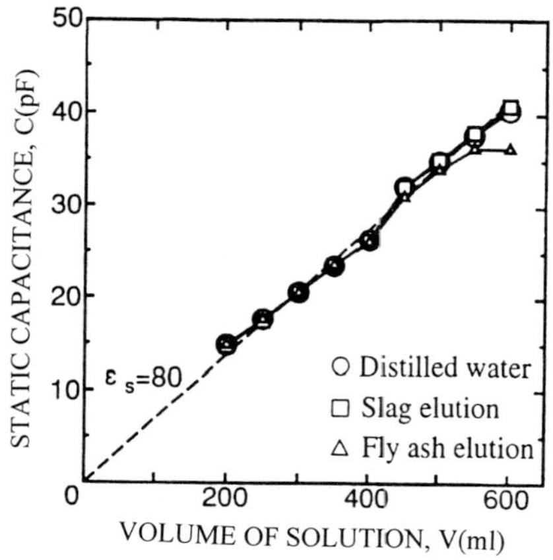

Figure 6 Capacitance of initial solution

Table 3 Permittivity and capacitance of initial solution

VOLUME V(ml)	PERMITTIVITY ε(nF/m)			STATIC CAPACITANCE C(pF)		
	Distilled water	Fly ash solution	Slag solution	Distilled water	Fly ash solution	Slag solution
200	0.740	0.745	0.745	14.8	14.9	14.9
250	0.700	0.708	0.708	17.5	17.7	17.7
300	0.687	0.687	0.687	20.6	20.6	20.6
350	0.674	0.674	0.671	23.6	23.6	23.5
400	0.660	0.658	0.663	26.4	26.3	26.5
450	0.718	0.693	0.718	32.3	31.2	32.3
500	0.704	0.682	0.700	35.2	34.1	35.0
550	0.693	0.662	0.687	38.1	36.4	37.8
600	0.683	0.607	0.675	41.0	36.4	40.5

and in Figure 6. The static capacitance of distilled water, fly ash solution, and slag solution increased in almost direct proportion to the volume of liquid in the tank, with the values being very similar for all three substances. Only the solution of fly ash showed a decrease of the order of several percent at large volumes. At the same time, we measured the static electric capacitance of solutions prepared from freshly obtained fly ash and from slag prepared 1 year before. The slag solution parameters on average were within 0.1% of those of the distilled water. The capacitance of the fly ash solution was 1 to 2% lower than that of the distilled water, which is closer than the other above-mentioned parameters. The curves in Figure 6 indicate that at low solution volumes, the static capacitance is slightly greater than a directly proportional relationship. This can be explained by the static capacitance of the air contained of water capacitance between the electrodes above the solution surface (of the order of 1 pF). The reason for the decreased capacitance of the fly ash solution observed at large volumes compared to that of distilled water is not quite clear. This phenomenon is probably caused by factors such as dissolution of some fly ash components, an increase in conductivity, an increase in leakage current, an increase in solution temperature, appearance of gas bubbles as a result of dissolution, and so forth.

Permittivity

From the capacitance C obtained by means of the LCZ-meter, the permittivity can be determined as

$$\varepsilon = C \times d / S = C\, d^2 / V$$
$$S = V(cm^3) / d(cm)$$

In order to avoid measurement errors, consolidated comparative results are shown in Figure 7. Here, d is the distance between electrodes (10.4 cm), S is the area (in cm^2) of the electrodes corresponding to the solution level, and V is the volume of the solution used in the test (in cm^3). The permittivity of the distilled water averaged over the entire volume was (0.70 ±0.04) $\times 10^{-9}$ F/m. Its value was nearly constant, with variations not exceeding 6%. This average value exactly corresponds to the specific permittivity ε_s=80. By contrast, the permittivity of the slag elution is generally within 1% of that of distilled water, while the permittivity of the fly ash solution is several percent lower than that of distilled water when the volume is greater than 400 ml. The permittivities of these three substances, expressed relative to the permittivity of distilled water, show that the permittivity of the fly ash eluate is lower than that of distilled water by several percent, while that of the slag eluate is on average less than 0.1% lower than that of distilled water and exhibited practically no charge over a period of 1 year.

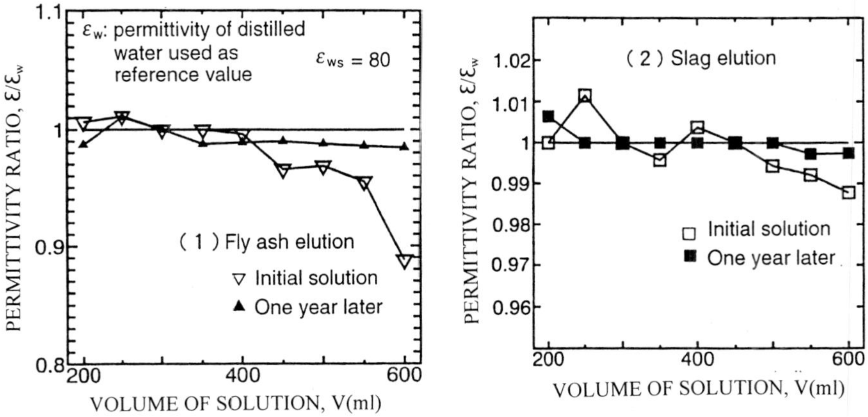

Figure 7 Permittivity ratio of solution

Specific conductance

Measurements of specific conductance, which can be used as a characteristic of water pollution, were obtained with a ring-electrode meter and are shown in Table 4 and Figure 8. The tested materials used for the measurements were the same as in previous experiments. The specific conductance of the fly ash eluate was 15,700 μS/cm, which is almost 15,000 times that of distilled water, and the conductivity of the slag obtained by plasma treatment was 24.6 μS/cm, which is of the same order as the conductivity of rainwater (10 to 30 μS/cm), or about 20 times that of distilled water. These results demonstrate that the leaching of pollutants from the slag is extremely low compared to that of fly ash (about 1/1000), thereby confirming the remarkable efficiency of plasma treatment.

Table 4 Measured electric conductivity

	DISTILLED WATER	FLY ASH SOLUTION	SLAG SOLUTION
Temp. of solution at the time of measurement	19.2 °C	19.4 °C	19.2 °C
Specific conductance	1.21 μS/cm	15,700μS/cm	24.6μS/cm

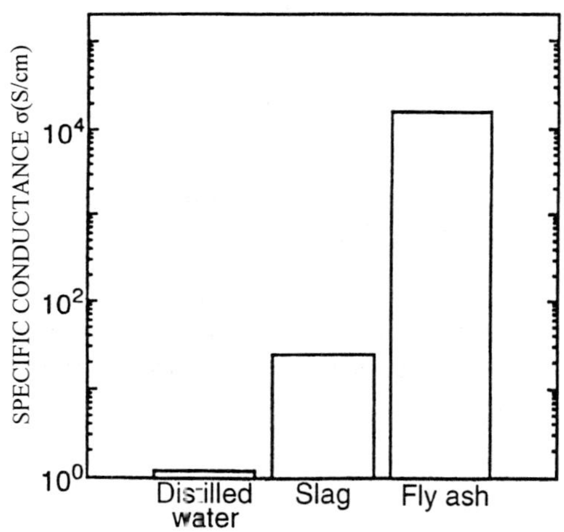

Figure 8 Measured values of electric conductivity

CONCLUSIONS

1. Incinerator fly ash can be effectively slagged by plasma treatment, resulting in a substantial volume reduction.

2. By plasma treatment, plasma molten slag has an excellent watertightness compared with original fly ash. And solubility test results in a distilled water met satisfaction of the drainage standards by the Prime Minister's Office in Japan.

3. In the more hard condition such as pH 4 in solubility test, its results also cleared the standards. It means that plasma molten slag which is buried under ground keeps watertightness even if it is exposed in the acid rain.

4. Comparison of the characteristics of eluates of fly ash and slag in distilled water reveals that the permittivity and static capacitance of these three substances differ from each other only slightly, within a few percent. Specifically, the permittivity of incinerator fly ash eluate compared to that of distilled water was several percent lower, while the permittivity of slag was of the order of 0.1% even after eluting for a period of 1 year, thus demonstrating good stability of the slag. Slag has good insulating properties; its electrical conductivity is only one-thousandth that of the incinerator ash.

ACKNOWLEDGEMENTS

The authors wish to thank the Furuta and Inaba laboratory's students who assisted in conducting experiments, and personnel of Nippon Special Machinery Co., Ltd. who designed and fabricated the equipment used in this study.

This research has been conducted under the research project "Treatment of Hazardous Waste & Reuse by Ultra-High temperature Plasma" at the Institute of Science and Engineering of Chuo University supported by the Ministry of Education of Japan.

REFERENCES

1. T.INABA, M.NAGANO, M.ENDO. Investigation of plasma treatment for hazardous wastes such as fly ash and asbestos. Trans. IEE Japan, Vol.117-D, No. 7, July 1997, pp. 831-838 (Japanese), Electrical Engineering in Japan, Vol-126, No.3, pp.73-82, 1999,Scripta Technica.

2. T.INABA, Y. WATANABE, M.NAGANO, T.ISHIDA, M.ENDO. Ar torch plasma characteristics and its application to waste treatment, Thin Solid Films,Vol.316(1998),pp.111-116.

A GEOCHEMICAL PERSPECTIVE ON THE FACTORS THAT CONTROL LONG-TERM EMISSIONS FROM MUNICIPAL SOLID WASTE INCINERATOR RESIDUES

C A Johnson

Swiss Federal Institute for Environmental Science and Technology (EAWAG)

Switzerland

ABSTRACT. One of the main environmental concerns with regard to the utilisation or the landfilling of municipal solid waste incinerator (MSWI) ashes is the long-term leaching and transport of heavy metals into the environment. The matrix or major components control leachate chemistry and provide sites for sorption and have a key role in controlling the mobility of heavy metals. Heavy metal concentrations are controlled by a number of mechanisms, most of which are pH dependent. Under the basic conditions encountered in fresh and cement-stabilised ashes, heavy metal oxyanions are more mobile than heavy metal cations. The system is well buffered and likely to keep concentrations of the latter low over centuries. The problem associated with heavy metals in construction materials is related more to uncontrolled dispersal into the constructed environment.

Keywords: Municipal solid waste incinerator (MSWI) bottom ash, Leachate, Geochemistry, Weathering, Emissions, Heavy metals

Dr C A Johnson took a degree in Chemistry at Imperial College, London and obtained a PhD in Applied Geochemistry from the same college. She carried post doctoral studies in Switzerland and the United States and now has a permanent position at the Swiss Federal Institute of Environmental Science and Technology. She works in the Department of Resource and Waste Management and heads a research team that investigates the long-term hazards associated with waste management practices, in particular landfilling. The group investigates the mechanisms that control the mobility of contaminants in the laboratory and the field.

INTRODUCTION

One of the main environmental concerns with regard to the utilisation or the landfilling of waste materials is the leaching and transport of contaminants into the environment. Municipal solid waste incinerator (MSWI) ashes pose a potential problem due to their relatively high heavy metal content and reactivity in contrast to geogenic materials. The question if and under which conditions these metals are released has been a central topic research in recent years (eg. 1, 2). It has been shown that heavy metal mobility can only be understood in context of hydrological, mineralogical and geochemical processes. Hydrological processes determine the extent of the leaching process and depend on the permeability, the homogeneity and the physical characteristics of the material. The geochemical properties of a material govern the availability of components for leaching and are therefore of crucial importance (Figure 1). The matrix or major components control leachate chemistry and provide sites for sorption and thus have a key role in controlling the mobility of heavy metals.

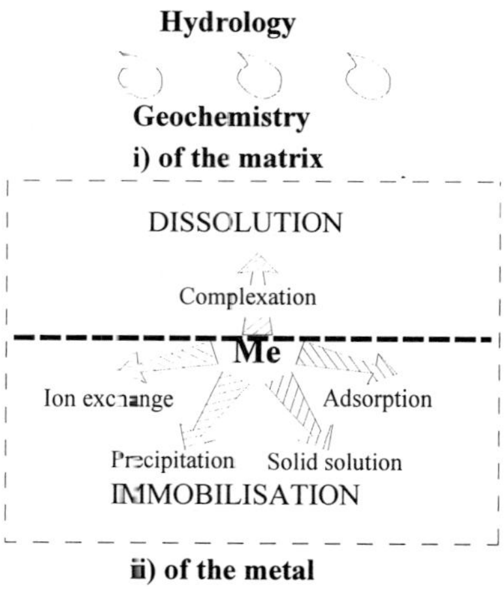

Figure 1 A schematic representation of factors that affect heavy metal mobility

Heavy metal concentrations are controlled by a number of mechanisms, schematically shown in Figure 1. By the formation of dissolved complexes total metal ion concentrations can be enhanced by orders of magnitude. The effect can only be quantified if all complexes are known. This means that anions such as Cl^-, SO_4^{2-}, CO_3^{2-}, HCO_3^-, OH^- and organic ligands have to be known in order to determine heavy metal cation concentrations. For heavy metal oxoanions the corresponding complexes with cations have to be determined. Mechanisms that limit solubility are associated with the solid phase. These include surface adsorption, cation exchange and incorporation into solid phases as solid solution. Such mechanisms are well known and described elsewhere (3). Quantification of these processes can only be determined by an understanding of the solid phase. The concentration of available sorption sites, the cation exchange capacity or the type and composition of solid solutions have to be known if the influence on the heavy metal concentration is to be determined. Only the precipitation of pure solid phases is independent of solid phase composition. The popularity of the use of pure solid phases as reference for the solubility of heavy metals is partly due to the simplicity of the model assumption.

Heavy metal concentrations in leachates of from recently landfilled MSWI residues are generally in the μg L^{-1} range. Heavy metal anions are present in the largest concentrations. In leachates of cement stabilised air pollution control (APC) residues CrO_4^{2-} and MoO_4^{2-} have concentrations in the hundred μg L^{-1} range. Molybdate is present in the highest concentrations of all heavy metals (approximately 300-400 μg L^{-1}) in leachate from MSWI bottom ash residues. Heavy metal cations tend to be in the low or sub μg L^{-1} range. The difference in concentration may be primarily ascribed to the pH of the leachate. The observed concentrations of some elements can be explained in terms of the processes shown in Figure 1. The mobility of the heavy metals may change as a function of time due to ageing processes within the MSWI residue matrix, that change leachate pH values or the sorptive properties of the matrix. In this paper the geochemical factors that potentially control the long-term emissions of heavy metals are examined. The focus is on MSWI bottom ash because it is quantitatvely and qualitatively the more important waste stream for use in construction materials.

GEOCHEMICAL CONTROLS OF HEAVY METAL SOLUBILITY

The mobility of heavy metals depends to a large degree on the geochemical properties of the element itself, but also on the pH of the leachate and on the properties of the solid phase. A detailled examination is not possible here. Instead a few illustrative examples will be shown.

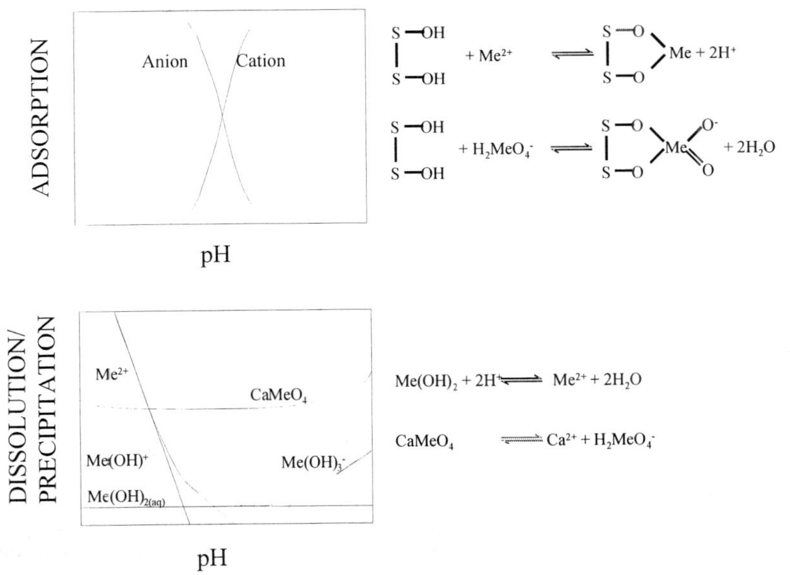

Figure 2 Schematic representation of the pH dependence of the sorption process (where S-OH is a surface hydroxyl group) and of precipitation.

The pH of the leachate is the most important factor in determining mobility, as illustrated in Figure 2. In the case of specific adsorption, or so-called inner sphere binding where metal ions coordinate with donor oxygen atoms from the surface, cations sorb in alkaline conditions, as encountered in MSWI ash leachates and anions sorb in neutral to acid pH conditions. This means that anions are likely to be mobile in ash deposits. The affinity of cations to surface complexation is related to their tendency form complexes. Lead, for example, hydrolyses at relatively low pH values and also sorbs at low pH values. Ideal surfaces for binding are (hydr)oxides, in particular Fe hydroxide. The solubility of heavy metals in relation to dissolution/precipitation reactions is also strongly pH dependent, in particular for heavy metal cations. Hydroxyl complexes and others enhance solubility in alkaline solutions and influence the solubility in acidic solutions. The solubility of oxyanions can be controlled by the precipitation of metallate salt and tends to be less pH-dependent.

Illustrative examples are shown for MoO_4^{2-}, WO_4^{2-}, Cu(II) and Cd(II). The selected oxyanions appear to be controlled by the precipitation of Ca metallate salts. The solubility of Cu(II) appears to be enhanced by organic ligands and Cd(II) solubility appears to be controlled by the precipitation of $CdCO_3$. It should however be pointed out that correlations between observed and predicted values should be corroborated with other evidence. This is a problem in heterogeneous wastes. It should also be pointed out that the agreement is poor for a number of metals, in particular those for which sorption could be important because it is almost impossible to estimate the concentration of surface sites.

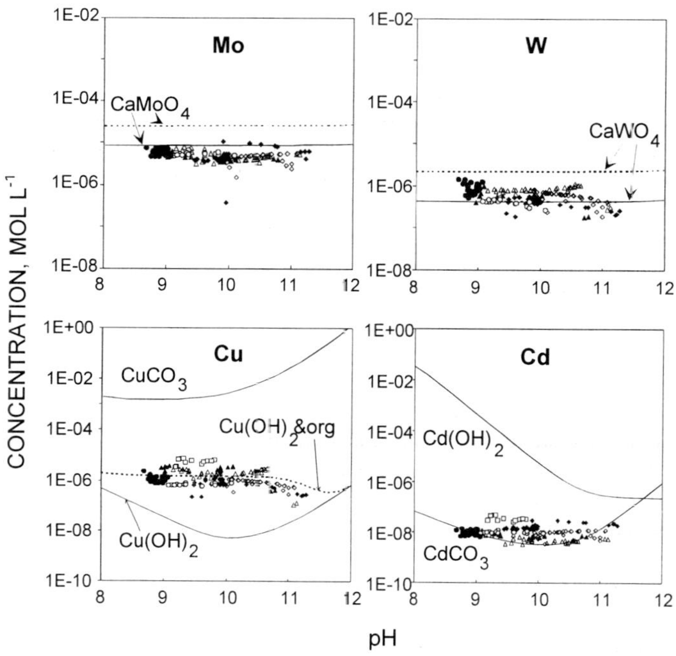

Figure 3 Heavy metal concentations from 7 sampling events at Landfill Lostorf, Switzerland (Nov.94, ◊; May95, ◆; Jun.95, □; Nov.95, Δ; Jun.96, ▲; Aug.96, ❍; Nov.96, ●) as a function of pH. plotted together with solubility curves of solids indicated in the figure (4). Charges are omitted for simplicity.

Changes in solution parameters such as pH, in Ca(II) concentration, ligand concentrations, or surface sites available for sorption affect heavy metal mobility. In the next section, reactions involving the matrix components that potentially effect both the short- and the long-term mobility of heavy metals are examined.

GEOCHEMICAL PROPERTIES OF FRESH MSWI BOTTOM ASHES

The major matrix components of bottom ash are Si, Ca, Al and Fe (5). Estimated as oxides they account for approximately 80% of the dry mass of bottom ash. Other major components are Mg, CO_3, Na, K, S, Cl in order of abundance. Organic components make up a further 1-2% of the MSWI bottom ash material. Due to the heterogeneity of the material, a complete quantification of the components in which these elements are bound is not possible. However some attempts have been made on selected samples. Studies on a few Swiss MSWI bottom ashes have shown that Fe, for example, is present to over 95% in metallic form, 40%, of which is present in the form of tin cans. The remaining 5% is present Fe(II) oxide corrosion products on the metallic surfaces and Fe silicates (6). In fresh MSWI bottom ash deposits, approximately 15-25% of Ca is present as basic soluble salts, possibly portlandite ($Ca(OH)_2$), ettringite ($3CaOAl_2O_3CaSO_4.32H_2O$) or Ca silicate hydrates, a further 15-25% is present as calcite ($CaCO_3$) and the rest is possibly bound in aluminosilicates, phosphates etc. (7). Speciation studies on S suggest that a significant proportion of S may be present as sulphide in fresh samples (approx. 50%). Samples aged 10a had only 15% was reduced (8).

Further information can be gained by investigating the inputs of waste materials to incineration. A detailed study on the waste flows in St. Gallen, Switzerland, carried out by (9) defines 14 waste categories and suggests sources for 10 elements. For the major components Glenck estimates that approximately 60% of Ca originates from organic material, pet litter and dust, while the remainder is associated with glass products and an unknown fraction. Approximately 75% of Si is bound in glass products and the remainder in paper and organic material. This suggests that the availability of a large proportion of Si is limited by the weathering of the glass products that are evident in MSWI bottom ashes. Sulphur appears in a number of products, but primarily in organic material and paper products. Phosphorus originates to 80% from organic material and approximately half the Cl originates from plastics.

Mineralogical studies have found that MSWI bottom ash is composed of fine ash material and melted components of which half have crystallised and small quantities of metallic components, ceramics and stones (10). The relative amounts are quite variable. Spectroscopic and microscopic studies have identified a large number of minerals, including quartz (SiO_2), gehlinite ($Ca_2Al_2SiO_7$), akermanite ($Ca_2MgSi_2O_7$), calcite, plagioclase feldspars ($(K,Na)(Al,Si)_4O_8$) and wollastonite ($CaSiO_3$) as the main crystalline components. (11, 12).

The components primarily involved in controlling heavy metal mobility are those that influence leachate composition. These are the most soluble components, such as NaCl. Field investigations have shown that soluble salt concentrations of the elements Na, K, Ca, Cl and SO_4 in the 10-100 mmol L^{-1} range are typical (2). Leachate solutions of samples are generally highly basic (pH 11-12.5) due to the presence of basic salts of the cations listed above. Under field conditions, CO_2 may neutralise leachate to produce pH values closer to 8. Geochemical modelling can be used to suggest solubility controlling phases for the major components. This is illustrated in Figure 4. Here field data from Landfill Lostorf (4) are compared to concentrations of Ca(II), SO_4^{2-}, Al(III) and Si(IV) predicted to be in equilibrium different solid phases. It should be noted that SO_4^{2-} has been selected because of its influence on Ca(II) solubility and that Fe(III) is insoluble in the basic and neutral pH range. The calculations suggest that gypsum ($CaSO_4$) and ettringite and gibbsite ($Al(OH)_3$) may be involved in controlling the solubility of Ca(II), SO_4^{2-} and Al(III). The role of Si-containing phases such as quartz or Ca silicate hydrate (CSH) is unclear. Whilst the correlation of the field data with the pH-concentration trend of the model prediction for CSH is poor, the agreement with the trend for quartz is questionable due to kinetic considerations.

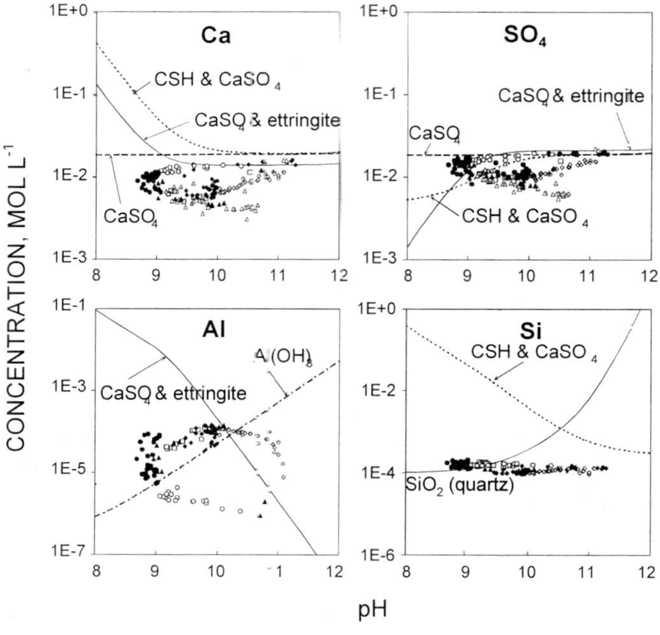

Figure 4 Concentrations of Ca, SO_4, Al and Si, from 7 sampling events at Landfill Lostorf, Switzerland (Nov.94, ◊; May95, ◆; Jun.95, □; Nov.95, Δ; Jun.96, ▲; Aug.96, ○; Nov.96, ●) as a function of pH. Also shown are calculated total concentrations in thermodynamic equilibrium with a given solid phase. Note the different scales of the y-axes (4)

LONG-TERM ALTERATION PROCESSES

Since the pH value of the leachate is the most important factor in determining heavy metal mobility, it is important to know which salts control pH and how they will change as a function of time. The acid neutralising capacity (ANC) of MSWI ash is the key parameter that can be used to predict the long-term release of heavy metal cations. It is the sum of all bases that are titrated with acid to an operational endpoint, and as such is an empirical parameter (7). Figure 5 shows the ANC of MSWI bottom ash taken from the same incineration plant. The fresh sample was taken directly from the Winterthur incineration plant, the older samples from borehole material from a landfill. The age was estimated from the depth of the sample. These values illustrate some typical features of ANC. Fresh samples are highly basic and their ANC is largely made up of soluble basic Ca salts such as portlandite, ettringite or perhaps Ca silicate hydrate. The ANC of older samples is predominantly $CaCO_3$ and it is almost always smaller.

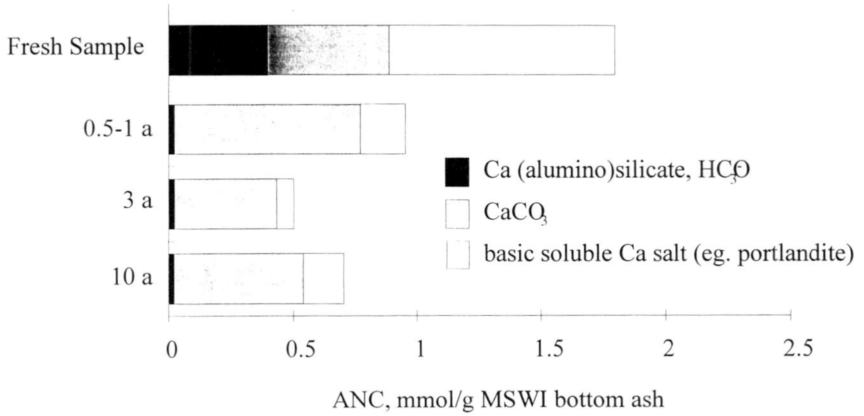

Figure 5 The ANC of MSWI bottom ash originating from the Winterthur incineration plant (Switzerland). The end point of titration was pH 7.5

The ANC is almost wholly comprised of Ca(II) salts and changes in ANC are primarily a result of changes in Ca mineralogy. However, a number of acid-base reactions involving other constituents of MSWI ash may be involved. One of the most important acid-base reactions is carbonation of basic Ca(II) salts involving atmospheric or microbially produced CO_2. In comparison to portlandite, Ca silicate hydrates or ettringite, $CaCO_3$ is relatively insoluble. A solution in equilibrium with portlandite will have pH value of approximately 12.5, while in equilibrium with $CaCO_3$ and an atmospheric partial pressure of CO_2 (0.00035 atm) the pH will be approximately 8.3. If the partial pressure of CO_2 is increased to 0.01 atm, as may be the case in soils, the pH can sink to 7.3. It should be noted, however, that carbonation does not influence the ANC.

Silicon is the predominant component of MSWI ashes, but the weathering of silicon containing components appears to be difficult to predict. There are indications that Si is not readily available, firstly because of its origin (to a large extent bottle glass), secondly the low solubility under high pH conditions and thirdly the lack of secondary mineral formation involving Si on a large scale. Whilst Ca silicate hydrates can be found in MSWI bottom ash deposits, they are only found sporadically. Clay formation, on the other hand, appears to be ubiquitous on the surface of glassy particles (13). It thus appears that Si containing minerals do not greatly affect ANC and that alterations may rather be seen in analogy to weathering processes in soils, the final products being silica (H_4SiO_4) and gibbsite ($Al(OH)_3$) (3)

$$\text{Me silicate} + HCO_3^- \longrightarrow MeCO_3 + H_4SiO_4$$

$$\text{Me Al silicate} + H^+ + H_2O \longrightarrow \text{Al silicate} + H_4SiO_4 + Me$$

$$\text{Al silicate} + H^+ \longrightarrow Al(OH)_3 + H_4SiO_4$$

where Me represents any cation. The weathering reaction consumes protons and thus has a buffering effect, but on a scale of centuries rather than decades.

Aluminium metal is not greatly altered during the incineration process and is only subsequently oxidised in moist alkaline conditions producing H_2.

$$Al + 3H_2O \longrightarrow Al(OH)_3 + {}^3/_2H_{2\,(g)}$$

This reaction is thermodynamically favourable and occurs within weeks. Larger metal particles can survive for several years and are probably protected by an oxide layer. Though the availability of Al can influence geochemical processes within deposits, it has no influence on the ANC.

Iron may have surface coatings of Fe(II/III) oxides and silicates, but the major oxidation reaction occurs after incineration, most probably within decades. The oxidation reaction has no effect on the acid neutralising capacity. $Fe(OH)_3$ is insoluble and has little buffering effect (until pH 3).

$$Fe^0 + {}^3/_4\,O_2 + {}^3/_2H_2O \longrightarrow Fe(OH)_3$$

A significant proportion of S may be present in reduced form. The oxidation reaction releases protons, but since the content of S is around 2-4 g/kg, the influence on the acid neutralising capacity is small (approx. 5%).

$$FeS + {}^9/_4O_2 + {}^5/_2H_2O \longrightarrow Fe(OH)_3 + SO_4^{2-} + 2H^+$$

Residual N in MSWI ash appears to present as ammonium. Nitrification can be microbially mediated under oxic conditions releasing protons. Assuming all N to be present as ammonium influence small around 5%

$$NH_4^+ + 2O_2 \longrightarrow NO_3^- + 2H^+ + H_2O$$

Biodegradation of organic carbon has a potential influence on the longevity of the buffer capacity primarily through the elevation of the partial pressure of CO_2. However too little is known about the kinetics of this process and that of gas exchange in a deposit. Unless the organic content is high enough it is unlikely that the influence will be significant.

The overall influence of oxidation reactions on the ANC appears to be small. The reduced ANC of aged samples may be due to the formation of less soluble, Ca containing minerals, possibly at the surface of the glassy particles. For the estimation of the long-term control of the leachate pH, it is advisable to adopt a conservative value. This is the initial $CaCO_3$ content of the MSWI ash. A more realistic value of ANC will lie between this and total ANC of the fresh sample, the extent of carbonation depending on external factors.

Secondary minerals will most probably formed within the first decades. These include $Fe(OH)_3$, $Al(OH)_3$ and possibly clay minerals. They provide surface or exchange sites for sorption. Their quantification is, however, very difficult. Salts and organic ligands that are in part responsible for enhancing solubility are most probably washed out in decades. Though difficult to quantify, it does appear as if the retarding influence of MSWI ashes increases with time.

The time scales of emissions can only be roughly estimated. The mobility of heavy metal oxyanions that are most probably controlled by Ca metallate precipitation can be estimated from the solubility of the appropriate salt and the total content of the element concerned. For heavy metal cations, where leachate pH plays a decisive role, the time required for the ANC to become depleted is the best guess for the time in which heavy metal cation emissions remain low. Though dependent on the dimensions and situation of the deposit, depletion occurs over very long time scales, perhaps thousands of years (14). It should nevertheless be pointed out, that preferential flow routes through the MSWI ash could be depleted long before this.

POTENTIAL USES AND OUTLOOK

The mixing of MSWI ashes with cement causes a significant increase in ANC directly proportional to the Ca content. It also lowers the permeability and porosity and further influences the mobility of a number of heavy metal cations and oxyanions. In addition to the mechanisms discussed in the previous section, heavy metals may be incorporated into cement minerals. There is evidence to show that Zn(II) can be incorporated into Ca silicate hydrate and that this can substantially reduce Zn(II) mobility (15). Against this, it should be pointed out that the more alkaline pH regime does generally increase Zn(II) solubility that this potentially cancels out the immobilising effect. The mobility of oxyanions could still be affected by Ca(II). Molybenum mobility appears to be controlled by the precipitation of $CaMoO_4$, but could also be incorporated into monosulphate (16). It could be concluded that the addition of heavy-metal containing MSWI ashes to cement most probably causes little problems with regard to heavy metal emissions, also in the long term. In all probability this is also the case. However, there are problems associated with incorporating such waste materials in construction.

Firstly, care has to be taken to avoid problems with the major constituents of MSWI ashes. The Al metal in the ash oxidises in the alkaline cement environment, producing H_2, which has an expansive effect. Salts such as sulphate and chloride can have a deleterious effect on concrete durability.

Secondly and most importantly, the heavy metal content of a concrete containing MSWI ashes can be a problem for secondary materials management when the construction is later demolished. This is particularly true for air pollution control residues, where Zn and Pb concentrations can be in the percent per weight range. Recycling of construction materials can lead to increasing metal concentrations when concrete containing MSWI ashes is reused. Further, the demolition materials containing wastes must be appropriately landfilled. In situations where different waste streams are mixed, the mobility of the heavy metals must be reassessed. Control of heavy metal fluxes can only then be effective if materials containing high concentrations of heavy metals are registered. The salient question is then whether or not the future generations are interested. The argument for use of MSWI bottom ash in construction is often that of saving resources. Using Switzerland as an example, this is not so. Only 1% of gravel usage could be saved in road construction in this way. It has often been stated that roads would be or are an alternative landfill. It is important to consider whether construction is an alternative landfill, a temporary solution or a really good construction material with specific desirable properties.

REFERENCES

1. BELEVI, H., STÄMPFLI, D.M. AND BACCINI, P.. Chemical behaviour of municipal solid waste incinerator bottom ash in monofills. Waste Management and Research, 1992, 10, 153-167.
2. HJELMAR O. Characterisation of leachate from landfilled MSWI Ash, Proceedings Int. Conf. Municipal Waste, Hollywood, Florida 1989. Minister of Supply and Services Canada, 1989.
3. STUMM W.. Chemistry of the solid-water interface. Wiley-Interscience, New York, 1992. pp. 427.
4. JOHNSON C.A, KAEPPELI M., BRANDENBERGER S., ULRICH A. AND BAUMANN W. Hydrological and geochemical factors affecting leachate composition in municipal solid waste incinerator bottom ash. Part II: The geochemistry of leachate from Landfill Lostorf, Switzerland. Accepted by the Journal of Contaminant Hydrology.
5. BACCINI P. AND BRUNNER P.H.. Behandlung und Endlagerung von Reststoffen aus Kehrichtverbrennungsanlagen. Gas Wasser Abwasser, 1985, 65, 403-409.
6. ZELTNER C.. Eisen in KVA-Schlacken. Diploma Study, ETH Zürich, Switzerland, 1992, pp. 56.
7. JOHNSON C.A., BRANDENBERGER S. AND BACCINI P. The acid neutralising capacity of municipal waste incinerator bottom ash. Environmental Science and Technology, 1995, 29, 142-147.
8. REDLE M. Sulfide In Kehrichtschlacken, Diploma Study, ETH Zürich, Switzerland, 1992, pp. 56.
9. GLENCK E.. Contribution des ménages urbains dans le métabolisme des biens et des éléments chimiques des filières de désapprovisionnement des biens de consommation solides. Ph D. Study, ETH Zürich, Switzerland, 1994, pp.182.
10. KIRBY, C.S. AND RIMSTIDT, J.D.. Mineralogy and surface properties of municipal solid waste ash. Environmental Science and Technology, 1993, 27, 652-660.
11. EUSDEN J.D., EIGHMY T.T., HOCKERT K., HOLLAND E. AND KIMBERLY M. Petrogenesis of municipal solid waste combustion bottom ash. Applied Geochemistry, 1999, 14(8), 1073-1091.
12. PFRANG-STOTZ G. AND REICHELT J. Charakterisierung und Bewertung von Muellverbrennungsschlacken aus 15 Muellverbrennungsanlagen unterschiedlicher Verfahrenstechnik. Muell und Abfall, 1999, 5, 262-268.
13. ZEVENBERG, C., VAN REEUWIK, P., BRADLEY, J.P., BLOEMEN, P. AND COMANS, R.N.J.. Mechanism and conditions of clay formation during natural weathering of MSWI bottom ash. Clays and Clay Minerals , 1996, 44, 546-552.
14. JOHNSON C.A. Chemische Eigenschaften und Langzeitverhalten der Müllschlacke. In Deponierung fester Rückstände aus der Abfallwirtschaft: Endlagerqualtiät am Beispiel der Müllschlacke, Baccini P. and Gamper B. (Eds.), VDF, Zürich, Switzerland, 1993, 35-54.
15. LUDWIG C., ZIEGLER F. AND JOHNSON C.A. Heavy metal binding mechanisms in cement-based waste materials. In: "Waste Materials in Construction, Putting Theory into Practice", Goumans, J.J.J.M., Senden, G.J., van der Sloot, H.A. (Eds.) Elsevier, Amsterdam, 1997, 459-468.

THE USE OF SOLUBLE PHOSPHATES TO STABILIZE HEAVY METALS IN FLY ASH

M R Lyons

Wheelabrator Technologies Inc.

United States of America

ABSTRACT. This paper discusses a patented and commercially proven process for chemically treating municipal solid waste combustion (MWC) fly ash to reduce the leaching of certain heavy metals. The process employs proprietary forms of soluble phosphate as the chemical treatment agent. The addition of phosphate to ash reduces the leaching of lead and other metals by converting soluble compounds, such as lead hydroxide [$Pb(OH)_2$], into more stable and insoluble mineral phases. Metals leaching is reduced by phosphate treatment under both laboratory leach tests and field disposal conditions. The reduction in metals solubility allows the treated ash to meet regulatory leaching limits and be safely disposed as a non-hazardous waste, or recycled. The paper discusses the chemical reactions that occur when fly ash is treated with soluble phosphate, presents leaching data from treated ash, outlines the operating history of the treatment process, and discusses the costs of the process when used at a MWC facility.

Keywords: Municipal waste combustion fly ash, Trace metals, Leaching, Treatment.

Mr Mark Lyons, is Project Vice President at Wheelabrator Technologies Inc. He is responsible for identifying and managing business development projects for Wheelabrator. His responsibilities include marketing the company's ash treatment and recycling technologies.

INTRODUCTION

The combustion of municipal solid waste in modern waste-to-energy facilities is now a widely practiced and accepted method of solid waste disposal in developed nations throughout the world. One of the reasons waste-to-energy has become widely accepted by environmental regulatory authorities and the general public is that modern WTE facilities strictly control air emissions from the combustion process. The design and operational practices of modern waste-to-energy facilities in fact produce so few air emissions that, on a pounds of total emissions per kilowatt hour basis, waste-to-energy facilities are cleaner sources of electricity than oil or coal-fired power plants [1].

THE GENERATION AND NATURE OF FLY ASH

The use of sophisticated air emissions control equipment at a WTE facility to clean the combustion process exhaust gas before it exits the smokestack produces several solid waste byproducts which must be properly managed. The operation of acid gas removal systems produces waste materials such as wet scrubber sludge or dry scrubber residues, depending on the design of the emission control device.

Electrostatic precipitators (ESPs) and fabric filters/baghouses remove particulate matter from the exhaust gas and thereby generate a solid waste material known as fly ash. ESPs and baghouses also collect dry scrubber residues carried over in the exhaust gas from semi-dry or dry acid gas scrubbing systems. All of these air emissions control residues will be collectively termed fly ash or fly ash/scrubber residue in this paper.

Fly ash collected at modern waste-to-energy facilities can contain many chemical substances: unreacted lime from dry or semi-dry scrubbers, salts such as calcium chloride and calcium sulfate produced by the reactions between scrubber injected lime and acid gasses, and smoke and soot particles. Also present in fly ash are very low concentrations of organic compounds formed during the waste combustion process, such as dioxins, and trace metals such as lead and cadmium. Some trace metals in the MSW fuel are volatized by the high temperatures of combustion, and become entrained in the furnace exhaust gas. Most of the volatized metals condense on the surfaces of fly ash particles, also entrained in the exhaust gas, as the gas cools on its way from the furnace to the smokestack. The condensed trace metals are collected along with the fly ash in the WTE facility's ESP and/or baghouse [2].

TRACE METAL LEACHING FROM FLY ASH

The primary environmental concern surrounding the proper disposal of MSW combustion fly ash is the potential for leaching of salts and trace metals from the ash when disposed of in a landfill. Rainwater or snowmelt which comes into contact with the landfilled fly ash could potentially dissolve some of the salts and trace metals in the ash. The metal and salt laden water, called ash leachate, could then migrate from the landfill and impact groundwater or surface water quality.

Several management strategies have been devised to control or minimize the leaching of trace metals from fly ash. One of the management options commonly practiced by the WTE industry is treatment of fly ash with one or more chemical agents to reduce the leaching of trace metals from the ash. This paper will now discuss in some detail one commercially proven and widely practiced chemical treatment process for reducing the leaching potential of certain trace metals in ash. The process uses various forms of phosphate to convert soluble trace metals in fly ash, such as lead, into new, more insoluble and geochemically stable mineral phases. This phosphate treatment method is commercially marketed as the WES-PHix® Process by Wheelabrator Technologies Inc. of Hampton, New Hampshire, U.S.A.

PROCESS DEVELOPMENT AND COMMERCIALIZATION

Laboratory Research

The WES-PHix phosphate treatment process for MWC ash was developed in 1986 by a Wheelabrator affiliate (Allied Signal Laboratories) under a research contract. The ash treatment research was undertaken in response to a permit requirement that the ash from a Wheelabrator waste-to-energy facility under construction in Claremont, New Hampshire must meet stringent state leaching test limits for eight metals, including lead and cadmium. Previous testing of ash from the company's operating facilities in other states indicated that untreated ash from the new facility might have difficulty meeting the New Hampshire leaching standards. The previous testing had shown that lead had a tendency to leach in concentrations above the state regulatory limit of 5 mg/L.

Since the New Hampshire waste-to-energy facility was under construction and not yet generating a fly ash/scrubber residue product, ashes from operating facilities of similar design were tested. Various combinations of fly ash, scrubber residue product, and bottom ash were tested during the experiments. Also, a variety of chemical treatment agents were tested for their effectiveness at immobilizing lead and cadmium.

Initial experiments with alkaline agents such as lime and cement were not effective at reducing lead solubility as measured by the applicable regulatory leaching test (the U.S. EPA Extraction Procedure Toxicity Test). The ineffectiveness of lime and cement were likely due to the high pH of the treated fly ash. Lead, being an amphoteric metal, becomes more soluble at a pH greater than about 12 [3]. Cadmium was successfully immobilized by lime and cement, however, because this metal is quite insoluble at a pH of greater than about 7.5.

Calcium sulfide (CaS) was also tested as a potential immobilization agent for lead and cadmium. Since CaS is an excellent buffer, it successfully controlled cadmium leaching, but the immobilization effects of sulfide on lead were overcome by the high pH of the treated fly ash/scrubber residue, resulting in excessive lead leaching.

Furthermore, some H_2S gas was generated during the sulfide experiments, which would present health and safety concerns if calcium sulfide were to be used in a full scale commercial ash treatment system.

Various forms of phosphate based treatment agents were also studied. Several forms were found to be highly effective at controlling both lead and cadmium leaching across a wide pH range. The addition of phosphate to both fly ash/scrubber residue and combined ash (fly ash plus bottom ash) resulted in the treated ash easily meeting the state leaching limits for lead and cadmium.

Furthermore, treatment with phosphate did not significantly change the physical haracteristics of the ash. The ash retained its free flowing, particulate nature after phosphate treatment, and did not solidify.

Some leaching test results from the original phosphate treatment experiments are shown below.

Table 1 Effects of phosphate treatment of fly ash/scrubber residue

METAL	UNTREATED	TREATED	REGULATORY LIMIT
Lead	11.8	0.23	5.0
Cadmium	1.27	0.25	1.0

Table 2 Effects of phosphate treatment on combined ash

METAL	UNTREATED	TREATED	REGULATORY LIMIT
Lead	12.0	0.31	5.0
Cadmium	2.82	0.70	1.0

The particle size distribution data in Table 3 demonstrates how the phosphate treated ash retained its free flowing, particulate nature during the experiments.

Table 3 particle size distribution of ash before and after phosphate treatment

PARTICLE SIZE, mm	PERCENT OF TOTAL	
	Untreated	Treated
<0.074	3.9	3.7
0.074 – 0.42	8.2	8.2
0.42 – 2	23.6	21.3
2 – 9.5	64.3	66.1

Commercialization of Process

The phosphate treatment process developed in the laboratory by Wheelabrator/Allied Signal was first used on a full scale, commercial basis at Wheelabrator's new waste-to-energy facility in Claremont in 1987. The process was started up simultaneously with the start-up of the facility. The treated ash generated by the facility was subjected to intensive testing during the first few weeks of facility operation. The treated ash was found to consistently leach lead and cadmium at concentrations well below the state regulatory limits. The treated ash from the Claremont facility has now been tested on a monthly basis for approximately twelve years, and it has never exhibited toxicity characteristics as defined by the State of New Hampshire or the U.S. EPA.

The laboratory research performed by Wheelabrator/Allied Signal in the mid-1980s led to a U.S. patent being granted for the phosphate treatment of MWC ash in 1988. Patents for the process have also been granted in Japan and Europe, including the United Kingdom.

PROCESS CHEMISTRY

The question of how phosphate treatment of ash actually reduces trace metals leaching has been extensively researched by Wheelabrator. In fact, this chemical treatment process is probably the most extensively studied ash stabilization technology in the world. One of the most comprehensive investigations was conducted on behalf of Wheelabrator by a consortium of universities, led by the University of New Hampshire, from 1995 to 1997. Other institutions involved in the research were the University of Massachusetts, the Massachusetts Institute of Technology, Louisiana State University and Bates College. This research investigated the mechanisms and reaction products of phosphate treatment for a variety of solid waste materials, including fly ash/scrubber residue, bottom ash, electric arc furnace dust, and contaminated soils.

The research confirmed that the addition of certain types of phosphate to these waste materials formed new mineral phases in the wastes which controlled metals leaching. The new mineral phases formed by phosphate treatment have extremely low solubilities and high geochemical stabilities in leaching environments.

The UNH research employed numerous sophisticated analytical methods to determine the chemical composition of untreated and phosphate treated ash. Neutron activation analysis (NAA), X-ray fluorescence (XRF), and X-ray photoelectron spectroscopy (XPS) were used to quantify elements. Field-emission scanning electron microscopy (FESEM) was used to examine particle size and morphology. Secondary ion mass spectroscopy (SIMS) was used to identify stabilization reactions. X-ray powder diffraction (XPPD) was used to identify bulk crystalline mineral phases. Leaching tests such as the Dutch Total Availability Test were also used to determine the effectiveness of phosphate treatment (Eighmy, 1996).

The research showed, for example, that untreated fly ash/scrubber residue from a Wheelabrator WTE facility with semi-dry scrubbers and fabric filters contained leachable lead, present in the form of lead hydroxide [$Pb(OH_2)$] and mendipite ($Pb_3O_2Cl_2$). Analysis of the phosphate treated residues showed the presence of newly formed crystalline phosphate minerals such as hydroxypryromorphite [$Pb_5(PO_4)_3Cl$]. Both the literature and the observed leaching behavior of these phosphate minerals confirm that they are geochemically stable and highly insoluble.

Table 4 lists some of the metal compounds detected in the untreated fly ash, and some of the metal phosphate minerals found in the treated residues.

Table 4 Some metal compounds detected in untreated and phosphate treated fly ash

	UNTREATED	TREATED
Lead Phases	$Pb(OH)_2$ $Pb_3O_2Cl_2$ Pb_2O_3	$Pb_5(PO_4)_3OH$ $Pb_5(PO_4)_3Cl$ $Pb_4O(PO_4)_2$
Cadmium Phases	$Cd(OH)_2$ $Cd\ P_2$	$Cd_5(PO_4)_3Cl$ $Cd_3(PO_4)_2$
Copper Phases	$Cu(ClO_4)_2$-$2H_2O$	$Cu_5(PO_4)_3OH$
Zinc Phases	$Zn_2P_2O_7$-$3H_2O$	$CaZn_2(PO_4)_2$ $Zn_3(PO_4)_2$

The UNH research concluded that phosphate treatment causes relatively unstable and soluble metal compounds to dissolve and react with the phosphate, forming nanometer-sized homogenous metal phosphate precipitates. There was no evidence of metal sorption to the fly ash particle surfaces. The newly formed metal phosphate minerals are much more geochemically stable and resistant to leaching than surface sorption of metals. The research confirmed previous studies, which concluded that phosphate reacts primarily with metals that are divalent metal cations in solution. The fraction of metals in ash available for leaching was significantly reduced by phosphate treatment. The UNH study showed that, as measured by the Dutch Total Availability Leach Test, lead leaching was reduced by 99.5%, copper by 58%, cadmium by 38%, aluminum by 75%, and zinc by 28% [4].

ADDITIONAL LEACHING DATA

Other studies of Wheelabrator's phosphate ash treatment process have been performed in recent years. One notable study was done by the U.S. EPA in 1993 [5]. The agency studied four proprietary treatment technologies and compared their effectiveness to a generic portland cement process. One of the treatment technologies evaluated was the WES-PHix process [6]. The EPA study concluded that WES-PHix was the most effective process of those evaluated for reducing the leachability of lead. Some select leaching data from the 1993 EPA Solidification/Stabilization Report are presented in the following two tables.

Table 5 Dutch Availability Test Results (mg/kg) for Fly Ash/Scrubber Residue

METAL	UNTREATED	CEMENT	PHOSPHATE	REDUCTION. %
Lead	1008	1692	3	99.7
Cadmium	133	88	91	32
Copper	197	238	96	51
Zinc	7913	7173	5107	36
Chromium	6	17	ND	100

The above data independently confirm the UNH research results. Phosphate reduced the leaching of all of the listed metals as measured by the total availability leaching test. Also, the EPA study showed that cement treatment actually increased the leaching of lead, chromium and copper. The EPA also performed a distilled water leaching test on untreated and treated fly ash. The results of this test are presented in the table below.

Table 6 Distilled Water Leach Test Results for Fly Ash/Scrubber Residue

METAL	UNTREATED	CEMENT	PHOSPHATE	REDUCTION. %
Lead	1079	187	10	99
Cadmium	ND	ND	ND	-
Copper	2	ND	ND	100
Zinc	127	19	14	89
Chromium	ND	ND	ND	-

The above data also indicates that phosphate is a more effective ash treatment agent than portland cement in a distilled water leaching environment.

PROCESS OPERATING HISTORY

As stated previously, the WES-PHix phosphate treatment process was first used commercially to stabilize MWC ash at the Wheelabrator Claremont WTE facility in 1987. Since that time, the process has been used at over 30 facilities in North America and at over 60 facilities in

Japan. Over 10 million tonnes of MWC ash have been successfully treated with the WES-PHix Process; approximately 9,000 tonnes of ash are treated each and every day with the technology. The process is currently used to treat fly ash/scrubber residue only, bottom ash only, and combined ash. Furthermore, no ash stream treated with the WES-PHix Process has ever exhibited toxicity characteristics as defined by the U.S. EPA. Leaching data from three facilities in the United States currently using phosphate to treat fly ash is presented in Table 6. This data has been provided to the author under the condition that the names of the facilities be kept confidential.

Table 7 TCLP Results (mg/L) for Phosphate Treated Fly Ash

	METAL	UNTREATED	TREATED	REGULATORY LIMIT
Facility A	Lead	46	0.49	5
	Cadmium	6.3	0.58	1
Facility B	Lead	11	<0.5	5
	Cadmium	6.3	<0.05	1
Facility C	Lead	24	<0.5	5
	Cadmium	<0.05	<0.05	1

In the U.S., phosphate treated ash is both landfilled and beneficially used. Several dedicated MWC ash recycling facilities use phosphate to stabilize trace metals in bottom ash or combined ash, and then the treated ash is sold as an aggregate for use as structural fill or in road construction projects. Phosphate treated combined ash is also used as a daily cover at sanitary landfills in several U.S. states.

PROCESS ADVANTAGES

Effectiveness

The use of phosphate to stabilize trace metals in MWC ash has been shown to be highly effective. Both university research and commercial process operating data confirm the short-term and long-term effectiveness of phosphate treatment. The 1993 U.S. EPA study showed phosphate to be the most effective process of those evaluated for stabilizing lead in MWC ash, as measured by leaching tests. The recent UNH research confirmed the formation of highly stable and insoluble metal phosphate minerals in MWC ash such as cadmium phosphate and lead chloropyromorphite. A separate UNH study has also provided strong empirical evidence that phosphate treatment is a highly effective long-term metal stabilization process for MWC ash [7]. The study involved placing both untreated and phosphate treated MWC combined ash in separate artificial monofills (lysimeters) on the university campus in 1988. The ash has been subjected to precipitation/infiltration, repeated freeze/thaw cycles, and wetting/drying phenomena for over ten years. All leachate generated by the lysimeters/monofills has been collected and analyzed for dissolved metals and other parameters. Leachate collection and analysis has occurred every 1-3 months, depending on the level of infiltration. The test results show that phosphate treatment has reduced lead leaching from the treated ash by one order of magnitude, compared to the untreated ash. Lead and other trace metals such as cadmium are typically below laboratory detection limits in the treated ash leachate. This study confirms that phosphate treatment reduces the long-term leaching of lead from MWC ash.

Process Cost

The cost of using phosphate to treat MWC ash compares very favorably to the cost of other treatment technologies such as portland cement. Phosphate treatment costs will vary depending on several factors, including the type of ash treated, the quantity of ash being treated, and the local price of phosphate. In the U.S., the cost of using phosphate to treat combined ash is approximately \$5 per tonne of ash treated. The cost to treat flyash/scrubber residue is higher, due to the higher fraction of leachable metals in the ash and the need to add greater amounts of phosphate to immobilize the metals. Phosphate treatment of fly ash in the U.S. typically costs between \$15 and \$25 per tonne. The capital cost of a phosphate treatment system will also vary, depending on the type of phosphate used and the existing ash handling equipment at the WTE facility. Treatment system capital costs in the U.S. typically are between \$100,000 and \$350,000.

Physical Characteristics of Treated Ash

Phosphate treatment of ash does not significantly change the physical characteristics of the MWC ash. The treated ash retains its free-flowing, granular nature, which facilitates subsequent handling of the material. Also, phosphate treatment does not significantly increase the weight of treated ash, thereby minimizing disposal expenses if the costs are based on weight. Furthermore, if a liquid form of phosphate is used, the volume of the treated ash is not increased. Liquid phosphate is also an effective surfactant, which enhances the dust suppression properties of water and helps control fugitive dust emissions from treated ash.

Process Reaction Kinetics

Another important advantage of phosphate treatment is the rapid reaction kinetics of the process. The formation of the insoluble phosphate mineral phases in the treated ash occurs within several minutes to several hours of treatment (Eighmy, 1996). The quick reaction time allows the treated ash to be immediately tested for regulatory compliance purposes and removed off-site for further management. This precludes the need for a large ash storage area at the WTE facility.

Process Performance Record

Phosphate treatment of MWC ash is now one of the most widely practiced ash stabilization technologies in the U.S. and Japan. Experience has shown the process to be extremely reliable. Over 10 million tonnes of ash have been successfully stabilized with phosphate.

CONCLUSIONS

Chemical treatment of MWC ash with phosphate is an effective method for reducing the leaching of trace metals from ash. Thousands of laboratory leaching tests have been performed on phosphate stabilized ash and have confirmed the effectiveness of the process. Phosphate treatment is also one of the most extensively studied ash stabilization technologies in the world. These studies have confirmed that phosphate stabilization is an environmentally sound method for reducing the leaching of certain trace metals from MWC ash. Phosphate treatment ensures that all types of MWC ash will meet regulatory leaching limits, and will facilitate the proper management of ash whether landfilled or beneficially used.

REFERENCES

1. AIMS COALITION, America's Newest Energy Source, 1994, pp 12-13.

2. INTERNATIONAL ASH WORKING GROUP, Municipal Solid Waste Incinerator Residues, Elsevier Science B.V., Netherlands, 1997, pp 263-312.

3. O'HARA, M. AND SURGI, M. Immobilization of Lead and Cadmium in Solid Residues from the Combustion of Refuse Using Lime and Phosphate, U.S. Patent No. 4,737,356, 1988.

4. EIGHMY, T. AND CRANNELL, B. Fundamental Research on the Mechanisms of the WES-PHix® Process: Studies on Wheelabrator Dry Scrubber Residue, University of New Hampshire, 1996, pp 17-59.

5. KOSSON, D., KOSSON, T. AND VAN DER SLOOT, H. Evaluation of Solidification/Stabilization Treatment Processes for Municipal Waste Combustion Residues. U.S. EPA, 1993, pp 324-337.

6. LYONS, M. The WES-PHix® Ash Immobilization Process, Proceedings of the 7th International Conference on Municipal Waste Combustor Ash Utilization, Resource Recovery Report, 1995, pp 156-168.

7. EIGHMY, T. AND CRANNELL, B. Final Report of the Field Lysimeter Leaching Experiments: Wes-pHix Amended Combinded Ash and Scrubber Residue, University of New Hampshire, 1998.

GAS GENERATION AT A MUNICIPAL WASTE COMBUSTOR ASH MONOFILL- FRANKLIN, NEW HAMPSHIRE

C N Musselman, W A Straub

CMA Engineers Inc

J N Bidwell, J E Carpenter

Enviornmental Risk Limited

J R Presher

Concord Regional Solid Waste/Resource Recovery Cooperative

United States of America

ABSTRACT. The paper describes the characteristics of gas generated within a residue landfill in Franklin, NH, USA accepting combined ash (bottom ash, fly ash, scrubber product from direct dry lime injection) from a 500 ton per day mass burn waste to energy facility in Concord, NH. Temperatures within the landfill mass were found to be up to 156°F, presumably due to exothermic chemical reactions within the residue. Methanogenic bacterial activity does not appear to play a major role in gas generation. Hydrogen gas was measured at concentrations in excess of the lower explosive limit. The hydrogen gas is postulated to be generated by reactions of elemental aluminum and other metals within the landfilled residue. Volatile organic compounds were present in concentrations at the low end of concentrations generally reported for MSW landfills. The information presented is likely representative of the gas characteristics within the landfilled mass, and in the head space within leachate collection piping systems.

Keywords: Municipal waste combustor residue, Ash landfills, Landfill gas, Hydrogen gas generation

Craig N Musselman is an environmental engineer and the president of CMA Engineers, Inc. of Portsmouth, NH. He has presented papers on the generation of gas in ash landfills, on the utilization of bottom ash in asphaltic concrete for roadway reconstruction and on landfill design topics.

William A Straub is an environmental engineer and the vice president of CMA Engineers of Portsmouth, NH. He has presented papers on landfill design, landfill gas, and the fate and transport of contaminants in solid waste landfills.

Jeremy N Bidwell and Joyce E Carpenter were, respectively, project manager and associate of Environmental Risk Limited, Bloomfield, CT.

James R Presher is the Director of the Concord, NH Regional Solid Waste/ Resource Recovery Cooperative.

INTRODUCTION

The characterization of landfill gas generated at municipal solid waste landfills has received significant attention in the United States in recent years. Generation of gas at municipal waste combustor (MWC) ash monofills is, however, generally assumed to be negligible and there is little, if any, published information available concerning the characteristics of gas generated at ash monofills. As ash landfills move towards closure in the future, and as progress continues in utilizing certain components of MWC ash residue in construction applications, it is important to identify the mechanisms by which gases may be generated in MWC ash so that appropriate design and material management decisions can be made.

The information reported upon herein was based upon the analysis of gas samples from leachate collection system cleanout pipes at a single lined MWC ash landfill (1, 2). The data were gathered because site operating staff had posed questions as to the contents of the gaseous emissions. The gas characteristics indicated the effects of exothermic chemical reactions in the landfilled ash which affected gas composition in ways which were unexpected, and not widely known.

PROJECT BACKGROUND

The lined MWC ash landfill in Franklin, New Hampshire is owned by the 27 member municipalities of the Concord, NH Regional Solid Waste/Resource Recovery Cooperative ("the Cooperative"). The landfill was initially constructed in 1988, and has operated continuously since 1989.

The landfill accepts combined bottom ash, fly ash and dry lime scrubber product from the 500 ton per day waste to energy (or Municipal Waste Combustor) facility in Concord, New Hampshire owned and operated by a subsidiary of Wheelabrator Technologies, Inc. The MWC facility processes about 177,000 tons per year of municipal solid waste and generates about 68,000 wet tons per year of combined ash. Bottom ash is discharged from the combustion units into wet ash troughs, from which the bottom ash is removed by means of drag chain conveyors. The facility's air pollution control system consists of pneumatic injection of hydrated lime directly into the exhaust gas ductwork for control of acid gas emissions, with fabric filters for control of particulates. The fly ash and dry lime scrubber product is treated with the proprietary WES-PHix process, and combined with the bottom ash on the bottom ash drag chain conveyor. The WES-PHix process involves the addition of phosphoric acid (H_3PO_4) to the fly ash to promote the formation of lead phosphates, in order to limit the solubility of lead in the ash.

The lined MWC ash landfill in Franklin, New Hampshire consists of a double lined landfill with a primary leachate collection system consisting of HDPE piping above the primary (or uppermost) liner. The leachate collection pipes discharge to leachate tanks at the low side of the site, from which leachate is pumped to a remote sewer connection. The leachate tanks are vented to the atmosphere.

Since the primary leachate collection pipes beneath the landfilled ash flow only partially full, air flows through the pipes with intake at the leachate collection tanks at the low side of the site. Air flows through the primary leachate collection pipes and discharges through the leachate system cleanout pipes located on the high side of the site, through a chimney" type effect.

The landfill is designed to eventually reach a final maximum depth of about 120 feet. In 1996, residue grades were about 45 feet above the landfill liners. In order to facilitate future leachate collection needs at higher elevations, two leachate collection pipes had been extended vertically through the fill. One of these two pipes had been perforated every several feet and wrapped with geotextile fabric. During a storm event, ash had been deposited in the perforated vertical pipe, which had clogged the bottom 1.5 feet of pipe where the vertical pipe connected to the leachate collection pipe. This clogged pipe had a water surface within the pipe, thus closing off convective air access to the leachate collection system. A cross section of this vertical pipe is presented as Figure 1. In this condition, this one vertical pipe inadvertently functioned as a typical landfill gas well. This clogged vertical pipe is noted as the "Phase 1 Vent".

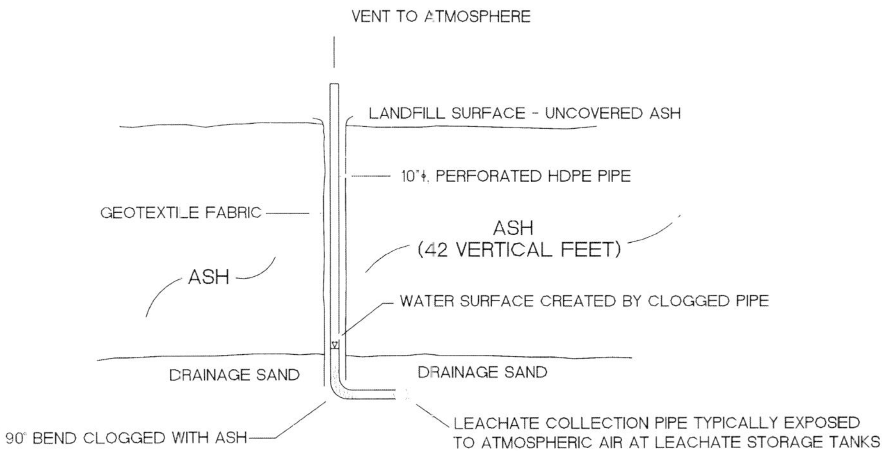

Figure 1 Phase 1 vent – Cross section

In 1995, two site staff members posed the question to the Cooperative as to the nature of the gas. In response to that question, the Cooperative began to gather data on the gas emissions.

The first evaluation was made by Cooperative staff using a handheld explosivity meter ("LEL meter"), on the various cleanouts and leachate collection pipe outlets (or "vents"). The meter's alarm was tripped at two of the outlets for elevated H_2S concentration, and the alarm for lower explosive limit ("LEL") exceedance was tripped for the Phase 1 Vent. This information led the Cooperative to decide to sample and analyze the gaseous emissions.

ASH CHARACTERISTICS

There are a number of factors pertaining to the specific ash handled at the Franklin site and the operating scheme at the landfill which may affect the generation of gas. These characteristics and the potential impacts of each are listed and discussed below.

Ash Combustible Content - A comprehensive ash sampling and analysis program conducted in 1990 and 1991 indicated that the bottom ash at that time had a combustible content averaging 6.4% as measured by Loss on Ignition tests (3). This combustible content may be considered moderate to high by US standards. This may have affected non-methane organic compound concentrations.

Lime Content - The direct dry lime injection system requires the addition of lime at 2 to 4 times the stoichiometric requirement for acid gas control. This rate is higher than at most plants with more stoichiometrically efficient spray dryer absorption systems. The ash at Franklin may have greater alkalinity and unreacted lime than is typical of other ash monofills. The primary leachate, however, has had a pH of 7 or below for the past seven years.

Phosphoric Acid Addition - Small amounts of phosphoric acid are added to the fly ash/scrubber product using the proprietary "WES-PHix" process. The addition of phosphoric acid has been considered in terms of its potential to have an effect on gas generation. Any effect is presumed minimal due to the neutralization of phosphoric acid by the lime.

Landfill Operating Scheme - The Franklin ash landfill is operated aggressively as a "dry tomb" type of facility in comparison to most US landfills. The limited amount of time (2-3 years), during which the landfilled ash has been subjected to percolation of water may have an effect on chemical reactions taking place in the landfill.

Other than those factors described above, the ash generated by the Concord, New Hampshire waste to energy facility is considered to be typical of most US facilities based on the type of facility, operation and ash characterization data available.

LANDFILL PIPING CONFIGURATION

The specific layout of the various vents is depicted schematically on Figure 2. The nomenclature for each gas sampling point, and the specific circumstances of each location are presented in Table 1.

GAS TEMPERATURE AND FLOWS

The landfill interior temperature was measured by lowering a weighted thermocouple into the Phase 1 and Phase 2 vents. Ambient air temperature at the time of sampling was 39°F.

The Phase 2 vent could be measured only 9 feet down the pipe due to the presence of a bend. The temperature was 91°F. Gas in this vent was diluted by atmospheric air from the leachate collection system below.

The temperature of undiluted gas in the clogged Phase 1 vent ranged from 132°F to 156°F. The temperature profile with depth is indicated on Figure 3. These temperatures were unexpectedly high and were higher than would typically be expected in an MSW landfill. Optimum temperatures for thermophilic anaerobic digestion of MSW range from 120°F to 135°F (4). It is postulated that the elevated temperatures are likely due to a variety of exothermic, primarily inorganic chemical reactions taking place in the landfilled ash.

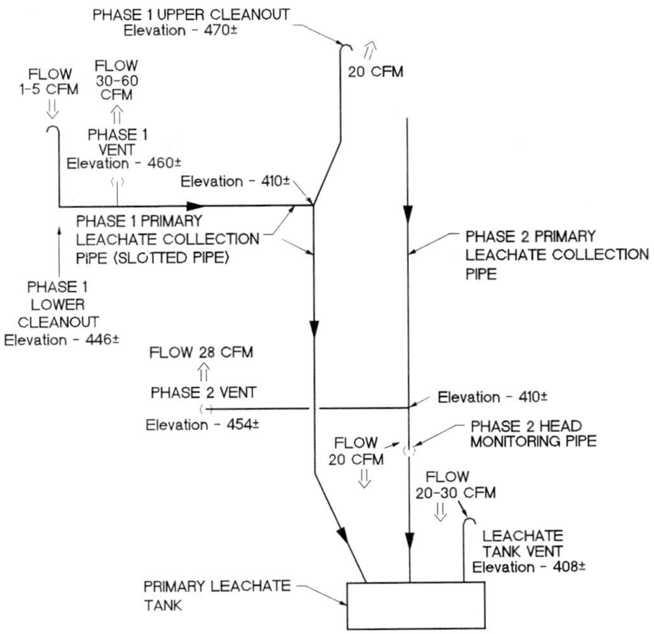

Figure 2 Leachate collection system schematic

Table 1 Gas sampling locations

SAMPLING POINT	TYPE OF PIPE	DEPTH OF ASH	AGE OF ASH
Phase 1 Vent	Perforated, Clogged at Depth (Similar in Function to Gas Well)	42 feet	0-4 years
Phase 2 Vent	Solid Wall, Open to Leachate System Below (Diluted by air flow from below)	46 feet	1-2 years
Phase 1 Upper Cleanouts	Solid Wall, Open to Leachate System Below (Diluted by air flow from below)	54 feet	4-8 years

Measured gas flows are as indicated on the piping schematic presented as Figure 2. A chimney effect is apparent as pipes open to the atmosphere at lower elevations indicated flow into the piping system and the vents connected to the piping system which outlet at higher elevations emitted gas at flows of 20 and 28 cubic feet per minute (CFM). Gas flows at the Phase I vent were estimated to be 30 to 60 CFM. This flowrate is similar in magnitude to the rate which might be experienced at steady state conditions in an active MSW landfill gas well. The zone of influence within the landfill mass of the Phase 1 vent is unknown.

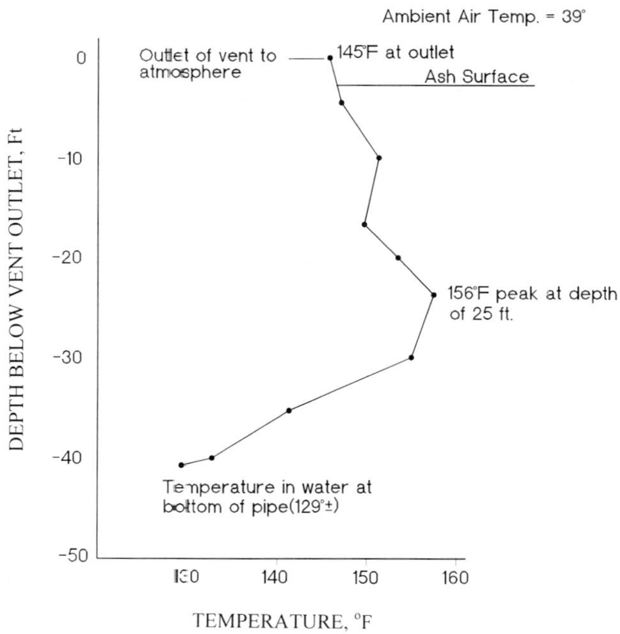

Figure 3 Gas temperature profile with depth

MWC LANDFILL GAS CHARACTERISTICS

The vents which had exhaust flow to the atmosphere were sampled and analyzed for atmospheric gases non-methane organic compounds (NMOCs) and sulfur compounds. Samples were gathered using evacuated SUMMA passivated (polished) canisters with samples withdrawn over a 60 minute duration, per Method TO14. Atmospheric gases were analyzed by ASTM Method D-3416. NMOCs were analyzed by gas chromatograph using Method TO12 (total NMOC), and by gas chromatography/mass spectroscopy for the full Method TO14 target compound list. Sulfur containing compounds were quantified through ASTM Method 5504.

The summary of the atmospheric gas analyses is presented in Table 2. The oxygen content in the Phase 2 Vent and the Phase 1 Upper Cleanout, both of which were open to atmospheric air in the leachate collection pipes beneath the landfill, were near typical atmospheric levels. The oxygen content in the Phase I Vent, which was clogged and functioning in a way similar to a gas well, was markedly depressed, but not absent. Methane was present in the Phase I Vent at a concentration of 780 ppm. This is low compared to the levels present in gases from MSW landfills (typically 500,000 ppmv, or 50%), and at a level well below the lower explosive limit (LEL) for methane of 50,000 ppmv. If methanogenic bacterial activity were at a high level, the O_2 concentration in the Phase 1 Vent would have been expected to approach zero.

Hydrogen was detected in the gas in the Phase 1 Vent. The hydrogen concentration in both Phase 1 Vent samples was measured to be 5.1%, which is in excess of the 4.0% LEL for hydrogen. The presence of hydrogen at these concentrations is likely the cause of the explosiveness alarm in the initial gas screening described earlier. The hydrogen is believed to be generated by oxidation reduction reactions of aluminum and other reactive metals, and/or other chemical reactions, taking place within the landfill mass, as discussed later herein.

Table 2 Summary of atmospheric gas analyses

LOCATION	OXYGEN (%)	NITROGEN (%)	CARBON MONOXIDE (PPM)	METHANE (PPM)	CARBON DIOXIDE (%)	HYDROGEN (%)
Phase 1 Vent						
first (can 3)	3.5	91	110	780	.008	5.1
second (can 4)	3.6	91	110	780	.008	5.1
Phase 2 Vent						
first (can 1)	21	79	<20	<20	.038	.087
second (can 2)	21	79	<20	<20	.038	.092
Phase 1 Primary upper cleanout						
first (can 5)	21	79	<20	30	.051	<.024
second (can 6)	21	79	<20	30	.050	<.024

Hydrogen sulfide concentrations were elevated in the Phase 2 Vent and in the Phase 1 upper cleanout. H_2S is typically a by-product of the anaerobic decomposition of organic matter. The H_2S concentrations are likely indicative of microbial activity within the landfill mass.

The NMOC concentrations are presented in Table 3. The NMOC concentrations were present at the low end of the range of concentrations reported in the literature (4) for NMOC concentrations in MSW landfill gas. Leachate from the landfill site rarely has detectable concentrations of volatile organic compounds with the exception of periodic low concentrations of acetone.

The concentrations of specific organic compounds were reported for two samples from the Phase 1 Vent and for one sample from the Phase 2 Vent. The results are presented in Table 4. The reported concentrations are all lower than presently applicable health and safety related standards and are generally less than those anticipated in MSW landfill gas (5).

Following the completion of the sampling and analysis described herein, the clog in the Phase 1 vent was removed hydraulically. Later LEL meter testing indicates that the Phase 1 and Phase 2 vents subsequently have similar characteristics.

DISCUSSION OF RESULTS AND IMPLICATIONS

Generation of Hydrogen Gas

The measured hydrogen gas may be generated by chemical reactions of elemental aluminum and other elemental metals in the presence of water. Typical oxidation reduction reactions may be as follows (6,7):

$$2\ Al^0 + 3H_2O = Al_2O_3 + H_2$$

$$Al^0 + 2\ H_2O = AlOOH + 1.5\ H_2$$

$$Al^0 + 3\ H_2O = Al(OH)_3 + 1.5\ H_2$$

Table 3 Summary of NMOC and sulfur gas analyses

LOCATION	NON-METHANE ORGANIC COMPOUNDS PPM)	HYDROGEN SULFIDE (PPM)	TOTAL REDUCED SULFUR (PPM)
Phase 1 Vent			
first (can/bag 3)	38	<0.004	0.290
second (can/bag 4)	40	<0.004	0.270
Phase 2 Vent			
first (can/bag 1)	3.3	11.0	11.0
second (can/bag 2)	4.4	15.0	15.0
Phase 1 Primary upper cleanout first (can/bag 5)	0.45	11.0	11.0
Second (can/bag 6)	0.48		

Table 4 Summary of Organic Compound Analyses (All reported in ppb)

COMPOUND	PHASE 1 VENT CAN 3	PHASE 1 VENT CAN 4	PHASE 2 VENT	TLV[2]-TWA (ACGIH[1])	STEL[3] (ACGIH[1])
Chloromethane	75	110	29	50,000	100,000
Methylene Chloride	130	<40	30	50,000	
Benzene	96	97	<16	10,000	300*
Toluene	100	93	<16	50,000	
Acetone	9,900	9,500	1,400	750,000	1,000,000
Tetrachlorethane	<48	<40	<24	25,000	100,000
2-Butanone (MEK)	840	720	170	200,000	300,000
Hexane	1,300	1,200	<65	50,000	
Ethanol	6,400	5,700	4,200	1,000,000	
Heptane	690	640	<65	400,000	500,000
Total Organic Compounds	29,661	22,310	6,179	N/A	N/A

1. ACGIH: American Conference of Governmental Industrial Hygienists
2. TLV-TWA: OThreshold Limit Value - Time Weighted AverageO (8 hour work day/40 hour work week, without adverse effect)
3. STEL: OThreshold Limit Value - Short Term Exposure LimitO (maximum 15 minute exposure)

The aluminum reactions are facilitated in the presence of either strong acids or strong bases, which may etch the surface of the aluminum, exposing more elemental aluminum to the reaction. In typical uses of aluminum subject to atmospheric exposure, an adherent protective film of aluminum oxide forms on the surface of the aluminum in pH conditions in the range of 4 to 9 (8). This protective surface film precludes further corrosion of the underlying aluminum. However, in the presence of strong acids or bases, this film may be destroyed and the hydrogen gas generating corrosion reactions can continue. When the protective film is destroyed (or the aluminum is "depassivated"), the exposed elemental aluminum reacts vigorously in the presence of moisture until all exposed aluminum is reacted. The protective oxide film reforms within seconds of exposure of elemental aluminum (9), which limits the length of time that aluminum corrosion continues to occur in atmospheric exposure.

In the landfill, the conditions under which corrosion occurs are more complicated. The aluminum is present both in bulk form and in fine particles, and the chemical conditions which affect corrosion vary locally throughout the landfill mass. Aluminum corrosion can be facilitated by the presence of chloride ions (10), which are present in abundance in MWC ash. The chloride ions affect the aluminum through pitting, exposing additional elemental aluminum to the corrosion reaction. The chemistry of aluminum is complex, and the reactions taking place in the heterogeneous landfilled residue are not well understood.

Aluminum is present in significant quantities in the ash. Bottom ash from the Concord, New Hampshire MWC facility has a range of aluminum content of 3.4 to 6.4% by weight. Aluminum is also typically present in fly ash in significant concentrations. It is anticipated that a significant percentage of the aluminum present in the ash exists in the elemental form having the potential for the hydrogen generating oxidation reduction reactions.

Hydrogen gas may also be generated by similar reactions with other elemental metals. The oxidation of elemental aluminum, zinc, chromium, iron, copper and other elemental metals can proceed in the proper environmental conditions, releasing electrons. The electrons, in the presence of water and, again, in the proper environmental conditions, can result in the generation of H_2 gas. The aluminum reaction is highlighted herein because of the redox potential of the aluminum reaction, the significant quantity of aluminum in the ash, and the likely presence of elemental aluminum in greater quantities than is likely for the other elemental metals. Oxidation reduction reactions of many elemental metals probably contribute to varying extents to the H_2 gas generation.

Hydrogen is also generated as a byproduct in certain organic decomposition reactions (i.e., in the later stages of anaerobic, non-methanogenic decomposition in an MSW landfill (4)), and can be generated as a byproduct in a variety of other chemical reactions.

It is considered unlikely that the addition of phosphoric acid plays a significant role in hydrogen gas generation at the landfill site. Although the Material Data Sheet supplied by the phosphoric acid supplier warns of the hydrogen gas generation potential when in contact with aluminum, this is due to the acidic nature of the phosphoric acid. The H_3PO_4 acid is neutralized at the MWC facility upon mixing with the highly alkaline scrubber product.

The unusually high alkalinity of the fly ash at the Franklin, New Hampshire landfill in all likelihood results in greater hydrogen gas generation than would be the case for an ash with less available lime. Hydrogen gas at other sites should still be anticipated to be present, although likely at a slower generation rate.

The hydrogen gas generating reactions of aluminum have been known for many years, and have been used in the manufacture of lightweight concrete with a cellular structure (11, 12).

Aluminum powder is added to cement and sand. In the presence of lime or other alkali, hydrogen bubbles are formed throughout the concrete, creating a more porous structure. The practice is more prevalent in Europe than in North America.

Anecdotal information is available regarding experience with the generation of hydrogen gas from MWC ash. Fly ash utilized as an aggregate substitute in asphalt cement in Germany developed gas bubbles during paving (13). The gas was analyzed and found to be primarily hydrogen. The generation of hydrogen gas and methane from fly ash and scrubber residue has been documented in laboratory studies in Germany (13, 14). Explosions in a fly ash system in Denmark were attributed to hydrogen gas (15, 16). Explosive conditions were subsequently prevented through ventilation of ash conditioning equipment.

The presence of hydrogen gas at levels above the LEL is not believed to be a major risk factor with respect to landfill operations. Landfill gas from MSW landfills often exhibits methane concentrations above the LEL and below the upper explosiveness limit, at concentrations where explosions could occur in certain circumstances. This risk is managed by proper gas handling in passive or active gas collection systems at closed landfills and by appropriate monitoring of gas concentrations at the site periphery to assure that subsurface migration to nearby structures does not occur. If the generation of hydrogen gas is confirmed at other ash landfill sites, similar design and operation precautions would likely be appropriate for ash landfills as well. Appropriate precautions should be taken when excavating into ash in confined spaces, or when drilling into old landfilled ash.

The potential for generation of hydrogen gas should be considered when evaluating ash utilization alternatives. Until further information is developed, the utilization of unencapsulated ash in close proximity to confined spaces such as structures may be inadvisable. Some researchers (11) have suggested that the aging of ash prior to utilization might be effective in reducing the potential for hydrogen gas release after utilization. The length of time required and the effectiveness of any reduction in subsequent gas generation are not yet known.

Ash Temperature

The elevated temperature determined within the landfill mass is most likely due to a variety of exothermic chemical reactions. Preliminary, theoretical thermodynamic calculations have indicated that neither the exothermic aluminum reactions described herein nor the lime hydration reactions would be expected, alone, to generate sufficient heat to bring the landfilled mass to the 156°F maximum temperatures measured. Other exothermic chemical reactions also undoubtedly play a role.

From a design standpoint, the elevated ash temperature is not significantly problematic. Temperatures in the ranges measured do not negatively affect most flexible membrane liners and geotextile materials typically used in landfill construction. The design of leachate collection pipes is somewhat temperature dependent in that the calculation of the pipe wall material's modulus of elasticity requires temperature input, depending upon the type of pipe used. Until valid information is developed from other sites indicating the prevalence of much lower ash temperatures, the use of 160°F as the temperature input in leachate collection pipe strength calculations, where required, may be advisable.

Provisions for Gas Management in Ash Landfill Closure Designs

It is apparent from the information reported upon herein that gas management provisions need to be incorporated in the design for closure of an MWC ash landfill. Design procedures similar to those utilized for the closure of smaller MSW landfills with passive gas venting systems may be appropriate. The installation of a suitable gas migration layer beneath impervious capping materials, and passive vents through impervious capping layers, should be, and typically are, incorporated in closure designs. Further information is needed regarding the movement of gas within landfilled ash to determine whether passive gas wells within the ash are needed or are otherwise of benefit.

Additional Information Needs

Information is needed from other ash landfills to confirm the temperature of ash at various depths, and to assess hydrogen gas generation rates and concentrations. Both the temperature and the hydrogen gas generation rates may be lower at other sites where less lime is available for hydration reactions. It should be noted that the hydrogen gas concentrations measured at any site are less a function of generation rate than they are a function of the site=s tendency to accumulate gases. The data reported upon herein provides no information regarding the nature of gas pressure and concentration gradients within an ash monofill.

The effect of aging the ash prior to utilization with regards to the subsequent generation of hydrogen gas also deserves investigation.

CONCLUSIONS

The following conclusions can be drawn from the information reported upon herein:

1. MWC ash disposed in an ash monofill may generate gas with NMOC concentrations on the low end of the typical range for NMOCs in MSW landfill gas.

2. Gas flowrates generated in an ash monofill may not be insignificant from a design perspective.

3. Hydrogen gas may be generated within the ash mass at an ash monofill from reactions of elemental aluminum and/or from other sources. Hydrogen gas concentrations may exceed the lower explosive limit.

4. Ash temperatures within an MWC ash monofill may be on the order of 156°F, due to reactions of elemental aluminum, lime hydration reactions and other exothermic chemical reactions. The potential for elevated temperatures should be considered in the design of landfill components.

5. Anaerobic non-methanogenic and methanogenic decomposition of waste does not appear to be a significant factor in the generation of gas at an ash monofill, despite the presence of an uncombusted fraction in the ash.

6. Lower ash temperatures and decreased rates of generation of hydrogen gas may be experienced at other ash landfills having less lime available for hydration reactions than was present in the ash at the Franklin, New Hampshire site. However, elevated temperatures and some hydrogen gas generation should be anticipated until experience indicates otherwise.

7. Passive gas vents, and effective gas migration layers beneath impervious caps, should be incorporated in designs for closure of ash monofills.

8. Appropriate precautions should be taken when excavating into ash in confined spaces, or when drilling into old landfilled ash.

9. The potential for generation of hydrogen gas should be considered in ash utilization programs. Until further information is developed otherwise, the utilization of unencapsulated ash in close proximity to enclosed spaces (i.e., structures) may be inadvisable. The potential benefit of aging of ash on reducing subsequent hydrogen gas generation in utilization merits investigation.

REFERENCES

1. "Vent Testing Program, Concord Regional Solid Waste/Resource Recovery Cooperative, Secure Residue Landfill, Franklin, New Hampshire"; Environmental Risk Limited, May, 1996.

2. "Report on Landfill Gas Testing, Franklin, NH Secure Residue Landfill"; CMA Engineers, Inc.; June, 1996.

3. EIGHMY, TT; GRESS, D.; ZHANG, X.; TARR, S; AND WHITEHEAD, I; "Bottom Ash Utilization Evaluation for the Concord, New Hampshire Waste-to-Energy Facility"; Environmental Research Group, University of New Hampshire; May, 1992.

4. "Air Emissions from Municipal Solid Waste Landfills - Background Information for Proposed Standards and Guidelines"; United States Environmental Protection Agency; Research Triangle Park, NC; March, 1991.

5. CARPENTER, JE AND BIDWELL, JN; "Air Emissions at a Municipal Solid Waste Landfill"; Environmental Risk Limited; Bloomfield, CT; in the Proceedings of the 17th Biennial Waste Processing Conference, American Society of Mechanical Engineers; 1996.

6. WEFERS, K. AND BELL, GM; "Oxides and Hydroxides of Aluminum, Technical Paper No. 19"; Alcoa Research Laboratories, 1972.

7. WEFERS, K. AND MISRA, C ; "Oxides and Hydroxides of Aluminum, Alcoa Technical Paper No. 19, Revised"; Alcoa Laboratories, 1987.

8. PAUL, S.; MITRA, P.K.; AND SIRKAR, S.C.; "Passivity Breakdown and PZC of Aluminum Chloride Water System"; in Corrosion, Journal of the National Association of Corrosion Engineers; March, 1993.

9. "Corrosion, Volume 1, Metal/Environment Reactions", Edited by Shreir, L.L.; Newnes Butterworth Publishers; 1976.

10. METIKOS-HUKOVIC, M. AND BABIC, R.; "Inhibition of the Hydrogen Evolution Reaction on Aluminum Covered by 'Spontaneous' Oxide"; in Journal of Applied Electrochemistry, Volume 24; 1994.

11. NEVILLE, A. M.; "Properties of Concrete, 3rd Edition"; Pittman Publishing, Inc.; Marshfield, MA; 1981.

12. NEUFELD, R.D.; VALLEJO, L.E.; HU, W.; LATONA, M.; CARSON, C.; AND KELLY, C.; "Properties of High Fly Ash Content Cellular Concrete"; in American Society of Civil Engineers Journal of Energy Engineering; April, 1994.

13. EIGHMY, TT; University of New Hampshire; Personal Communication; 1996.

14. OBERSTE-PADTBERG, R.; SEWEDEN, K.; AZur Freisetzung von Wusserstoff aus Mörteln mit MVA-Reststoffen@; Wasser, Luft and Boden 6:61-62; 1990.

15. "An International Perspective on Characterization and Management of Residues from Municipal Solid Waste Incineration, Volume 1"; the International Ash Working Group; July, 1995.

16. HJELMAR, O.; AStofudraskniny fra flyveaskefra affaldsforbraendingsanlaeg.@ Repport til Miljøstyrelsen. VKI, Hørsholm, Denmark, 1993.

17. SURGI, M. R.; Memorandum Re: "Thermodynamics and Potential Reactions Causing Elevated Temperatures in Ash Monofills"; Analytical and Environmental Services, Inc., Glencoe, IL; May 31, 1996 (unpublished preliminary calculations).

PILOT AND FULL SCALE WASHING OF INCINERATOR BOTTOM ASH

J K Boddum

J Skaarup

AFATEK A/S

Denmark

ABSTRACT. An investigation was performed with the purpose of evaluating suitable methods of improving the quality of municipal solid waste incinerator bottom ash. Full scale washing in the quench tank of an incinerator plant and pilot scale washing/separation at a soil cleaning facility has been tested. Emphasis has been laid on the emission of salts. The effect of washing was evaluated by a leaching test CEN prEN 12457, procedure A. Criteria for success has been forthcoming limit values proclaimed by The Danish Environmental Protection Agency. The effect of separating fines and organic material was evaluated by determination of total content in the produced fractions. Quench tank washing was found unfit to reduce the leaching of salt sufficiently. The effect of increased water supply in the quench tank was negligible. Washing on a pilot scale soil cleaning facility reduced the emission of sodium and chloride to meet the limit values. Sulphate proved to be difficult to remove. The limit value was not reached. Arsenic, cadmium, chromium, lead and zinc was found enriched in the fine fraction <100 μm.

Keywords: MSWI bottom ash, Quench tank washing, Soil cleaning facility, Leaching

Mr Jens K Boddum, is M.Sc. in Environmental Engineering from the Technical University of Denmark. Employed as development engineer at AFATEK A/S.

Mr Jørgen Skaarup, is Managing Director of AFATEK A/S and M.Sc. with speciality in Chemistry and Environmental Engineering. 20 years experience within management of the Environment, both in public and private sector. Main working fields have been. Waste Water Treatment, Ground Water Pollution, Site Remediation of Soil Contamination, Soil Cleaning and Reuse of Municipal Solid Waste Incinerator Bottom Ash.

INTRODUCTION

Thermal treatment with energy recovery accounts for the disposal of approximately 80% of the municipal solid waste produced in Denmark. This causes the production of in the order of 500.000 tons of incinerator bottom ash per year. Traditionally the bottom ash is utilized in various construction works, as a substitute for gravel. In 1995-97 app. 80% of the bottom ash was reused, whereas the remaining 20% was landfilled [1].

Since 1983, the utilization has been regulated by a decree issued by the Danish Environmental Protection Agency [2]. In this, requirements are set to the total content of cadmium, lead and mercury. Furthermore alkalinity and pH are subject to demands. Failure to comply with these limits, primarily the lead content, account for the above-mentioned 20% landfilled MSWI bottom ash.

The Danish EPA is currently reviewing this decree, and recently a draft for a new decree was issued [3]. In compliance with other European countries, the review introduces new means of evaluating possible environmental impact from the use of a material. The leaching is to be determined by a batch test CEN prEN 12457, procedure A, proposed by the European Committee of Standardization, CEN [4]. The test determines the leaching at a liquid to solid ratio (L/S) of 2 l/kg and limits are set to concentration of components in the eluate. Combined with limits to the total content of certain components, the residue is categorized in one of four categories. The limit values are given in Table 1. Results of the proposed leaching test on Danish MSWI bottom ashes are also given for comparison.

Table 1 Limit values categorizing residues to be utilized in construction work. Total content in mg/kg. Leaching in batch leaching test at L/S = 2 l/kg. Concentrations in μg/l, except chloride, sodium and sulphate in mg/l. Leaching from MSWI bottom ash is given for comparison

COMPONENT	CATEGORY 1&2A Total content	CATEGORY 1&2A Leaching	CATEGORY 2BLeaching	MSWI BOTTOM ASH LEACHING	
				I/S Vest[1]	I/S Amager[2]
Chloride		150	3000	830	2100
Sodium		100	1500	500	1100
Sulphate		250	4000	1700	940
Arsenic	20	8	50	3.4	2
Cadmium	0,5	2	40	0.1	0.3
Chrome, Tot.	500	10	500	-	0.7
Chrome (VI)	20				
Copper	500	45	2000	6.2	4
Lead	40	10	100	<0.3	21
Mercury	1	0,1	1	-	0.03
Nickel	30	10	70	1.3	2
Zink	500	100	1500	-	15

1. Analysis of bottom ash from the incinerator plant I/S Vestforbrænding [5]
2. Average of analysis of 6 samples from the incinerator plant I/S Amagerforbrænding [6]

The residues are thereby categorized as follows:

Category 1: Residues that meet the demands of total content and the emission of category 2a. May be used in common construction work.

Category 2a: Residues that exceed the limit value of total content but respect the leaching values of category 2a. May be used in most construction works provided a top lining securing against contact is applied. Furthermore limitations to the extend of the layer are given.

Category 2b: Residues that exceed the limit value of category 2a leaching, but meet the values of category 2b leaching. May be used in roads, paths and in wire pits under an impermeable top lining (concrete, asphalt). Use under foundations and floors are allowed too.

Category 3: Residues that exceed the limit value of category 2b leaching. May not be reused in construction work without special permission.

Table 1 shows that untreated MSWI bottom ashes are categorized as 2b products, mainly because of the leaching of salts. Generally, the leaching of metals from the slag does not represent a problem. In some cases though, leaching of copper and lead are found to exceed the limits of category 2a, presumably due to their amphoteric properties [7]. The leaching of lead from bottom ash from the incinerator plant I/S Amagerforbrænding is an example. Out of six analyses, two exceeded the limit value, the leaching was 78 and 28 μg/l.

The restrictions connected with category 2b limits the utilization of bottom ash to app. 40% of the reuse of today. The alternative is deposition at a cost of app. 100 EUR/tons, including tax. This gives a strong incentive to improve the quality of the bottom ash. It is estimated that bringing the bottom ash in category 2a will make it possible to maintain a reuse percentage of 80 - 90% of today s. To respect the limits of category 2a, the salts sodium, chloride and sulphate needs to be fixated or removed.

This paper presents a work done to establish suitable methods of improving the MSWI bottom ash quality. The primary object has been reduction of the leaching of salts, in order to meet the environmental demands set forth by the authorities. Secondly, a determination of possible side effects of the washing methods, namely reduction of content of heavy metals and organic material, has been performed. This was done to allow for an overall evaluation of the quality improvement by the tested methods.

EXPERIMENTAL

In removal of soluble salts, washing is the obvious choice. This investigation evaluates two different methods of washing. In the literature it is often suggested, that simple in-plant measures i.e. washing in the quench tank be applied in order to reduce the leaching of salts [8,9]. Very few investigations deal with full-scale washing, but rely on batch tests performed on a laboratory scale. Through a full-scale experiment, the efficiency of a simple quench tank washing has been determined.

Danish legislators have proclaimed a stringent set of limit values concerning the leaching of salts. Especially the limit values regarding sulphate are expected to cause difficulties, presumably due to the low solubility of gypsum [10]. It was therefore expected that a more thorough washing process might be needed, in order to meet these values. Thus it was decided to attempt washing on a pilot-scale soil cleaning facility, giving the opportunity of a more complete wash. This method also allows for a separation of the treated material. In particular the fraction of fines is of interest. Some investigations suggest that certain heavy metals be enriched in the fine fraction [11]. This means, that removal of the fine fraction could lead to a less environmentally hazardous product. Furthermore, the physical properties of the bottom ash are improved by the separation of fines [12], potentially making it a more attractive product to the contractors. Reduction of the amount of organic material is also desirable from a contractor point of view.

ANALYSIS

All laboratory work has been performed by The Danish Water Quality Institute (VKI).

Leaching Test

The leaching is determined by CEN prEN 12457, procedure A [4]. This is a one stage batch test at L/S=2 l/kg. The leaching media is demineralised water adjusted to pH=4 by addition of HNO_3. The sample is crushed to a diameter less than 4 mm. 100 g of sample is added 200 ml of leachant in a polyethylene bottle. The bottle is agitated for 24 hours. After 15 minutes of settling, the eluate is filtered using a 0.45 μm membrane filter. The pH and electric conductivity is measured in the eluate. Afterwards the eluate is acidified to pH=2, using HNO_3, and kept for analysis. The content of calcium, potassium, sodium, chloride and sulphate in the eluate was determined.

Total Content

For determination of total content, samples are digested and analysed according to DS 2210 [13]. This includes partial digestion of 1 g sample, using 7 M HNO_3 and autoclaving for app. 0.5 hour.

Measured parameters are pH, alkalinity, loss on ignition (LOI), calcium, potassium, sodium, total chloride, total sulphate, arsenic, chromium, cadmium, copper, nickel, lead and zinc

QUENCH TANK WASHING

The quench tank washing experiment was performed at the municipal solid waste incinerator plant I/S KARA in Roskilde, Denmark. The facility has four oven lines, two Vølund 3-grate systems, each with a capacity of 3 tons of waste pr. hour and two Vølund rotary ovens, each with a capacity of 7 tons pr. hour. The washing was performed at one of the 3-grate ovens.

Description

The oven has a capacity of 3 tons pr. hour thereby producing app. 0.6 tons of bottom ash pr. hour. After the burnout on the third grate, the ashes fall into a quench tank. This serves to extinguish any remaining combustibles, causes larger pieces to fracture and reduce dust emission. From the tank, the ashes are transported by a plate conveyor onto a vibrating conveyor. This leads to an ash storage bunker. The quench system is illustrated in Figure 1.

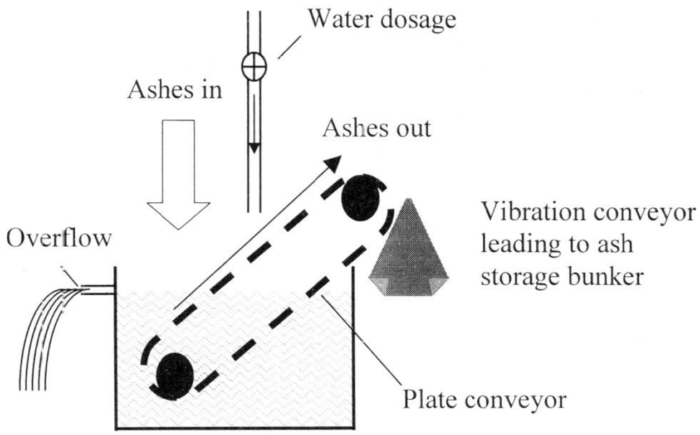

Figure 1 Draft of the quench tank at I/S KARA, with the performed modifications

During normal operation, the quench tank receives water from a reservoir, that contains a mixture of process water and rainwater. The varying origin of the water results in a fluctuating quality of the water used for quenching the ashes.

Modifications enabled a bypass of the recycled water and dosage of clean water through a pipe instead. The pipe was mounted with an adjustable valve and a gauge to control the water flow into the tank. Water in excess was disposed of via an overflow pipe, mounted at normal operations water level.

Experiments

The bottom ash was washed at 4 L/S ratios, 0.5, 1, 2 and 4 l/kg. At normal operations, the ratio is app. 0.5 l/kg and using less than this amount of water was considered unwise. At 4 l/kg the quantity of water used was near the capacity of the system.

No data on the waste composition and amounts at the time of the experiments were available, but no irregularities were reported.

On basis of residence time and flow, the necessary equilibration time before sampling was calculated.

Sampling

For each L/S ratio app. 100 kg of ash was sampled. The samples were taken from the vibrating conveyer over a 12-hour period, with app. 2 hours intervals. The sample was then thoroughly mixed and divided gradually, until app. 10 kg of sample remained. This was stored in a sample container until analysis. For comparison a sample of ash produced during normal operations was collected using the same sampling procedure. Finally a sample was taken directly from the third grate during a routine shutdown of the oven.

During washing at each of the four L/S ratios, a water sample was taken from the overflow, for determination of the salt content. A sample of the water used during normal operations and of the clean water was also taken.

Results and Discussion

Leaching from the sampled ashes are illustrated for the critical parameters chloride, sodium and sulphate in Figure 2 below.

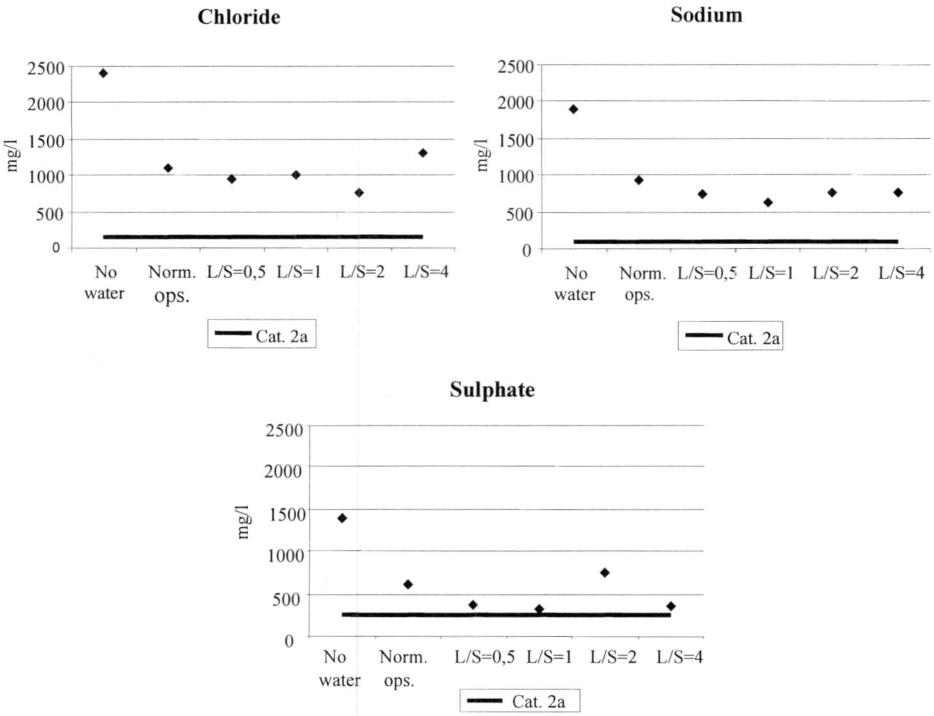

Figure 2 Leaching of salts from untreated and quench tank washed MSWI bottom ash. Limit values of category 2a are also shown

The figure shows that quite a significant washing effect is obtained even by quenching with recycled water under normal operations. However, the figure clearly illustrates, that no or very little effect is achieved by substitution of recycled water with clean water. Also the effect of intensifying the waterflow through the tank is shown to be negligible if not non-existing.

It is also illustrated that the limit values of category 2a are not reached for any of the critical parameters. Surprisingly sulphate is the component closest to meeting the limit value.

An explanation for the lacking effect of the substitution of recycled water with clean water is needed. Table 2 below lists the salt concentrations in the washing water and recycled water.

The salt concentration in recycled water is somewhat higher than in clean water. But when relating them to the salt concentration of the washing water; the difference in concentration is minor. This could explain the very little effect of substituting recycled water with clean water. Another reason for the lacking effect of washing with clean water might be that no measure has been taken to secure a thorough mixture between water and ashes. As shown on Figure 1, the water is applied on the belt conveyor and runs off in the opposite end of the quench tank. This was chosen to enable the best possible mixture. It is important that the water does not run off on the surface towards the overflow. The water entering the tank being colder than the water already in the tank should prevent this from happening. It is possible that a mechanical installation creating a better mixture in the tank could improve the washing effect.

Table 2 Analysis of salt content in water samples

PARAMETER	UNIT	CLEAN WATER	RECYCLED WATER	WASHING WATER L/S=0.5	WASHING WATER L/S=1	WASHING* WATER L/S=2	WASHING WATER L/S=4
PH		8.4	8.3	12.5	12.4	12.4	12.4
Alkalinity	mmol/l	4.6	31	50	35	29	30
Calcium	mg/l	110	96	870	1200	730	800
Potassium	mg/l	4.7	110	1200	900	140	730
Chloride	mg/l	36	270	2900	2700	650	1700
Sodium	mg/l	15	140	1500	1000	340	670
Sulphate	mg/l	79	140	170	330	130	210

*This sample diverts significantly from the others. No immediate explanation could be found

PILOT SCALE WASHING ON SOIL CLEANING FACILITY

An experiment has been performed on a pilot scale soil cleaning facility. The facility is designed by Australian Defence Industries Ltd. In 1998 KK Miljøteknik A/S bought the facility in order to gather experience on the subject soil cleaning. AFATEK A/S has had access to the facility for a period, and experiments have been performed to test its suitability for MSWI bottom ash washing.

Description

The facility has a capacity of app. 500 – 1000 kg pr. hour. Figure 3 shows the layout of the facility.

The material is feed through an apron feeder. From this the material is transported into a course scrubber by a belt conveyor. Water is added in the scrubber via a valve with gauge. The flow of material is controlled by adjusting the speed of the belt conveyor. The course scrubber enables a good contact between material and water and provides a homogenous mixture for further treatment. The volume of material in the coarse scrubber at operation, is app. 0.1 m^3. The dissolution of salts is expected to take place here.

From the coarse scrubber, the material falls into a spiral classifier via a 12 mm sieve. The spiral classifier transports the sand and gravel fraction to a container.

The overflow from the classifier (light material and fines) is screened on a 100 µm vibrating screen. This serves to separate the organic material from the filler fraction. The material that passes through the sieve is pumped to a thickener. This conditions the slurry of water and filler before it is filtrated in a chamber filter press.

By the treatment the material is separated into three fractions:

- Sand and gravel (>100 µm) 85-95%
- Light/organic material 0-5%
- Filler (<100 µm) 5-10%

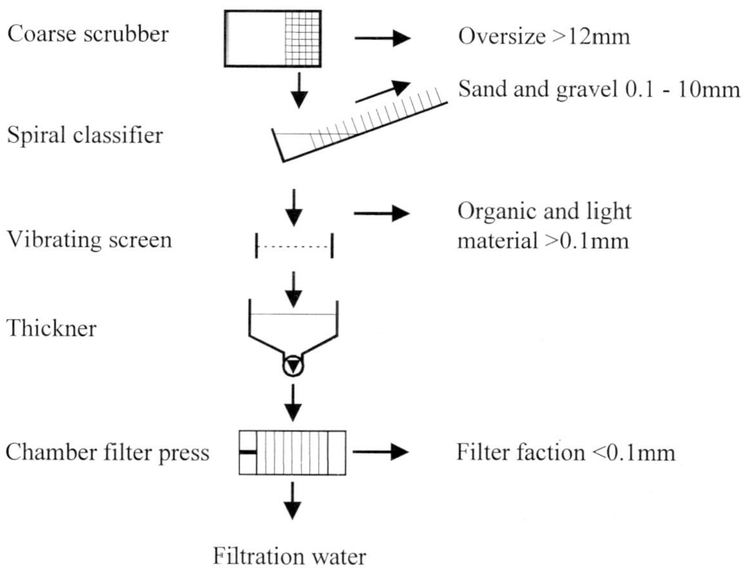

Figure 3 Schematic description of the washing process used at pilot scale soil cleaning facility

Experiments

MSWI bottom ash from the incinerator plant I/S Vestforbrænding has been washed on the facility. The ashes was app. 1 month old when washed the first time. Metals were separated from the ash by a belt magnet. Since larger particles made it difficult to maintain a constant flow to the coarse scrubber, the ash was pre-screened over a 10 mm screen.

Initially the ash was washed at three L/S ratios, 1.5, 4 and 10 l/kg. In order to secure an identical residence time at each L/S ratio, the total flow Q (ash and water) was fixed at 1 m^3/hour.

An experiment of stepwise washing was performed. The same ash was used, but app. two month later. This meant that an effect of ageing should be taken into consideration. The ash was washed three times at a L/S ratio of app. 1.5 l/kg in each step. The total flow Q was the same as in the previous experiment.

Sampling

Upon wash, samples were taken from the sand and gravel fraction, the filler and the organic fraction. App. 10 kg of sample was taken. Samples of the water after filtration and of the clean water were also taken. In the 3-step wash samples were taken from the sand and gravel fraction after each step. After the final step samples of organic material and filler was taken along with samples of filtration water and clean water.

Results and Discussion

Leaching

In Figure 4A and B is illustrated the results of leaching test on the sampled ashes for the critical parameters chloride, sodium and sulfate.

Figure 4A shows that there is a significant effect of washing the ash. The largest effect is seen after washing at lowest L/S ratio. Increasing water addition from 1.5 to 4 l/kg has an effect whereas the effect of increasing L/S from 4 to 10 l/kg seems insignificant. In Figure 4B a significant reduction in leaching can be realised after each step.

Figure 4 show that it is possible to reduce the content of leachable chloride and sodium to meet the limits of category 2a in a one step wash at a L/S ratio of 2-4 l/kg. In a multiple step washing at L/S=1,5 l/kg pr. step, 2-3 steps, corresponding to L/S=3 – 4.5 l/kg are required to reach the same result. Therefore it is difficult to argue that a multiple step washing at lower L/S ratio is preferable to a single step washing at higher ratio.

Sulphate leaching from the treated material does not meet the limits set forth by Danish legislators. Upon washing at a L/S ratio of 10 l/kg, the concentration in the leaching test is 400 mg/l.

One notices a significant difference between leaching of sulphate from the untreated ash analyzed after one and three months. This could be caused by inhomogeneity of the ash. Another possibility is that it is an effect of ageing. The pH of the untreated ash decreased from 10.8 to 9.4 over the two month period. This clearly indicates that ageing has taken place. Stekeete et al. [9] describes the mobilization of sulphate as one unfavourable effect of ageing

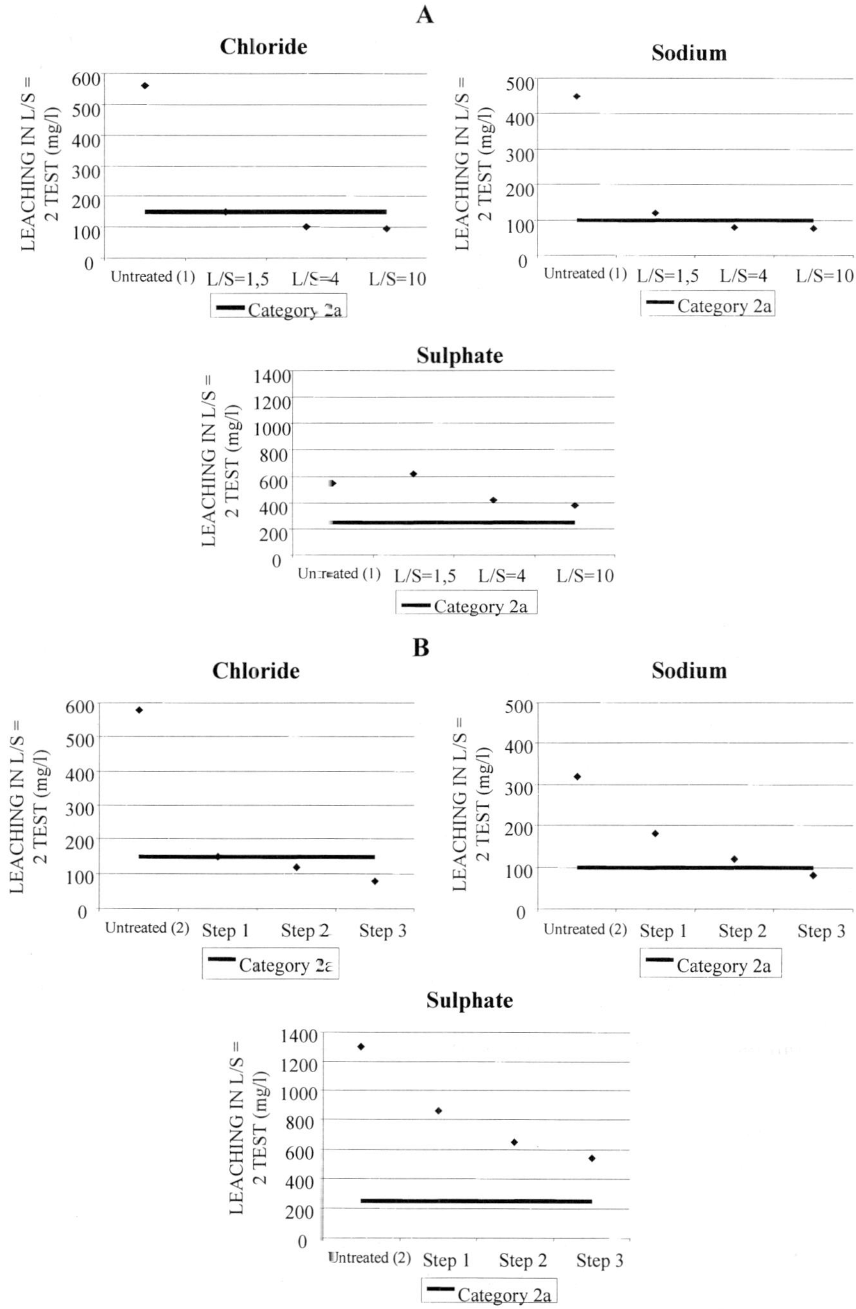

Figure 4 Result of leaching test on MSWI ashes washed at A: three L/S ratios on a pilot scale soil cleaning facility. B: three steps each at a L/S of 1.5 l/kg on a pilot scale soil cleaning facility

Total Content

Analysis of the total content in the fractions produced by the washing show that some components are enriched in the fine fraction. Figure 5 show the distribution between fractions for chloride, sulphur, calcium, cadmium, lead and zinc.

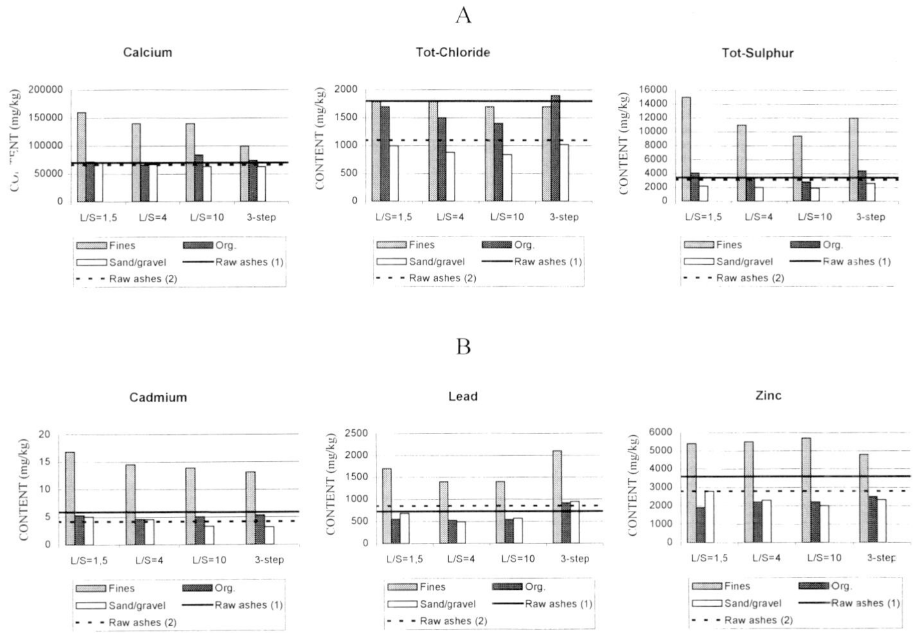

Figure 5 Distribution of calcium, chloride and sulphur (A) and cadmium, lead and zinc (B) between fines (<100 µm), organic material and sand/gravel (>0,1 mm) in treated bottom ashes. Content in sand and gravel in the 3-step wash represent the average of the three samples

Particularly sulphur and calcium are found in high concentrations in the fine fraction. This might indicate the presence of anhydrite or gypsum. It means that the reduction in sulphate leaching from the sand and gravel is not only caused by dissolution of sulphate, but also by the separation.

The total content of salts in the fines constitutes of salts in the ash and in the residual water. However measures of the salt concentration in the filtrated water show the content to be insignificant compared to the content in the ash.

With respect to heavy metals, Figure 5B shows a significant overrepresentation of cadmium, lead and zinc in the fine fraction. Similar results are found for chromium and arsenic, whereas copper and nickel show no such characteristics.

The occurrence of heavy metals is not considered a problem with respect to leaching, since the high natural pH of the ash results in a low solubility of the metals. However a reduction of the metal content is always desirable and the results indicate that separation of the fines produce a less environmentally hazardous product.

Figure 6 below shows the loss on ignition in the three fractions produced by the washing process. The loss on ignition can be taken as a guideline for the content of organic material.

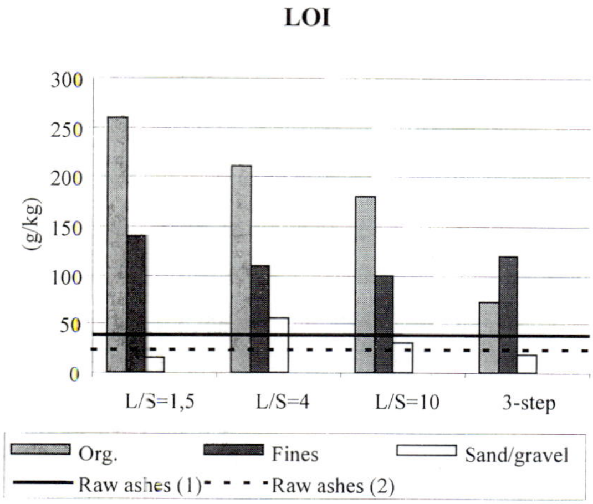

Figure 6 Loss on ignition in the produced fractions

The figure shows a LOI up to 25% in the light/organic fraction. In the fines the LOI is about 10-15%. An explanation could be the presence of particular carbon. Consequently the LOI should be reduced in the sand and gravel fraction by the separation. The figure proves this to be true except for washing at L/S=4 l/kg. It is assumed that this divergence is caused by sample inhomogenities. Provided this is true, the washing process also serves to produce a material with better construction characteristics.

CONCLUSION

It has been shown that quench tank washing of MSWI bottom is insufficient when the objective is to comply with upcoming Danish legislation. Virtually no effect was seen when substituting recycled water with clean water. Increasing the water flow through the quench tank had no effect on the leaching.

Pilot scale washing on a soil cleaning facility was found suitable for treatment of incinerator bottom ash. Leaching of chloride and sodium was reduced to respect the expected very stringent limit values. Sulphate however proved to be difficult to remove. A reduction to app. 400 mg/l in leaching test was the best result achieved and required a L/S ratio of 10 l/kg. Limit value is expected to be 250 mg/l. Separation of fines accounted for part of the reduction in leaching of sulphate, since sulphate was found enriched in the fine fraction.

The size separation of the ash was successful. The LOI in the sand and gravel was reduced, as was the content of heavy metals.

REFERENCES

1. MILJØSTYRELSEN. Orientering fra Miljøstyrelsen, Nr. 1, 1999

2. MILJØSTYRELSEN. Bekendtgørelse nr. 568 af 6. december 1983 om genanvendelse af slagger og flyveaske, 1983

3. MILJØSTYRELSEN. Høringsudkast til Bekendtgørelse om genanvendelse af restprodukter og jord til bygge- og anlægsarbejde, 1999.

4. EUROPEAN COMMITTEE FOR STANDARDIZATION. Draft CEN prEN 12457, July 1996.

5. LUDVIGSEN, K. Anvendelse af ubundne slagger fra affaldsforbrænding som vejbygningsmateriale, AFATEK A/S, Roskilde, 1995.

6. HOLM, P.E., LEHMANN, N.K.L., ASMUSSEN, O.W. AND HJELMAR, O. Undersøgelsen af stofudvaskningen fra slaggeprøver fra Amagerforbrænding, VKI, 1999.

7. IAWG: CHANDLER, J.A., EIGHMY, T.T., HARTLÉN, J., HJELMAR, O., KOSSON, D.S., SAWELL, S.E., VAN DER SLOOT. H.A. AND VEHLOW, J. Municipal Solid Waste Incinerator Residues. Studies in Environmental Science 67. Elsevier B.V. 1997.

8. SCHNEIDER, J., VEHLOW, J. AND VOGG, H. Improving the MSWI Bottom Ash Quality by Simple In–Plant Measures. In Goumans, van der Sloot and Aalberts (Editors): Environmental Aspects of Construction with Waste Materials, Elsevier Science B.V. 1994. pp 605-620

9. LAMERS, F.J.M. AND BORN, J.G.P. Upgrading Techniques for the Quality Improvement of Municipal Waste Incinerator Residues. In Goumans, van der Sloot and Aalberts (Editors): Environmental Aspects of Construction with Waste Materials. Elsevier Science B.V. 1994. pp 645-654

10. STEKETEE, J.J., DUZIJN, R.F. AND BORN, J.G.P. Quality Improvements of MSWI Bottom Ash by Enhanced Aging, Washing and Combination Prosesses. In Goumans, Senden, van der Sloot (Editors): Waste Materials in Construction: Putting Theory into Practice, Elsevier Science B.V. 1997. pp 13-23

11. HJELMAR, O. AND VEHLOW, J. Sammenligning af slagger fra forskellige typer affaldsforbrænding, fase 1. "Den Blå Serie" nr. 36. I/S Vestforbrænding, 1995

12. GOHLKE, O. AND WESTINER, E. Bautechnische Aspekte der Wäsche von Müllverbrennungsschlacken, Müll und Abfall, Vol. 2, 1999. pp 74-77

13. DANSK STANDARD. DS 2210, Vandundersøgelse. Metaller i vand, slam og sediment. Bestemmelse ved flammeløs atomabsorptionsspektrometri. Elektrotermisk atomisering i grafitovn. Generelle principper og retningslinier. 1982.

VALORISATION OF MSW INCINERATION RESIDUES IN BELGIUM: A DEVELOPING MARKET

K C Vrancken

B Laethem

V Dutré

VITO Grondstoffen

Belgium

ABSTRACT. The market for valorisation of incinerator residues in Belgium has not come to full development. Nevertheless, various initiatives are taken on both regulatory and industrial side. Projects have been defined in Wallonia and Flanders, dealing with both mechanical treatment and immobilisation. The bottom ash from the Brussels region is treated following the Dutch certified system. The Flemish Regulation on Prevention and Management of Waste (VLAREA) provides a legal basis for the application of waste as a secondary raw material. Wallonia has set up a new regulatory framework for bottom ash valorisation. In this paper a survey of the current state-of-the-art on valorisation and regulation of incinerator residues in Belgium is given. This is put in perspective with the practices and legislation in the surrounding countries.

Keywords: Bottom ash, Valorisation, Regulation, Legislation, Leaching

Dr K C Vrancken, is working as a project co-ordinator on waste and secondary raw materials. He has been working as R&D and training professional in environmental engineering industry and joined Vito in 1997. His current work involves the study of leaching properties and valorisation options for secondary raw materials on behalf of the industry and/or the Flemish Government. He is involved in several European projects on characterisation and valorisation of residues.

B Laethem, is head of the Raw Materials Centre of Expertise of Vito. He has been engaged in many Vito-projects on characterisation, evaluation and standardisation of secondary raw materials on behalf of the industry and/or the Flemish Government. He is representing Belgium within the European Recycling Forum, working group C (Research and innovation) and is involved in different governmental working groups on secondary building materials and grind substitutes.

Dr V Dutré, is a project leader on waste and secondary raw materials. She is engaged in studies on leaching mechanisms of contaminants from candidate secondary raw materials, development of leaching tests, studies on the environmental impact of utilisation of secondary raw materials. She is a member of various CEN TC 292 Working Groups on characterisation of waste.

INTRODUCTION

In Belgium about 50% of the municipal solid waste is incinerated. Grate furnaces have been constructed since the '70's. Over the last 5 years, major investments have been made in the expansion of flue gas cleaning systems. The sector has been under a lot of public pressure concerning flue gas emissions, even after this. This apparently has distracted the attention from the valorisation of residues. Whereas bottom ash reuse was practised in the surrounding countries, all Belgian incinerator residues have been disposed until 1999. Only in the last two years, the valorisation has been initiated and the sector faces a rapid growth.

The development of a market for bottom ash valorisation is triggered by the regional regulatory framework. The type of treatment needed and the possible fields of application are determined by the legislation. The three Belgian regions (Flanders, Wallonia and Brussels) are facing a different legislation. A second important market factor is the demand from neighbouring countries. The Netherlands, Germany and France have working treatment units. These look for additional capacity trough import of Belgian ashes. Thus, the exported material finds application under a different legislation. Comparison of the legislations shows that there are clear differences.

In this article, we will put the developments on the Belgian market in a European perspective. In order to do so, an inventory is made of the Belgian bottom ash production and the regulation and valorisation practices are compared.

INVENTORY

Assessing valorisation options starts from knowledge of the amount of material present. Table 1 gives the actual production of bottom ash and the valorisation percentage. Data for Belgium are compared to those for The Netherlands, France and Germany.

Flanders

Flanders currently has an MSW incineration capacity of 1.072.000 tons/y in 11 plants. An extra capacity of 150.000 tons/y is planned by the year 2001. The theoretical incineration capacity will therefore amount up to 1.222.000 ton/y. The corresponding production of bottom ash can be estimated to be in-between 270.000 and 340.000 tons/y.

In 1995 the MSW incinerators produced 196.748 tons of bottom ashes. The full bottom ash production is currently landfilled or used as a covering material at the disposal site. The Flemish government has the intention to reach 50% valorisation of the bottom ash by the year 2001 [1].

Wallonia

The Walloon region has a MSW incineration capacity of 480.000 ton/y in 4 incinerator units. For technical reasons, only two of these work at full capacity. This resulted in 72.635 tons of bottom ash in 1995. At full capacity, the bottom ash production would be 120.000 ton/y. The Walloon Waste Management Plan foresees a capacity increase to 800.000 ton/y by the year2002 [2]. Nevertheless, this plan may be changed due to a moratorium on the construction

Table 1 Bottom ash production and valorisation in some West-European regions

	FLANDERS (1999)	BRUSSELS Region(1998)	WALLONIA (1999)	NETHERLANDS (1998)	FRANCE (1997)	GERMANY (1999)
Incineration capacity	1.072.00	500.00	480.000	4.500.000	12.700.000	12.000.000
Production of botton ashes	196.748*	130.000	72.635*	975.000	2.500.000	3000000*
% valorisation of bottom ashes	-landfill	-(external) valorisation	-road construction (Thumaide) external valorisation	-large-scale public works	-road construction -mining (Caen)	-road construction
	* 1995		* 1995			* 1995

of new incinerators, issued recently by the new government. The same waste management plan foresees an increasing valorisation of incinerator residues, growing from 25% in 2000 to 75% in 2002. This must be facilitated with the new regulation, issued in the decree of the Walloon Government of 20 May 1999.

Brussels

The Brussels region has one MSW incineration plant, with a total capacity of 500.000 ton/y. The yearly production of bottom ash is 130.000 ton/y. These are transported to the Netherlands, where they are treated and valorised under product certification.

European Context

For the whole of Belgium, a maximal production of 590.000 ton/y can be envisaged in the near future. In absolute figures, this is low as compared to the surrounding countries. The Netherlands, France and Germany produce from 1 to 3 million tons. The established treatment market in these countries has a pulling effect on the Belgian flow. Untreated material is wanted for treatment and treated material may be demanded for large scale works. However, the legislative frameworks show distinct differences.

LEGISLATIVE FRAMEWORK AND VALORISATION

Flanders

The general framework for the application of secondary raw materials has been defined in the Decree on Waste Materials (Afvalstoffendecreet, April 1994). Secondary raw materials have been defined here as waste materials that comply with specific requirements, concerning composition and/or use, that guarantee application without any environmental risk. A list of candidate secondary raw materials and the criteria that have to be met have been defined in the Flemish Regulation on Waste Prevention and Management (Vlaams Reglement inzake Afvalvoorkoming en -Beheer), shortly VLAREA [3]. Waste materials which comply with the requirements, cease to be considered as waste from the moment they have been delivered at the site of the user. The user needs to have the permission or possibility to apply the material as a secondary raw material. Control on the effective valorisation of the residues is followed up administratively by means of a certificate of use. In this way, VLAREA is one of the first legislations in Europe, that defines secondary raw materials in more general legal terms.

Pre-treated incinerator bottom ash is classified as a candidate secondary building material. This can be either in granular or bonded form. VLAREA sets limits for the environmental quality of a candidate secondary raw material. It stipulates that the quality of residues is determined by the concentration of 8 heavy metals (As, Cd, Cr, Cu, Hg, Pb, Ni, Zn) and organic compounds. Requirements for the concentration of heavy metals are considered as target values. This implies that materials exceeding the target value, can be used, albeit under controlled conditions. Concentration values for organic compounds are strict limit values. Additional limit values are imposed for the leaching of heavy metals. Leaching should be tested using standard procedures. Granular materials are tested using a column test following the Dutch NEN7343 [4]. Monolithic or shaped materials are tested using a diffusion tank test, following the Dutch NEN7345 [5].

In recent laboratory testing it appeared that the current bottom ash quality, without further treatment, does not meet the VLAREA limits. Cu, Pb and Zn are the main critical compounds. Therefore, VLAREA only classifies pre-treated incinerator bottom ash as a candidate secondary building material. This can be either in granular or bounded form. Pre-treatment can involve washing processes, size fractionation and ageing to stabilise the product and to decrease leachability. The low leaching limit for Cu (0.5 mg/kg) appears to be the decisive factor for valorisation. Treated MSWI-bottom ash can be applied in embankments, as a road base course or in concrete masonry blocks.

The initiation of VLAREA has opened the market for bottom ash valorisation. This has gradually attracted the interest of industry. At present the main industrial players in the waste treatment field are setting up initiatives in this new field.

- A wet mechanical treatment unit is under construction at Beveren (Antwerp). This installation with a capacity of 120.000 ton/y will be started up in the summer of 2000.
- In another treatment project at Grimbergen (Brussels), bottom ash is sieved and deferrised and applied for ground works at the same site. This plant gets input from both Flemish and Walloon incinerators
- Centralised collection and transport for treatment of the ashes in German units is considered by one of the larger waste handling firms.
- The application of the bottom ash in bound materials is studied in several industrial research projects.

These initiatives stimulate the market for valorisation of the bottom ash.

Brussels

The Brussels Region does not have a working legal framework for valorisation of bottom ash. Valorisation of the bottom ashes is effectuated trough a Dutch treatment plant. The product has obtained the Dutch quality certificate and is taken up by the local market.

Wallonia

The decree of the Walloon government of 20 May 1999 establishes a list of materials to be used as products. The decree also includes limit values and test methods to evaluate the environmental quality of the material. MSW incinerator bottom ash is present on the list. A distinction is made according to the type of valorisation: materials for civil works (i.e. as a road base material) and compounds in the production of finished products (i.e. use in concrete blocks etc.). The list states that bottom ash should be pre-treated using sieving, metal separation and ripening. Additionally, the bottom ash should not be mixed with fly ash or boiler ash. The use in civil works is specified as application as a road base material. In this type of valorisation a certificate of use needs to be obtained.

Before use as a material for civil works, the environmental quality of the ash should be tested using the NEN7343 column leaching test. This applies for both ashes in granular form and ashes mixed with hydraulic binders. Limit leaching values are given for a set of 13 metals and 4 anions. For organic compounds, the total content is limited. On basis of the results of these tests a certificate of use is issued. Regular quality control is performed using DIN-S4 batch mixing tests. The limit values for this quality test differ from those for the column test.

If the ash is used in the production of 'derivatives resulting from mixing with a hydraulic binder' no leaching tests are required. This category applies for e.g. the production of concrete blocks.

The definition of the limit values for the above-stated regulation was based on practical tests, performed at the bottom ash treatment unit at Thumaide. This treatment unit has been in operation since several years, but the end product was not reused in practice. The product obtained a certificate of use under the actual legislation and is now valorised by a concrete producer as road base material.

The treatment consists of a size separation and magnetic separation. A 0-20mm product is produced. This is subjected to a 18 weeks maturation in a closed hall. The unit has an annual production of 20.000 tons. This will increase after a planned expansion of the incinerators. Thus, all the ash of the Thumaide incinerator is valorised.

The other 3 Walloon incinerators bring their deferrised bottom ash to a centralised treatment centre in Flanders.

The new Walloon government, that was installed in July 1999, expressed the intention to change the current legislation on secondary materials. This might have an effect on bottom ash valorisation as well.

The Netherlands

The valorisation of bottom ashes is regulated by the Building Materials Decree (BMD) (Bouwstoffenbesluit). This decree deals with all stony materials in outdoor constructions. Thus, it is a true product standardisation and not a waste regulation.

Leaching is tested using the Dutch standard column and tank tests (NEN 7343 and NEN 7345, respectively). On basis of limit values for composition and leaching, two categories of materials are defined. Category 1 (N1) materials may be used without any isolation measures. Application will not lead to an exceeding of the immision limits. Category 2 (N2) materials may only be applied using isolatory measures. These have to prevent or reduce contact with rain and ground water. MSW bottom ashes are classified as a special category. Products from this category can only be used in specific applications, taking specific isolation measures. In practice, bottom ash is applied in embankments or as a road base course, with prevention of contact with rain water.

A product quality certificate is issued for plants that comply with specific requirements of good practice and product quality. The amount of certified bottom ash has increased from 200 kton in 1991 to 830 kton in 1998 [6]. With respect to the BMD limits, the leaching of Cu, Mo and Sb appears critical. All other elements largely comply with N2-quality and in some cases with N1-quality [6,7]. In order to reach this quality, bottom ashes are treated in specialised units having : sieving, size reduction, ferro and non-ferro separation, ripening in open air or with irrigation. Further treatment methods, causing accelerated ripening, are being studied [8].

France

The valorisation of incinerator residues was initiated by the Article 14 of the Ministerial Decree of 25 January 1991. The regulation on treatment of bottom ashes is given in a circular of the Minister of the Environment of 9 May 1994.

Bottom ash is classified on basis of physical and chemical characteristics in 3 categories: category V (valorisation) : with limited leachable fraction; category M (maturation): with intermediate leachable fraction; category S (stockage) : with high leachable fraction. Products of category V may be used in road construction. Category M material has to be treated (ripened) or landfilled. Category S materials has to be landfilled. The leaching test is performed in three successive steps, according to the French standard NF X 31-210. The limit values have been defined on basis of the actual bottom ash quality and current applications. They have not been based on a notion of environmental impact.

Application of the category V material is possible in the construction of roads and parkings or in compact elevations of at least 3m height. The contact with rain, ground or surface water has to be limited. In Caen bottom ash is used in mining applications.

Germany

The regulation for valorisation of residues is defined in the Technical Rules and the LAGA-Merkblatt. Additionally, technical quality criteria are defined. For MSW bottom ash, the regulation is given in the Länderarbeitsgeselschaft Abfall (LAGA)-Merkblatt : 'Entsorgung von Abfällen aus Verbrennungsanlagen für Siedlungsabfälle'. This document states limit values and analytical procedures for MSW bottom ash. The LAGA-Merkblatt defines leaching values as strict limit values and composition limits as target values. The prescribed leaching test is the DIN 38 414 S4. The LAGA-Merkblatt is a general directive for the whole Federal Republic of Germany. The various states can apply a modified (more rigid) version. In practice, this causes a very differentiated valorisation policy. This is reflected in the valorisation ratios, ranging from 0% in Bayern to 100% in Nord-Rine Westphalia. However, a general tendency towards federal harmonisation does exist.

The LAGA limits can be met by MSW bottom ash after mechanical treatment, ripening and wet treatment (irrigation or washing). Ashes are mainly used in embankments and as a road base material. On a smaller scale they are applied in hardened roads, cycling paths and walkways.

EVALUATION

A full comparison of the valorisation potential in various regions can only be made after comparison and evaluation of the limit values that have to be met. These determine the treatment that should be applied and ultimately the cost for reaching the recycling goals.

Table 2 gives the limit values applicable for reuse as a granular material. In the last column, the average values for Dutch bottom ash, as measured from 1993 to 1997 is given. Data have been taken from all regular quality controls as performed by the national organisation of waste managers (VVAV) at all the MSW incinerators and from the National Institute for Environmental Protection (RIVM) [9].

Test Procedure

Besides the actual limit value, also the applied test procedure should be considered. The leaching of granular materials is tested in either column tests or batch mixing tests. The main parameters of these leaching tests are the liquid/solid-ratio (L/S) and the particle size used. Most countries use leaching test with /S = 10 l/kg. In France, a three-step procedure is used, with an L/S=10 in each consecutive step. For comparison of the emission limits, the French limits should therefore be divided by a factor 3, though this is only an approximation.

Limit values for results of the DIN-S4 procedure are expressed in mg/l, these are displayed in *italics*. Conversion to mg/kg-units is made through a multiplication with the L/S-ratio of 10. Nevertheless, there is no straight correlation between DIN-S4 and NEN7343 leaching behaviour. This is mainly due to the fact that the DIN-S4 test is performed on material without particle size reduction. In the column test, the material is crushed to a 0-4mmm particle size. This affects the surface/volume ratio and the natural pH of the material. Both parameters affect the leaching behaviour of the various elements in a variable way.

Limit Values

From comparison of the tabulated limit values the following conclusions may be drawn:

The Flemish parameter set defines leaching as well as concentration limits for metals and it states concentration values for a broad range of organic compounds. The number of metals has been confined to 8 (As, Cd, Cr, Cu, Hg, Pb, Ni, Zn). As stated above, the total concentration limits for metals are target values. The German regulation has a comparable set of 7 heavy metals (excluding As).

According to the French legislation only a limited number of compounds have to be tested. The rigidity of these limit values as compared to the other legislations is very variable from one parameter to the other.

Table 2 Legislation for reuse of MSW bottom ash as granular construction material, compared to average BA quality in the Netherlands for 1993-97

		Flanders VLAREA Granular materials		**Wallonia** Arrêté 20/05/99		**the Netherlands** BMD Granular materials			**France** category V	**Germany** LAGA Merkblatt		**Dutch treated BA** 1993-1997
		leaching (mg/kg)	total content (mg/kg)	certification (mg/kg)	quality control *(mg/l)*	category N1 leaching* (mg/kg)	category N2 leaching* (mg/kg)	total content (mg/kg)	leaching (mg/kg)	leaching *(mg/l)*	total content (mg/kg)	mg/kg
Leaching test		NEN7343		NEN7343	*DIN-S4*	NEN7343	NEN7343		AFNOR X31-210	*DIN-S4*		NEN7343
L/S		10		10	*10*	10	10		3 x 10	*10*		10
particle size (mm)		<4		<4	*<32*	<4	<4		<4	*<32*		<4
Metals												leaching
Antimony	Sb	-	-	0,3	*0,2*	0,05	0,43	-	-	-	-	0,22
Aluminium	Al	-	-	2000	*2000*	-	-	-	-	-	-	
Arsenic	As	0,8	250	0,8	*0,1*	0,88	7	-	2	-	-	0,022
Barium	Ba	-	-	-	-	5,5	58	-	-	-	-	0,6
Cadmium	Cd	0,03	10	0,03	*0,1*	0,032	0,066	-	1	*0,003*	20	0,003
Chromium	Cr	0,5	1250	0,5	-	1,3	12	-	1,5	*0,2*	2000	0,08
Cobalt	Co	-	-	0,25	*0,1*	0,42	25	-	-	-	-	0,05
Copper	Cu	0,5	375	5,0	*2*	0,72	3,5	-	-	*0,3*	7000	3
Mercury	Hg	0,02	5	0,02	*0,02*	0,018	0,079	-	0,2	*0,001*	-	0,001
Lead	Pb	1,3	1250	2,2	*0,2*	1,9	8,7	-	10	*0,05*	6000	0,07
Molybdenum	Mo	-	-	1,8	*0,15*	0,28	0,91	-	-	-	-	1,52
Nickel	Ni	0,75	250	1,8	*0,2*	1,1	3,7	-	-	*0,04*	500	0,13
Selenium	Se	-	-	-	-	0,044	0,1	-	-	-	-	0,01
Tin	Sn	-	-	-	-	0,27	24	-	-	-	-	0,04
Titanium	Ti	-	-	2,4	*2*	-	-	-	-	-	-	
Vanadium	V	-	-	-	-	1,6	32	-	-	-	-	0,23
Zinc	Zn	2,8	1250	4,0	*0,9*	3,8	15	-	-	*0,3*	10000	0,09
Other Inorganic Compounds												
Potassium	K	-	-	1700	-	-	-	-	-	-	-	
Bromide	Br^-	-	-	-	-	2,9	41°	-	-	-	-	7,6
Chloride	Cl^-	-	-	6000	*500*	600	8800	-	-	*250*	-	2615
Cyanide (free)	CN^-	-	-	-	-	0,013	0,076	-	-	*0,02*	-	0,01
Cyanide (tot)	CN^-	-	-	0,2	*0,46*	0,067	0,38	-	-	-	-	0,048
Fluoride	F^-	-	-	20	*5*	13	100	-	-		-	14,1
Sulphate	SO_4^{2-}	-	-	4000	*1000*	750	22000	-	10000	*600*	-	5058

Table 2 Legislation for reuse of MSW bottom ash as granular construction material, compared to average BA quality in the Netherlands for 1993-97 cont

	Flanders VLAREA Granular materials		Wallonia Arrêté 20/05/99		the Netherlands BMD Granular materials			France category V	Germany LAGA Merkblatt		Dutch treated BA 1993-1997
	leaching (mg/kg)	total content (mg/kg)	certification (mg/kg)	quality control (mg/l)	category N1 leaching* (mg/kg)	category N2 leaching* (mg/kg)	total content (mg/kg)	leaching (mg/kg)	leaching (mg/l)	total content (mg/kg)	mg/kg
Monocyclic Aromatic Hydrocarbons			total content	total content							total content
Benzene	-	0,5	x	-	-	-	1,25	-	-	-	0,01
Ethylbenzene	-	5	x	-	-	-	1,25	-	-	-	0,01
Styrene	-	1,5	-	-	-	-	-	-	-	-	
Toluene	-	15	x	-	-	-	1,25	-	-	-	0,01
Xylene	-	15	x	-	-	-	-	-	-	-	0,01
BTEX total	-	-	sum x = 2,1	-	-	-	-	-	-	-	
Polycyclic Aromatic Hydrocarbons											
Anthracene	-	-	-	-	-	-	10	-	-	-	0,07
Benzo(a)anthracene	-	35	-	-	-	-	50	-	-	-	0,13
Benzo(a)pyrene	-	8,5	y	-	-	-	10	-	-	-	0,1
Benzo(ghi)perylene	-	35	y	-	-	-	50	-	-	-	0,08
Benzo(b)fluoranthene	-	55	y	-	-	-	-	-	-	-	
Benzo(k)fluoranthene	-	55	y	-	-	-	50	-	-	-	0,27
Chrysene	-	400	-	-	-	-	10	-	-	-	0,23
Fenantrene	-	30	-	-	-	-	20	-	-	-	0,3
Fluoranthene	-	40	y	-	-	-	35	-	-	-	0,28
Indeno(1,2,3cd)pyrene	-	35	y	-	-	-	50	-	-	-	0,09
Naftalene	-	20	-	-	-	-	5	-	-	-	0,06
PAH's total (sum 10)	-	-	sum y = 4,3	-	-	-	75	-	-	-	1,6
Other Organic Compounds											
Extractable organo-halogens (EOX)	-	3	7	7	-	-	3	-	-	3	0,08
Pesticides (without Chlorine)	-	0,5	-	-	-	-	0,5	-	-	-	0,1
Pesticides (with Chlorine)	-	0,5	-	-	-	-	0,5	-	-	-	0,3
Hexane	-	1	-	-	-	-	-	-	-	-	
Heptane	-	25	-	-	-	-	-	-	-	-	
Mineral oil	-	1000	1500	1500	-	-	500	-	-	-	101
Octane	-	90	-	-	-	-	-	-	-	-	
PCDD/PCDF			-	-	-	-	-	-	-	$(0,6\text{-}30)*10^{-6}$	
Polychlorobifenyls (PCB)	-	0,5	0,2	-	-	-	0,5	-	-	-	0,14
Total Organic Carbon	-	-	-	-	-	-	-	500	-	1%	

* for h= 0,7m

° excemption arrangement17/12/1998

The Walloon legislation defines limits for a broader range of metals, comparable to the Dutch BMD. Main differences are the presence of Al and Ti, and the absence of Ba, Se, Sn and V. A different set of limits is used for certification and quality control, with application of a different leaching test.

Concerning anions, VLAREA states no limit values. Wallonia states limits for 4 anions at a level between Dutch N1 and N2. LAGA also foresees limits for anion leaching.

Bottom Ash Quality

From the data in Table 2 the bottom ash quality after treatment can be compared to the various legislative frameworks. This comparison shows that for organic compounds the material stays well below the limits. In the thermal process of incineration organic compounds are fully destroyed. Comparison of leaching values to Belgian legislation shows that the material does not comply with the VLAREA Cu-limit. The Cu leaching value of 0.5 mg/kg appears very strict, and lower than limits in the other countries. In order to comply with this limit, the bottom ash will have to be size separated and only the coarser fraction can be valorised. For the fine fraction further immobilisation through hydraulic binders has to be considered.

The material does meet the Walloon legislation for metal leaching, but does not comply for sulphate leaching. This may be solved by washing of the material.

Comparing the BA quality to the Dutch legislation, Sb and Mo are critical as well. The Dutch government has initiated research programs to solve this problem through additional washing and rapid ageing of the material. Meanwhile, the Dutch bottom ash is valorised under a specific category, i.e. using special isolatory measures.

From these data it is clear that bottom ash treatment in Flanders faces a severe Cu leaching limit. The Walloon regulation provides less severe limits, but for a broader range of metals. Anion leaching is limited as well The sulphate limit is close to the average Dutch BA quality.

CONCLUSION

The Belgian market for bottom ash treatment and valorisation is growing rapidly. This has been initiated by new legislation as well as by demands from the surrounding regions. Part of the Belgian ashes are exported at present to treatment units in the Netherlands. Interest from Germany does exist.

The evolution and development of bottom ash valorisation in practice faces the problem that the different regions in Belgium and countries in Europe have a non-uniform legislation. This may cause a transport of residues for valorisation to those regions with the less severe regulation. Furthermore, the application of different leaching tests hampers a true intercomparison of regulations. This shows the need for a further harmonisation of legislation and test methododology on a European scale.

REFERENCES

1. Afvalstoffenplan, OVAM, 1997-2001, 1997

2. Horizon 2010, Plan Wallon de Déchets, Office Wallon des Déchets, 1998

3. Vlaams Reglement inzake Afvalvoorkoming en -beheer, Het Belgisch Statsblad, April 16th, 1998.

4. NEN 7343, Leaching characteristics of building and solid waste materials, Leaching tests: Determination of leaching behaviour of inorganic components from powder and granular building and waste materials, NNI, 1995.

5. NEN 7345, Leaching characteristics of building and solid waste materials, Leaching tests: Determination of leaching behaviour of inorganic components from shaped building materials, monolithic and stabilized waste materials, NNI, 1995.

6. Adviescentrum Reststoffen, Jaarverslag, 1998

7. F.J.M. LAMERS, J.W. VAN DEN BERG, J.G.P. Born, in Waste Materials in Construction: Putting Theory into Practice, Elsevier, The Netherlands, 1997, 735.

8. J.G.P. BORN, VVAV, Kwaliteitsverbetering van AVI-bodemas door wasprocessen en de combinatievan wassen en versnelde veroudering, report VVAV96005TR.R, 1996 .

9. TH. G. AALBERS ET.AL., Bouwstoffen Nader Bekeken, Eburon, Delft, the Netherlands, Bijlage 1, 1998

MECHANISM AND KINETICS OF THE WEATHERING OF MSWI BOTTOM ASH GLASS

C Zevenbergen
Vermeer Infrastructure Development B.V.
L P van Reeuwijk
International Soil Reference and Information Centre
J Keijzer
IWACO B.V.
Netherlands
J Bradley
Georgia Institute of Technology
United States of America

ABSTRACT. Combustion residues such as municipal solid waste incinerator (MSWI) bottom ash and coal fly ash are predominantly composed of high temperature solids. In an atmospheric environment these solids are metastable and will alter to form thermodynamically stable assemblages of minerals. Results of our research have revealed that a substantial fraction of MSWI bottom ash consists of reactive glasses and reactive thermal decomposition products of (native) clay minerals. Its large surface area, initially high 'active'-Al and associated Si content, and high pH predispose the ash to form abundant clay minerals within a relatively short course of time. Amorphous, hydrous aluminosilicates are found to be intermediate reaction products and are important repository of heavy metals. Upon weathering and depending on the local environmental conditions these intermediate reaction products may transform and crystallize into clay minerals (e.g. illite) or zeolites. However, the latter has not yet been substantiated by direct observations.

Keywords: MSWI bottom ash, Glasses, Transmision electron microscopy, Weathering, Clay formation, Heavy metal leaching

Dr C Zevenbergen, is Director Environment of Vermeer Infrastructure Development B.V., Hoofddorp, the Netherlands. His present interest is in the long-term leaching behaviour of contaminated sites, waste materials in constructions, and landfilled wastes.

Dr L P van Reeuwijk, is associate professor and Head of the laboratory of the International Soil Reference and Information Centre, Wageningen, the Netherlands. His present interest is in quality management and in volcanic ash soils and zeolites

Prof Dr J Bradley, is Professor Material Science at the Georgia Institute of Technology and and executive Director of MVA, Inc., Norcross, Georgia, USA. He is an internationally recognized scientist in the field of electron microscopy.

Mr J Keijzer, is analytical chemist at the laboratory of IWACO B.V., Rotterdam, the Netherlands. His present interest is in waste characterization.

INTRODUCTION

In order to manage ash disposal and utilization, a fundamental understanding of the long-term leaching processes is necessary. Recent studies, using geochemical modeling in conjunction with conventional petrographical and advanced microanalytical techniques, have significantly contributed to our present understanding of the long-term geochemical alterations processes on MSWI bottom ash deposits [1-7]. Municipal solid waste incinerator (MSWI) bottom ash is predominantly composed of high-temperature solids which, in an atmospheric environment, will alter to form thermodynamically stable assemblages of minerals. It is more and more recognized that weathering of bottom ash occurs by processes and sequences similar to those documented for soils of volcanic origin. Both ash types are composed mainly of glassy constituents.

In this paper recent work carried out by the authors on MSWI bottom ash glass characterization and weathering is summarized.

MORPHOLOGY AND MINERALOGY

In contrast to coal fly ash and MSWI fly ash, the ash particles of MSWI bottom ash exhibit a high degree of angularity (see Fig. 1). Morphological characteristics of the particles depend both on their exposure conditions and on their chemical composition. The largest particles (>0.5 mm) exhibit the most heterogeneity and diversity and may contain incompletely transformed residual minerals and (manufactured) glasses. Among the smaller, predominantly glassy particles, particles with crystalline domains are frequently observed, which indicate that during incineration part of the native clay minerals decomposes resulting in a poorly ordered structure but without forming a rigid glass. This implies that temperatures within the incineration chamber remain locally much lower than the melting temperature of clay minerals (<< 800 °C). Vesicular particles, which are usually abundant, have been exposed to higher temperatures in the combustion zone. Spherical glassy particles are infrequently observed in the submicron size range. Automated particle analysis using a scanning electron microscope (SEM) revealed that over 40 % of the surface area of a typical bottom ash sample consists of CaSi, AlSi, CaAlSi, and AlkaliSi particles and 30% of the surface are Si particles (see Fig. 2)[6]. The first type of glassy particles are likely to be most vulnerable to weathering and thus can be expected to play an important role in the bulk reactivity of bottom ash. Based on these findings it can be concluded that bottom ash contains particles with different composition and thermal history, which as a consequence show different weathering rates.

X-ray diffraction results indicate that the most abundant minerals in weathered bottom ash are quartz, calcite, magnetite, hematite, gypsum, ettringite, akermanite, gehlenite and cristobalite [2, 7]. However, these crystalline compounds make up only a minor fraction of the mass of the total ash [4]. The majority of these crystalline phases are likely to be preserved from the original ash rather than formed during weathering. Of secondary origin are: calcite, gypsum, ettringite and possibly part of hematite.

GLASS WEATHERING

The weathering behaviour of obsidian, a naturally occcuring volcanic glass with a composition very similar to the that of MSW bottom ash, has been extensively investigated, since the extent

Figure 1 Backscattered electron image of bottom ash particles

of surface hydration of obsidian offers a method of dating prehistoric artifacts. Obsidian-like synthetic glasses are also used as containment media for nuclear wastes. Advanced electron microscopy has given insight in the processes at the surface of these glasses and revealed that glass weathering occurs in one or a combination of several ways: (i) preferential dissolution of network modifiers leaving a residual, leached layer, (ii) total or congruent dissolution, and (iii) reaction with the surrounding solution to form new crystalline or amorphous compounds on the glass surface. Although the mechanism of glass weathering is well understood, our knowledge of the kinetics is still very limited due to generally low alteration rates and uncertainties concerning the natural conditions during alteration. In comparison with natural systems bottom ash is

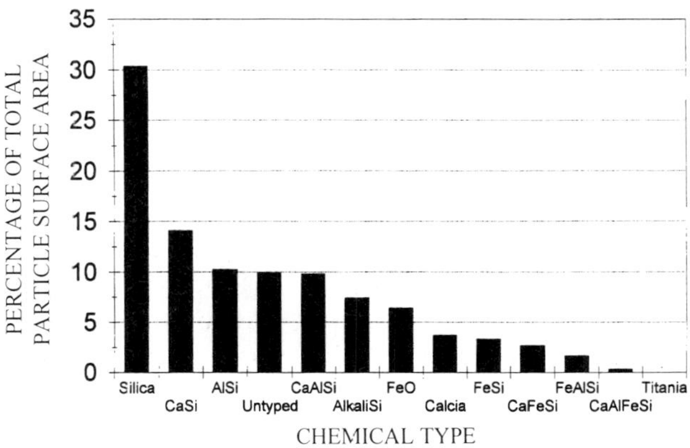

Figure 2 Particle surface area of a typical bottom ash sample
(particle types: silica => 75% Si, MSi =>10% M, > 10% Si <10% any other element, MAlSi = >10% M, >10%Al, >10%Si, <10% any other element, etc.)

unique in that it is composed of glasses with different compositions, which in a deposit or in a concrete product are subject to the same weathering environment. This emplies that bottom ash is an ideal material to study fundamentals of glass weathering phenomena.

The formation of secondary phases from glasses is well documented for soils of volcanic origin [8]. The type of secondary phases which form in these soils, and which are likely to be formed in bottom ash deposits, depend largely on the pH of the solution and the leaching rate as schematically presented in Figure 3. Both very low and very high pH accelerate glass weathering by rapidly dissolving the silicate structure. Zeolites, smectites, and illites may originate from volcanic glass either gradually via an amorphous aluminosilicate phase or by precipitation. When there is no leaching, glasses hydrolyze without the reaction products being transported. Depending on the content of alkalies and alkaline-earths in the glass, the contact solution may become saline and alkaline. Under these conditions zeolites may form. Experimental formation of clays and zeolites from glasses indicates greater reactivity of basaltic-type compositions relative to high-silica glasses [9].

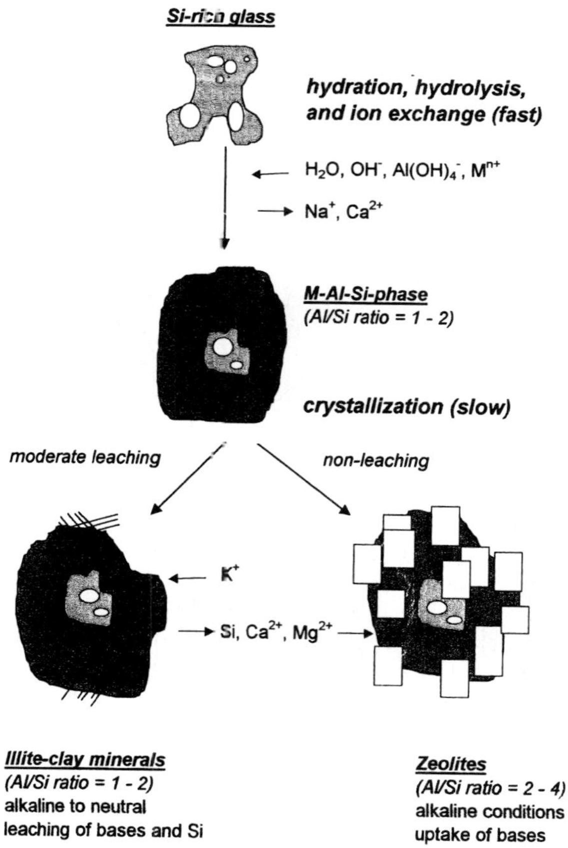

Figure 3 Neoformation of clays and clay-like materials from glass.

TEM examination of bottom ashes which have been exposed to weathering have revealed widespread decomposition of the glass phase and formation of secondary aluminosilicates [10.11]. For example, examination of 4-year old ash shows formation of amorphous hydrous aluminosilicates (allophane) as the first reaction product of glass weathering. These secondary aluminosilicates are rich in Ca and have a low Al/Si-ratio (e.g. Si-rich) relative to allophane in volcanic ash and are found as discrete flakes and as weathered rims on the surfaces of glass particles (see Fig. 4). In 12-year old ash derived from an open, well drained ash deposit, we have found direct evidence of neoformation of well-ordered primitive clay (illite). (see Fig. 5)[4]. The quantity of this clay mineral is, however, too low to detect by X-ray diffraction analysis. We used acid oxalate extraction for the quantitative estimation of the 'active' Al (Al_{ox}) and associated Si (Si_{ox}) fraction of bottom ash and coal fly ash samples derived from deposits with different age. This is the fraction which is associated with amorphous hydrous aluminosilicates such as allophane and imogolite, and Al humus complexes. The results are plotted in Figure 6 and are compared with data of unweathered natural glassses. These data reveal that (i) the (amorphous) clay content of the weathered synthetic ashes are significantly higher than that of the unweathered natural glasses en (ii) the rate of clay formation in bottom ash seems to be higher than that in coal fly ash. We speculate that the latter is a result of the lower temperatures during combustion which results in an incomplete transformation and decomposition of the native materials

In contrast to volcanic ash, MSWI bottom ash contains large amounts of alkaline earth oxides and, during initial weathering, solutions in contact with bottom ash generally develop extremely high pH values by reaction with oxide components, leading to rapid glass dissolution. Due to uptake of CO_2 from the atmosphere (carbonation) and microbial respiration, the pH of bottom ash decreases and stabilizes at values around 8.3 *[5,7]*. Since neutralization reduces the solubility of Al and Si, this pH decrease promotes (further) precipitation of these constituents from the ash pore water solution.The initially very high pH and its neutralization during weathering distinguishes bottom ash and most other indutrial ashes from volcanic ash and is responsible for the observed widespread and rapid formation of non-crystalline clays in the weathered bottom ashes.

In order to understand the underlying mechanism and conditions of neoformation of clay minerals from MSWI bottom ash glasses, we have examined the compositional and mineralogical properties of alteration products of weathered bottom ash glasses in more detail. TEM observations in conjunction with experimental leaching data indicate that clay formation from bottom ash glasses occurs by reaction among dissolved species as well as by reaction with an altered layer. Depending on the glass composition, twe have observed
wo different mechanisms of glass weathering: (i) *in situ* transformation on surfaces of silica rich glass particles and (ii) congruent dissolution followed by precipitation outside the glass spheres on surfaces of aluminosilicate glass particles. These secondary aluminosilicates tend toward the same composition irrespective of the composition of the parent glass. These observations lead to the conclusion that the type of clay mineral formed is determined by the external chemical environment (e.g. pH) rather than by the glass composition and alteration mechanism.

IMPACT ON HEAVY METAL LEACHING

It has long been recognized that altered aluminosilicate layers formed on natural and synthetic glasses play a significant role in the retention of trace elements during leaching [12]. We have

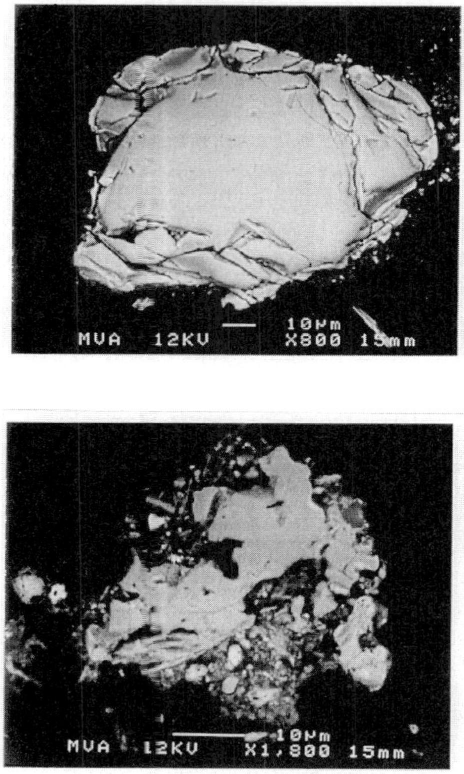

Figure 4 Backscattered images of bottom ash glass grains in (a) fresh bottom ash, and (b) in weathered bottom ash. A thick alteration rim of cly has developed on the weathered glass grain.

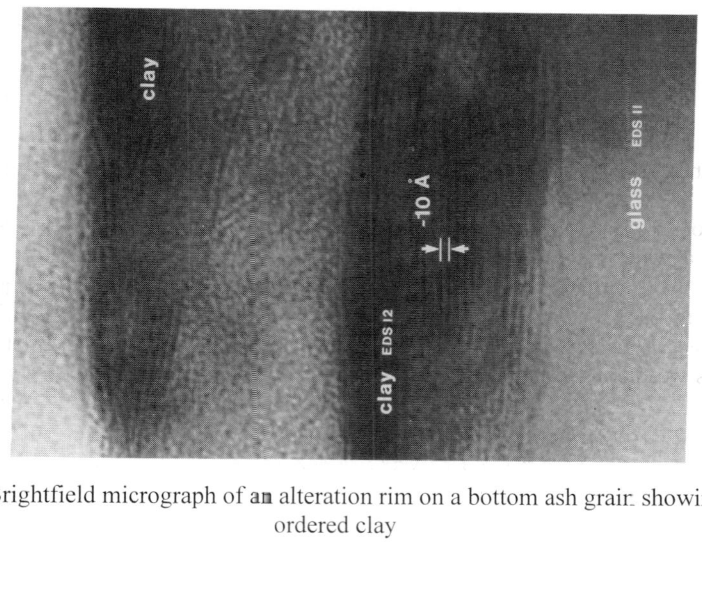

Figure 5 Brightfield micrograph of an alteration rim on a bottom ash grain showing well ordered clay

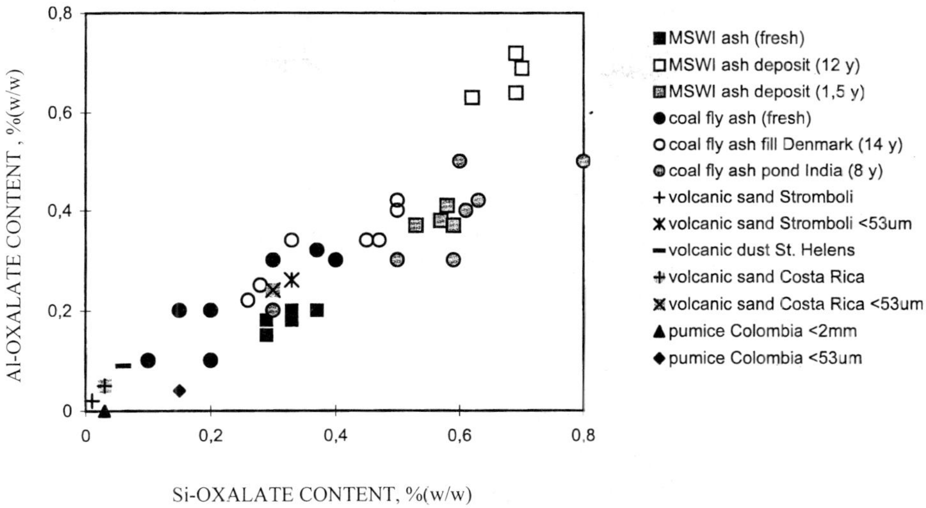

Figure 6 Al and Si oxalate extractable fraction (in % w/w) from fresh and weathered bottom ashes and other synthetic and natural glasses.

noted by direct observations using EDS element mapping that heavy metals such as Zn and Cu are at least to some extent incorporated into newly formed clays in weathered bottom ash. Zn-rich amorphous aluminosilicates (with Zn contents of more than 10 % w/w) have recently been observed in a MSWI botom ash deposit by Meima [13]. Clay formation in bottom ash depositsmay, therefore, play an important role in the mobilisation/retention profiles for heavy metals.

CONCLUDING REMARKS

Although there is some literature on the occurrence of glasses, there has been little emphasis on the processes of glass weathering in thermal residues such as MSWI bottom ash. Micro-analytical observations reveal that the bottom ash matrix, which consists primarily of glasses that cover a wide compositional range, is not chemically inert but 'thermally activated'. In comparison with natural systems, bottom ash is unique in that it is composed of glasses with different composition which are subject to weathering in the same environment. Hence, bottom ash deposits provide an ideal setting for fundamental studies on glass alteration processes in general. Its weathering features are analogous to those observed during weathering of natural mineral assemblages and glasses. Glass weathering may have important implications for the geochemistry of contaminants in a MSWI bottom ash disposal environment and must therefore be taken into account in assessing the environmental impact of these ashes.

REFERENCES

1. COMMANS, R.N.J., VAN DER SLOOT, H.A. AND BONOUVRIE, P.A. Geochemical reactions controlling the solubility of major and trace elements during leaching of municipal solid waste incinerator residues. Proc. Int. Conf. Municipal Waste Combustion, March 30-April 2, 1993, Williamsburg, VA, J. Kilgroe, ed., AWMA, Pittsburg, PA.

2. EIGHMY, T.T., EUSDEN, J.J., MARSELLA, K., HOGAN, J., DOMINGO, D., KRZANOWSKI, J.R. AND STAMPLI, D. Particle petrogenesis and speciation of elements in MSW incinerator bottom ashes. Environmental aspects of construction with waste materials. Goumans, J.J.J.M., Van der Sloot H.A. and Aalbers Th.G. (eds.) Elsevier Science B.V., pp. 111-136, 1994.

3. INTERNATIONAL ASH WORKING GROUP (IAWG). An international perspective on characterization and Management of residues from Municipal solid waste Incineration, ECN-report, 77 pp., 1994.

4. KIRBY, C.S. The aqueous geochemistry of municipal solid waste ash. Ph.D. Thesis, Virginia Polytechnic Institute and State University, 1993.

5. MEIMA, J, AND COMANS, R.N.J. Geochemical modeling of weathering reactions in municipal solid waste incinerator bottom ash. Environmental Science and Technology , 31, 1269-1276, 1997.

6 ZEVENBERGEN, C. VANDER WOOD, T., BRADLEY, J.P., VAN DER BROECK, P., ORBONS, A.J. AND VAN REEUWIJK, L.P. Morphological and chemical properties of MSWI bottom ash with respect to the glassy constituents. Hazardous waste & Hazardous Materials, 11, pp. 371-383, 1994.

7. ZEVENBERGEN, C, AND COMANS, R.N.J. Geochemical factors controlling the mobilization of major elements during weathering of MSWI bottom ash. In: Environmental aspects of construction with waste materials. Goumans, J.J.J.M., Van der Sloot H.A. and Aalbers Th.G. (eds.) Elsevier Science B.V., pp. 179-194, 1994.

8. WADA, K. Minerals and mineral formation in soils derived from volcanic ash in the tropics. Sci. Geol., Mem., 85, pp. 69-78, 1990.

9. KIRKMAN, J.H. AND MCHARDY, W.J. A comparative study of the morphology, chemical composition and weathering of ryolitic and andestic glass, Clays and Clay Minerals, 15, pp. 165-174, 1980.

10 ZEVENBERGEN, C, BRADLEY, J.P., VANDER WOOD, T., BROWN, R.S., VAN REEUWIJK, L.P., AND SCHUILING, R.D. Microanalytical investigation of mechanisms of MSWI bottom ash weathering. Microbeam Analysis, 3, pp. 125-135, 1994.

11 ZEVENBERGEN, C., VAN REEUWIJK, L.P., BRADLEY, J.P., BLOEMEN, P., AND COMANS, R.N.J. Mechanism and conditions of clay formation during natural weathering of MSWI bottom ash. Clays and Clay Minerals, Vol. 44, No, 4, pp. 546-552, , 1996.

12. PETIT, J.C., DRAN, J.C. AND TROTIGNON, L.. Mechanism of heavy element retention in hydrated layers formed on leached silicate glasses. Mat. Res. Symp. Proc. Vol. 127, pp. 33-40., 1989.

13. MEIMA, J.A. Geochemical modelling and identification of leaching processes in MSWI bottom ash, Ph.D Tthesis, University Utrecht, 1997.

MUNICIPAL SOLID WASTE INCINERATOR RESIDUES: PHASE COMPOSITION AND DISTRIBUTION OF ORGANIC AND INORGANIC POLLUTANTS BETWEEN THE OUTPUT STREAMS

B Hentschel

C Marb

Bavarian State Office for Environmental Protection

Germany

ABSTRACT. Based on recent studies at six municipal solid waste incinerators in Bavaria, the different sorts of residues (process wastes) are characterized with respect to phase composition, pollutant content and physical properties. The temporal variations of these characteristics are described and the respective ranges of data are given, which are similar for all incinerators, regardless of differences in technical design and waste input. The overall approach, including the sampling of all output streams (solids and gas) and the determination of mass streams and volume flow rates, allows the calculation of total pollutant outputs of modern waste incinerators and the description of pollutant distributions within the system. It is demonstrated, how bottom ash quality can be improved, applying higher process temperatures (oxygen enrichment of primary air), while changes in volume flow rates of primary and secondary air including the partial substitution of the secondary air by steam injection didn't show an effect on the characteristics of the residues.

Keywords: Municipal solid waste incineration, Process wastes, Phase composition, Pollutant distribution, Pollutant loads, Operating conditions.

Dr-Ing B Hentschel, is a scientist at the Waste Technology Centre of the Bavarian State Office for Environmental Protection, Augsburg, Germany. He is working in the field of thermal waste treatment technologies, especially with respect to the chemical and mineralogical characterization of the different solid residues, produced in the course of thermal treatment and air pollution control, as well as the assessment of MSWI bottom ash as a secondary raw material.

Dr-Ing C Marb, is head of the Waste Technology Centre of the Bavarian State Office for Environmental Protection, Augsburg, Germany. He specialises in waste composition and disposal methods like composting of biological wastes, thermal treatment of municipal solid wastes or sewage sludge and remediation of contaminated soils respectively. His current interests include the development of low-waste and non-waste technologies as future disposal methods.

INTRODUCTION

Waste incineration is a common method of municipal solid waste treatment in many industrial countries of the world. In Bavaria 81 % of the 2.7 Mio. Mg of the residual waste (municipal solid waste, part of the industrial waste, bulky waste) produced in 1997 [1] were thermally treated in 19 municipal solid waste incinerators (MSWI) and 1 pyrolysis plant. Given this situation and considering the changes in waste composition and in the legal demands (air pollution control, operating conditions, quality of the residues, etc.), there is a continuous need for an up-to-date assessment of

- the environmental impact (especially concerning the total pollutant output of municipal solid waste incinerators and the pollutant loads of the various output streams) as well as
- the physical and chemical characteristics of the different process wastes (bottom ash, fly ash, etc.) with respect to a possible utilization as secondary raw materials.

About 73 % of the 547000 Mg of bottom ash (incl. metal scrap) produced in 1997 [1] were processed and recycled (landfill technique, road construction, etc.). For this main process waste, which is generally suitable for utilization with respect to environmental demands in Germany, there is still little known about the actual quality fluctuations or temporal variations in pollutant contents due to the heterogeneity of the waste input. But especially this information is of key importance to the assessment of bottom ash as a construction material with a defined and predictable quality and to the discussion of new waste treatment technologies.

SAMPLING AND ANALYTICAL METHODS

Plant Description

From 1995 to 1999 six municipal solid waste incinerators with different kinds of grate firing systems (rocking grate, reciprocating grate, etc.) were sampled and studied: three of them (I, II and VI) designed according to the regulations of the 17th ordinance regarding enforcement of the Federal Emissions Protection Law (17. BImSchV) [2] and three plants (III to V) constructed before 1990, that had to be retrofitted. Table 1 summarizes some important technical characteristics of the six MSWI (I - VI) investigated, including the design of the different air pollution control (APC) systems.

With respect to the air pollution control strategies plants with a preliminary dust removal unit (electrostatic precipitator, cyclone, etc.) can be distinguished from those systems, where dust removal is integrated into the chemical units.

Sampling

As a basis for the analytical studies representative samples of all output streams of the MSWI had to be obtained. This means that besides the definition of adequate sample sizes, various fundamental problems had to be addressed, for example separation effects (particularly for coarse grained materials like the bottom ashes), spatial and temporal variations caused by the

Table 1 Technical characteristics of the six municipal solid waste incinerators

PLANT	START OF OPERATION	WASTE INPUT MG/H	SAMPLING	WASTE GAS CLEANING UNITS	
I	1995	7.6	1996	electrostatic precipitator (ESP)	3-stage scrubber - wet ESP - SCR - fabric filter
II	1994	16.8	1995	electrostatic precipitator (ESP)	3-stage scrubber - SCR - fabric filter
III	1989	6.5	1997	- - -	spray absorber - fabric filter - 2-stage scrubber - wet ESP
IV	1982	12.7	1995	- - -	spray absorber - fabric filter - SCR
V	1984	9.6	1995	waste gas cooler - cyclone	evaporation cooler - fabric filter - SCR
VI	1994	6.5	1999	multicyclone	spray absorber - fabric filter - 2-stage scrubber

heterogeneous waste input, periodical fluctuations due to plant operation characteristics (e.g. periodical cleaning procedures for air pollution control aggregates), etc. Based on these considerations, experiences from the mining industry, dealing with heterogeneous bulk goods for a long time, and legal demands (e.g. LAGA guidelines for sampling of solid wastes [3,4]) a detailed sampling strategy for each class of solid process wastes was developed.

At each plant one-week sampling campaigns were carried out. Eight individual samples were taken each day over 8-hour periods and combined and homogenized in the evening to obtain daily composite samples (DS). At the end of the week the total amount was again homogenized to extract weekly composite samples (WS). In this way six samples were obtained per plant and week (plant VI: only three daily composite samples). The size of the composite samples ranged from 2 kg for the very fine grained fabric filter dusts to about 1 Mg for the heterogeneous bottom ashes, which were classified, sorted (removal of metal scrap and large unburned particles), and divided into the grain size fractions 0 - 8 mm („fine“) and 8 - 32 mm („coarse“) immediately after the collection at the wet slag removal unit. The bottom ash fraction larger than 32 mm was rejected. All composite samples were shipped to the laboratories for further processing (drying, crushing, milling, etc.) to finally obtain the samples for analysis.

Dirty gas samples were taken either immediately after the boiler or in the case of plant I and II behind the preliminary dust removal unit (ESP). Clean gas was sampled just before the stack according to the relevant VDI regulations. Discontinuous samplings (suctions) were carried out isokinetically for 30 min. (inorganic) and 8 hrs. (organic) respectively.

All solid process wastes studied at the six incinerators are summarized in Table 2. Sampling locations (S) as well as ranges of mass streams of the solid process wastes and volume flow rates of the dirty and clean gas determined in the course of the sampling campaigns are given in Figure 1.

Table 2 Solid process wastes

PLANT	SOLID PROCESS WASTES
I	bottom ash - boiler ash - ESP dust - filtration residue - activated carbon
II	bottom ash - grate screenings - boiler ash - ESP dust - filtration residue - fabric filter dust
III	bottom ash - grate screenings - boiler ash - fabric filter dust
IV	bottom ash - grate screenings - boiler ash - fabric filter dust
V	bottom ash - grate screenings - fabric filter dust
VI	bottom ash - boiler ash - cyclone dust - fabric filter dust

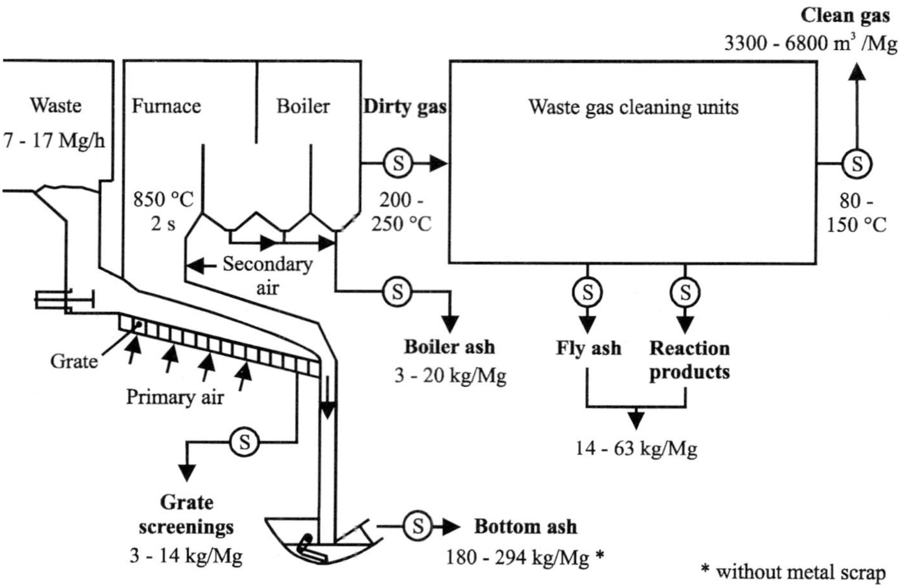

Figure 1 Sampling locations (S) and output streams related to the waste input

Analytical Methods

Physical characteristics

As a basic characterization loss on ignition (ashing at 550 °C), particle size distribution (sieve analysis), surface area (BET), and porosity (Hg-intrusion) were determined.

Chemical analyses

The analytical program included the determination of environmentally relevant pollutants like heavy metals (Pb, Cu, Zn, Cd, Hg, etc.) and halogens by ICP/OES, AAS and/or RFA, and of organic components like polycyclic aromatic hydrocarbons (PAH), polychlorinated dibenzo-p-dioxins and -furanes (PCDD/PCDF), and polychlorinated biphenyls (PCB) by HRGC/MS

(PAH, PCB) and by HRGC/HRMS (PCDD/PCDF), both in SIM mode. Clean-up, fractionation and final separation were carried out by standard methods described in [5]. For PAH 24 components in the gas and 16 components (acc. to EPA 610) in the solids, for PCDD/PCDF 17 congeners (acc. to 17. BImSchV [2]) and for PCB 6 congeners (acc. to AbfKlärV [6]) were determined. The total organic carbon (TOC) was measured by combustion followed by IR-spectrometry.

Phase analyses

Mineralogical phase analyses were done by x-ray powder diffraction using full pattern refinement for quantification of crystalline phases and by light microscopy including point-counting methods to determine the relative amounts of crystalline and amorphous (ash, glass, etc.) phases. The total amorphous content of the bottom ashes was also determined by x-ray diffraction (full pattern refinement and addition of a spike material (ZnO, KCl)).

RESULTS AND DISCUSSION

Phase Composition, Burnout and Physical Characteristics

Due to the considerable heterogeneity of the different materials, all parameters analyzed showed large fluctuations during the one week sampling periods. The coarse bottom ash fraction (8 - 32 mm) was generally characterized by better burnout parameters (loss on ignition, total organic carbon content) and significantly lower contents of organic and inorganic pollutants. In Figure 2 ranges of total organic carbon (TOC) contents are given for the daily and weekly composites of the bottom ash fine fractions (loss on ignition: 1.0 - 4.4 mass-%) exemplarily; the TOC of the coarse fractions ranged from 0.02 - 1.2 mass.-% (loss on igni- tion: 0.3 - 3.6 mass-%).

The different kinds of process wastes showed characteristic phase compositions, see Table 3. Bottom ashes and grate screenings were dominated by silicates, carbonates, oxides, and large amounts of amorphous components (fine fraction: 62 - 77 mass-%, coarse fraction: 70 - 82 mass-%), while in the bottom ashes anhydrite ($CaSO_4$) was found as a further main component (22 - 56 mass-% of crystalline material). The dusts from the electrostatic precipitators of plants I and II resembled the boiler ashes, but showed much larger amounts of halite (NaCl) and sylvite (KCl), while phase compositions of the fabric filter dusts largely depended on the APC systems and sorbents (lime, calcium hydrate, etc.) used.

By microscopic methods (point-counting) it could be confirmed, that silicate glass phases produced by partial melting formed the main part of the amorphous components (fine fraction: 27 - 43 vol.-%, coarse fraction: 28 - 53 vol.-%) of bottom ashes and that total porosities (open and closed pores) varied between 3 and 10 vol.-% in both fractions. The fine fractions also showed surface areas per unit volume (BET) significantly higher (2.2 - 6.7 m^2/g) than those of the coarse fractions (0.5 - 2.2 m^2/g). Single Hg-intrusion analyses (bottom ashes of plant II) gave total open pore volumes of about 0.2 cm^3/g for the fine and 0.1 cm^3/g for the coarse fractions

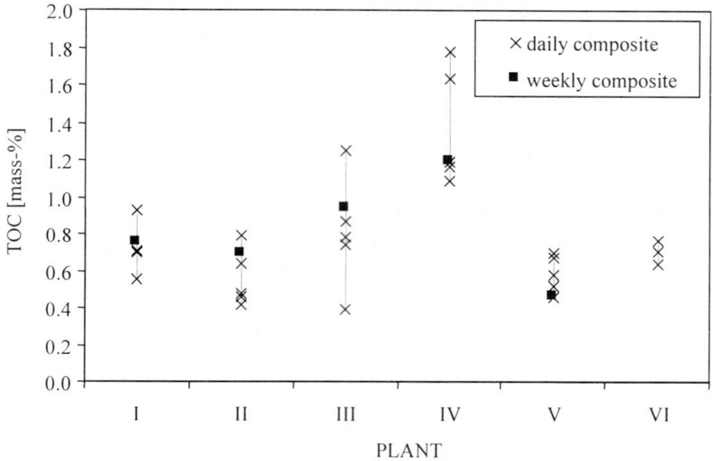

Figure 2 Total organic carbon contents of bottom ashes (fine fraction) adapted from [7]

Table 3 Phase composition of the solid process wastes (main mineral phases)

COMBUSTION	AIR POLLUTION CONTROL		
Bottom ash / grate screenings	**Boiler ash**	**ESP dust**	**Fabric filter dust**
silicates: quartz, melilites, pyroxenes, feldspars	**silicates:** quartz, melilites, pyroxenes, feldspars	**silicates:** quartz, melilites, pyroxenes, feldspars	**silicates:** quartz, melilites, pyroxenes, feldspars
oxides: spinels, hematite, perovskite, rutile, CaO	**oxides:** spinels, perovskite, rutile, periclase	**oxides:** spinels, perovskite, rutile, CaO	**oxides:** spinels, perovskite, rutile
carbonates: calcite, dolomite	**carbonates:** calcite, dolomite, magnesite	**carbonates:** calcite, dolomite, magnesite	**carbonates:** calcite, dolomite
sulfates: anhydrite	**sulfates:** anhydrite	**sulfates:** anhydrite	**sulfates:** anhydrite
hydroxides: portlandite	**chlorides:** halite, sylvite	**chlorides:** halite, sylvite	**chlorides:** halite, sylvite
phosphates: apatite			**hydroxides:** portlandite, CaClOH

Pollutant Distribution and Total Output

Ranges of concentration

The inorganic pollutants showed characteristic distribution patterns between the different process wastes in accordance with their respective volatilities. Mercury, cadmium, zink, lead, and chlorine were enriched in the residues of the air pollution control systems, while copper was found to a large extend in the residues of the actual combustion process (bottom ashes, grate screenings). The results are exemplified for cadmium and copper in figures 3.

The combustion residues (plants I to V) also showed higher concentrations of PAH than the APC residues, which were enriched in PCDD/PCDF and PCB (Table 4). Concentration of organic pollutants in bottom ashes and grate screenings were generally very low. Burnout parameters (loss on ignition, total organic carbon) didn't indicate correlations with the concentrations of organic pollutants, except for the PAH of the bottom ashes, which correlated with the respective TOC. Concentrations of organic pollutants measured in the dirty and clean gas of plants I to V are summarized in Table 5 for comparison.

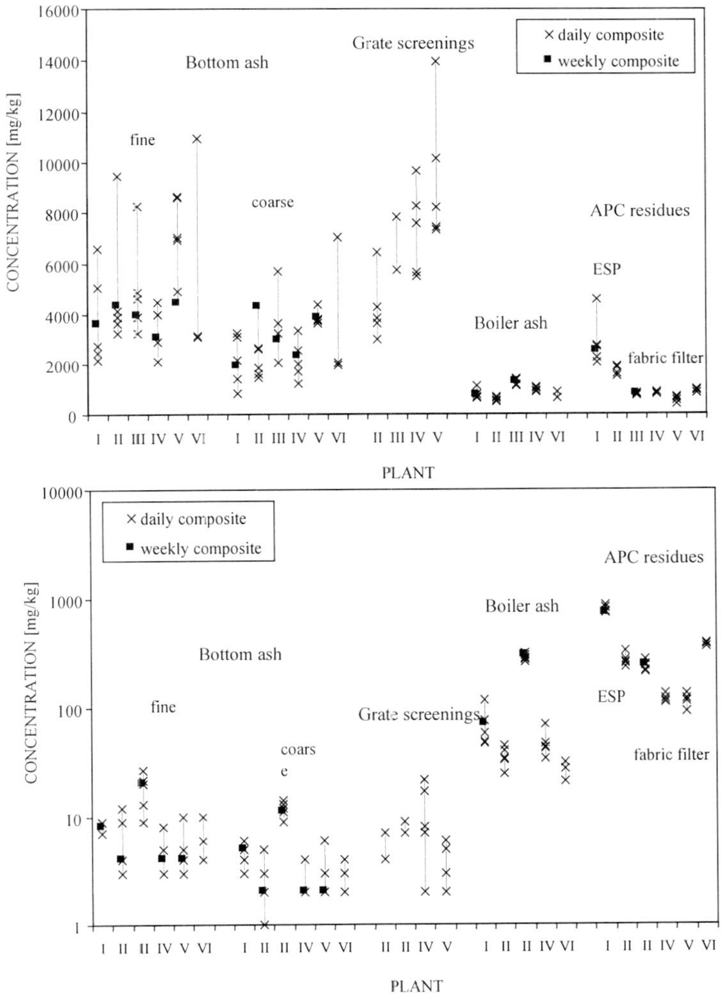

Figures 3 Copper (above) and cadmium (below): ranges of concentration in the solid process wastes adapted from [7]

Table 4 Ranges of concentration of organic pollutants in the solid process wastes [7]

			BOTTOM ASH Fine	BOTTOM ASH Coarse	GRATE SCREENINGS	BOILER ASH	ESP DUST	FABRIC FILTER DUST
PAH	Mean	µg/kg	746	329	2034	142	89.1	546
	Min		83.4	29.2	191	2.3	28.6	6.9
	Max		2150	1900	14100	1430	273	4010
	Median		631	127	860	21.0	43.0	122
PCDD/ PCDF	Mean	ng I-TEQ/kg	14.8	5.3	9.0	289	518	1368
	Min		3.6	0.7	2.5	87.4	311	290
	Max		30.9	11.7	24.7	1125	1010	3706
	Median		14.0	5.2	5.8	253	455	1210
PCB	Mean	µg/kg	21.7	34.1	8.8	13.3	13.9	15.8
	Min		1.4	0.4	1.0	0.3	1.3	0.5
	Max		52.8	102	34.0	36.0	62.6	41.8
	Median		24.3	15.7	3.9	3.4	1.5	15.3

Table 5 Ranges of concentration of organic pollutants in the dirty and clean gas [8]

			DIRTY GAS	CLEAN GAS
PAH	Mean	ng/m^3	201	42.4
	Min		26	25
	Max		1110	67
	Median		134	37.7
PCDD/ PCDF	Mean	ng I-TEQ/m^3	3.59	0.008
	Min		1.20	0.001
	Max		5.90	0.030
	Median		4.10	0.004
PCB	Mean	ng/m^3	33.1	10.7
	Min		1.2	0.1
	Max		82	43
	Median		34	2.5

Distribution and total output

From the concentrations in the solid process wastes and the flue gas as well as the mass streams and volume flow rates determined simultaneously, the total output per Mg of waste input of various pollutants and the specific contribution of each output stream to the total pollutant load was calculated for each plant. Assuming that all possible output streams were taken into account, the pollutant content of the waste input can be estimated from the mass balance and the total of the output streams. In Table 6 the total output loads calculated for several elements are compared with published data [9,10]. The ranges determined in this study lie in the lower part or slightly below the ranges given in [10] for average municipal solid wastes in Germany. This observation illustrates the influence of changes in waste-management concepts during the last years on the characteristics of MSWI process wastes.

For the elements under consideration no significant differences between the individual plants (I - VI) due to differences in plant design (grate technology, furnace design, air pollution control systems, etc.) or to differences in waste input could be observed with respect to element distribution and total output.

Table 6 Total output loads (per Mg of waste input) calculated for several relevant elements and specific contribution of each output stream (%), compared to data from [9,10]

RESIDUES	LEAD	COPPER	ZINK	CADMIUM	MERCURY	CHLORINE
Bottom ash	32 - 65	84 - 98	30 - 69	7 - 26	1 - 9	8 - 28
acc. to [9]:	70	-	60	12	2	8 - 13
Grate screenings	3 - 10	4 - 5	1 - 3	< 1	< 1	1 - 2
acc. to [9]:	-	-	-	-	-	-
Boiler ash	3 - 13	1 - 3	5 - 15	6 - 10	< 1	2 - 9
acc. to [9]:	1	-	2	3	< 0.3	-
Σ APC residues	29 - 53	2 - 9	26 - 55	64 - 86	90 - 98	63 - 89
acc. to [9]:	30	-	38	83	70 - 80	-
Clean gas	<< 0.1	<< 0.1	<< 0.1	<< 0.1	0.1 - 6	< 0.1 - 0.3
acc. to [9]:	< 1	-	< 1	< 2.5	< 10	< 0.1
Total output [g/kg$_{waste}$]	**0.5 - 0.9**	**0.6 - 1.4**	**1.1 - 1.9**	**0.006 - 0.015**	**0.0005 - 0.002**	**3.3 - 10.1**
Waste input acc. to [10]:	0.6 - 2.5	0.45 - 2.5	0.44 - 3.5	0.01 - 0.04	0.002 - 0.007	5 - 15

In case of the organic pollutants an estimation of the composition of the waste input is not possible, because organic compounds are destroyed (furnace) and in some cases formed (e.g. de novo formation of PCDD/PCDF along the gas path) within the system. Nevertheless, the calculation of the total output of different organic pollutants can serve as a basis for the evaluation of the environmental impact of MSWI plants. In Table 7 the total output of PCDD/PCDF (I-TEQ) and the specific contribution of each output stream is given for plants I to V. The results demonstrate, that the total PCDD/PCDF output of the more recent (Table 1) plants I and II, designed according to 17th BImSchV [2], are significantly lower than those of plants III to V, constructed before 1990, which had to be retrofitted. More than 90 % of the total PCDD/PCDF load was concentrated in the boiler ash and the APC residues, while the clean gas didn't show any significant contribution to the total loads.

In case of the PCB bottom ashes as the main process wastes carried more than 70 % of the total output load (plants I and II: 2 - 4 mg/Mg$_{waste}$, plants III to V: 10 - 15 mg/Mg$_{waste}$). This was due to the large ranges of PCB concentrations with similar mean values (Table 4) in all output streams and the large mass stream of the bottom ash.

Table 7 Total output loads (per Mg of waste input) calculated for PCDD/PCDF (I-TEQ) and specific contribution of each output stream [%]

PLANT	TOTAL OUTPUT µg I-TEQ/Mg_{waste}	BOTTOM ASH %	GRATE SCREENINGS %	BOILER ASH %	APC RESIDUES %	CLEAN GAS %
I	10	10	-	31	59	< 1
II	18	9	-	8	83	< 1
III	64	8	< 1	1	91	< 1
IV	83	2	< 1	12	87	< 1
V	95	1	< 1	-	98	< 1

Influence of the Operating Conditions

In the course of additional one week sampling campaigns at plants III, IV and VI the influence of operating conditions on the characteristics of the process wastes, especially on the quality of the bottom ash, was studied.

At plant III an increase in the mean surface temperature of the moving bed in the main combustion zone from 1060 °C to 1200 °C could be achieved by an oxygen enrichment of the primary air from 21 to 28 vol.-% and the reduction of the volume flow rates of primary and secondary air (both from 18000 to about 13000 m^3/h) combined with a partial flue gas recirculation. The higher process temperatures led to an improvement (reduction) of the parameters characterizing the bottom ash burnout (loss on ignition and total organic carbon content) [11], which is shown in Figure 4 for the TOC, and to bottom ashes with significantly increased amounts of coarse grained particles. A reduction of the furnace temperature of about 100 K in the main combustion zone caused by an increase in the volume flow rate of the combustion air (plant VI) seemed to have little effect on the process wastes. Mean TOC contents of the bottom ashes were slightly lower (fine: 0.7 mass-%, coarse: 0.4 mass-%) for the regular operating conditions than for the experimental changes (fine: 0.9 mass-%, coarse: 0.5 mass-%), while other parameters (e.g. loss on ignition or heavy metal contents in the bottom ash) didn't show significant differences between the operating conditions. Because not all of the analyses (especially for the organic components) are finished at present, a final interpretation can't be given so far. Changes of volume flow rates of primary and secondary air including the partial substitution of the secondary air by steam injection (plant IV) also didn't show an effect on the characteristics of the residues.

CONCLUSIONS AND OUTLOOK

The studies presented in this paper give a survey of the chemical, physical and mineralogical characteristics of all sorts of MSWI process wastes, including the total pollutant output of modern MSWI plants in Bavaria and the relative contribution of each output stream to the total. Regardless of the technical applicability, the environmental impact of recycling projects or of thermal waste treatment itself will decide on the general acceptance of this technology

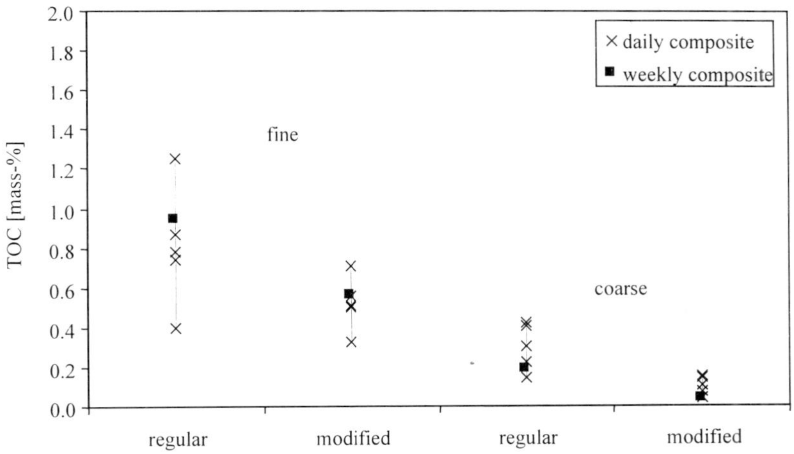

Figure 4 Total organic carbon contents of bottom ashes for regular and modified operating conditions at plant III [7]

in public opinion as well as in the construction industry. For this reason and considering changes in the technical, economical, and legal background, continual efforts are necessary to provide an up-to-date data base with respect to all relevant environmental questions connected with the technology of thermal waste treatment.

Based on these data, the technical suitability of MSWI bottom ash as a secondary raw material for large scale construction projects is studied in a new research project, carried out in cooperation with Max Aicher GmbH & Co, Freilassing, Germany, a construction company working in the field of bottom ash recycling for many years. The project, supported by the Bavarian State Ministry for State Development and Environmental Affairs, especially focuses on the geotechnical properties (compression, optimum water content, mechanical stability, volume stability, leachability, permeability, etc.) of large bottom ash bodies of different grain size fractions and is mainly based on experiments at technical scale bottom ash heaps (about 4000 Mg bottom ash each), fully equipped with sensors (temperature, water content, pressure, etc.) as well as actual recycling projects (foundations, dams, etc.). It aims at the fundamental characterization and improvement of present recycling applications, and the development and study of new fields of bottom ash recycling.

ACKNOWLEDGEMENTS

This paper is based on fundamental investigations of the Bavarian State Office for Environmental Protection as well as recently published data [7,11,12] from a research project (May 1995 to July 1998), carried out in cooperation with the Bavarian Institute for Waste Research (BIfA GmbH), Augsburg. We would like to acknowledge the financial support provided by the German Environmental Foundation (DBU), Osnabrück, and the Bavarian State Ministry for State Development and Environmental Affairs, Munich.

REFERENCES

1. BAYERISCHES LANDESAMT FÜR UMWELTSCHUTZ. Abfallwirtschaft - Hausmüll in Bayern - Bilanzen 1997, München, 1998.
2. ANONYMOUS. Siebzehnte Verordnung zur Durchführung des Bundes-Immissionsschutzgesetzes (Verordnung über Verbrennungsanlagen für Abfälle und ähnliche brennbare Stoffe - 17. BImSchV) vom 23.11.1990, BGBl. I, p 2545, ber. p 2832
3. LAGA: Richtlinie zur Entnahme und Vorbereitung von Proben aus festen, schlammigen und flüssigen Abfällen, PN 2/78, in: Schriftenreihe chemische Analytik und Umwelttechnologie, Bayer. Landesamt für Umweltschutz, Heft 2, München, 1979. pp 7 - 20
4. LAGA: Richtlinie für das Vorgehen bei physikalischen und chemischen Untersuchungen im Zusammenhang mit der Beseitigung von Abfällen - PN 2/78 K - Grundregeln für die Entnahme von Proben aus Abfällen und abgelagerten Stoffen, in: Müllhandbuch, Nr. 1859, Lfg. 2/84. Erich Schmidt Verlag, Berlin, 1983
5. HAGENMAIER, H, BRUNNER, H, HAAG, R, KUNZENDORF, H J, KRAFT, M, TICHACZEK, K AND WEBERRUSS, U: Stand der Dioxin-Analytik. VDI-Berichte. 634, pp 61 - 89
6. ANONYMOUS. Klärschlammverordnung (AbfKlärV) vom 15.4.1992, BGBl. I, p 912, geändert durch VO v. 6.3.1997 BGBl I. p 446
7. KNORR, W, HENTSCHEL, B, MARB, C, SCHÄDEL, S, SWEREV, M, VIERLE, O AND LAY, J-P. Rückstände aus der Müllverbrennung - Chancen für eine stoffliche Verwertung von Aschen und Schlacken. Initiativen zum Umweltschutz 13. Erich Schmidt Verlag, Berlin, 1999
8. RIEDEL, H, HENTSCHEL, B AND THOMA, H. Organic pollutants in stack gases compared to solid emissions of municipal solid waste incinerators. Organohalogen Compounds. Vol. 40, 1999. pp 465 - 468
9. CHANDLER, A J, EIGHMY, T T, HARTLEN, J, HJELMAR, O, KOSSON, D S, SAWELL, S E, VAN DER SLOOT, H A AND VEHLOW, J. Municipal solid waste incinerator residues. The International Ash Working Group (IAWG). Studies in Environmental Science 67. Elsevier, Amsterdam, 1997.
10. RAT VON SACHVERSTÄNDIGEN FÜR UMWELTFRAGEN. Sondergutachten des Rates von Sachverständigen für Umweltfragen vom September 1990 „Abfallwirtschaft“, Drucksache 11/8493, Deutscher Bundestag, 11. Wahlperiode, Wiesbaden.
11. HENTSCHEL, B, MARB, C, SCHÄDEL, S AND SWEREV, M. Untersuchungen zum Ausbrand von Schlacken aus thermischen Abfallbehandlungsanlagen mit Rostfeuerung. Beihefte zum European Journal of Mineralogy. Vol. 10, 1, 1998. p 126
12. MARB, C, HENTSCHEL, B, VIERLE, O, THOMA, H, DUMLER-GRADL, R, SWEREV, M, SCHÄDEL, S AND FIEDLER, H. PCDD/F in bottom ashes from municipal solid waste incinerators in Bavaria, Germany. Organohalogen Compounds. Vol. 32, 1997. pp 161 - 166

INREC^{PLUS} – THE NEXT STEP IN WASTE INCINERATION ASH RECYCLING

A Selinger
Ch Wieckert
V Schmidt
B Carcer
H Rüegg
G Mechtersheimer
ABB Alstom Power
Switzerland

ABSTRACT. Ashes from waste incineration need to be treated before they can be recycled, i. e. for use in civil construction, due to their toxic potential.

Bottom ash, accounting for 80-90% of the solid residues from grate incineration, can be described as a mixture of inert materials, inorganic binders and metal scrap. After appropriate treatment, all main components are well-suited for recycling. The InRec^{plus} process is based on **dry** bottom ash discharge with subsequent sorting. Compared to conventional wet bottom ash processing, more metals in a better quality are recovered, as corrosion and cementation is prevented. In addition, the main mineral fraction is not only low in metals, but can also be further cleaned from organic carbon, salts and heavy metals with simple mechanical means.

Fly ash of waste incineration needs thermal treatment, before reuse in civil construction can be considered. With two complementary thermal processes, InRec^{plus} covers a wide range of ash vitrification demands: Not only any combination of ash input material is covered, but also different fuels, throughput ranges and end product quality requirements.

Finally, InRec^{plus} features a patented method to turn the glassy melt into foam slag, a valuable lightweight material with numerous applications in civil construction, creating revenues which reduce the overall treatment costs.

Keywords: Bottom ash, Fly ash, Dry sorting, Vitrification, Foam slag

Dr A Selinger, Chemist, is product manager of the ash treatment systems.

Dr Ch Wieckert, Physicist, is a development engineer.

Dr V Schmidt, Chemical engineer, is responsible for the residues treatment research project.

B Carcer is process engineer in R&D.

H Rüegg, Mechanical Engineer, is R&D manager of the waste-to-energy business area unit.

Dr G Mechtersheimer, Physicist, is a technical project manager.

All authors are employed by ABB Alstom Power in Switzerland.

INTRODUCTION

In developed economies, dumping of untreated waste is not longer seen as a sustainable solution. Thermal treatment is increasingly favoured complementary to other recycling efforts. After incineration, 1000 kg of municipal waste are turned into energy, cleaned off-gas and some 280 kg of solid residues; typically 240 kg bottom ash and 40 kg fly ash and waste water treatment solids.

Without landfill of untreated waste, these residues will become the major solid waste effluent of an economy. Total recycling of these ashes would therefore turn landfills for municipal waste obsolete - a big and challenging step towards sustainability.

However, any ash recycling effort must pass two crucial requirements:

- protection of the environment from potentially toxic ash contents
- within given legislation, the costs have to be competitive.

Environmental problems may arise from the ashes content of heavy metals, organic toxics like dioxins and soluble salts [1].

Currently, bottom ash from wet discharge systems is widely and successfully used - mainly for road construction - in countries like France, Germany, Netherlands and Denmark [2]. However, there are several areas where improvements are desirable:

- a lower metal content would reduce gas formation and prevent volume instabilities
- the quality of separated metals is rather poor due to corrosion and contamination
- salt content may cause leaching problems.

Fly ash from waste incineration is currently not recycled in a sustainable way within Europe, but landfilled after various types of treatment.

THE INREC CONCEPT

Dry Ash Discharge

Traditionally, the bottom ash is dropped from the grate into a water sealed discharger. In the InRec concept, that is replaced by a dry ash discharge system. Depending on the subsequent ash sorting, the ash can be discharged through a double-flap lock, which provides sealing to the atmosphere. In this case, the sorting process includes handling of the coarse fraction. Alternatively, a roller grate can be used to separate the coarse material within the ash discharge system. In this case, only the main fraction with particle sizes of up to 40 mm is discharged dry, while the coarse fraction is dumped into a water bath as usual. Both options have been installed in a Swiss incineration plant and were successfully tested for several month. A new waste incinerator constructed for Tokyo Metropolitan/Japan will use this system in the near future.

Dry Sorting of Ash

The sorting of the dry ash is performed with well proven mechanical components. In a first step, the fine fraction (i. e. particles smaller than 2mm) is separated from the remainder. This way, the dust load for the following steps is minimised. In addition, the fine fraction concentrates components less suitable for construction.

Iron is removed by means of a magnet drum and an overhead magnet, while an eddy current separator is used to extract the non-ferrous metals from the bottom ash. To separate metals which are sintered to mineral material and to crush larger aggregates down to gravel size, an impactor is used.

This way, besides a small oversize fraction, four main fractions are obtained:

- 50-60% main mineral fraction (particle size i. e. 2-40 mm) for construction purposes
- 15-20% ferrous metals
- 1-2% non-ferrous metals, of which typically $^2/_3$ is aluminum and $^1/_3$ is copper and brass
- 15-30% fine fraction (for further treatment, see below)

All the mechanical components of the sorting process have been tested full-scale with InRec material.

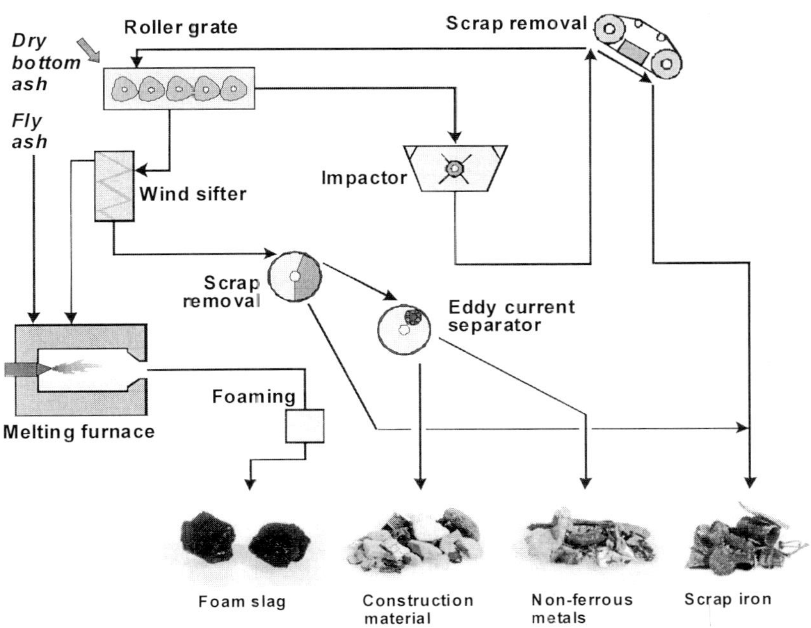

Figure 1 Dry sorting combined with optional vitrification and foam slag production

Vitrification

The fine fraction, as well as the boiler and the filter ash, may be treated in many different ways, depending on legislation and the desired quality. In many countries, fly ash can be landfilled after treatment i. e. with chelating agents. The fine fraction, due to its solidifying hydraulic potential, can be landfilled directly after mixing with water.

However, the heavy metal and dioxin content of these materials will remain unchanged, and recycling is not possible. For further quality improvement, thermal treatment is a mandatory step.

By vitrification, a residue is transformed into three or four fractions:

- a glassy slag (typically major fraction),
- a secondary fly ash, rich in salts like NaCl, but also containing most of the heavy metal compounds (e.g. chlorides from Pb, Zn and Cd)
- an off-gas, containing most of the sulfur as SO_2 and part of the chlorine as HCl.
- plus, if all bottom ash is vitrified, an additional metallic fraction mainly from Fe, but containing also most of the Cu.

Full bottom ash vitrification, however, is currently pushed exclusively in Japan.

Compared to other disposal methods for handling the toxic content in residues (like e.g. stabilisation with cement and landfill, washing, sintering) vitrification offers major advantages:

- destruction of organic toxics like dioxins
- added value from usage of the main product
- recovery of valuable heavy metals (depending on current economical context)
- effective inclusion of residual heavy metal into the silicate matrix of the main product (no leaching)
- improved handling of the material due to its fluidity (compared e.g. to sintering).

Starting in 1987, ABB Alstom Power has performed various steps on laboratory, pilot and on industrial scale for the development of complementary vitrification processes. Table 1 summarises main features of the two processes promoted today. Additionally an arc furnace process for very large throughputs was tested ("AshArc").

The fully electrically heated Deglor is the only ash melting process, which is able to continuously melt fly ashes on a commercial scale in Japan. The furnace design is based on electric glass oven technology [3]. The process is suited for treating also smaller amounts of residues. Bottom ash fractions can only be processed after effective pre-separation of the metal fraction (like in InRec).

Table 1 Main features of ABB Alstom Power vitrification processes

Type of furnace	Rotary kiln	Glass oven (Deglor)
Input	Fly ash; Bottom ash; Mixtures thereof	Fly ash; Residues from flue gas treatment; Bottom ash fine fraction
Heat source	Burners with O_2, optionally feeding through burner	Ceramic top heaters, bath electrodes (resistive heating)
Primary energy	Oil/Gas/Coal etc. O_2	Electricity
Residence time	1-2 hours	1 day
Redox level	adjustable	oxidative
Temperature	1400 °C (oxidative), 1500 °C (reductive)	1300 °C
Throughput range	0.5 – 5 t/h	0.2 – 1 t/h
Development status	Numerous 50 kg test runs; 2 pilot plants (300kg)	Commercial plant in Japan for fly ash plus flue gas cleaning products since 1996 (200 kg/h + additives)

A major advantage of the recently developed rotary kiln process is its great flexibility with respect to input, fuel and redox level. Numerous variants were tested, treating fly ashes, bottom ashes and mixtures thereof. Successfully applied heating fuels included coarse coal swimming on the molten slag (burned using O_2 lances), several gaseous fuels and also dried sewage sludge. Feeding was performed as well through the burner or separately; the burners were operated with various oxygen to fuel ratios ranging from very reductive to oxidising. Feeding through the burner is a very effective technology in terms of heat input and mass transfer due to the huge surface area as long as the individual particles do not cluster. Figure 2 shows schematically one variant of the rotary kiln process.

Foam Slag

Economic evaluation of vitrification process costs show, that transformation of the molten slag into a valuable products is absolutely necessary in Europe in order to be competitive to disposal costs. For this reason, as well as for ecological reasons (recycling), we produced a number of different product from the molten slag and evaluated the respective recycling potential: Mineral wool, cement additive, gravel substitute for road construction and ceramic tiles. Finally, our efforts were focused on the production of foam slag for the following reasons:

- it can be produced directly from the out-flowing melt, using an add-on technology
- sale profits are considerably high (at least 30 Euro /m^3)
- a huge market is available

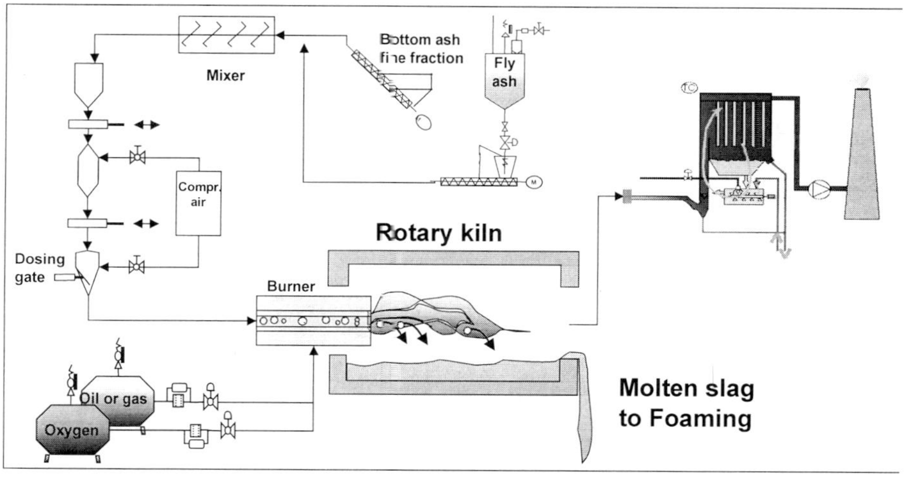

Figure 2 Rotary kiln melting process variant: Fly ash and fine fraction of bottom ash are transported in dense flow directly into the flame of an O_2 burner heating a rotary kiln.

A similar product, made from cullet, is already successfully marketed by a Swiss company.

Main uses include

- thermal insulation, fireproof up to high temperatures
- substitute for gravel, where light-weight material is needed

PRODUCT QUALITY: LAB AND PILOT PLANT RESULTS

Dry Sorting of Ash

As mentioned, dry bottom ash enables separation of the fine particles, different from wet bottom ash. This opens a large potential for quality improvement, as shown in Figure 3, where the relative concentration of several ash components are shown as a function of the particle size.

It can be seen, that by separation into a fine particle fraction and a gravel size fraction, separating i. e. at 2 mm grain size will result in a gravel fraction with reduced toxic potential:

TOC (Total organic carbon) represents the content of potentially toxic organic compounds. Chloride and sulphate tend to leach, which is not desired. Reduction of the chemically active lime reduces the alkalinity of the main fraction. As lead and zinc become insoluble at lowered pH values, their leaching is thus effectively prevented.

Unfortunately, not all toxic components can be separated into the fine fraction, due to agglomeration (sintering) of fine particles as one reason. Therefore, different combinations of mechanical treatment these ashes have been investigated during this project, to find the most

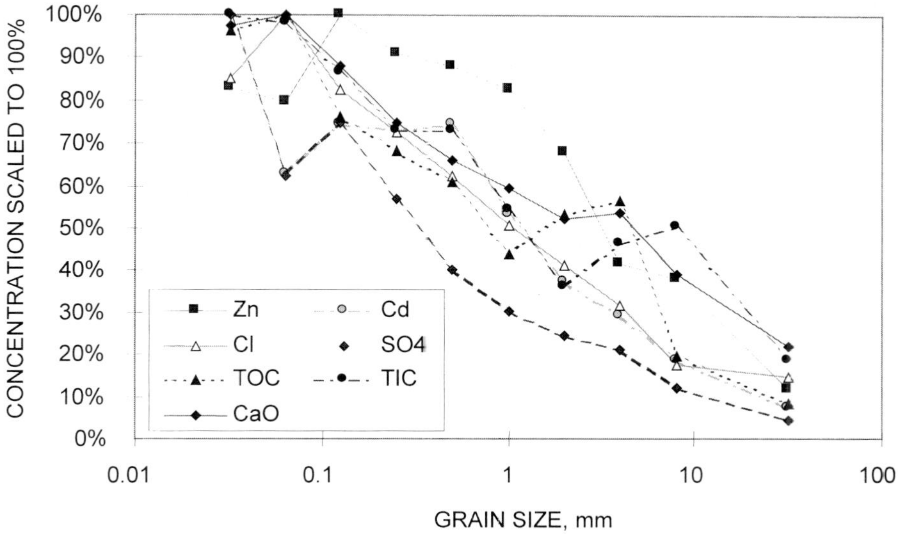

Figure 3 Fine fraction enrichment of seven ash components

effective set-up. The following table shows the mass balance and heavy metal content of the main bottom ash fractions produced during a pilot test campaign of dry extraction and sorting systems:

Table 2 Heavy metal content of bottom ash and InRec fractions, compared to the Swiss TVA legislation for inert materials

		BOTTOM ASH TOTAL < 32 mm	SCRAP IRON (NO COARSE SCRAP INCL.) < 32 mm	NON-FERROUS METALS < 32 mm	MINERAL FRACTIONS		CH-TVA
					Fine part < 2 mm	Gravel size 2..32 mm	Inert Material
Mass	W-%	100%	12%	1.7%	28%	58%	
Pb	mg/g	1.5			3.5	0.9	0.5
Zn	mg/g	3.0		50	4.5	1.6	1
Cu	mg/g	6.7		200	4.5	3.5	0.5
Cd	mg/g	0.008			0.020	0.005	0.01
Ni	mg/g	0.10			0.15	0.09	0.5

In this case, only fine particle separation and removal of ferrous and non-ferrous metals was applied. The very low heavy metal levels defined in Swiss legislation for inert materials are not met. However, in all other European countries, heavy metal leaching rather than content is regulated.

Of the large number of available technologies and possible combinations of mechanical sorting equipment the following process has been identified as best in terms of separation performance at reasonable costs:

1. Separation of over-sized material with a roller grate.

2. Separation of a light fraction with a so-called zig-zag-sifter containing light ash, unburned carbon, salts etc.

3. Magnetic separation of ferrous metals. A combination of at least two systems, i.e. over-band magnet plus head drum is recommended.

4. Eddy-current separation of non-ferrous metals.

At this point, the bottom ash quality is high enough to allow for immediate recycling as construction material according i.e. to the German LAGA Z2 quality standards.

Depending on the amount of oversize ash, but also on the desired use of this ash, we recommend oversize scrap removal combined with impact milling, as shown in Figure 1. This way, most of the oversize is recovered as scrap or gravel.

Further steps can be applied to further reduce the heavy metal content of the main fraction:

5. Milling (breaking) of bottom ash to sizes < 1 mm or less. Metals (Cu, Zn) are separated by sieving due to their ductile behaviour.

6. Optional advanced separation steps based on density differences between metals and mineral matter, including the possibility of more efficient but also more costly wet sorting systems.

If step 5 is included, the process allows to separate ca. 75% of the total copper content and over 60% of the total zinc content into the non-ferrous metals fraction. The relative amount of copper in the different fractions is shown in figure 4.

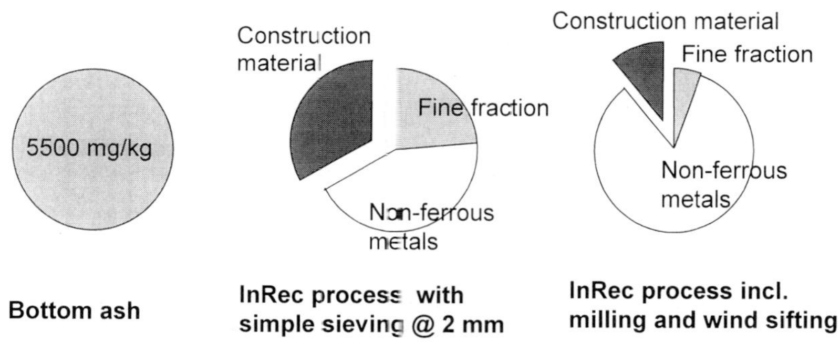

Figure 4 Distribution of copper into the different InRec fractions

The remaining construction material fraction still contains heavy metals in concentrations roughly twice what is required for the Swiss TVA inert material quality (numbers in parenthesis); with 1200 mg/g Cu (500), 1500 mg/g Zn (1000) and 1300 mg/g Pb (500).

Vitrification

Oxidative vitrification of fly ash in a glass furnace ("Deglor") produces a very homogenous, leaching resistant glass product [5]. A commercial Deglor plant for the vitrification for fly ash and residues from flue gas cleaning is in operation in Japan since 1996. Over 2000 t of fly ash have been vitrified. Furnace lifetimes of about 1.5 years have been reached. Basically all the Cd and Pb and about half of the Zn and Cu is transferred to the secondary fly ash. Since residual heavy metals in the slag are chemically bound in the silicate matrix, the glassy slag meets all international leaching test requirements.

Although it does not significantly influence the leaching rates, for true sustainable and unlimited reuse it may be advantageous to further reduce the *content* of heavy metals in this product. Also, for reuse in Switzerland it is necessary to reduce the heavy metal content below the above mentioned limits for inert materials. In oxidative vitrification processes (Deglor or rotary kiln in oxidative mode) for treatment of fly ash the Zn limit and for treatment of bottom ash the Zn and Cu limit cannot be met. no is included in the silicate matrix, which results in a strongly reduced evaporation [6].

The respective low heavy metal contents can be realised using a reductive melting process. #

The main chemical reactions responsible for heavy metal depletion are

$$M \text{ x } ZnO + CO(g) <==> M + Zn(g) + CO_2(g)$$

and

$$M \text{ x } CuO + CO(g) <==> M + Cu + CO_2(g),$$

where M x ZnO and M x CuO represents the glassy slag matrix (mixture of oxides of Si, Al, Ca, Na, K, Fe etc.) with the included heavy metal oxides.

According to these equations, Zn is evaporating and leaves the furnace with the off-gas, while the metallic copper is transferred into a metal phase below the glassy melt which has to be present for effective Cu content reduction.

The retention time required to achieve the desired reduction in heavy metal content is strongly dependent on the temperature, on the transport processes to the melt surface and on the redox conditions. The rotary kiln is a suitable component to realise an effective mixing of the melt, leading to a rapid transport to the surface. Figure 5 gives the results of a set of experiments performed on a 50 kg scale at 1500 °C to determine the speed of the Zn-evaporation as a major input for the design of the process. It shows the Zn-content in the melt as function of the treatment time under four different reductive conditions. The respective melts had been produced from fly ash under oxidising conditions leading to a "final" Zn-content of about 1%, as a start value for the evaporation test. It can be concluded, that the reduction of the Zn-content can be realised with a characteristic (exponential) time constant of about 30 minutes. This is much faster than in basically unmixed reactors [7].

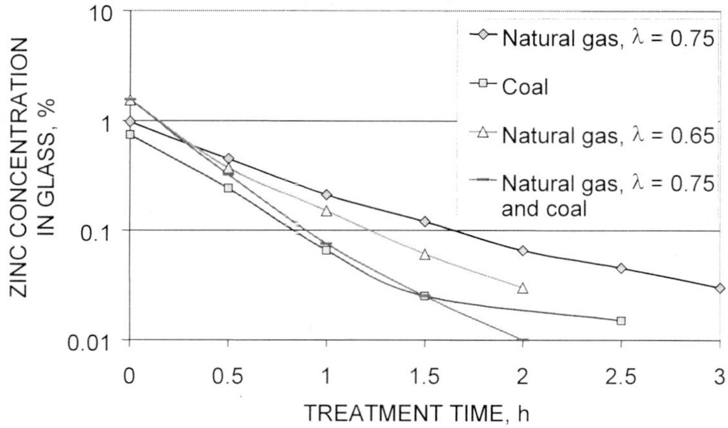

Figure 5 Zn depletion with time at 1500 °C for different conditions (λ = oxygen to fuel ratio in the burner)

We performed similar experiments for Cu reduction with a metal phase. Based on these and other experiments we have all necessary data available required for the design of InRec^{plus}-vitrification plants basically for all types of inputs and fuels.

Foam Slag

We have developed a process to produce a foam directly from a flow of molten slag. The basic principle is to disperse the slag stream using an intense high velocity gas stream and to add a foaming agent in form of a fine powder to the dispersed slag. This mixture foams on a surface and the resulting light foam is frozen due to simultaneous cooling. The resulting foam slag fulfils the requirements for use as light-weight aggregate for thermal insulation purposes in terms of low density (below 500 kg/m^3), high mechanical strength, closed pores and so on.

Based on numerous tests we derived the chemical and physical requirements of a foaming process resulting in foam slag of the desired quality. Depending on the composition and properties of an outflowing slag flow, conditioning with customised additives may be required to prepare the slag for foaming. In addition to Deglor or InRec^{plus} slags, the process may be applied to numerous existing furnaces producing slag flows, as well.

MASS BALANCE AND COST STRUCTURE OF AN INREC REFERENCE SYSTEM

For a waste incineration plant treating 200'000 tons per year (25 t/h; 600 t/d), a mass balance like displayed in Figure 6 will result. The numbers indicate mass flow in t/y. For simplification, fuel and additive consumption is not included. For vitrification of 1000 kg of mixed ash, 65 kg of oil and 180 Nm^3 of oxygen will be consumed. For flue gas cleaning and foaming, ca 50 kg of lime and other additives are necessary.

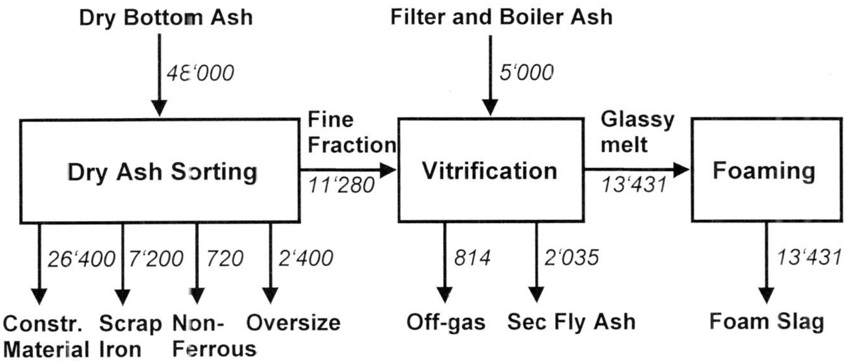

Figure 6 Mass flow of an InRec^{plus} reference system. Numbers indicate tons/year

The estimated cost and revenue scheme for an average European market, based on the same 200'000 t/y reference plant is given in Table 3. The investment cost for a plant including all InRec^{plus} process steps will be ca 10 MEuro. The resulting capital costs are included in the table, as well as personnel, operation, maintenance, energy and additives.

Table 3 Cost and revenue scheme for a reference InRec^{plus} plant in Euro (€).

PROCESS/PRODUCT	T/Y	EURO/T	KEURO/Y	REMARK
Dry Ash Sorting	48000	18	864	Bottom ash
Vitrification	16280	200	3256	Fly ash + fine fraction
Foaming	16280	25	407	directly from melt
Constr. Material	2400	0	0	assmpt.: no revenue
Scrap Iron	7200	-40	-288	revenue !
Non-ferrous Metals	720	-250	-180	valuable product !
Oversize Fraction	2400	10	24	cheap landfill or reuse
Sec. Fly Ash	2035	200	407	haz. waste landfill
Foam Slag	13431	-60	-805	revenue !
Total mixed ash	53000	69.51	3684	Compare to landfill

All numbers depend on the region and the ever changing market conditions and should only be considered as indication. It is clearly shown that vitrification is the major cost factor, although only less than $^1/_3$ of the total ash is vitrified. Therefore, we recommend to consider vitrification only for the fraction of the ash with high toxic potential. For much larger plants, specific costs will be considerably lower, as the capital costs are most important.

CONCLUSIONS

Overall, the InRec^{plus} approach optimises ash treatment economics and ecology likewise:

- recovery of metals in unprecedented quality
- customised generation of gravel from bottom ash
- vitrification only for selected fractions of special concern
- low vitrification cost due to improved rotary kiln technology
- foam slag revenues further decrease treatment costs.

However, the current trends in Europe have created a paradox situation, which basically prevents advanced ash treatment and reuse on most markets:

If more waste is incinerated, the waste landfills get less business. This has created a fierce price competition on the waste and residue market. True recycling of ashes and ash components is only competitive if it is kept very simple or if the treatment plants are already depreciated.

Additionally, storage of untreated ashes in old mines is classified as reuse in Germany. At the same time, the volume of these cavities as well as the competition grows, which turned formerly expensive mine filling now very cheap. Most of the surrounding countries are also taking advantage of this disposal opportunity.

Hopefully, the terms "reuse" and "recycling" will become more clearly defined in the future. Ultimately, advanced ash treatment has to become a corner stone of waste management, if sustainability remains an important goal of modern societies.

ACKNOWLEDGEMENTS

Part of this work has been supported by the Swiss National Science Foundation within the Priority Programme Environment, which is greatly acknowledged.

REFERENCES

1. J. J. J. M. GOUMANS, H. A. VAN DER SLOOT AND TH. G. AALBERS (Eds.), "Environmental Aspects of Construction with Waste Materials", Studies in Environmental Science 60, Elsevier Amsterdam (1994).

2. J. J. J. M. GOUMANS, G. J. SENDEN AND H. A. VAN DER SLOOT, "Waste Materials in Construction - Putting Theory into Practice", Studies in Environmental Science 71, Elsevier Amsterdam (1997).

3. J. IORI, J. BALG AND CH. WIECKERT, "Detoxification of municipal waste incineration residues by vitrification", ABB Review 6/7 (1995), 9-16.

4. A. SELINGER, V. SCHMIDT, CH. WIECKERT, CH. STEINER UND H. RÜEGG, "Advanced Treatment of Waste Incineration Ashes", in A. Barrage und X. Edelmann (Eds.), R'99 Recovery, Recycling, Re-integration Congress Proceedings, EMPA Dübendorf (1999) S. II.234-II.239.

5. D. TRABER, U. MÄDER, U. EGGENBERGER, F.-G. SIMON AND CH. WIECKERT, "Phase Chemistry Study of Products from the Vitrification Processes AshArc and Deglor", Glass Sci. Technol. 72 (1999), 91-98

6. S. STUCKI AND A. JAKOB, "Thermal treatment of incinerator ashes: Factors influencing the evaporation of $ZnCl_2$", Waste Management 17, 231-236 (1997).

7. C. ZELTNER, "Petrologische Evaluation der thermischen Behandlung von Siedlungsabfällen, Diss. ETH Zurich Nr. 12688, 1998.

TRAPPING OF CHLORIDE, SULFATE AND HEAVY METALS

O Barbery

A Ghodsi

University of Brussels

Belgium

ABSTRACT. Bottom ashes coming from house waste incineration (MSWI) could be reused as under-layer for road construction if they satisfy enforced standards limiting heavy metals, chlorides and sulfates content in the leachate. We set up two processes for the inerting of MSWI : the first one consists in using anion traps, while the second one is an encapsulation. The leaching tests used are the German standard DIN 7348-S4 and the Dutch standard NEN 7343. The results shows that the finest fraction (less than 2 mm) constitutes generally only 20% of the total, but it contains the majority of pollutants. That's why we decided to study this fraction. After separation of the MSWI, the fraction < 2 mm is encapsulated with a reactive silicone. In this way, we satisfied the Walloon decree and the heavy metals are totally blocked. The first methodology allows to decrease chlorides and sulfates content in the leachates. It consists in mixing small quantities of Zn, Mg and Al hydroxycarbonate with MSWI. This molecule traps chlorides and sulfates within its structure. Both processes are easy to carry out and allows to obtain low values of heavy metals, chlorides and sulfates content in the leachates.

Keywords : MSWI, Bottom ash, Hydrotalcite, Heavy metal.

Ir Olivier Barbery, is a civil engineer in chemistry from the University of Brussels. After studying the synthesis of hydrotalcite for heat stabilisation, he works with Watco and the Walloon Region on a research program concerning the stabilization of chlorides, sulphates and heavy metals in the bottom ashes. Now he is working as technical responsible at CRIC, Research Centre for the Cement Industry in Belgium.

Ir Alain Ghodsi, is a civil engineer in chemistry from the University of Brussels. He's First Attaché at the Environmental Ministry for the Walloon Region of Belgium, responsible for research. He's also industrial collaborator at the ULB's Civil Engineering Institute.

INTRODUCTION

The production of wastes has become a significant economic and environmental issue. Waste production increases yearly. Disposal may be harmful to human health, by causing land, air or groundwater pollution.

Incineration is a common method used to treat urban wastes. The incineration reduces the volume of waste and produces residues and especially bottom ashes (10% of the initial waste) that could be recycled in the construction sector thanks to their composition.

Figure 1 Bottom ashes

Bottom ash is a grey granular material, containing a lot of pieces of glasses (Figure 1). The chloride, sulphate and heavy metals content present environmental problems. The main difficulty is to inert these bottom ashes.

In Belgium, waste management is within the competence of the Regions.

To prevent the use of bottom ash, the Walloon Region has promulgated a decree giving threshold values of heavy metal and anions to be satisfied by these ashes.

The project aims at studying and providing solutions to improve leaching behaviour of bottom ashes. This study tackles specifically with the presence of chlorides, sulphates and heavy metals.

MATERIALS AND EXPERIMENTAL PROCEDURE

Leaching tests

A lot of leaching tests have been developed. To be in accordance with the Walloon decree, we have selected two tests:

- The Dutch test, NEN 7343
- The German test, DIN 7348-S4

The Table 1 gives the threshold value imposed by this decree for each leaching tests.

Table 1 Threshold value

	DIN S4	NEN 7343
Cl^-	500 mg/l	6000 mg/kg
$SO_4^=$	1000 mg/l	4000 mg/kg
As	0.1 mg/l	0.8 mg/kg
Cd	0.1 mg/l	0.03 mg/kg
Cr	0.1 mg/kg DM	0.5 mg/kg
Co	0.1 mg/l	0.25 mg/kg
Cu	2 mg/l	5.0 mg/kg
Hg	0.02 mg/l	0.02 mg/kg
Pb	0.2 mg/l	2.2 mg/kg
Mo	0.15 mg/l	1.8 mg/kg
Ni	0.2 mg/l	1.8 mg/kg
Zn	0.9 mg/l	4 mg/kg

(DM : dry material)

The decree recommends two tests:

- The column test (Dutch test) [1]
- The German test [2]

The *column test* consists in the circulation of a solution of desionized water and nitric acid at a pH of 4 through a column containing the volume m0 of bottom ashes. The bed height of bottom ashes must be at least 4 times the internal column diameter (i.e. 50 mm). The maximum flow of the solution is 50 ml/hour. The leachate is collected at the column top in 7 fractions. The Table 2 shows the collection of fractions.

Table 2 Fraction in the column test

FRACTION	FRACTION VOLUME (L)	L/S (l/kg)
K1	$(0.1 \pm 0.01) \times m_0$	0.1 ± 0.01
K2	$(0.1 \pm 0.01) \times m_0$	0.2 ± 0.02
K3	$(0.3 \pm 0.03) \times m_0$	0.5 ± 0.05
K4	$(0.5 \pm 0.05) \times m_0$	1.0 ± 0.1
K5	$(1 \pm 0.1) \times m_0$	2.0 ± 0.2
K6	$(3.0 \pm 0.3) \times m_0$	5.0 ± 0.5
K7	$(5.0 \pm 0.5) \times m_0$	10 ± 0.02

The total ratio liquid /solid equals 10.

The concentrations are obtained by the formula:

$$C_{i,k} = (L_k / m_0) \times (C_i / \mu) \quad (1)$$

Where

L_k = volume of the fraction k in litre
m_0 = mass of the dry bottom ashes in kg
C_i = concentration of the i species in µg/l
μ = correction factor = 1000 µg/mg

For each element, the fraction's sum gives the amount of pollutant in the bottom ashes.

The *German test* consists in mixing 100 gr of bottom ashes with 1 litre of desionized water. The ratio L/S must be equal to 10. The mix is agitated 24 hours with minimum one round each minute. After 24 hours, the mix is filtered and the leachate is analysed by the conventional method. This test is easier to perform than the column test.

MSWI

The Bottom ashes used in this study have been produced by the incineration of wastes coming from a city. The bottom ashes were analysed by fluorescence X using a semi-quantitative analysis. These results give an indication of the potentiality of a compound to be polluting (see Table 3)

Table 3 Composition of the bottom ashes by x-ray fluorescence

% IN WEIGTH	RESULT 1	RESULT 2	RESULT 3	AVERAGE
Cl	0.614	0.632	0.627	0.624
As	0	0	0	0
Cd	0	0	0	0
Cr	0.0576	0.0524	0.0556	0.055
Co	0	0	0	0
Cu	0.183	0.176	0.17	0.176
Hg	0	0	0	0
Pb	0.704	0.732	0.602	0.679
Mo	0	0	0	0
Ni	0.031	0.031	0.031	0.031
Zn	0.214	0.211	0.217	0.214

The two leaching tests have been performed on these bottom ashes. The results are given in Tables 4 and 5.

The results obtained with the two leaching tests are quite similar. In both tests, the lead contents are beyond the threshold values. The copper contents obtained with the Dutch test are four times higher than the threshold values. That can be explained by the use of an acidic solution for the Dutch test. Indeed, as shown by the Figure 3, the pH has a great influence on the solubility of heavy metals. The chlorides and sulfates contents obtained are close to the limits.

Table 4 Results of the German test on the bottom ash without treatment

PPM	RESULT 1	RESULT 2	RESULT 3	AVERAGE	LIMITS
Cl^-	290	270	284	282	500
$SO_4^=$	600	580	589	590	1000
As	<0.1	<0.1	<0.1	<0.1	0.1
Cd	<0.1	<0.1	<0.1	<0.1	0.1
Cr	<0.1	<0.1	<0.1	<0.1	0.1
Co	<0.1	<0.1	<0.1	<0.1	0.1
Cu	1.78	1.76	1.76	1.77	2
Hg	0.02	<0.02	<0.02	<0.02	0.02
Pb	0.86	0.8	0.83	0.83	0.2
Mo	<0.15	<0.15	<0.15	<0.15	0.15
Ni	0.05	0.05	0.05	0.05	0.2
Zn	0.15	0.15	0.15	0.15	0.9

Table 5 Results of the Dutch test on the bottom ash without treatment

MG/KG	RESULT 1	RESULT 2	RESULT 3	AVERAGE	LIMITS
Cl^-	4967.6	4189.2	3977.5	4378	6000
$SO_4^=$	1277.3	822.7	938.6	1013	4000
As	<0.5	<0.5	<0.5	<0.5	0.8
Cd	0.04	0.021	0.017	0.026	0.03
Cr	0.256	0.152	0.151	0.186	0.5
Co	<0.2	<0.2	<0.2	<0.2	0.25
Cu	15.233	10.34	12.12	12.56	5.0
Hg	<0.02	<0.02	<0.02	<0.02	0.02
Pb	12.02	5.97	11.45	9.81	2.2
Mo	0.135	0.094	0.104	0.11	1.8
Ni	0.617	0.602	0.321	0.51	1.8
Zn	0.38	0.44	0.491	0.44	4

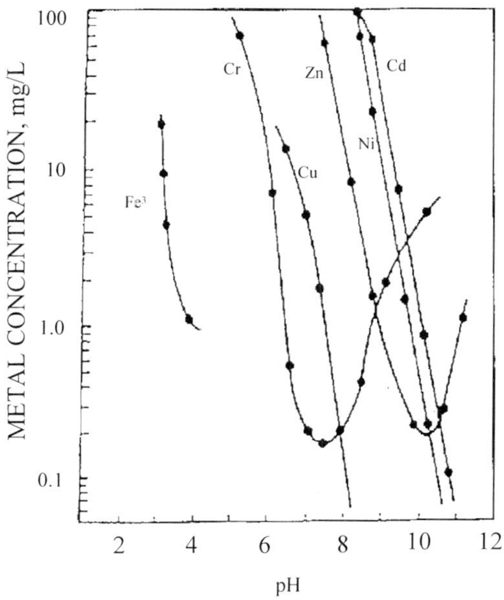

Figure 3 Solubility of heavy metals function of the pH

Sampling

As bottom ashes are a heterogeneous material, the sampling is difficult to realise. The bottom ashes composition depends on a lot of parameters like the season, the months, the place where the incinerator is, etc.

Several sampling methods have been presented by different authors. Two methods have been selected for this study: the Gy's method and the sampling by quartation.

The Gy's [3] method consists in evaluating visually the x components present in the ashes and their concentration. X sampling weight are estimated by a complex formula. The higher and the lower weights are eliminated and the weight of sample to be taken have to be chosen between the x-2 weights. This method is quite difficult to perform and subjective as it depends on a visual estimation.

Another method consists in taking 60 kg of MSWI, in three fractions of 20 kg, in the incinerator. These fractions are divided into four parts of 15 kg, from which two parts are selected and mixed together to obtain 30 kg. The process is repeated to obtain a mix of 15 kg, in which a sample of 300 gr is taken. This method is easier to realize than the Gy's method.

RESULTS

Trapping of chlorides and sulphates [4-5]

Chlorides and sulphates are highly soluble compounds. The idea developed in this study is to trap chloride and sulphates in the structure of a mineral compound. The mineral used is an hydrotalcite compound, composed of layers of Zn, Mg and Al hydroxide in piles separated by mobile intermediary layers of carbonates ($CO_3^{=}$).

This compound has positively-charged brucite-like layers in which some of the divalent cations have been replaced by trivalent anions to obtain compounds like

$$[M^{II}_{1-x}M^{III}_{x}(OH)_2]^{x+}(A^{n-})_{x/n}.z\,H_2O \qquad (2)$$

The hydrotalcite particles are coated with stearine which is an hydrophobic compound Figure 4. This coating allows the hydrotalcite to be protected against aggressive attack without blocking the anions diffusion The hydrotalcite like compound used in this research is a product which is used for thermic stabilisation of PVC.

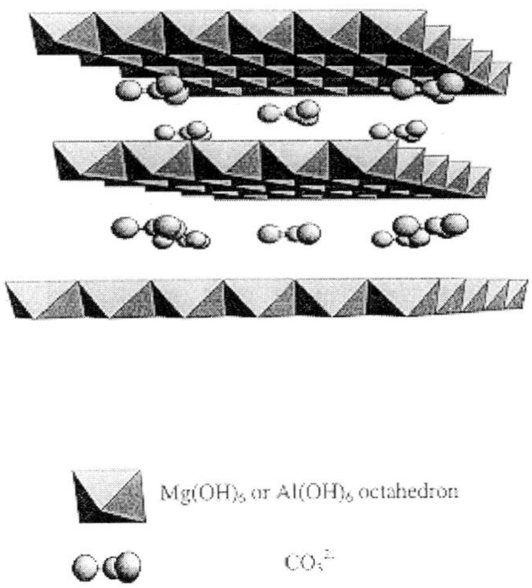

Figure 4 Hydrotalcite like layer

Before realising the two leaching tests, 1% in weight of hydrotalcite is homogeneously mixed in the dry bottom ashes

The sulfates and chlorides content obtained with the German test (DIN S4) are given in the Table 6. The results obtained with the Dutch test are similar. As the hydrotalcite is an anion trap, it doesn't influence the content of heavy metals in the leachate.

Table 6 Bottom ashes with 1%in weight of hydrotalcite. Dosage of chloride and sulfate by HPLC in the leachate by DIN S4

ANIONS	RESULT 1	RESULT 2	RESULT 3	AVERAGE	LIMITS
Cl^-	210	180	195	195	500
$SO_4^{=}$	360	345	346	350	1000

The chlorides and sulphates content are decreased by around 50 %. These results confirm that these anions could have been trapped in the hydrotalcite structure.

Trapping of heavy metals

As using hydrotalcite doesn't influence the heavy metals content in the leachate, it was decided to coat the bottom ashes by a hydrophobic compound, a polymer of methylhydrogenosiloxane. The monomer of methylhydrogenosiloxane reacts with the groups OH or H at the surface of bottom ashes. 1% in weight of siloxane is mixed with the bottom ashes. The mix is placed in a steam room at 60 °C for about 8 hours to obtain a perfect coating by reticulation.

This method was performed on the ungraded bottom ashes. The heavy metals contents obtained in the leachates are similar to those obtained on the non-treated bottom ashes. One hypothesis to explain this observation is that most of the pollutants are found in the finest particles (< 2 mm), which constitute only 20% of the bottom ashes. As a high quantity of siloxane is consumed by the biggest particles, the finest one is not properly coated.

To confirm this hypothesis, the bottom ashes were sifted and separated in two groups, i.e. one part <2 mm and the other > 2 mm. Each group was analysed by x-ray fluorescence. The results are given in the next tables.

Table 7 X-ray fluorescence on the fraction < 2 mm of bottom ashes

% IN WEIGTH	RESULT 1	RESULT 2	RESULT 3	AVERAGE
Cl	0.59	0.62	0.56	0.59
As	-	-	-	-
Cd	-	-	-	-
Cr	0.07	0.09	0.08	0.08
Co	-	-	-	-
Cu	0.66	0.55	0.56	0.59
Hg	-	-	-	-
Pb	0.13	0.15	0.14	0.14
Mo	-	-	-	-
Ni	0.15	0.15	0.15	0.15
Zn	0.3	0.3	0.3	0.3

The results shows that the fraction < 2 mm contains about 90% of the pollutants. Therefore, this fraction needs to be coated, while the fraction > 2 mm doesn't need any treatment.

The method was thus performed on the particles of bottom ashes <2mm. The results obtained with the two leaching tests are given in the next tables (tables 9 and 10).

Table 8 X-ray fluorescence on the fraction > 2 mm of bottom ashes

% IN WEIGHT	RESULT 1	RESULT 2	RESULT 3	AVERAGE
Cl	0.183	0.188	0.187	0.186
As	-	-	-	-
Cd	-	-	-	-
Cr	0.05	0.05	0.05	0.05
Co	-	-	-	-
Cu	0.32	0.31	0.33	0.32
Hg	-	-	-	-
Pb	0.06	0.07	0.08	0.07
Mo	-	-	-	-
Ni	0.05	0.06	0.05	0.06
Zn				

Table 9 DIN S4 on the fraction of bottom ashes < 2 mm + 1 % of siloxane

PPM	RESULT 1	RESULT 2	RESULT 3	AVERAGE	LIMITS
Cl^-	305	322	315	314	500
$SO_4^{=}$	708	715	709	711	1000
As	<0.1	<0.1	<0.1	<0.1	0.1
Cd	<0.1	<0.1	<0.1	<0.1	0.1
Cr	<0.1	<0.1	<0.1	<0.1	0.1
Co	<0.1	<0.1	<0.1	<0.1	0.1
Cu	<2	<2	<2	<2	2
Hg	<0.02	<0.02	<0.02	<0.02	0.02
Pb	<0.2	<0.2	<0.2	<0.2	0.2
Mo	<0.15	<0.15	<0.15	<0.15	0.15
Ni	<0.2	<0.2	<0.2	<0.2	0.2
Zn	<0.9	<0.9	<0.9	<0.9	0.9

Table 10 NEN 7343 on the fraction of bottom ashes < 2 mm + 1 % of siloxane

MG/KG	**RESULT 1**	**RESULT 2**	**RESULT 3**	**AVERAGE**	**LIMITS**
Cl^-	865	889	868	874	6000
$SO_4^{=}$	1135	1149	1160	1148	4000
As	<0.8	<0.8	<0.8	<0.8	0.8
Cd	<0.03	<0.03	<0.03	<0.03	0.03
Cr	<0.5	<0.5	<0.5	<0.5	0.5
Co	<0.25	<0.25	<0.25	<0.25	0.25
Cu	<5.0	<5.0	<5.0	<5.0	5.0
Hg	<0.02	<0.02	<0.02	<0.02	0.02
Pb	<2.2	<2.2	<2.2	<2.2	2.2
Mo	<1.8	<1.8	<1.8	<1.8	1.8
Ni	<1.8	<1.8	<1.8	<1.8	1.8
Zn	<4	<4	<4	<4	4

With both tests, the heavy metals contents are below the detection limits and thus below the threshold value of the Walloon decree.

The chlorides contents obtained with the Ducth test are decreased by 75%. But concentrations obtained with the German test are similar to those of the non-treated bottom ashes (table 4). The German test is a dynamic test. An abrasion between the particles take place, what could damage the coating. The coating of biggest particles could be preferentially broken, what could liberate the chlorides. This assumption could explain the difference observed between the two tests.

CONCLUSIONS

This study proposes solutions to inert bottom ashes in order to allow their use in road construction. The main difficulty is the high content of chlorides, sulfates and heavy metals found in the leachate of these bottom ashes.

To realize this study, we use two leaching tests: the German test and the Dutch test. This research was carried out in two parts.

The first part tackles with the presence of chlorides and sulfates in the leachate. The solution proposed in this paper consists in trapping the chlorides in the structure of a mineral compound, hydrotalcite. The contents obtained with the two leaching tests shows are decreased by around 50%. But, this product doesn't influence the content of heavy metals.

The second part of this paper deals with the heavy metals content in the leachate. It was suggested to coat the bottom ashes with a hydrophobic compound, siloxane. It was shown that 90% of the pollutants are found in the finest particles (< 2 mm). What explained that coating the ungraded bottom ashes doesn't reduce the content in heavy metals. A separation of the particles have to be performed and only the finest particles have to be coated. The coating of these particles was realised and the results shows that this method allows the content of heavy metals to be reduced below the threshold values of the Walloon decree.

It can be concluded that the two methods proposed give encouraging results. They are quite easy to perform and relatively cheap. Indeed, hydrotalcite can be produced by using salts of Zn, Mg and Al and carbonates. The pH of the solution must be equal to 11.5. The price of this product is around 1.5 US$/kg while the siloxane is more expensive (about 50 US$/kg). The solution to perform have to be chosen depending on the bottom ashes composition and the enforced regulations.

ACKNOWLEDGMENTS

The authors would like to acknowledge the financial support provided for the project by the Walloon Region as parts of a First project (n°9713618 - VISA 97 / 52341) and the Valomac company.

REFERENCES

1. NEN 7343, Leaching characteristics of solid earthy and stony building and waste materials. Leaching tests. Determination of the leaching of inorganic components from granular material with the column test.", 1er edition, february 1995, ICS 13.030-70;91.100, Nederlands Normalisatie-instituut.

2. "DIN 7348- S4, Leaching characteristics of solid earthy and stony building and waste materials. Leaching tests. Determination of the leaching of inorganic components from granular material with the column test.", Germany.

3. "Théorie de l'échantillonage", Jean-Louis Pineau, Le coin des spécialistes, 1975, pp59-62.

4. "Synthesis and Characterization of hydrotalcite-like compounds containing V3+ in the layers, and of their calcination products", F.M. Labajos, V. Rives, Inorganic Chemistry 35 (1996), pp 1154-1160.

5. "Hydotalcite – like double layer Structure", R. Allman, Institute fur kristallographie.

TREATMENT OF APC RESIDUES FROM MSW INCINERATORS

O Hjelmar

H Birch

J B Hansen

DHI – Institute of Water and Environment

Denmark

ABSTRACT. A process for treatment of semidry and dry APC system residues from MSW incinerators based on an initial aqueous extraction followed by re-suspension and stabilisation of the filter cake with carbon dioxide and/or phosphoric acid was developed in 1994 - 1997 and was shown to produce a material with vastly improved disposal properties. This paper summarises some of the earlier results and presents some preliminary results of a new ongoing R&D project aimed at optimising and preparing the process for full-scale implementation and extending the applicability of the process to include fly ash and wet scrubber sludge. The results of leaching tests performed on residues treated at a new large-scale pilot plant indicate that good results may be obtained for most contaminants using carbon dioxide and phosphoric acid, both for semidry residues and fly ash. Using CO_2 alone also yields good results, but the leachability of Pb is higher at low pH values. Further work has been initiated to improve the stabilised residues and reduce the leaching of SO_4^{2-}, Hg and some oxyanions.

Keywords: Air pollution control system residues, Waste incineration, Treatment, Extraction, Stabilisation, Disposal, Leaching, Trace elements, Salts, Environmental impact.

Mr O Hjelmar, is Chief Engineer at DHI-Institute of Water and Environment in Hørsholm, Denmark. He specialises in waste characterisation, development, application and interpretation of waste leaching tests, waste treatment technology, and development of sustainable utilisation and landfilling strategies. He is Director of the Waste Research Centre (C-RES), a joint R&D co-operation project between universities, R&D institutions and industry for development of strategies, treatment methods, cleaner technologies, testing tools and models, which can enhance sustainable utilisation and landfilling of waste.

Mr H Birch, is a senior Chemical Engineer at DHI-Institute of Water and Environment in Hørsholm. He specialises in industrial and chemical process technology related to treatment and reuse of water and waste streams, cleaner technology and savings of energy and resources. He is also associated with C-RES.

Dr J B Hansen, is an Environmental/Chemical Engineer at DHI-Institute of Water and Environment in Hørsholm, Denmark. She finished her ph.d. thesis on mobility of heavy metals in landfill leachate at the Technical University of Denmark in 1999. Since then she has worked with characterisation and treatment of waste and validation of pre-standard leaching tests. She is also associated with C-RES.

INTRODUCTION

Solid residues from air pollution control (APC) systems at municipal solid waste (MSW) incinerators constitute a serious and increasing waste management problem in several industrialised countries. The problem of environmentally safe management of these residues, particularly the residues from the semidry and dry acid gas cleaning processes and the fly ash, has not yet been solved in a satisfactory and sustainable manner. The residues are in many cases simply stored indefinitely in big bags or they are landfilled under conditions, which in the long term cannot prevent unacceptable impacts on the environment.

The APC residues, including fly ash, are in many countries as well as in EU legislation classified as hazardous waste due to their high contents of trace elements/heavy metals and soluble salts. Some of the APC residues (those from the dry/semidry lime injection processes) are strongly alkaline due to a substantial content of excess lime, and the high pH favours the leaching of several contaminants, particularly lead. Therefore, treatment is or will be necessary and/or required prior to landfilling (or utilisation) of most APC residues from MSW incinerators.

Substantial efforts have been spent developing treatment processes, which can render these residues more suitable for disposal or even enable beneficial use of the residues. The fact that several of these processes for treatment of dry and semidry APC residues include or are preceded by a washing or extraction step reflects the general acknowledgement that the large content of soluble salts constitutes a major obstacle to the attainment of an environmentally stable treatment product. One treatment process, which resulted in strongly improved environmental properties of the treated residues, was developed and its feasibility proven in a laboratory and pilot scale study, which was started in 1994 and finalised in 1997 [1,2]. This paper summarises some of the earlier results and presents some new preliminary results of an ongoing programme for further development of that process.

THE RESIDUES

The most common APC system residues from MSW incinerators are those from the so-called dry and semidry lime injection processes and the wet scrubbing process. In some countries the wet and the semidry processes are combined to yield only solid residues.

The residues from the dry and semidry processes consist of a mixture of fly ash, unreacted excess lime and reaction products, primarily calcium chloride, from the acid gas removal system. The material is collected as a fine powder in electrostatic filters or fabric filters. The difference between the dry and the semidry processes are that the lime is injected or sprayed into the flue gas as a powder in the dry process and as a slurry in the semidry process. The dry process is usually somewhat less effective than the semidry process and for the same concentration of acid gases in the flue gas a higher stoichiometric ratio of lime is required to reach the same acid gas removal efficiency as the semidry process would. The residues from the dry process therefore generally have a higher content of excess lime than the residues from the semidry process. Sometimes the fly ash is collected separately upstream of the acid gas cleaning system. In those cases two streams of solid waste are collected: one consisting of fly ash and another consisting of reaction products (mostly soluble salts) and excess lime. The separately collected fly ash typically contains 20 to 25 percent (by weight) of readily soluble salts, mostly chlorides but also sulphates of sodium, potassium and calcium. The admixtures of dry/semidry residue and fly ash typically contain 30 to 35 percent (by weight) of the same soluble salts, but dominated by calcium chloride (or calcium hydroxychloride, CaOHCl, H_2O). With few exceptions, and most notably among these mercury, most of the trace elements/heavy metals present in the residues are mainly associated with the fly ash matrix [3].

In the wet process the fly ash is always pre-collected upstream of the wet scrubber(s). In addition to the fly ash, the process yields a small stream of sludge from treatment of the scrubber effluent and a stream of treated wastewater containing most of the soluble salts. The purpose of the wastewater treatment process is to reduce the content of trace elements/heavy metals in the wastewater to a low level prior do discharge to the sewer or to a recipient. The sludge contains the precipitation agents (often TMT, trimercaptotriazine), trace elements/heavy metals, gypsum, carbonates, hydroxides and other salts. Gypsum is sometimes collected in a separate stream. In some cases, the wastewater treatment sludge is mixed with the fly ash prior to landfilling.

The leachability of some of the components from the fly ash, the dry/semidry residues (with fly ash included) and one example of wet scrubber sludge mixed with fly ash is illustrated in Table 1. The data are results of accelerated laboratory leaching tests carried out as combined column and batch leaching tests (first a column test representing a liquid to solid ratio (L/S) of 0-2 l/kg, then the material was removed from the columns and subjected to a serial batch test representing L/S = 2-5 l/kg and 5-25 l/kg). The table shows that the leachability of salts and most of the trace elements is highest from the dry semidry residues and the fly ash and substantially lower from the mixture of wet scrubber sludge and fly ash.

Table 1 Amounts of selected compounds (in mg/kg) leached from APC residues from MSW incinerators in combined column and batch leaching tests at an accumulated L/S = 25 l/kg. TDS = total dissolved solids. NVOC = non-volatile organic carbon. After [3].

COMPONENT	FLY ASH	DRY/SEMIDRY RESIDUES	WET SCRUBBER SLUDGE + FLY ASH
TDS	210,000 – 230,000	290,000 – 380,000	140,00000
Ca	9,100 – 24,000	61,000 – 109,000	17,000
Na	23,000 – 30,000	12,000 – 17,000	21,000
K	28,000 – 50,000	17,000 – 29,000	21,000
Cl^-	89,000 – 106,000	116,000 – 200,000	56,000
SO_4^{2-}	13,000 – 41,000	470 – 3100	32,000
As	0.10 – 0.19	< 0.02 – 0.04	0.28
Cd	0.051 – 35	0.03 – 0.44	< 0.0006
Cr	0,053 – 0.12	< 0.6 – 2.4	0.52
Cu	0.025 – 0.31	0.13 – 22	< 0.004
Hg	< 0.013	< 0.003	< 0.001
Mo	7,1 – 14	1.1 - 3.0	12
Ni	< 0.13	< 0.2	< 0.2
Pb	17 – 370	220 – 3,400	< 0.0011
Zn	< 1.4 – 580	45 – 340	0.15
NVOC	14 – 46	71 - 780	78

THE TREATMENT PROCESS

The process (henceforth referred to as the VKI process) has been developed for treatment of residues from the dry and semidry APC processes (with fly ash admixed with the reaction products) but it may also be applied to the residues from the wet scrubbing process (fly ash and sludge). It is based on an initial aqueous extraction of the residues and a subsequent stabilisation of the washed filter cake with carbon dioxide and/or phosphoric acid. The carbon dioxide source could e.g. be the cleaned flue gas. A diagram of the process is shown in Figure 1.

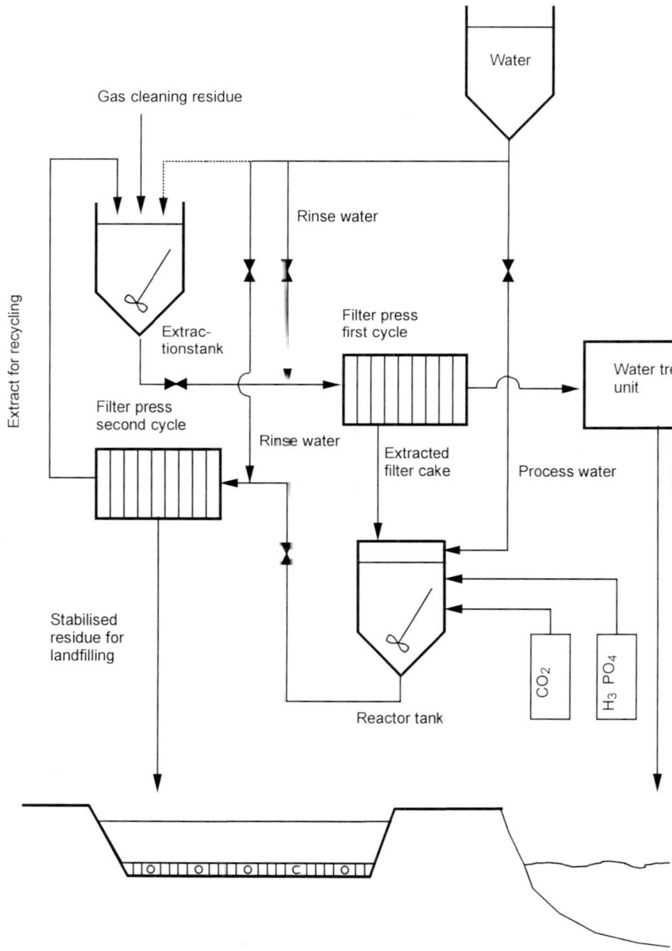

Figure 1 Diagram showing the principle of the VKI treatment process for APC system residues from MSW incinerators.

The process was targeted at producing at treated product with a good acidity buffering capacity, a neutral to slightly alkaline pH in contact with water and a low content of soluble salts and trace elements/heavy metals. This was largely achieved during the initial process development period: The original pH of 11.6 of the semidry APC residue was reduced to 10.3 after treatment, and testing of the treated residue exhibited a very substantial reduction of the leachability of salts and most heavy metals/trace elements as compared to the untreated residue. An acid neutralisation capacity of 2.0 eqv/kg (titration to pH = 7.0) was retained. The results indicate that these residues in all probability can be landfilled safely. It has further been shown that the wastewater from the process can be treated by state-of-the-art techniques (pH adjustment and TMT addition followed by flocculation and separation) and subsequently discharged to a municipal wastewater treatment plant or a marine recipient [1,2].

The effectiveness of the treatment may be illustrated by a scenario calculation based on the results of a batch leaching tests performed on untreated and treated APC system residue from a semidry process. The scenario is a landfill of 100 m x 100 m surface area, average depth = 10 m, containing 130000 tonnes (dry weight) of semidry APC residue. The annual rate of infiltration of precipitation is estimated at 250 mm, and the results, which are shown in Table 2, are the total amounts of contaminants released with the leachate from untreated and treated residues, respectively, over a period of 100 years. The scenario is a very simplified one, and ageing effects etc. are not taken into account. The annual amount of leachate produced under these conditions is 2500 m^3.

Table 2 Estimates of the amounts of selected components released with the leachate from a landfill containing 130000 tonnes of semidry APC residues over a period of 100 years. Landfill size: 100 m x 100 m x 10 m.

PARAMETER	UNTREATED RESIDUE TONNES	TREATED RESIDUE TONNES
Chloride	23000	7.8
Sulphate	350	26
Ca	9400	9.9
Na	2400	4.2
K	5100	3.1
As	< 0.02	< 0.0003
Cd	0.005	0.00003
Cr	< 0.003	0.016
Cu	0.94	< 0.0003
Ni	< 0.01	< 0.0003
Pb	200	0.00012
Zn	6 2	0.016

From Table 2 it can be seen that the treatment substantially reduces the expected release of several of the contaminants (most notably chloride, Ca, Na, K, Pb, Cd and Zn) whereas the release of others is reduced less thoroughly (e.g. sulphate) or even increased (Cr). Although the absolute release of Cr (and other oxyanions) is not very large, one of the objectives of the ongoing R&D project is to improve the stabilisation of these components.

In addition to the batch leaching tests referred to above, samples of untreated and washed semidry APC system residue from an MSW incinerator and residue treated by CO_2 alone and by PO_4^{3-} and CO_2 in combination were subjected to pH-static leaching tests at L/S = 5 l/kg. The results, which are presented in Figure 2 for a few components, illustrate the effectiveness of both treatments on the leachability of these components as a function of pH. Although both treatments reduce the leachability of the components substantially, low leachabilities of Pb at low pH values are only achieved using the combined carbon dioxide/phosphoric acid treatment.

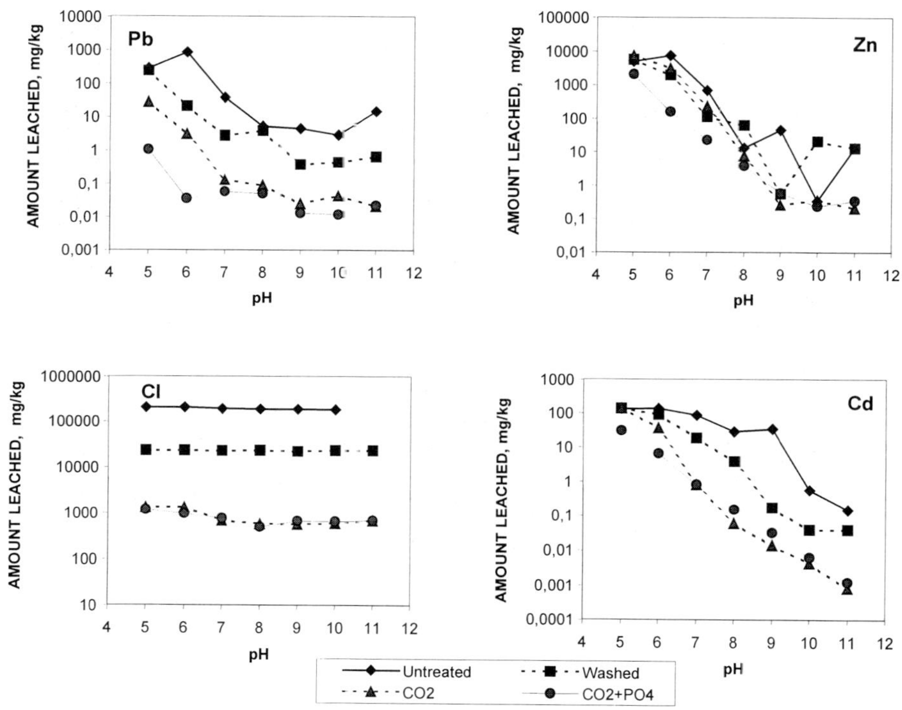

Figure 2 Results of pH-static leaching tests on untreated, washed and treated residues from the semidry APC system at a MSW incinerator. The tests were performed at L/S = 5 l/kg.

THE ONGOING R&D PROJECT

The Pilot Scale Plant

The VKI process is being further developed and optimised in an ongoing R&D project, which was initialised in 1998 and will be completed in 2000. The overall objective of the ongoing project is to prepare the process for full-scale implementation. The project is focused on further improvement of the stabilised product, minimisation of the consumption of process water through internal recycling of various water streams, optimisation of process parameters and reaction conditions, minimisation of the consumption of additives/testing of alternative additives, stabilisation of particularly soluble trace elements and development of an improved understanding of the chemical processes upon which the stabilisation method is based. For this purpose, a large pilot treatment plant has been constructed at the largest MSW incinerator in Denmark, Vestforbraending. The batch pilot plant which can treat 200 to 800 kg of residue per batch is being used for the experimental work and to demonstrate the effectiveness of the treatment method when applied to residues from the dry/semidry APC processes as well as to fly ash and sludge from the wet scrubbing process. A picture of the batch pilot scale treatment plant is shown in Figure 3.

Figure 3 The pilot scale treatment plant at Vestforbraending. The tall tank behind the stairways is the reactor tank, and the filter press is visible on the platform in the upper right corner of the picture. The extraction tank is located one floor below this level.

Expriences and Preliminary Results

The first year of the project period (1998) was spent designing and constructing the pilot treatment plant, and several months of the first half-year of 1999 were spent testing the functionality of the plant and performing the necessary corrections and improvements.
Once the initial technical problems were overcome, the pilot plant functioned as intended and several tests runs were performed on batches of semidry APC system residue from Amagerforbraending and fly ash collected upstream of the wet scrubbers at Vestforbraending.

The initial test runs were used to optimise the sequence of stabiliser addition and minimise the dosages of carbon dioxide alone and carbon dioxide and phosphoric acid in combination as well as to optimise the end-point pH (the reaction control is based on pH measurements) and the retention time in the reaction tank. An end-point pH of 7.0 was chosen. The sequence of addition of stabilising agents did not appear to be critical. Particular efforts were spent trying to minimise phosphoric acid consumption since this is more expensive than carbon dioxide. The carbon dioxide was added as simulated flue gas as a mixture of 10% pure CO_2 and 90% atmospheric air. The feasibility of using ash from incineration of sewage sludge as an alternative source of phosphoric acid is also being investigated. Sewage ash typically contains the equivalent of 25% (by weight) of phosphate, which must be treated with strong acid to become available as ortho-phosphate. 50 to 150 kg of sewage sludge ash would be required to stabilise 1 tonne of semidry APC residue. The initial tests gave promising results except for Cd, which became more leachable than when commercial phosphoric acid was used.

The stabilised products from the test runs were subjected to pH-static leaching testing at L/S = 10 l/kg and pH = 5 and 7, respectively. They were also subjected to a 2-stage batch compliance leaching test (a slightly modified version of the draft prEN 12457-3 prepared by CEN/TC 292). Some of the results of the leaching tests on untreated and treated semidry residues are presented in Figure 4 (pH-static tests) and Table 3 (batch compliance tests). Results of pH-static leaching tests performed on untreated and treated fly ash are shown in Figure 5.

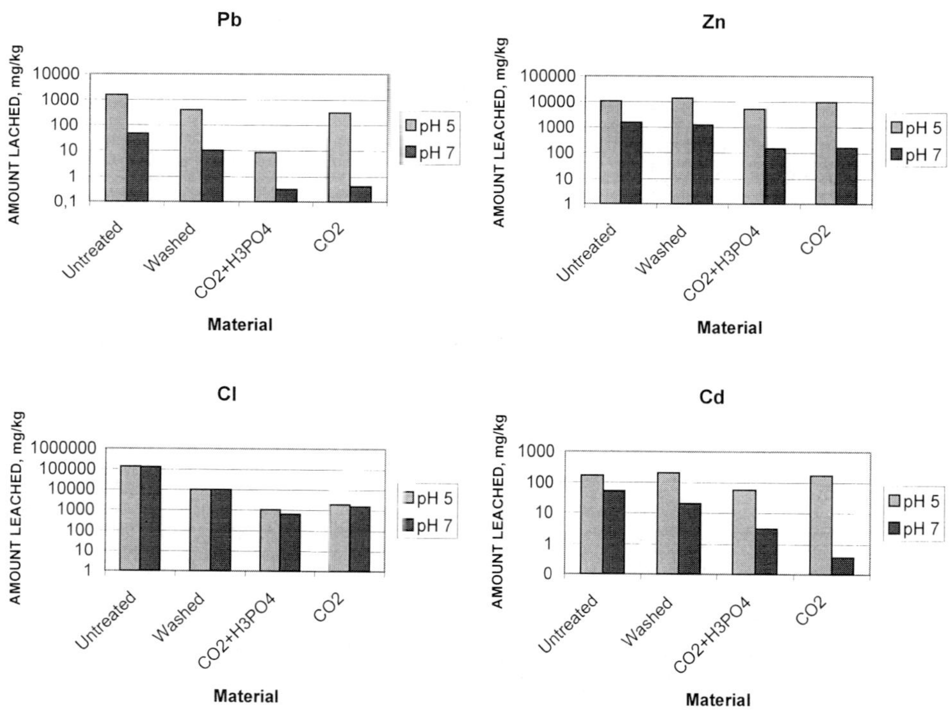

Figure 4 Results of pH-static leaching tests performed at L/S = 10 l/kg on untreated, washed and treated residues from the semidry APC process.

Table 3 Results of 2-stage compliance batch leaching test. The results shown are accumulated leached amounts at L/S = 10 l/kg. The test is slightly modified (using a first step of L/S = 5 l/kg instead of 2 l/kg). The table shows pH values in both eluates

PARAMETER	UNIT	CEN DRAFT COMPLIANCE BATCH LEACHING TEST, PREN 12457-3 (1999) PH AND ACCUMULATED LEACHED AMOUNTS FOR L/S = 10 l/kg			
		Untreated Residue	Washed Residue	Treated Residue CO_2+H_3PO_4	Treated Residue CO_2
pH		11.8/12.5	12.5/12.6	9.6/9.5	9.8/9.8
Chloride	g/kg	140	7.4	0.49	1.2
Sulphate	g/kg	21	15	15	15
As	mg/kg	0.10	< 0.10	0.01	0.01
Cd	mg/kg	0.09	0.04	0.008	0.004
Cr	mg/kg	0.34	0.75	1.8	2.6
Cu	mg/kg	3.5	0.90	0.07	0.04
Hg	mg/kg	0.22	0.08	0.19	0.29
Ni	mg/kg	< 0.05	< 0.04	0.008	0.009
Pb	mg/kg	1800	490	0.07	0.11
Zn	mg/kg	57	29	0.30	0.22

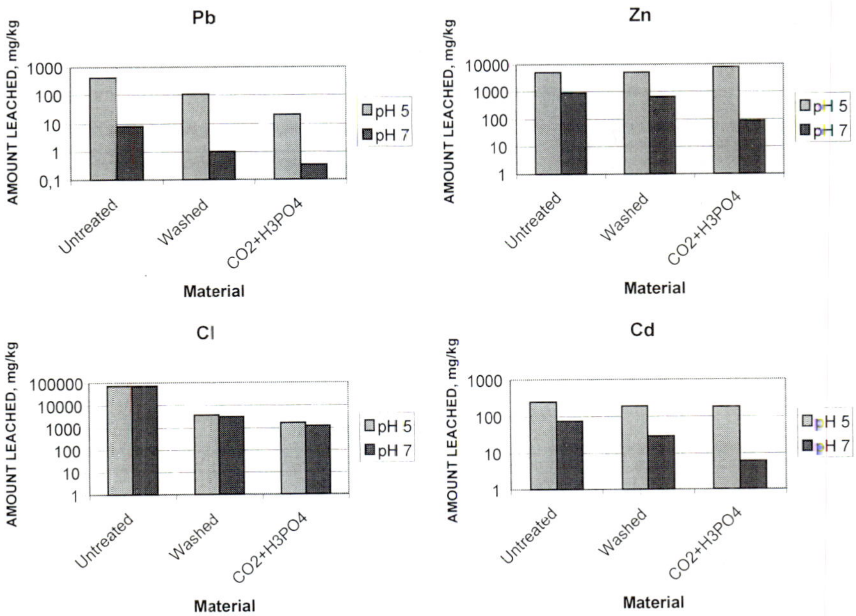

Figure 5 Results of pH-static leaching tests performed on MSWI fly ash at L/S = 10 l/kg.

From Table 3 it can be seen that substantial reductions of the leachability of most contaminants (except for sulphate, Hg and Cr) are achieved with both treatments of semidry APC residue. Further work has been planned to reduce the leachability of sulphate, Hg and Cr (and other oxyanions). Figure 4 confirms the results shown in Figure 2, namely that the addition of phosphoric acid is required to maintain a low leachability of Pb at lower pH values. Table 3 shows that the target values for pH have been achieved successfully. The results in Figure 5 indicate that the fly ash behaves and responds to the treatment in a manner, which is very similar to that of the semidry APC system residues.

CONCLUSIONS

It has previously been demonstrated that a process for treatment of semidry and dry APC system residues from MSW incinerators based on an initial aqueous extraction followed by re-suspension and stabilisation of the filter cake with phosphoric acid and/or carbon dioxide can produce a material with vastly improved disposal properties. A 2.5 year R&D project was initiated in 1998 with the objective to optimise and prepare the process for full-scale implementation, and to extend the applicability of the process to include fly ash and wet scrubber sludge. Most of the process optimisation and adjustment work is being carried out on a large-scale batch pilot treatment plant, which has been constructed at a large MSW incinerator in Copenhagen. The quality of the stabilised residues are evaluated on the basis of the results of pH-static leaching tests and batch leaching tests. Preliminary results indicate that good stabilisation results can be achieved for most components using carbon dioxide and phosphoric acid, both for semidry residues and fly ash. Using CO_2 alone also yields good results, but with higher leachability of Pb at low pH values. Further work is being carried out to reduce the leaching of sulphate and Hg as well as Cr and other oxyanions. Test results indicate that it may be feasible to use sewage sludge ash as a source of phosphoric acid.

ACKNOWLEDGEMENTS

The ongoing R&D programme is financed by the two largest MSW incinerators in Denmark, I/S Amagerforbraending and I/S Vestforbraending, through their joint landfill operation company AV Miljoe.

REFERENCES

1. HJELMAR, O. and BIRCH, H. (1997): Treatment of air pollution control residues from MSW incinerators prior to landfilling. In: Proceedings Sardinia 97, Sixth International Landfill Symposium, S. Margharita di Pula, Cagliari, Italy, 1997, pp. 535-544.

2. HJELMAR, O., HOLLAND, D., POULSEN, B., DALAGER, S., BIRCH, H., ROSE, N., PETERSEN, J.H., WESTERMANN, J., PEDERSEN, J., ASMUSSEN, O.W., RASK, P. SKOVBORG, P., HOLM, P.E., OLSEN, P.B., JENSEN, F. A process for treatment of residues from dry/semidry APC systems at municipal solid waste incinerators. EFP-94 1323/94-002. Final report prepared by VKI, FLS miljoe a/s and I/S KARA, 1997.

3. HJELMAR, O. Leaching properties of fly ash from MSW incinerators. Report prepared for the Danish EPA by VKI. Hoersholm, 1993 (in Danish).

PART TWO:

USE OF INCINERATOR BOTTOM ASH IN FILL AND HIGHWAY WORKS

PROCESS INTEGRATED CARBONATION OF BOTTOM ASHES

R Koralewska

M Faulstich

Technical University of Munich

Germany

ABSTRACT. The goal of the investigations which are described in the paper was to define and test the process conditions of an integrated wet chemical treatment of raw bottom ashes in a bottom ash discharger. This should lead to an improved quality of raw bottom ashes. Results indicated that easily soluble salts and fine particles were removed by the washing procedure. In laboratory experiments, the concentrations of heavy metals in leachates after treatment could be lowered significantly by the addition of gaseous carbon dioxide. However, in some pilot-scale experiments the concentrations of carbon dioxide in the process fluids were too low, and hence the expected stabilization of heavy metals was not achieved. The chemical reactions of the carbon dioxide mostly happen in the washing solution. Stabilization could be reached only with a large amount of additive. After wet chemical treatment all raw bottom ashes exhibited a significant improvement of technical construction material characteristics. This procedure may be applied without incurring substantial investment costs by retrofitting existing incineration plants to allow on-site washing of raw bottom ashes. In addition the carbon dioxide from the cleaned flue gas can be used to accelerate the aging processes.

Keywords: Wet chemical treatment, Washing procedure, Carbonation, Leaching properties, Heavy metals, Salts

Dr R Koralewska, was a Research/Teaching Fellow at the Institute of Water Quality Control and Waste Management of the Technical University of Munich. His research interests included the treatment procedures of bottom ashes. Now he is project-engineer in the research and development department of the Martin GmbH for Environmental and Energy Technology, Munich.

Professor M Faulstich, is professor and head of the Department of Waste Management at the Institute of Water Quality Control and Waste Management of the Technical University of Munich. His investigations include concepts and technologies for the analysis, evaluation and implementation of prevention, re-use, treatment (thermal/ biological) and disposal of waste and residues on the basis of material flow and life cycle analyses.

INTRODUCTION

The incineration of household waste and related industrial waste in Germany results in approximately 3 million Mg of bottom ash yearly. After incineration a mixed residue consisting of ash, glass, ceramics, metals, unburned material and natural stones is produced. The main chemical contents of bottom ashes are oxides of silicon and aluminum but also oxides of alkali and alkaline earth metals which contain parts of inorganic pollutants. Examples are the heavy metals copper, chromium, nickel, zinc, lead, cadmium, mercury and their compounds. The heterogeneity of bottom ash is determined by the changing composition of household waste.

Waste regulations mandate that re-use of this residue be given priority. Re-use does not usually occur, largely because technical construction material and environmental characteristics either do not allow or strongly restrict utilization. Bottom ash in comparison to currently used construction materials demonstrates high pollutant solubility, low expandability and high frost sensitivity. In order to improve these parameters, a minimum three-month ageing is necessary. But the processes of carbonation and hydration reactions that take place during the ageing do not go to completion even after several months [1]. With regard to the mineralogical composition four types of bottom ashes can be distinguished [2] (Table 1).

Table 1 Types of bottom ashes [2]

TYPE	DEFINITION
Grate ash	material discharged from the end of the grate
Raw bottom ash	grate ash discharged via a quench tank (bottom ash discharger)
Bottom ash	raw bottom ash which had been processed and stored for at least three months
Aged bottom ash	raw bottom ash which had been stored for several years

In the bottom ash discharger the grate ash contacts water for the first time. It is a dry, hot and very porous material with a high potential for chemical reactions (hydration). It is the first step of the ageing process and raw bottom ashes undergo with respect to their mineral composition further mineralogical processes as alteration and new formation of mineral phases. The most important reaction of the bottom ashes during storing outside is the conversion of portlandite to calcite under influence of carbon dioxide of the air. This process changes the leaching properties and technical construction material characteristics in a high degree.

The goal of our project was two-fold: to develop a simple and cost effective integrated wet chemical treatment process and to test and compare which of the technical construction material and environmental characteristics of raw bottom ashes could be improved. Additional goals were a reduction in the solubility of heavy metals and an acceleration of hydration and carbonation reactions.

In Figure 1 the concept of an integrated wet chemical treatment is provided. On one hand as much as possible easily soluble pollutants (heavy metals, salts) and fine particles should be removed by a washing procedure with water. On the other hand the remained pollutants (heavy metals) should be chemically and / or physically stabilized by the specific addition of gaseous carbon dioxide. For this reason, raw bottom ashes were washed with water and in addition of gaseous carbon dioxide in one and the same process in laboratory and pilot-scale experiments.

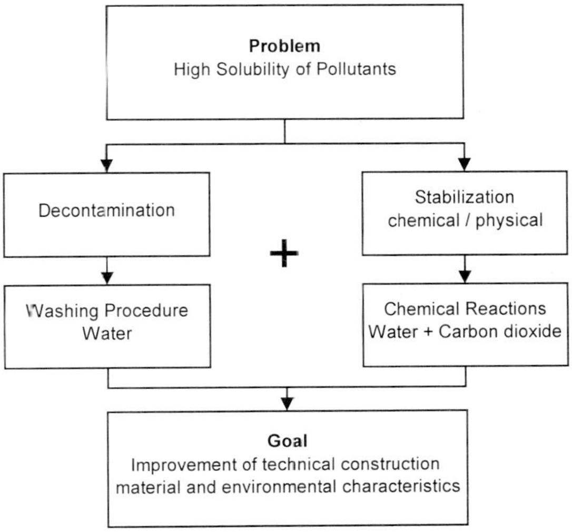

Figure 1 Wet chemical treatment

It was found that easily soluble pollutants and fine particles were removed. The chemical reactions of carbon dioxide mostly happen in the washing solution. Stabilization of heavy metals could be reached and the concentrations of heavy metals in leachates could be lowered. The on-site washing of raw bottom ashes could be retro-fitted in an incineration plant with low expenditure.

METHODS

Sampling and Conditioning

Dry grate ashes were removed continuously from two Bavarian incineration plants (A and B). Therefore the bottom ash discharger was driven without any water. The screening curve of the grate ashes of each incineration plant was determinated and afterwards the 10 Mg of grate ashes were separately divided into grain fractions 0/5, 5/10, 10/15, 15/25 and 25/40 mm by a screening drum. One third of the amount of the grate ashes was discarded as oversize fraction and coarse metal waste. The different grain fractions were filled in containers (volume 20 litres), flushed with nitrogen gas, and sealed as air-tightly as possible. It was thus possible to investigate each fraction or to remix the screening curve of the grate ash for each incineration plant at any time. The gaseous nitrogen was necessary to slow down any uptake of moisture. A leachate analysis one year after sampling showed that there was no significant change in the grate ashes as a result of ageing.

For each grate ash, a comprehensive characterisation of the technical construction material and chemical (solids and leachate analyses according to the LAGA leaflet [3]) characteristics took place. For the preparation of the solids analysis ferrous metals, non-ferrous metals, and uncombusted material were removed by a magnet and by hand and quantified.

Laboratory experiments

First experiments of wet chemical treatment were carried out in a glass reactor to investigate the reactions of the grate ashes during washing with water and in contact with gaseous carbon dioxide during this washing procedure. Therefore 750 g of remixed grate ash of grain size of 0/25 mm was filled into a glass column and percolated by the recirculated washing water at a L/S ratio of 2:1. Plastic rings were necessary to guarantee an unrestrained flow of the washing fluid. Gaseous carbon dioxide was blown into the raw bottom ash through a glass pipe which was installed into the material at the beginning of each experiment.

Pilot-scale experiments

These investigations were carried out in a Martin bottom ash discharger of a scale of 1:10 with 150 kg grate ash per experiment. Therefore the different grain fractions were remixed. Variables of the washing procedure were the quantity of washing water, the addition of gaseous carbon dioxide and the hydraulic residence time. The washing water was pumped through nozzles on the top of the inlet chute and recirculated at a L/S ratio of 3,5:1. Fine particles of the raw bottom ashes were retained by sieves and a filter out of the washing fluid. Gaseous carbon dioxide (100%) was blown into the bottom ash discharger directly. The difference to the laboratory experiments were the continuous refill of grate ash by hand and discharging of treated raw bottom ash by the discharging ram. To investigate the situation in an upscale bottom ash discharger of an incineration plant, investigations without active washing in the water bath took place. Water was only added to replenish any losses due to discharge of water along with the raw bottom ash.

The effects of the washing procedure and the addition of gaseous carbon dioxide at various concentrations on the composition of washing solutions, the leachate composition, the constitution of treated raw bottom ash and retained fine particles were analyzed. In addition, selected technical construction material characteristics were determined. Some treated raw bottom ashes were stored outside for three months and analyzed in the same way.

For some treated raw bottom ashes and bottom ashes titration to a pH value of 4,0 took place. Therefore the bottom ashes were crushed to grain size $\leq 0{,}1$ mm. 10 g of this material were stirred with 100 ml of deionized water and 0,2 ml/min of 1 M HNO_3 was added continuously. The goal was the qualitative comparison of different treated raw bottom ashes by their buffer curves.

RESULTS AND DISCUSSION

Laboratory Experiments

The concentration of heavy metals in leachates of treated raw bottom ashes B is provided in Figure 2. It is shown a significant decrease of lead and zinc after addition of gaseous carbon dioxide. The pH value of the leachate was lowered from 12,5 (water) to 10,5 (CO_2 3,0 l/min) which is the main reason for the lower contents of these amphotere heavy metals. The chemical reactions of the carbon dioxide happened mostly in the washing solution which was found as flocculation in the glass reactor.

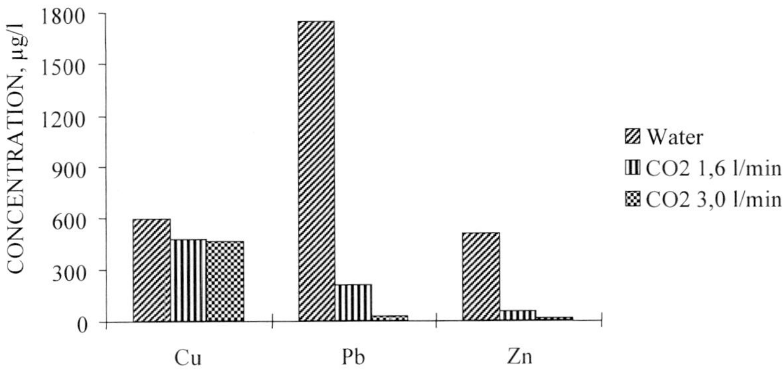

Figure 2 Contents of heavy metals in leachates of treated raw bottom ashes B (laboratory experiments)

Easily soluble salts (chlorides, sulfates) were leached out by the washing procedure. There was an increase in mobilization of sulfates and no further increase of chlorides by the addition of gaseous carbon dioxide. In the leachates the concentrations of chlorides were lowered by a physical stabilization. The porous surface was closed by increased formation of calcite. The carbon dioxide treatment led to cementation of the treated raw bottom ashes.

Pilot-scale Experiments

In Figure 3 contents of heavy metals in leachates of CO_2-treated raw bottom ashes B are provided. The results of washing procedures are shown in comparison to water bath experiments with the same amount of carbon dioxide added. An increasing amount of carbon dioxide led to a significant decrease of the concentration of heavy metals in leachates. The investigations in a water bath using the same amount of gaseous carbon dioxide showed that the heavy metals were leached out of the treated raw bottom ashes at a lower rate.

The expected stabilization of heavy metals was not achieved by using gaseous carbon dioxide in all pilot-scale washing procedures. One reason was the high ratio of water to raw bottom ash. This ratio was chosen to leach out as much easily soluble salts as possible and to remove fine particles. So the concentration of carbon dioxide in the continuously exchanged washing solution was lower in comparison to water bath experiments. Another reason was the reaction of the carbon dioxide with the calcium in the washing solution directly. Calcium is one main element in solids and washing solutions.

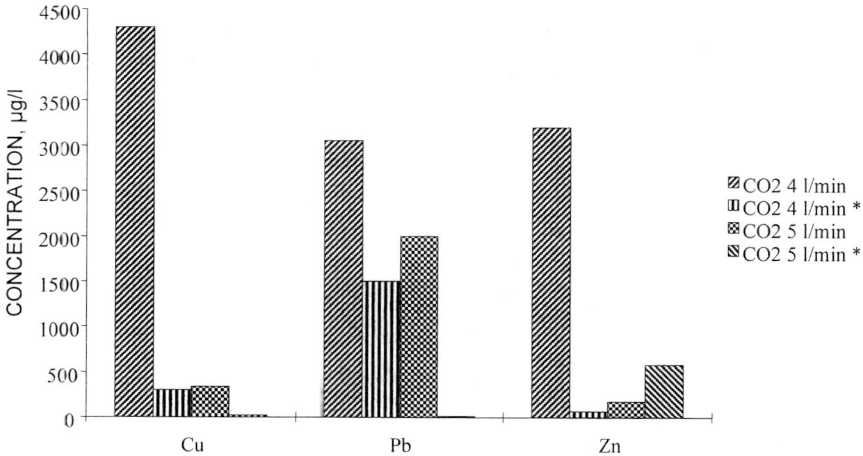

Figure 3 Contents of heavy metals in leachates of CO_2-treated raw bottom ashes B (pilot-scale experiments, * water bath)

Easily soluble salts (chlorides, sulfates) and fine particles were leached out by the washing procedure and there was no significant increase in mobilization of salts by the addition of carbon dioxide. The investigations with the water bath provided that easily soluble salts were leached out at a lower rate, the concentration in the leachates of the treated raw bottom ashes was significantly higher and there was no extraction of fine particles because the wash solutions were not exchanged.

The pH value of the recirculated washing fluid during the wet chemical treatment reached 11 to 12 very fast and was only lowered when a great amount of carbon dioxide was offered. In the CO_2-treated raw bottom ashes a cementation was found.

After aging of some treated raw bottom ashes for three months an additional significant reduction of the concentration of lead and zinc in leachates was shown. This is a result of the reaction of the bottom ashes during storing outside under influence of carbon dioxide of the air. Especially the conversion of portlandite to calcite lowers the pH value of the leachates to 10,5. The parallel new formation of mineral phases results in an increased concentration of sulfate in the leachates.

The contents of salts in leachates of CO_2-treated raw bottom ash A and fine particles A from the washing procedure (Figure 4) corroborate that chlorides and sulfates are mainly found in fine particles [4]. The finest particles had the greatest content in the leachates. Also the content of the heavy metals lead and zinc in solids and the loss on ignition in the fine particles were significantly higher than in treated raw bottom ashes.

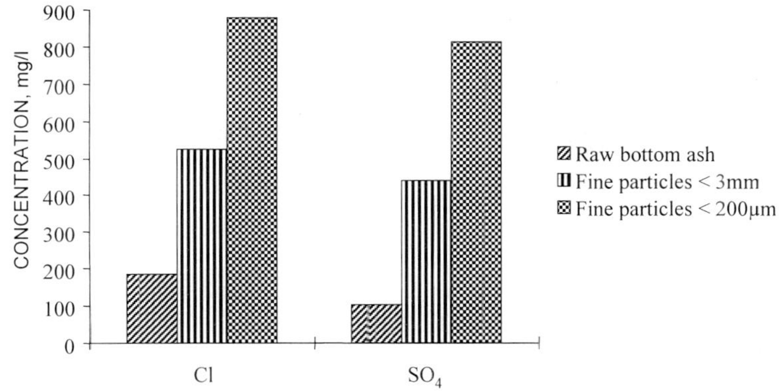

Figure 4 Contents of salts in leachates of CO_2-treated raw bottom ash A and fine particles A from the washing procedure (pilot-scale experiments)

A comparison of the titration of different treated raw bottom ashes B is provided in Figure 5. All treated raw bottom ashes show the typical plateaus of carbonates at pH 10,3 and 6,3. The water-treated raw bottom ash reaches pH 4 first. OH^--ions were leached out and there was less new formation of calcite. The CO_2-treated raw bottom ashes showed the best buffering properties. Only at low pH values the bottom ash (stored for three months) has a higher buffer capacity because of silicate phases which formate during aging [4].

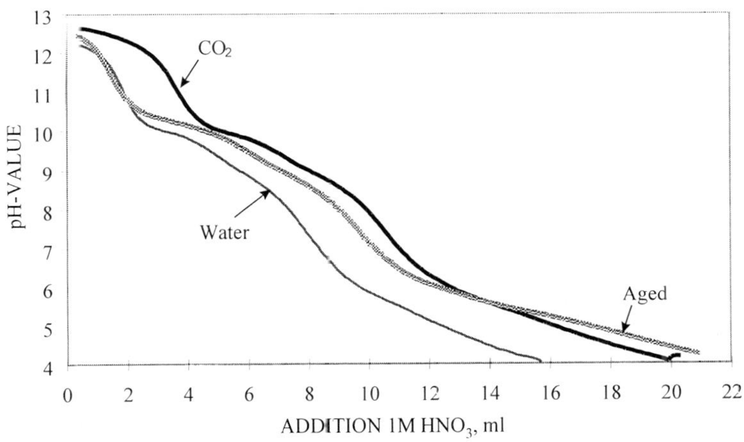

Figure 5 Titration of different treated raw bottom ashes B (pilot-scale experiments)

Technical Construction Material Characteristics

The characterization of the technical construction material characteristics was carried out by the Institute of Building Materials of the Technical University of Munich [5,6]. According to these investigations, the freeze-thaw resistance of the grate ash and treated raw bottom ash samples was sufficient. The corresponding total amounts of scaled material were clearly below the limit of 9 M.-% specified by TL HMVA-StB 95 [7].

The soundness of the grate ashes and treated raw bottom ashes A and B was determined following the technical delivery conditions for the quality control of steelworks slag in road construction [8]. The favourable effect of wet chemical treatment on the soundness of the bottom ashes is clearly apparent (Figure 6).

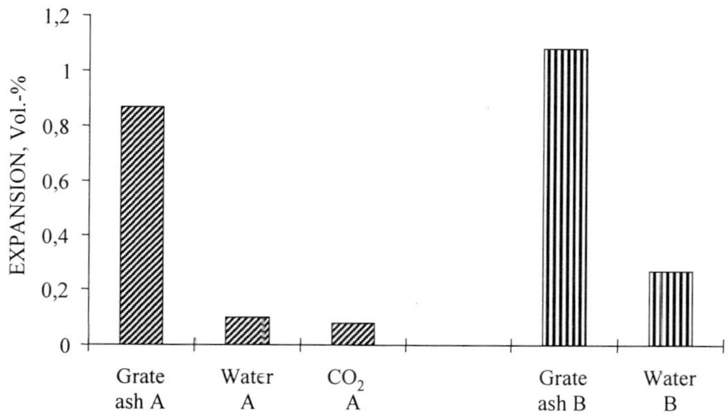

Figure 6 Soundness of grate ashes and different treated raw bottom ashes A and B (pilot-scale experiments)

After wet chemical treatment all raw bottom ashes exhibited a significant improvement of technical construction material characteristics. However, the main reason for these effects is in the first step not the addition of gaseous carbon dioxide but the separation of fine particles and the mixing of the raw bottom ashes with the fluid by the washing procedure. Reactive components undergo mineralogical processes which will not happen in the construction material. The conversion of portlandite to calcite and the beginning of cementation under influence of added carbon dioxide is a further effect.

CONCLUSIONS

Results indicated that easily soluble salts and fine particles were removed by the washing of raw bottom ashes. The concentrations of heavy metals in leachates after treatment could be lowered significantly by the addition of gaseous carbon dioxide. The expected stabilization of heavy metals could be achieved by acceleration of ageing processes, mostly by changing the pH value in the leachates of the treated bottom ashes. However, the chemical reactions of the carbon dioxide mostly happen in the washing solution. Stabilization can be reached only with a large amount of additive. After wet chemical treatment all raw bottom ashes exhibited a significant improvement of technical construction material characteristics.

This procedure may be applied without incurring substantial investment costs by retrofitting existing incineration plants to allow on-site washing of raw bottom ashes. In addition the carbon dioxide from the cleaned flue gas can be used. The washing solutions could then be used in the washer system of the flue gas cleaning facility. It is important to note that any treatment applied must not interfere with subsequent handling of the material.

ACKNOWLEDGEMENTS

The authors would like to acknowledge the Institute of Building Materials of the Technical University of Munich and the Martin GmbH for Environmental and Energy Technology, Munich for their substantial contributions to this research project. The financial support was provided by the Bavarian Ministry of Regional Development and Environmental Affairs.

REFERENCES

1. JAROS, M.; HUBER, H. Emissionsverhalten von MVA-Schlacke nach forcierter Alterung. Waste Reports No. 06. Waste Management, Universität für Bodenkultur, Wien, 1997.

2. PFRANG-STOTZ, G.; REICHELT, J. MWSI Bottom Ashes: Mineralogy, Leachability and Technical Aspects. Waste Reports No. 08. Waste Management, Universität für Bodenkultur, Wien, 1998. pp 27-35.

3. LAGA LÄNDERARBEITSGEMEINSCHAFT ABFALL. Anforderungen an die stoffliche Verwertung von mineralischen Reststoffen / Abfällen: Technische Regeln, 1994.

4. FÖRSTNER, U.; HIRSCHMANN, G. Langfristiges Deponieverhalten von Müllverbrennungsschlacken. Abschlußbericht BMBF-Verbundvorhaben Deponiekörper, Teilvorhaben 1, Hamburg-Harburg, 1997. 259 pp.

5. KORALEWSKA, R.; DORNER, H.; FAULSTICH, M. Bottom Ash from Household Waste Incineration - Treatment, Conditioning and Applications. Proceedings of R´99 World Congress, Genf, 1999. pp II.193-II.198.

6. WESTINER, E. Untersuchungen zur Raumbeständigkeit von hydraulisch gebundener Hausmüllverbrennungsasche. Ph.D. Thesis, Lehrstuhl für Baustoffkunde und Werkstoffprüfung, Technische Universität München, 1998.

7. FGSV FORSCHUNGSGESELLSCHAFT FÜR STRAßEN- UND VERKEHRSWESEN. Technische Lieferbedingungen für Hausmüllverbrennungsasche im Straßenbau. TL HMVA-StB 95, Köln, 1995.

ENVIRONMENATAL IMPACTS OF WASTE UTILISATION: DEVELOPMENT OF RISK-RELATED LIMIT VALUES FOR LEACHING OF CONTAMINANTS IN REGULATORY TESTING

O Hjelmar

DHI – Institute of Water and Environment

P E Holm

Royal Veterinary and Agricultural University

T H Christensen

Technical University of Denmark

Denmark

ABSTRACT. This paper describes a methodology, which in contrast to most existing regulatory systems provides a direct relationship between the risk posed to the groundwater by various utilisation scenarios for mineral waste types and the results of leaching tests performed on the waste. The concept has been tested on two mineral waste types and soil with promising results. The model is based on flux considerations and does not take attenuation mechanisms into consideration. Only impacts on the groundwater quality are considered in this model, and it must be supplemented with other relevant criteria to ensure a satisfactory protection of the environment and human health against the potential negative effects of utilisation of inorganic mineral waste materials and soils. In this paper, leaching data for coal fly ash are used to exemplify scenario calculations.

Keywords: Granular mineral waste materials, Soil, Utilisation, Risk-related limit values, Leaching tests, Regulatory system, Environmental impact, Groundwater, Scenario.

Mr O Hjelmar is Chief Engineer at DHI-Institute of Water and Environment in Hørsholm, Denmark. He specialises in waste characterisation, development, application and interpretation of waste leaching tests, waste treatment technology, and development of sustainable utilisation and landfilling strategies. He is Director of the Waste Research Centre (C-RES), a joint R&D co-operation project between universities, R&D institutions and industry for development of strategies, treatment methods, cleaner technologies, testing tools and models, which can enhance sustainable utilisation and landfilling of waste.

Dr P E Holm is an associate professor at the Department of Soil Chemistry at the Royal Veterinary and Agricultural University in Copenhagen, Denmark. He is a specialist in environmental chemistry, geochemistry and microbiology. His research experience includes waste and soil characterisation, risk assessment, phytoremediation and soil chemistry.

Dr T H Christensen is professor of Environmental Engineering at the Department of Environmental Science and Engineering, Technical University of Denmark in Lyngby, Denmark. Most of his research activities have focused on the environmental aspects of waste disposal on land including sludge and compost application to land, heavy metal cycling in agricultural soils, industrially contaminated soils, landfills, groundwater pollution, waste treatment and remediation technology. He is one of the partners of C-RES.

INTRODUCTION

The potential utilisation of waste products such as inorganic residues from waste combustion/energy production or industrial processes are in many countries subject to regulations that require that the waste products are characterised and that they meet specified criteria, if permission for utilisation is to be granted. The characterisation required is often based on the performance of a specific leaching test, and the criteria to be met are usually specified in terms of associated limit values for the concentrations (e.g. mg/l) of various components in the eluate produced or the amounts (e.g. mg/kg of waste product tested) of various components leached. From an administrative perspective, this appears to be a satisfactory management tool. From a scientific and environmental perspective, however, the situation is often less satisfactory because of the general lack of relationship between the actual risk of an environmental impact and the limit values. This is due to the pragmatic way in which limit values are often established.

This paper describes an alternative approach which establishes a relationship between the result of a leaching test carried out on a granular waste product and the environmental risk posed by various civil engineering utilisation applications of the waste product. Consequently, by using this methodology, it is possible to set limit values, which are related to the potential environmental impact. The description of a relationship between a utilisation scenario and a potential elevation of the concentration(s) of one or several leached contaminants in the groundwater at a downstream point of extraction for drinking water purposes allows the setting of limit values to be divided into two steps: a series of objective calculations and a few decisions based on environmental policy. In this case, the policy decisions consist of setting maximum allowable values of the increases of various contaminants in the groundwater at the point of extraction (at the well). The more objective calculations consist of establishing a relationship, firstly between the result of the leaching test and the potential flux of contaminants out of the utilised material, secondly between the flux of contaminants and the elevation of the contaminant level in the groundwater. Scenario calculations are used for this purpose.

DESCRIPTION OF THE MODEL

The Physical Scenario

The (inorganic) granular waste product which is utilised e.g. as a filling material in an embankment or a road ramp is described in a simplified manner as a box with surface area A (m^2), height H (m) and bulk dry density d (t/m^3). The material is placed on or below the surface of the ground above an aquifer. The general annual rate of infiltration of precipitation in the geographic area is I (m). See Figure 1. The maximum acceptable concentration of a given component in the leachate which appears at the bottom of the box as a result of the percolation of infiltrated precipitation through the material is defined as follows:

$$C_V \quad = \quad GE \times D \times P \times IR$$

where

GE is the maximum acceptable elevation of the concentration of the contaminant in question in the groundwater from the aquifer below the site at the point of extraction;

D is the ratio between the surface area of the groundwater catchment area considered and the surface area of the utilised material or, alternatively, a dilution factor indicating how many times the leachate from the utilised material is diluted from the point of entry into the groundwater to the downstream point of extraction of drinking water from the aquifer;

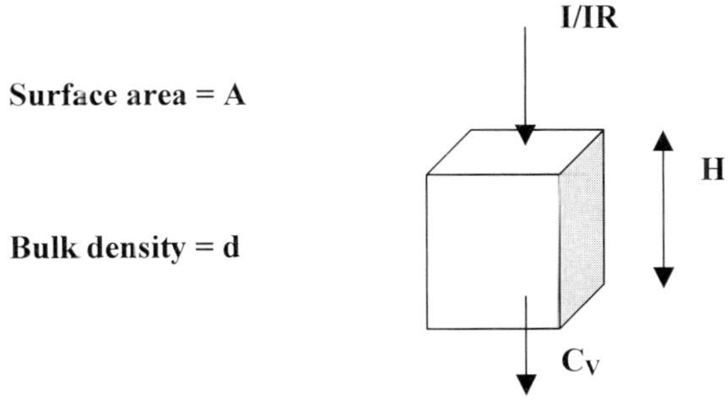

Figure 1 Simplified physical description of a utilisation application containing a granular inorganic material

P is a priority factor which provides the option of placing higher priority on some aquifers than others, e.g. based on its value as a drinking water resource;

IR is an infiltration reduction factor, which expresses the ratio between the general rate of infiltration, I, in the area and the rate of infiltration into (or percolation through) the utilised material which may be reduced by a top cover (increased surface run-off, increased evapo-transpiration). The amount of water percolating through the material will thus be I/IR.

This simplified model does not directly take into account the biogeochemical processes in the unsaturated zone below the site and in the aquifer, which would delay and retain a given leachate component during the transport of the leachate from the bottom of the "box" of utilised material to the drinking water extraction point. These attenuation processes therefore constitute an unquantified extra margin of safety.

The Criteria Related Timeframe, T_C

The leaching of several components, including e.g. salts, from inorganic waste materials and contaminated soil will often show high initial concentrations in the leachate with subsequently decreasing leachate concentrations as the leaching progresses and the liquid to solid ratio (L/S) increases. The concentration of other components (e.g. lead) in the leachate will be controlled mainly by other factors such as pH or redox potential. Since the model is based on the consideration of the flux of contaminants rather than contaminant concentrations, it seems reasonable to use an average value of C_V for a given component over a period of time, T_C, referred to as the criteria related timeframe. This implies e.g. that it is accepted that C_V for the component in question is exceeded during the initial phase of the leaching whereas the concentration of the component during a later phase will be lower than C_V, ensuring that the average concentration of the component in the leachate over the entire period T_C does not exceed C_V. It will most likely be necessary to assign different values of T_C to different groups of components. Components, which are leached quickly and which are mobile in the unsaturated zone and the aquifer (e.g. many salts), should be assigned relatively low values of T_C (e.g. 1-3 years), whereas higher values of T_C (e.g. 10 –25 years) can safely be assigned to less mobile components which are leached relatively slowly (e.g. several trace elements). The leaching of certain components, e.g. arsenic, are often solubility controlled in such a way that constantly increasing concentration levels with increasing L/S may be observed for a very long period of time. For such components T_C must be assigned with great care and the model should be applied with caution.

Specific Acceptable Mass, M_{SA}, and Accumulated Leached Mass, ALM_C

If the flux of a given component leached from the box of utilised material is considered, the specific acceptable mass of that component per unit weight of utilised material, M_{SA}, (e.g. expressed in mg/kg) which can be leached during the time period T_C may be calculated by substituting the equation defining C_V above into the expression used to calcuate the flux of the component out of the utilised material. This leads to the following equation:

$$M_{SA} = GE \times D \times P \times T_C \times I/(d \times H)$$

The result of a leaching test, which is usually accelerated, may for granular inorganic materials often be expressed as an amount of leached components per unit weight of material as a function of L/S or at a given L/S ratio. Assuming ideal conditions, L/S (e.g. expressed as l/kg = m^3/t) can be related to time for the physical scenario under consideration by the following equation:

$$T = (L/S) \times d \times H/(I/IR)$$

where T is the time corresponding to the value of L/S used. If the same time perspective is applied to the physical scenario describing a utilisation situation for a material and to the interpretation of the result of an accelerated leaching test on the same material, the equation shown above can be used to establish a relationship between the specific acceptable mass, M_{SA}, leached from the utilised material within the criteria related timeframe, T_C, and the result of the leaching test. The impact on the groundwater quality will not exceed an excess concentration corresponding to GE at the point of drinking water extraction as long as M_{SA} is not exceeded. The criteria related value of L/S corresponding to T_C for a given component can be determined from the following equation:

$$(L/S)_C = T_C \times I/(IR \times d \times H)$$

Under idealised conditions where the leached amounts of a given component as a function of L/S based on the results of leaching tests are available and where this description can be assumed to simulate the actual leaching conditions, the following condition must be satisfied to ensure that GE is not exceeded at the point of extraction of drinking water from the aquifer:

$$ALM_{Ci} < M_{SAi}$$

where ALM_{Ci} is the accumulated leached mass of component i found experimentally at the accumulated L/S value $(L/S)_C$ (see Figure 2) and where M_{SAi} is the calculated specific acceptable mass of component i leached from the utilisation scenario.

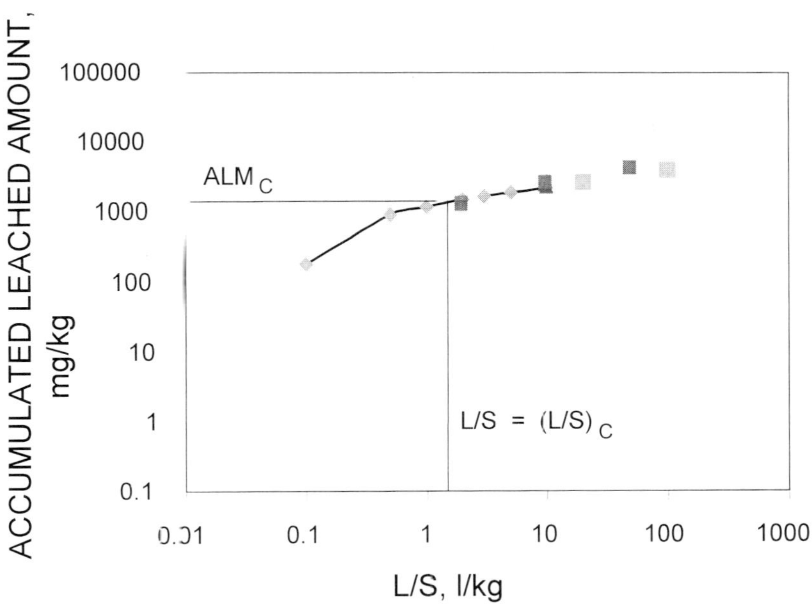

Figure 2 Graphical determination of the criteria related accumulated leached amount, ALM_C.

SCENARIO CALCULATIONS: TWO EXAMPLES

To illustrate the use of the concept and the model, they are applied to two different utilsation scenarios: an embankment with soil cover and road utilisation scenario (subbase under a paved road). In both cases the waste material is coal fly ash. The scenarios, which are very simple, are characterised by the parameters shown in Table 1. For both scenarios dilution factors of 10, groundwater priority factors of 1 (high priority), annual precipitation infiltration rates of 300 mm and coal fly ash bulk densities of 1.5 t/m^3 are assumed. It is further assumed that the embankment has an average height of 5 m and a top cover which reduces the rate of infiltration by a factor of 2, whereas the road subbase has a thickness of 0.5 m with a surface pavement which reduces the rate of infiltration by (at least) a factor of 10. The flow regime is assumed to be percolation in both cases.

Two partly leachable components of the coal fly ash, potassium (K) and chromium (Cr), are considered. The relatively abundant and mobile salt ion, K^+, is assigned a criteria related timeframe of T_C = 3 years, whereas Cr, which will be moderately mobile as the chromate ion, is assigned a T_C = 10 years. The following maximum acceptable elevations of the concentrations in the groundwater at the extraction point, GE_{Max}, are chosen: 150 mg/l (K) and 0.05 mg/l (Cr). The calculated values of the specific acceptable leached amounts, M_{SA}, and the criteria related L/S values, $(L/S)_C$, are shown in Table 2.

In Figure 3, K and Cr leaching results of lysimeter, column and batch tests on a coal fly ash are presented [1,2].

Table 1 Definition of sample scenarios

PARAMETER	UNIT	EMBANKMENT SCENARIO	ROAD SUB-BASE SCENARIO
Dilution factor, D	none	10	10
Groundwater priority factor, P	none	1	1
Rate of infiltration of precipitation, I	m/year	0.3	0.3
Dry bulk density of coal fly ash, d	t/m^3	1.5	1.5
Infiltration reduction factor, IR	none	2	10
Height of embankment/subbase, H	m	5	0.5

Table 2 Calculation of M_{SA} and $(L/S)_C$

SCENARIO	PARAMETER	POTASSIUM	CHROMIUM
Embankment	M_{SA}	180 mg/kg	0.20 mg/kg
Road subbase	M_{SA}	1800 mg/kg	2.0 mg/kg
Embankment	$(L/S)_C$	0.06 l/kg	0.20 l/kg
Road subbase	$(L/S)_C$	0.12 l/kg	0.40 l/kg

From the data in Figure 3, the accumulated leached mass, ALM_C, corresponding to the values of $(L/S)_C$ shown in Table 2 can be determined. The results are presented in Table 3, which also shows the final comparison of ALM_C and M_{SA} for each scenario and each component. The criteria are fulfilled for both scenarios for K ($ALM_C/M_{SA} < 1$) but not for Cr ($ALM_C/M_{SA} > 1$ for both scenarios). If the thickness of the subbase layer in the road scenario is reduced to 0.2 m, ALM_C/M_{SA} values of 0.04 (K) and 0.88 (Cr) are obtained, thus fulfilling the criteria for this scenario.

These results are indicative of the ability of the model to identify leachable components with potentially detrimental effects on downstream groundwater quality. Field observations have actually revealed significantly elevated concentrations of Cr (up to 0.5 mg/l) in groundwater a few hundred metres downstream of a coal fly ash monofill [1].

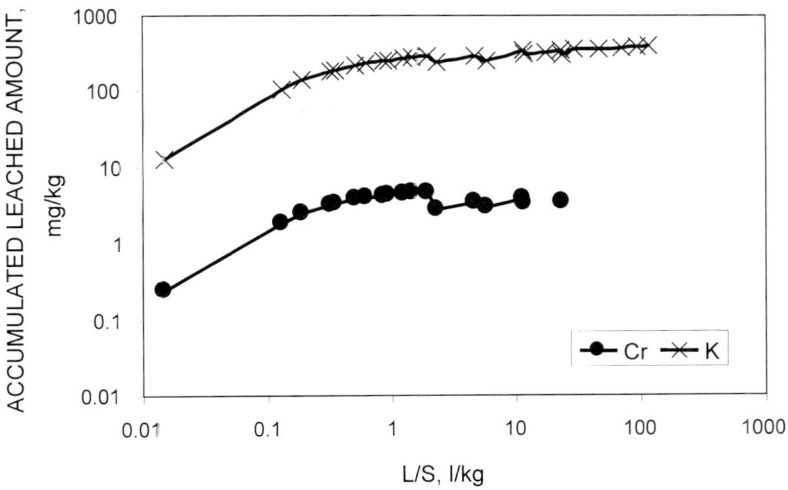

Figure 3 Results of lysimeter, column and batch leaching tests on a coal fly ash covering a range of L/S from 0.015 l/kg to 117.5 l/kg

Table 3 Calculation of ALM_C and check on fulfilment of criteria ($ALM_C/M_{SA} < 1$)

SCENARIO	PARAMETER	POTASSIUM	CHROMIUM
Embankment	ALM_C	104 mg/kg	2.5 mg/kg
Road subbase	ALM_C	100 mg/kg	3.6 mg/kg
Embankment	ALM_C/M_{SA}	0.58	13
Road subbase	ALM_C/M_{SA}	0.06	1.8

GENERAL EXPERIENCES WITH THE MODEL

In addition to the scenario calculations shown above, the model has been subjected to more comprehensive testing using the following values of GE (mg/l) and T_C (years): Chloride (250, 1 or 3), sulphate (220, 1 or 3), Na (150, 1 or 3), K (280, 1 or 3), As (0.009, 10), Cd (0.005, 25), Cu (0.1, 25), Cr (0.05, 10), Hg (0.001, 25), Ni (0.015, 25), Pb (0.009, 25), Se (0.01, 10) and Zn (0.04, 25). The chosen values of GE are to a certain degree dependent on the background concentration levels in the groundwater. A number of utilisation scenarios corresponding to the following parameter values and ranges were considered: F (1 – 10), P (1), IR (1 – 10), I (0.2 – 0.7 m/year), d (1.5 t/m^3), E (0.3 – 10 m) using leaching data on coal fly ash, incinerator bottom ash and soil. The detailed results are described in [3]. Some of the general findings and experiences are described below.

The calculated value of $(L/S)_C$ varies substantially from one scenario to another, and in many cases data allowing a precise determination of ALM_C at the desired $(L/S)_C$ are not available. In such cases the conservative (environmentally safe) approach is to use the closest larger value of L/S for which data are available. The danger in using data corresponding to a too large value of L/S is that the evaluation of the material and the scenario in question may be too restrictive: It will not always be possible to determine whether an apparent violation of the criteria $ALM_{Ci} < M_{Sai}$ is real or is caused by an unrealisticly high value of ALM_{Ci} because the accumulated leached amount of the component is described at a value of L/S which is too high. If, on the other hand, the criteria are complied with despite the use of a too high value of L/S, an extra safety margin is added. The alternative, always to use "taylor-made" leaching results, would most likely be too impractical for routine application. The solution will probably be to use a combination of very detailed knowledge of the leaching behaviour of groups of materials (fundamental characterisation) and related behaviour of the same materials in simpler routine leaching tests (compliance tests), possibly combined with some well proven extrapolation techniques.

Many of the most common utilisation scenarios lead to relatively small values of $(L/S)_C$, often below 1.0 l/kg and 0.1 l/kg, and leaching data corresponding to such low values can only be obtained from column or lysimeter leaching tests. This was also the case with the two coal fly ash scenarios, see Table 2.

Apart from the above mentioned problems and some inconsistencies caused by the dependency of $(L/S)_C$ on the height of utilised material when it is attempted to define limit values corresponding to uncontaminated soil, the concept described in this paper appears promising and useful. The inconsistencies can be overcome using an empirically based convention. The concept has proven a very powerful tool for pinpointing and excluding utilisation of materials and application of utilisation scenarios which are unfavourable in terms of leaching of specifically harmful components in significant amounts and designs which increase the flux of contaminants over a significant period of time. It also has the advantage of providing more than one option for improvement of the scenario designs and the waste material properties (both hydraulic design changes and chemical stabilisation changes are e.g. possible). Various general or specific restrictions may be imposed on particular types of utilisation applications of particular types of waste material in accordance with the performance of the material and the scenario when the system is applied.

In contrast to many other existing or proposed regulatory systems, this concept creates a direct relationship between the risk to groundwater posed by a particular application of a particular type of waste and the performance of that material in a leaching test system with associated limit values. It should be noted that other tests and criteria are necessary if the potential impact of the material due to e.g. direct exposure or fugitive dust emissions are to be included.

CONCLUSIONS

A methodology has been developed which provides a direct relationship between the risk posed to the groundwater by various utilisation scenarios for various mineral waste types and the results of a leaching test performed on the waste. The concept has been tested on two mineral waste types and soil with very promising results. It appears that the major weaknesses of the method, the requirement of leaching data at relatively low L/S values, and some inconsistencies which may occur when potentially uncontaminated soil is tested, can be overcome relatively easily. The model is based on flux considerations and does not take attenuation mechanisms into consideration. Only impacts on the groundwater quality are considered in this model, and it must be supplemented with other relevant criteria to ensure a satisfactory protection of the environment and human health against the potential negative effects of utilisation of inorganic mineral waste materials and soils.

ACKNOWLEDGEMENT AND DISCLAIMER

The development of the concept described in this paper was partly funded by the Danish Environmental Protection Agency, DEPA [2], but the opinions expressed are those of the authors and no official endorsement from DEPA should be inferred.

REFERENCES

1. HJELMAR, O., HANSEN, E.AA., LARSEN, F. AND THOMASSEN, H. Leaching and soil/groundwater transport of contaminants from coal combustion residues. Final report to EFP, Elkraft A.m.b.A. & EC (DGXII) prepared by VKI. Energiministeriets Forskningsudvalg for produktion og fordeling af el og varme. Miljoe og restprodukter, VKI, Hoersholm, Denmark, 1991.

2. HJELMAR, O., CHRISTENSEN, T.H. AND LUDVIGSEN, K. Investigation of the feasibility of classifying residues in relation to utilisation. Report for the Danish Environmental Protection Agency, VKI, Hoersholm, Denmark, 1996 (in Danish).

3. HJELMAR, O. HOLM, P. E., LEHMANN, N.K.L., ASMUSSEN, O. AND ROSE, N. Background for utilisation of contaminated soil and residues. Miljoeprojekt nr. 415, Danish Environmental Protection Agency, Copenhagen, Denmark, 1998 (in Danish).

A PHYSICAL EVALUATION OF MUNICIPAL SOLID WASTE GRATE ASH AS AN AGGREGATE SUBSTITUTE IN ASPHALTIC CONCRETE

S Karpinski
HDR Engineering

X Zhang
Heritage Research Group

V Janoo
US Army Corps of Engineers

D Gress
T Eighmy
University of New Hampshire
United States of America

ABSTRACT. The Laconia Bottom Ash Paving Project examined the feasibility of using municipal solid waste grate ash as an aggregate substitute in a pavement binder course. A section of RT.3 in Laconia, New Hampshire was paved with grate ash as an aggregate substitute in asphaltic binder course. The road was instrumented to measure temperature, moisture contents, depth of frost, traffic, and strains. In-situ testing involved Falling Weight Deflectometer (FWD) to determine layer moduli. The results of the testing were used to develop American Association of State Highway and Transportation Officials (AASHTO) layer pavement design coefficients. The performance of the grate ash as an aggregate substitute was found to be identical to that of the aggregate in the control mix. The performance illustrated that grate ash may be used as an aggregate substitute in asphaltic concrete to replace natural aggregate.

Keywords: Recycling bottom ash, Grate ash, MSW, Asphaltic Concrete, Aggregate Substitute, Field performance.

Mr Stanley Karpinski, is an engineer for HDR Engineering, Inc. Norfolk, Virginia, USA

Dr Xishun Zhang, is Vice President of Heritage Research Group, Indianapolis, IN, USA

Dr David Gress, Professor, Environmental Research Group, Department of Civil Engineering, Durham, New Hampshire, USA.

Dr Taylor Eighmy, Research Professor and Director of the Environmental Research Group in the Department of Civil Engineering, Kingsbury Hall, University of New Hampshire, Durham, NH, USA

Dr Vincent Janoo, Research Civil Engineer, Cold Regions Research Environmental Laboratory, US Army Corp of Engineers, Lime, New Hampshire, USA.

INTRODUCTION

The concept of using municipal solid waste combustion residues as construction materials is not new in the U.S. In the 1940's, the ash residue from some plants was utilized for embankment and subbase construction at various sites in Pennsylvania and New York [1]. In the 1960's and 1970's, several studies were performed and demonstration projects were built to evaluate the technical feasibility of using ash residue as a total or partial aggregate replacement in paving mixtures. The most comprehensive effort addressing the potential for using ash residue in various highway applications was completed by the Federal Highway Administration in 1976 [2]. Several field test sections were constructed. The performance of these test sections varied from "excellent" to "poor" during the periods of evaluation. A few studies have looked at using fused or treated residues as aggregate substitutes [3-4]. Work by Chesner et al. [5] has shown that bottom ash from the Southwest Brooklyn, New York combustor can be used as an aggregate substitute. Wiles and Shepherd have recently published a very comprehensive resource document relative to the beneficial use and recycling of Municipal Waste Combustion Residues [6].

OBJECTIVES OF RESEARCH

The objectives of this research were to evaluate the physical properties of municipal solid waste combustion grate ash as an aggregate substitute in order to make a direct comparison against a control section by paving a section of highway. The mix design for the project was based on previous work performed at the University of New Hampshire using gyratory test method procedures ([7-9]. The optimum asphalt contents being 5.1 and 7.0 percent for the control and test sections respectively. Detailed information relative to the Laconia project can be found elsewhere [10-12].

METHODS AND MATERIALS

Laconia Bottom Ash Paving Project

The Laconia bottom ash paving project is a 609.6 m section of Route 3 located in Laconia, New Hampshire, USA. Road construction occurred in late May 1993. The road design called for a 25 mm wearing course, 50 mm binder course, and a stabilized base course. The road was divided into two sections, a control section and a test section consisting of 50% bottom ash, both meeting the standards for a Type B binder base course mix. The control and test sections were 350.5 m and 259.1 m in length respectively. Both sections conformed to New Hampshire Department of Transportation guidelines for asphaltic concrete. The site was prepared by milling the existing asphaltic concrete and transporting it offsite for future use. Concrete pavement underneath was broken up and removed from the site. With the subbase exposed, holes were excavated and the temperature and moisture probes were installed.

The removal of the old paving materials resulted in loss of elevation requiring new base material to be brought in to restore the site to the original elevation. The asphaltic concrete that was previously removed was brought back and spread over the new base material and blended together with a reclaimer to create an asphalt stabilized base. After the blending was completed, it was found that several points, particularly the southern portion of the northbound lane, did not match the required elevations. Bringing in recycled asphalt material and spreading it in the affected areas accomplished elevation control. This resulted in this area of the base being thicker and having a higher asphalt content.

The asphalt concrete binder courses were placed using a short base rubber tire paver and compacted with rollers until a constant density was obtained by using a nuclear density device. The binder base had a uniform thickness of 50.8 mm, with the exception of the transition section. This region, 58.4 m in length has a binder thickness of 101.6 mm to allow for falling weight deflectometer (FWD) testing, strain gauges to be placed, and for removal of the asphaltic concrete for laboratory testing which required specimens with a thickness greater than 50.8 mm. After the binder courses were placed, a 25.4 mm wearing course of 100 percent natural aggregate was placed over the entire length of the project.

Instrumentation

Instrumentation was placed in the subgrade and the asphaltic base in both sections. The instrumentation was chosen to monitor the environmental conditions on the subbase, base, and asphaltic concrete. In the southbound lane, a hole was dug in which two identical sets of instrumentation trees and moisture probes were placed, one on each side of the transition. The hole was then backfilled and another hole dug in both the Northbound and Southbound lanes for induction loops. Strain gauges were then placed on top of the subbase material prior to the paving so they would be bonded to the bottom of the of the binder course layer. With the exception of the strain gauges, all instrumentation was located along the centerline of the lane.

Field Testing

The instrumentation trees contained thermocouples for temperature, resistance gauges for depth of frost, and steel rods connected to a time domain reflectometer (TDR) to measure moisture content. Induction loops were installed to record traffic. Strain gauges were installed to record the strain at the bottom of the asphaltic concrete during FWD testing. Data were collected every 60 minutes for a period of two years beginning in July, 1993 and ending in September, 1995.

FWD testing was performed over the entire length of the project at 30.5 m intervals except for the thickened section, where testing was performed every 1.52 m. The FWD testing was performed a total of six times: once on the base course, once on the binder course without the wearing surface, and four times after the wearing course was applied. Strain gauge testing was also performed with the FWD during the final four measurements.

Laboratory Testing

Laboratory testing of the asphaltic concrete consisted of three standard ASTM tests and one non-standard test. Tests were performed on specimens removed from the project site at 0 and 15 months after placement. The standard tests performed were: ASTM C496 Splitting

Tensile Strength of Concrete Specimens [13], ASTM D4123 Standard Test for Indirect Tension for Resilient Modulus of Bituminous Mixtures [14], and ASTM D1559 Standard Test for Resistance to Plastic Flow of Bituminous Mixtures Using Marshall Mix Design Apparatus [15]. The non-standard test was for thermal stress. This test was developed by the Strategic Highway Research Program (SHRP) to evaluate the resistance of the asphaltic concrete to cracking resulting from low temperatures.

RESULTS AND DISCUSSIONS

Field Testing

The field testing of the environmental conditions is important in determining the effect that they have on pavement performance. From the TDR probes, it was determined that the moisture contents between the test and control probes were approximately the same. The moisture contents at each depth also remained constant throughout the entire monitoring period except during the winter months when the material was frozen. A frozen period is recorded as a decrease in the moisture content due to the phase change of the water.

The moisture content increased with depth as was expected due to migration of water toward the surface caused by capillary rise, evaporation, and condensation. The data also illustrated that the water table never rose above the lowest moisture probe located at 1219 mm below grade. The temperatures during the monitoring period were approximately the same for each year. The temperature is important in that as it decreases, material stiffness increases. The material can also approach the glass transition temperature (T_g) of the asphalt cement, which is a low temperature unique to each material. Below this temperature, the material becomes brittle, hard, and more resistant to shear stress, and cracking of the pavement can occur due to applied loads and thermal shrinkage. During the two years of monitoring, cracks were not observed in either section.

The temperature and resistance gauges measured the effect of temperature and the state of moisture in the subgrade materials. The winter of 1993-94 had a depth of frost of 1346 mm below grade versus the winter of 1994-95, which had a depth of 889 mm below grade for both the test and control sections. The temperature is also important as frozen subgrade layers affect FWD testing.

The total amount of vehicles that passed over the road since construction was approximately 6.1 million vehicles. Traffic during the summer months increased the daily traffic by approximately 3,000 vehicles per day. The amount of trucks that have traveled on the road was approximately 427,000, based on a typical New Hampshire traffic distribution of 7.0% of the traffic being trucks as per the Transportation and Planning Department of the New Hampshire Department of Transportation.

The strain gauges were used to determine if the stiffness of the pavement had increased over time. With the gauge attached to the bottom of the pavement, an increase in stiffness of the pavement, either due to temperature or aging, would result in a lower strain reading. Both the control and test section exhibited the same reactions due to temperature. The gauge readings remained the same for the duration of the test program indicating the pavement had not aged significantly by showing increased stiffness.

The FWD data were used to evaluate the elastic properties over time and to evaluate the response of the pavements due to seasonal changes in temperature. The first testing was performed on the base and subgrade materials and the results showed the materials to be approximately equal for the entire length of the road. The second test was performed on top of the binder courses and the decrease in measured peak deflection was similar 1.27 mm in the Southbound lane and 1.78 mm in the Northbound lane. All subsequent deflections were performed on top of the wearing surface.

Wearing course tests were performed at different times during the year in order to determine the effect that the environment had on the pavement system. The temperature difference from May to November caused reduced deflections due to increased stiffness of the pavement. The remaining two tests were performed during the "spring thaw". The March 6, 1994 testing was performed while the subgrade was still frozen to 610 mm. Testing was then performed one month later when the subgrade was not frozen. The effect of the "spring thaw" was identified by increased deflection profiles recorded on these two days. The increased deflections are the result of increased moisture in the subgrade materials. The effect of the 101.6 mm thickened section binder course was very apparent during the "spring thaw." Deflections in the thickened section increased only 0.025 mm between the March 6 and April 5, 1994 test dates. The 50.8 mm sections showed an average increase in deflections of 0.075 mm, three times as much as the thicker section. This shows the structural benefit of a thicker binder course.

The FWD data were also used to determine the elastic modulus of the pavement sections. From the thickened section, test points were chosen based on uniformity of the base material. From these points, the FWD data were used to backcalculate the modulus values. A program called WESDEF was used for the process. Tables 1 and 2 represent the average modulus values from the back calculations. The backcalculation of the modulus values show the same behavior as was found in the analysis of the FWD deflections. The binder modulus values increase during the spring due to the low temperature and subgrade materials beneath the pavement being frozen. Modulus values increased on the April 5, 1994 test date for the Northbound lane, but the Southbound lane values remained approximately the same or decreased. These data were variable due to the "spring thaw" effect. The modulus values for the pavement would be expected to increase if the stiffness of the base and subbase materials decreased. This happened here as the material underneath the Northbound lane became weaker due to the increased moisture.

The Southbound lane modulus values are similar to those found on the November 5, 1993 test date. This indicated the subgrade materials under the Southbound lane returned to their original equilibrium states and that neither the binder or subgrade material had changed over the course of one winter.

The final test date October 13, 1994, produced modulus values, which decreased significantly from the 5 November 1994 test date. Both the WESDEF results and FWD deflection analysis indicated that the base material had increased in strength and stiffness over the period of one year. The increase in base material stiffness would have the result of

Table 1 Average modulus values

A) CONTROL SECTION		
	Northbound	Southbound
Test Date	Average Modulus, MPa	
November 5, 1993	6819±2110	8040±2010
March 6, 1994	7440±3540	14270±3320
April 5, 1994	8180±4570	8540±2900
October 13, 1994	3680±617	4140±549
B) TEST SECTION		
	Northbound	Southbound
Test Date	Average Modulus, MPa	
November 5, 1993	8970±1050	5980±1005
March 6, 1994	9860±1427	7620±642
April 5, 1994	13560±1654	6320±2430
October 13, 1994	4980±861	7680±3250

reducing the modulus value of any layer on top of it, in this case, the pavement. The data indicated that the decrease in binder moduli was not a function of the binder materials, but of the subgrade materials. The change in moduli values over time was attributed to the changing state of the subgrade materials, especially the subbase materials. Subbase materials are difficult to characterize as stated by Parker, Jr. [16] due to changing moisture contents in the subsurface materials. In situ layer material properties, such as modulus and thickness values, are difficult to determine and any variation in moisture conditions affects the final answer.

Layer Coefficients

The ultimate goal of the field testing was to determine the layer coefficients to be used in the AASHTO method for designing a pavement. The structural number in the AASHTO method is based on the following equation:

$$SN = a_1 D_1 + a_2 D_2 + a_3 D_3$$

Where SN is the structural number, a_x is the layer coefficient and D_x is the layer thickness. In design, the structural number of the system is known along with the layer coefficients. The layer thicknesses needed to obtain the desired structural number are easily determined. The focus of the FWD testing was to define layer coefficients for the control and test binders in order to compare both binders and to allow future design of pavements using the grate ash material. Layer coefficients were calculated using Rhode's Method [17] and the AASHTO design method [18].

Layer coefficients were determined from FWD test dates on November 5, 1993 and October 13, 1994. The May 8 , 1993 test was excluded as the wearing course had not yet been applied, as were the March 6 and April 5, 1994 tests due to the effects of the "spring thaw". Layer coefficients calculated using Rhode's method are presented in Table 2. The AASHTO coefficients are presented in Table 3. Layer coefficients from Rhode's method showed the control section was consistent over time. The test section, however, changed slightly over the course of a year. In the Northbound lane, the layer coefficient decreased, while in the Southbound lane, the layer coefficient increased. This was expected as the layer coefficient varies with a change in binder modulus, which was seen during the testing period.

Table 2 Layer coefficients by Rhodes method

A) CONTROL SECTION

Station	November 1993	October 1994
Northbound		
160+12	0.45	0.40
160+22	0.41	0.39
160+52	0.38	0.37
160+62	0.36	0.36
Average.	0.39	0.39
Southbound		
160+22	0.43	0.45
160+52	0.49	0.43
160+62	0.48	0.48
Average	0.46	0.46
Average	0.46	0.46

B) TEST SECTION

Station	November 1993	October 1994
Northbound		
159+12	0.44	0.39
159+22	0.40	0.38
159+32	0.43	0.36
159+42	0.42	0.42
159+62	0.42	0.39
Average	0.42*	0.39*
Southbound		
159+22	0.36	0.40
159+32	0.35	0.37
159+42	0.36	0.39
159+62	0.38	0.38
159+72	0.36	0.36
Average	0.36*	0.38*

* Difference between years is statistically significant at the 95% confidence interval.

The AASHTO method yielded equal layer coefficients for both the Northbound and Southbound control and test sections. The high modulus values of the asphaltic concrete resulted in the maximum value allowed by AASHTO, which is 0.44. For design purposes there is no difference between the control and test layer coefficients.

Table 3 Layer coefficients by the AASHTO Method

A) CONTROL SECTION		
Station	November 1993	October 1994
Northbound		
160+12	0.44	0.44
160+22	0.44	0.44
160+52	0.44	0.44
160+62	0.44	0.44
Average	0.44	0.44
Southbound		
160+22	0.44	0.44
160+52	0.44	0.44
160+62	0.44	0.44
Average	0.44	0.44

B) TEST SECTION		
Station	November 1993	October 1994
Northbound		
159+12	0.44	0.44
159+22	0.44	0.44
159+32	0.44	0.44
159+42	0.44	0.44
159+62	0.44	0.44
Average	0.44	0.44
Southbound		
159+22	0.44	0.44
159+32	0.44	0.44
159+42	0.44	0.44
159+62	0.44	0.44
159+72	0.44	0.44
Average	0.44	0.44

Laboratory Testing

The results of the indirect tensile tests illustrated the effect of temperature on asphaltic concrete. The indirect tensile strength increased by a minimum factor of four over a temperature range of 35°C. Statistical analysis, using a paired t-test, showed that the variation between the two mixes was small and that they were therefore equal.

The results of the resilient modulus testing produced values that were lower than those calculated from FWD testing, although the two values can not be compared directly. Lower values are typically found during this test due to the sample being unconfined during testing. The pavement that was tested in the field is placed in a state of triaxial stress, which affects the displacements and results in increased stress. Statistical analysis, again using a paired t-test, revealed the two mixes to be equal at all temperatures.

The field core data results after 15 months showed that the densities of the control and test mixes have not yet reached their final densities or air void contents. The difference between the laboratory mixes, batch plant, and field core values was due to loss of temperature during transport reducing the viscosity of the asphalt cement making compaction more difficult. It is very reasonable that the field cores have not achieved their final design values, as there has not been sufficient traffic over two years to cause complete compaction.

The Marshall stability values were greater for the batch plant testing during initial pavement construction than the field cores after 15 months of service life, with the test mix having larger values than the control. The increased stability values were possibly due to the higher temperature during laboratory compaction than during field compaction, which allowed increased compaction of the laboratory samples. The increased angularity of the grate ash in the test mix would be expected to have an higher stability value than the control mix. Even though the mix designs are for a binder course, they both exceed the minimum stability of 2,220 N specified by the Asphalt Institute for all mix designs and 6,670 N for a base course as specified by the Georgia Department of Transportation [19,20]. The NHDOT does not have a specification for a binder course, but specifies a value of 3340 N for a wearing course [21].

Flow values of the field cores were higher than the samples made at the batch plant, with the test mix having larger readings. The reduction in compaction of the field cores compared to the batch plant samples was the likely cause of the difference in the flow values. The larger flow values could be the result of a larger amount of fines in the test mix created during the mixing process. Flow values specified by the NHDOT, GaDOT, and Asphalt Institute all specify a flow value of between 8 and 16 [19-21], again with the NHDOT value being for a wearing course.

The results of the thermal stress test show that the control and test mixes both behave the same when exposed to low temperatures. The thermal stress test is used to find the bonding strength between the asphalt cement and the aggregate as affected by cold temperatures. The samples removed in May 1993 had an average cracking temperature of 5°C higher than the samples removed in August 1994. Aging of the asphalt cement is the most likely cause for the increase in cracking temperature. The May samples were stored for several months before testing which may have allowed O_2 access to react with the asphalt cement and therefore increasing its stiffness.

CONCLUSIONS

Based on the results presented above, the following conclusions were drawn from the research:

1. The test section modulus values at the conclusion of testing were larger than the modulus values of the control section: Northbound lane 4980 MPa versus 3680 Mpa, and Southbound lane 7680 Mpa versus 4140 Mpa respectively.

2. The "spring thaw" has the effect of placing additional stress upon the system requiring the pavement to provide more support to replace that lost in the subgrade materials.

3. The response of the pavement to physical loading is the same for the control and test sections.

4. Laboratory testing for tensile strength, resilient modulus and thermal stress revealed that the two mixes perform equally regardless of temperature.

5. The in situ mixes have not reached final gyratory testing machine mix design properties for density and air voids as of the conclusion of this work.

6. The AASHTO layer coefficients were found to be the same for both the control and test sections. For design purposes, the maximum value allowed by AASHTO, 0.44 is recommended for the ash base binder layer coefficient.

The results and conclusions indicate that the test and control mixes behave the same. Field testing showed a slight difference between the test and control mixes, but this can be attributed to differences in the subgrade materials, not in the binder base course. Laboratory testing of tensile strength, resilient modulus, and thermal stress revealed that the two mixes performed the same at any temperature. A final conclusion is that the bottom ash behaves as a natural aggregate and can be use as a replacement for natural aggregate in asphaltic concrete. Grate ash requires additional asphalt due to the higher porosity as compared to conventional aggregate.

ACKNOWLEDGEMENTS

The authors would like to acknowledge the financial support provided by the Concord Regional Solid Waste/resource Recovery Cooperative, Wheelabrator Environmental Systems, the U.S. Department of Energy via the National Renewable Energy Laboratory, and the U.S. EPA via Rutgers University. Special thanks goes to the many individuals from various agencies that were associated with the project

REFERENCES

1. PINDZOLA, D. and R. J. COLLINS. Technology for Use of Incinerator Residue as Highway Material - Identification of Incinerator Practices and Residue Sources. Report No. FHWA-RD-75-81 (Interim Report), Federal Highway Administration, 1975.
2. COLLINS, R. J., R. H. MILLER, S. K. CIESIELSKI, E. M. WALLO, M. J. BOYLE, D. PINDZOLA, and J. TROPEA. Technology for Use of Incinerator Residue as Highway Material. Report FHWA/RD-77/151, 1976.
3. SNYDER, R.R. Evaluation of Fused Incinerator as a Paving Material. Report FHWA-TS-88-229, 1980.
4. MEHAN, D. J. and W. F. Hooper. Processed Ash Demonstration Project. Proceedings of the Second Annual International Specialty Conference (Municipal Waste Combustion), AWWA, Pittsburgh, Penn., 1991, pp.627-641.
5. CHESNER W. H., R. J. COLLINS, and T. FUNG. An Assessment of the Potential Suitability of Southwest Brooklyn Incinerator Residue in Asphaltic Concrete Mixes. Final Report, Chesner Engineering, P.C. 2171 Jericho Turnpike, Commack, New York 11725, 1989.
6. WILES, C., and SHEPHERD, P., Beneficial Use and Recycling of Municipal Waste Combustion Residues a comprehensive resource document, National Renewable Energy Laboratory, U.S. Department of Energy, April 1999, pp 1-126.
7. EIGHMY, T.T., GRESS, D.L., ZHANG, X., TARR, S. and WHITEHEAD, I., Interim Report-Bottom Ash Utilization Evaluation for the Concord, New Hampshire Waste-to-Energy Facility, Durham: University of New Hampshire, 1992.
8. ZHANG, X., The Utilization of a Municipal Solid Waste Combustion Bottom Ash as a Paving Material. PhD. Dissertation, University of New Hamspshire, 1994.
9. ZHANG, X., GRESS, D., KARPINSKI, S., and EIGHMY, T.T., Utilization of Municipal Solid Waste Combustion Bottom Ash as a Paving Material, Transportation Research Record No. 1652 Vol 1, Seventh International Conference on Low-Volume Roads 1999, Baton Rouge Louisiana, pp 257-263.
10. The Laconia, New Hampshire Bottom Ash Paving Project, Volume 1: Environmental Testing Report, Environmental Research Group University of New Hampshire, Durham New Hampshire, National Renewable Energy Laboratory, US Department of Energy, March 1996 NREL/TP-430-230959.
11. The Laconia, New Hampshire Bottom Ash Paving Project, Volume 2: Environmental Testing Data Appendices Report, Environmental Research Group University of New Hampshire, Durham New Hampshire, National Renewable Energy Laboratory, US Department of Energy, March 1996 NREL/TP-430-230960.

12. The Laconia, New Hampshire Bottom Ash Paving Project, Volume 1: Physical Performance Testing Report, Environmental Research Group University of New Hampshire, Durham New Hampshire, National Renewable Energy Laboratory, US Department of Energy, March 1995 NREL/TP-430-230961.

13. AMERICAN SOCIETY OF TESTING MATERIALS, ASTM C496-90, Splitting Tensile Strength of Concrete Specimens, 04.02.

14. AMERICAN SOCIETY OF TESTING MATERIALS, ASTM D4123-82 Standard Test for Indirect Tension for Resilient Modulus of Bituminous Mixtures. 04.03.

15. AMERICAN SOCIETY OF TESTING MATERIALS, ASTM D1559-89, Standard Test for Resistance to Plastic Flow of Bituminous Mixtures Using Marshall Mix Design Apparatus, 04.03.

16. PARKER, Jr., F., "Estimation of Paving Materials Design Moduli from Falling Weight Deflectometer Measurements," Transportation Research Record, No. 1293, Washington, D.C., 1991, pp. 42-52.

17. RHODE, G.T., "Determining Pavement Structural Number from FWD Testing," Transportation Research Record, No. 1448, Washington, D.C., 1994, pp. 61-68.

18. AMERICAN ASSOCIATION OF STATE HIGHWAY AND TRANSPORTATION OFFICIALS, AASHTO Guide for Design of Pavement Structure, Washington, D.C.: American Association of State Highway and Transportation Officials, 1993.

19. ASPHALT INSTITUTE, Mix Design Methods for Asphalt Concrete, MS-2, College Park: The Asphalt Institute, 1974.

20. GEORGIA DEPARTMENT OF TRANSPORTATION, Standard Specification-Construction of Roads and Bridges, Atlanta, Ga: GaDOT, 1983.

21. NEW HAMPSHIRE DEPARTMENT OF TRANSPORTATION, Standard Specifications for Road and Bridge Construction, Concord, NH: NHDOT, 1990.

MSW BOTTOM ASH – EXPERIENCES IN THE NETHERLANDS

P Leenders

Insulinde Recycling Amsterdam

Netherlands

ABSTRACT. In the last 10 years more than 6 million tons of MSW Bottom Ash have been used in the Netherlands. Most applications were in highway embankment-construction projects as a fill material and industrial building sites as a construction material below high-level floors in large warehouses. Due to environmental regulations MSW Bottom Ash is used in large scale applications, i.e. more than 100,000 tons per project, up to projects with 1,000,000 tons. With a continuous production of MSW Bottom Ash, some 1,000,000 tons each year, on one side and intermittent realization of large scale projects on the other hand, it was experienced in the Netherlands that intermediate specialized companies are necessary to achieve a high degree of application. In this paper the experiences of a project with 1,000,000 tons of MSW Bottom Ash are described. Attention is given to design aspects, quality assurance of the bottom ash, costs and the position of the intermediate company.

Keywords: MSW bottom ash, Application, Embankment, Properties, Quality assurance, Intermediate company.

Mr Peter Leenders is a member of the board of Insulinde Recycling Amsterdam. He has been involved in applications of secondary materials for more than 15 years. He has worked as manager for the Dutch Waste Processing Association and as environmental policy advisor for Public Works Department Rotterdam. He is a member of several working groups on road building materials, environmental aspects of materials and waste processing policy.

INTRODUCTION

In the Netherlands some 5 Mt of waste incineration capacity is available per annum from 11 MSW incinerators (Table 1). The largest MSW incinerator in the Netherlands, AVR Rijnmond, produces more than 220,000 tons of MSW Bottom Ash per annum. In total some 1.0 Mt of MSW Bottom Ash is produced each year. All the bottom ash is treated (sieving, crushing and sorting out ferrous and non-ferrous metals). In the plant MSW Fly Ash is separated from the bottom ash.

Table 1 MSWI Plants in the Netherlands and amounts of waste incinerated in 1998

INCINERATOR	INCINERATION, kton
GAVI Wijster	413
AviTwente	288
ARN	237
AVIRA	297
HVC	450
GDA Amsterdam	778
AVR Rijnmond	884
AVR Rotterdam	385
GEVUDO	194
Watco Roosendaal	49
AZN	561

Quality control of the MSW Bottom Ash has been in use at all MSW incinerators since the end of the eighties. At present the vast majority of the treated bottom ash is produced with a quality KOMO certificate based on guidelines and legal demands placed on the environmental quality of the ash.

Referring to the Building Materials Decree, MSW Bottom Ash is classified as a category 2 or special category material. The leaching behaviour of the bottom ash, when compared with the allowable immission of contaminants in the natural soil, is such that applications of bottom ash are allowed only when the material is capped with a (nearly) impermeable cover on top and a position sufficient above the groundwater level.

Bottom ash has been used in large scale civil projects since 1985 (the Caland windbarrier in Rotterdam, 660,000 tons). The first large scale road embankment using bottom ash was in 1989 (Highway 15 Rotterdam, 470,000 tons). In recent years, bottom ash has also been used in highway embankments (Highway 27, 1.0 Mt and N-50, 200,000 tons), sound barriers (Vleuten) and in industrial building sites below high-level floors in large warehouses (Amsterdam, Heerenveen, Ridderkerk, Moerdijk). Furthermore, bottom ash is used as a construction material beneath the final cover of landfills. The sacle of some applications is given in Table 2.

Intermediate specialized companies like Insulinde Recycling have proven to be important to reach a high degree of application of MSW Bottom Ash in large scale projects, considering the continuous production on one side and the intermittent realization of the projects on the other hand.

This paper describes some experiences in the Highway 27 project and the position of the intermediate company in large scale applications of bottom ash.

Table 2 Large scale applications of bottom ash in the Netherlands

APPLICATION	BOTTOM ASH USED, kton
Caland wind barrier	660
Highway 15 Rijnmond	470
Motorway S20	180
Railway track Amsterdam	95
Highway 4/44	120
Container Terminal	150
Highway 27 Flevoland	1,000
Highway 50 Kampen	200

HIGHWAY 27

Project information

The project is situated in Flevoland, in earlier days part of the Zuiderzee, in the central part of the Netherlands. The total length of the new highway is 19 km. The highest point of the construction is 16 m above sea level.

The construction with MSW Bottom Ash started in 1994 and was completed in 1997. The height of the embankment varied between 1 m and 10 m. Inaccordance with the environmental regulations the bottom ash was covered after reaching the prescribed density by a sand-bentonite cover, thickness 0.20 m.

The new highway was officially commissioned in December 1999.

Design phase

In the design phase it was decided which parts of the project were suitable for using MSW Bottom Ash. Criteria for that were the position of the ground water table and the minimum height of a bottom ash embankment that is cost-efficient with respect of the costs of environmental measures. This minimum height is 1 – 2 m and depends mostly on the width of the embankment.

The Directorate General for public works and watermanagement (RWS) demanded that the costs of application of the bottom ash, environmental measures included, should not be more expensive than the application of natural sand. A calculation model was used to determine the different cost factors using the bottom ash. These included the measures of maintenance and monitoring during the life-time of the project.

In the design phase it was shown that a substantial amount of bottom-ash could be used: far more than 1.0 Mt in total. This amount was almost twice the yearly production of bottom ash in the Netherlands at that time. For this reason RWS decided to supply the bottom ash themselves to the contractors for the different parts of the project. In this way it was possible to have sufficient time for a careful design, gathering information of experiences in previous applications and, last but not least, getting the assurance that the amount of bottom ash would be available in the amounts wanted on the right moment during construction.

In this phase RWS was contacted by Insulinde Recycling, an intermediate and independent company with knowledge and experience in supplying bottom ash for application in large scale civil projects. The amount of bottom ash needed could be supplied and guaranteed by Insulinde due to several service-contracts of Insulinde with a number of MSW incinerating facilities.

The main differences between a design of the embankments using sand and using MSW Bottom Ash are:

- a drainage layer of sand before applying the bottom ash to be sure that with respect of settlements to be expected in the lifetime of the construction there is always a distance of more than 0.5 m between the bottom ash and the ground water table;
- a sand-bentonite capping on top and on the slopes of the bottom ash with a minimum thickness of 0.2 m;
- a 1.0 m layer of sand on top of the capping before the road foundation was constructed; this layer is necesssary for protection of the capping and to have sufficient space for ducts, cables, lamp-posts, etc.;
- near concrete constructions the embankment of bottom ash was stopped at a distance of at least 15 m to prevent possible negative effects of chlorides and sulphates from the bottom ash on the concrete construction;
- the monitoring of the ground water table and ground water quality for at least 10 years.

The environmental authorities were informed at this stage of the proposed application of bottom ash. A permit was not needed. Due to the amount of bottom ash needed and the limited storage possibilities for bottom ash at the site of the MSW incineration plants, it was necessary to have two temporary storage sites located near the project. The sites were situated near a canal. For these storage sites an environmental permit was needed and due to environmental regulations a HDPE liner was used to prevent contamination of the subsoil by the bottom ash or leachate from the bottom ash stored. The temporary sites were dismantled at the end of the construction phase.

The proposed application of MSW Bottom Ash was communicated during the design phase to the communities nearby. An exposition with information on the project and the use of bottom ash was available nearby the project.

Materials

The MSW Bottom Ash used in the project, in total nearly 1.0 Mt, originated from 4 different MSW incinerations plants in the Netherlands: in Rotterdam (67 %), Amsterdam (31 %) and Nijmegen (2 %). The bottom ash was transported to the site by ship (58 %) and by truck.

About 30 % of the total amount was stored during the construction phase in the temporary storage sites.

All bottom ash complied to the demands of the KIWA National Guidelines for Certification BRL 2307.

Some 210.000 tons of LD-slags from the Hoogovens steelindustries were also used in the project. This material was mostly used to achieve a smooth surface on the MSW Bottom Ash prior to the sand-bentonite cover. The LD-slags were transported completely by ship to the project. The quality of the slags was determined to be category 2 material according to the Building Material Decree.

The sand-bentonite for the capping was produced by Insulinde Recycling mixed-in plant on the site. The mix was delivered in the project by truck.

Construction phase

The MSW Bottom Ash was delivered in the project in amounts of 2,000 – 2,500 tons per day, with a maximum of 16,000 per week The capacity of the temporary storage sites was 150,000 tons each.

The bottom ash was laid in 0.4 m layers before compaction. The compaction rate of the bottom ash was monitored with the sand-replacement method. The average compaction measured was 102 %. The demand was an average of 98 % with a minimum of 93 %.

In the summer of 1994 there was local erosion of the compacted bottom ash due to heavy rainfall over a short period of time. This happened where the capping was not yet in place. To prevent erosion and the following contamination of the adjacent soil surface it was decided that for the rest of the project the compacted bottom ash was covered weekly with a thin layer of bitumina-emulsion.

During construction of a specific part of the embankment it was measured that the settlement of the subsoil was larger than that predicted. The cause of this proved to be a historic old river-bed in the underground not discoverd in the design phase. Differences in settlement of 0.8 m over a distance of 60 m were measured. To prevent a possible contact between the MSW Bottom Ash and the ground water it was decided to remove the sand-bentonite capping and to excavate the bottom ash, some 20,000 tons, up to the sand drainage layer. A thicker sand layer was brought in and the bottom ash and the sand bentonite were replaced. During this operation it was found that the original sand bentonite was not damaged due to the difference in settlements and that the bottom ash could be excavated without extreme efforts even after compaction.

The sand-bentonite for the project was produced with a maximum of 545m^3 per day. The material was compacted with a dynamic roller on horizontal surfaces and a dynamic plate on the slopes. Compaction rates of 98 % horizontal and 95 % on the slopes were reached. A demand on the hydraulic conductivity of 1 E-10 m/s was measured. In total some 36,000 m3 of sand bentonite was produced and applied.

Monitoring phase

The ground water level and quality at 9 places along the project were measured monthly during the construction phase and yearly since 1997. The quality was measured with the anorganic parameters copper, zinc, arsenic and molybdenium. No significant changes have been found.

Cost aspects

To calculate the costs of the application of MSW Bottom Ash in this project the following aspects were taken into account:

- spreading and compacting the bottom ash
- temporary covering with bitumina-emulsion
- costs of temporary storage sites
- production, spreading and compaction of sand bentonite
- extra efforts of personal in the design and in the construction phase
- monitoring during construction and during 10 years afterwards
- the alternative of natural sand delivered on site

This cost-calculation showed that the application of MSW Bottom Ash in this project was lower than the cost of natural sand by an average of 46 % (min. 28 % and max. 54 %).

Learning points of the project

The following learning points of the project have been identified:

- extra attention is needed in the design phase for possible changes in the ground water level during construction; the possibilities of constructing extra drainage measures are to be considered;
- in the design phase a good historic research of the local situation of the underground is essential; if necessary intensify the underground investigation locally;
- beware of the possibility of erosion of bottom ash not yet compacted due to heavy rainfall, especially in the case of a large horizontal surface and steep and long slopes; temporary covering may be necessary
- a sand layer on top of the sand-bentonite capping may act as a waterdrainage and prevent a soaked cover layer;
- a guarantee on the availability of enough bottom ash is necessary; the logistics of large amounts of bottom ash to be delivered in a short period are very important; temporary storage sites near the project can be very useful;

Role and position of the intermediate company

In the Netherlands is has been proved that the intermediate company is an important factor in achieving a high degree of application of MSW Bottom Ash.

The main parties involved in using bottom ash are:

- the public authorities and private companies as owners/builders
- the producers of bottom ash (MSW incineration plants)
- the building contractors
- the regulating authorities
- the intermediate companies for bottom ash

With exception of the intermediate company, the application of bottom ash is for neither party a 'core-business'.

Main activities, knowledge and experience of the intermediate companies are:

- technical consulting in each phase of the project
- acquisition of projects suitable for bottom-ash
- commercial service contracts with several MSWI plants; sometimes treatment and certification are included in the activities
- contracts for guaranteed supply to the users of the bottom ash
- transportation and temporary storage of bottom ash from different MSWI plants for large scale projects
- consulting and compliance on rules and regulations
- quality assurance and part of the certification procedures.

These activities require extensive knowledge and experience. Besides that a good working relationship with all parties involved is essential. For, especially the complex environmental regulations on the use of bottom ash in projects ask for intensive activities in different phases of construction: design, transportation, application and after-care.

CONCLUSIONS

The extensive experiences in the Netherlands with large scale applications of MSW Bottom Ash have proven that, even with strict environmental demands, it is possible to use bottom ash in a responsible way. However, the application in large scale embankment projects asks for careful design and intensive logistic planning. In large scale projects cost reductions of the costs of natural sand of up to 50 % are possible. The knowledge and experiences of specialized intermediate companies is essential to achieve a maximum application of all the bottom ash produced.

ACKNOWLEDGEMENTS

The author would like to thank the Directorate General for Public Works and Watermanagement (RWS) and the Dutch Waste Processing Association for their continuous and stimulation efforts to contribute to the high degree of application of MSW Bottom Ash in the Netherlands.

REFERENCES

1. EVALUATIE RAPPORT HERGEBRUIK AVI-BODEMASSEN, Aanleg Rijksweg A27 Eemnes-Almere, RWS Directie Ijsselmeergebied, January 1998 (in Dutch).

2. JAARVERSLAG 1998 ADVIESCENTRUM AVI-RESTSTOFFEN, ACR 1999 (in Dutch).

3. AFVALVERWERKING IN NEDERLAND, Gegevens 1998, AOO/RIVM/VVAV, July 1999 (in Dutch).

4. TREATMENT & BENEFICIAL USE OF MSWI RESIDUES IN THE NETHERLANDS, State of the art 1997, NOVEM 1997.

5. NATIONALE BEOORDELINGSRICHTLIJN AVI-BODEMAS, BRL 2307, KIWA 1999 (in Dutch).

6. BUILDING MATERIALS DECREE, Ministerie VROM, Ministerie V&W, 1995 (in Dutch).

7. BEREKENINGSSYSTEMATIEK VOOR DE ACCEPTATIEPRIJS AVI-BODEMAS IN GROOTSCHALIGE OPHOGINGEN, RWS/VVAV, 1997 (in Dutch).

MUNICIPAL SOLID WASTE INCINERATION (MSWI) BOTTOM ASH: CHARACTERISTICS AND THE USE IN ROAD CONSTRUCTION

G Pfrang-Stotz

Institute of Technical Chemistry/Thermal Waste Treatment

J Reichelt

University of Karlsruhe

Germany

ABSTRACT. Within the framework of a research project mineralogical, chemical and structural engineering methods were applied to raw bottom ashes as well as to processed bottom ashes stored for three months, all originated from 15 German municipal solid waste incineration plants operated on different process technologies. The MSWI bottom ashes consist mainly of silicates, oxides and carbonates. Secondary phases occuring are amorphous glass phases, several iron minerals and feldspars. Accessory and trace substances detected include metals, alloys as well as a great variety of sulfates. The mineral phases of the MSWI bottom ashes undergo alterations which means that while they are stored hydration reactions, solidification reactions through formation of calcite, sulfate reactions, salt formation reactions, corrosion reactions of iron and aluminium as well as solution reactions occur. The chemical analysis pointed out that the MSWI bottom ashes contain silicon, aluminium, iron and calcium as the principal elements. In addition, sodium, potassium and traces of heavy metals are present Accordingly, MSWI bottom ash is a calcium-aluminium silicate mixed with an iron fraction. The anions detected are mainly chlorides and sulfates. The chemico-physical leaching behaviour, which was studied in the German DEV-S4 elution tests, in the Dutch column and availability tests and in lysimeter experiments, depends on the amount of water available, the pH, the buffering capacity, and the prevailing mineralogical state. The structural engineering properties are determined by the mineral phases present. High contents of stable silicates and oxides produce a positive effect whereas high concentrations of sulfates and chlorides have a negative impact on the properties. Besides, the stability is controlled decisively by the degree of carbonatization.

Keywords: MSWI bottom ashes, Mineralogy, Environmental compatibility, Structural engineering properties.

Dr G Pfrang-Stotz is working at the Karslruhe Research Center at the Institute of Technical Chemistry/Thermal Waste Treatment. Her research interests includes the mineralogical composition and the environmental compatibility of MSWI bottom ashes and other industrial byproducts.

Dr Ing J Reichelt is working at the Institute of Road Construction at the University of Karlsruhe/Germany. He is a project manager of several research projects in the field of road construction. Aim of these projects are the possibilty of using of natural aggregates and industrial byproducts as construction materials.

INTRODUCTION

In the Federal Republic of Germany wastes should, in the first line, be avoided and, in the second line, be recycled or utilized for energy production [1]. When MSWI bottom ash is utilized in road construction as granular base material it has to comply with the structural engineering requirements, and its environmental compatibility has to be ensured.

With regard to the mineralogical composition four types of bottom ashes can be distinguished:

1. grate ash: material discharged from the end of the grate
2. raw bottom ash: grate ash discharged via a quench tank
3. bottom ash: raw bottom ash which had been processed and stored for at least three months
4. aged bottom ash: raw bottom ash which had been stored for several years

In order to evaluate these requirements, mineralogical, chemical and structural engineering methods were applied to raw bottom ashes as well as to processed bottom ashes stored for three months, all originating in 15 municipal solid waste incineration plants operated on different process technologies.

PRODUCTION AND PROCESSING OF MSWI BOTTOM ASHES

In the combustion chamber of a municipal solid waste incineration plant dry grate ash is produced on the grate at fuel bed temperatures of about 850° C. The grate ash is discharged via a quench tank (water bath). The resulting product is the raw bottom ash. After several weeks of storage in order to reduce the moisture content and the alkalinity, the raw bottom ash is processed to material for use in structural engineering.

Processing consists in most cases of screening and magnet separation. The products obtained are bottom ashes, 0/32 mm grain size, ferrous metals and rejects. The bottom ashes are stored for at least three months in order to improve their ecological and structural properties.

MINERALOGY

The MSWI bottom ashes are a multi-component system of slag fractions, glass, ceramics, metals, other constituents (e. g. rock) and unburned residues Figure1.

With a view to the crucial structural engineering and environmentally relevant properties of bottom ashes the mineralogical composition of the largest group of materials - the slag fraction - commands particular interest. The crystalline structure and the types of bond of the crystals in the mineral phases present are responsible for the physical and chemical properties of a material.

The MSWI bottom ashes consist mainly of silicates, oxides and carbonates. Secondary phases occuring are amorphous glass phases, several iron minerals and feldspars. Accessory and trace substances detected include metals, alloys as well as a great variety of sulfates Table 1.

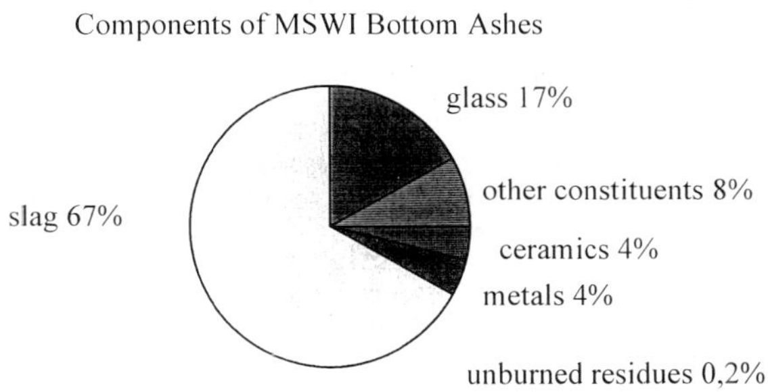

Figure 1 Components of MSWI bottom ashes

Table 1 Mineralogy of MSWI bottom ashes

Silicates:	Gehlenite	$Ca_2Al_2SiO_7$
	Åkermanite	$Ca_2MgSi_2O_7$
	Kalifeldspar	$(Na,K)[AlSi_3O_8]$
	Plagioclase	$NaAlSi_3O_8$- $CaAl_2Si_2O_8$
	Diopside	$Ca(Mg,Fe)Si_2O_6$
Oxides:	Quartz	SiO_2
	Magnetite	Fe_3O_4
	Maghemite	γ-Fe_2O_3
	Haematite	α-Fe_2O_3
Carbonates:	Calcite	$CaCO_3$
	Magnesite	$MgCO_3$
Hydroxides:	Portlandite	$Ca(OH)_2$
Sulfates:	Anhydrite	$CaSO_4$
	Bassanite	$CaSO_4x0,5H_2O$
	Gypsum	$CaSO_4x2H_2O$
Salts:	Halite	$NaCl$
	Sylvite	KCl
Elemental:	Iron, Aluminium, Copper	

The formation process of the grate ashes in the combustion chamber does not reach thermodynamic equilibrium with respect to the primarily formed mineral phases. This means, after discharging out of the combustion chamber via a quench tank the primarily formed raw bottom ashes are not thermodynamically stable products according to their mineralogical composition. Under ambient conditions they undergo with respect to their mineral composition further mineralogical processes.

In the first three months of storing mineralogical alteration corrosion processes (hydration reactions, solidification reactions, sulfate reactions, salt formation reactions, reactions of iron and aluminium, solution reactions) take place and stable bottom ashes are formed Figure 2.

- Quenching of the grate ash
- Uptake of water: about 20 to 35 wt.%
- Alteration/new formation of mineral phases

⇒ Fixing of glass fractions
⇒ Conversion: calcium oxide to portlandite

$$\underset{\text{calcium oxide}}{CaO} + H_2O \rightarrow \underset{\text{portlandite}}{Ca(OH)_2}$$

⇒ Conversion: anhydrite to bassanite

$$\underset{\text{anhydrite}}{CaSO_4} + \tfrac{1}{2} H_2O \rightarrow \underset{\text{bassanite}}{CaSO_4 \cdot H_2O}$$

⇒ Solution of the readily soluble salts

- Hydratation reactions (fast reaction)
- Solidification reactions

⇒ portlandite to calcite

$$\underset{\text{portlandite}}{Ca(OH)_2} + CO_2 \rightarrow \underset{\text{calcite}}{CaCO_3} + H_2O$$

⇒ Direct precipitation of calcite
⇒ Formation of calcium silicate hydrate phases (CSH)?

- Sulfate reactions

⇒ Conversion: anhydrite to bassanite

$$\underset{\text{Anhydrite}}{CaSO_4} + \tfrac{1}{2} H_2O \rightarrow \underset{\text{bassanite}}{CaSO_4 \text{ x } 0{,}5\, H_2O}$$

⇒ Direct precipitation of bassanite

- Salt formation reactions
- Solution reactions

Figure 2 Major mineral reactions taking place during the first three months of storage

After three month of storing the bottom ashes are fit for utilization in road construction (unbounded and hydraulically bounded material). This means the mineralogical processes determine the environmentally relevant and the structural engineering properties of MSWI bottom ashes.

CHEMICAL COMPOSITION OF MSWI BOTTOM ASHES

A comparison of the chemical bulk composition of MSWI bottom ashes expressed in terms of weight concentrations of oxides shows that the contents of the principal elements silicon, calcium, iron and aluminium have larger ranges. In the first place these differing ranges of the bulk composition can be refered to the different waste input as well as alteration reactions. The elements silicon, calcium and aluminium build up the mineral phases which represent the main mineralogical composition.

The mineralogical composition of MSWI bottom ashes and the contents of the principal elements deviate from the contents found in natural rocks and industrial by-products as blast furnace slags or electro oven slags. MSWI bottom ashes differ from acidic magmatites (silica concentration: 66 or more wt.%; granite, rhyolite) due to lower SiO_2- and higher CaO contents, but have a similar chemical bulk composition to basic magmatites (silica concentration: 45 to 52 wt.%; gabbro, diabas, basalt, nepheline). The CaO contents of the MSWI bottom ashes, however, are higher compared to the acidic magmatites. The blast furnace slags differ from MSWI bottom ashes in the first place by higher contents of CaO wheras electro oven slags show higher contents of CaO and lower contents of SiO_2 Figure 3.

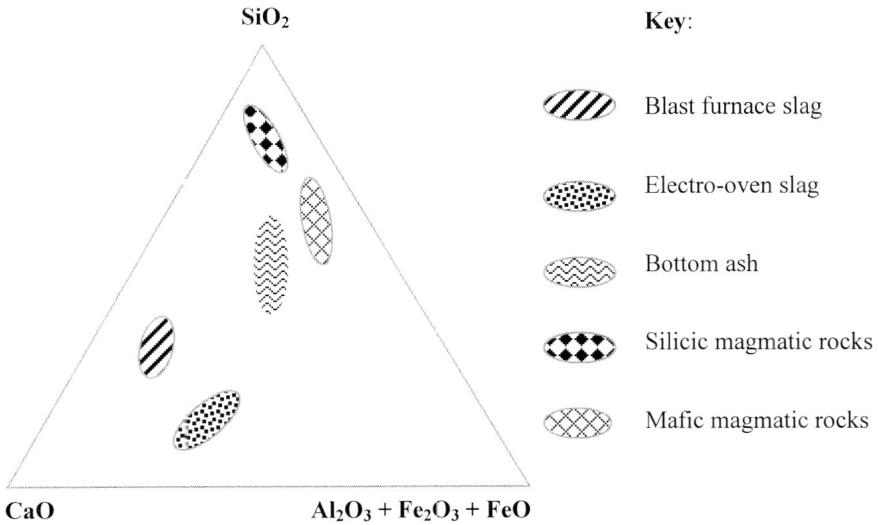

Figure 3 Chemical composition of MSWI bottom ashes

ELUTION BEHAVIOUR

The concentration of harmful specimen in the bottom ashes is no measure of the environmental impact (polution potential). More important are the leaching properties and the mobility of the components out of the mineral phases. The environmental impact depends upon the present minerals, the stage of alteration and the mobility or leaching properties of the component in question and thus is mainly controlled by:

- the speciation
- matrix bond (fixation in the matrix)
- the alkalinity of the material
- the actual pH of the leachate
- the complexing properties of the leachate
- the temperature

In order to characterize the solution behaviour the following 3 leaching tests were applied:

- the German DEV-S4 [2] test is used to check the compliance with German regulating standards (e.g. LAGA [4]) and to compare the leaching properties with those of industrial by-products [3]
- the column test [5] is used for the determination of the short- and medium-term leaching behaviour
- the availability test [6] is used to determine the leaching potential of each of the components of the bottom ashes

Lysimeter tests were performed on fresh raw bottom ash in order to be able to characterize the time resolution of actual release and to determine approximate values of the contamination of the percolates.

DEV-S4 Test

The DEV-S4 test is a non pH controlled extraction test where the sample is shaked in water during 24 hours. The liquid-solid ratio is 10.

The statistical interpretation of a quality controlled MSWI bottom ash shows Table 2.

- The largest fraction of cations in the solution is made up by calcium, sodium and potassium. The anions are sulfate and chloride.
- The largest standard deviations are refered to the elements calcium, sodium, potassium, chloride and sulfate. These elements build up the soluble salts halite (NaCl) and sylvite (KCl), the sulfates anhydrite ($CaSO_4$) and bassanite ($CaSO_4x0,5H_2O$) as well as carbonates, e.g. calcit ($CaCO_3$). The high leaching rates of these elements are attributable to those mineral phases, which react very sensitively to modifications of the physical-chemical environment.
- The arithmetic means of the concentrations of the heavy metals, chloride, sulfate and cyanide are clearly below the standards of LAGA.

Table 2 Elution data (DEV S4) of raw bottom ashes (F) and three months stored bottom ashes

MSW Incineration Plant		A		B		F	
Parameter	Unit	F	A	F	A	F	A
Calcium	mg/l	492	165	660	270	621	95
Sodium	mg/l	93	137	66	81	55	78
Potassium	mg/l	72	102	27	48	22	38
Iron	µg/l	62	53	74	48	75	64
Zinc	µg/l	328	113	716	284	393	154
Copper	µg/l	358	204	222	154	442	118
Lead	µg/l	1779	26	2635	1488	801	54
Chromium	µg/l	57	86	42	84	47	91
Nickel	µg/l	41	43	94	60	79	70
Sulfate	mg/l	318	294	90	13	991	258
Chloride	mg/l	228	225	174	146	292	133
Fluoride	mg/l	4	4	4.4	3.7	6.2	0.5
pH		12	11	12	12	12	11
El. Conductivity	µS/cm	5640	2035	7730	3650	6962	1236

Availability Test

The availability test has been designed to supply a measure for pollutially leachable fraction of an element under worst environmental conditions. The availability test is a pH controlled extraction test performed at pH 7 (maximum leaching of the soluble salts) and 4 (high leaching rates of heavy metals) [7].

The test results are Table 3.

- Calcium and potassium show the highest leaching rates.
- The leaching of heavy metals follows the sequence zinc, lead, copper, nickel and chromium.

Column Test

The column test is a dynamic flow-through test. Increasing volumes of water (pH = 4) flow through the column which is filled with the test material. Analyses of the eluates at the quantity of testing material results in a temporal splitting of the elution process. Various liquid-solid ratios (max. L/S = 5) give an indication of the resolution of the leaching process.

The test results are Figure 4,5.

- Raw bottom ashes show heavy metal concentrations which are varying between 0.2 and about 5 mg per kg.

- A comparison with the elution behaviour of stored bottom ashes shows that the curves are similar, but the measured elution contents of most of the heavy metals are much lower.The final accumulated elution rates are between 0.1 and 0.8 mg per kg.This shows clearly the positive effect of three months storing before utilization the bottom ash as material in unbounded layers in road construction.

Table 3 Availability Test: Maximum Leachable Quantities in mg/kg

ωES [mg/kg]	K	Ca	Zn	Cu	Pb	Cr	Ni	Sulfate	Chloride
Raw Bottom Ash C	2,37	46,65	1,76	0,08	0,72	0,01	0,02	39,13	3,67
Bottom Ash C	1,99	44,02	1,56	0,12	0,058	0,01	0,02	15,15	3,4
Raw Bottom Ash D	1,96	37,91	0,01	0,2	0,289	0,01	0,01	6,31	4,13
Bottom Ash D	1,25	25,36	0,51	0,09	0,036	0,01	0,01	2,88	3,78
Raw Bottom Ash G	1,32	23,93	0,41	0,08	0,09	0,004	0,005	6,77	2,39
Bottom Ash G	2,36	23,43	0,45	0,11	0,13	0,01	0,01	6,57	2,01
Raw Bottom Ash I	0,62	32,5	0,54	0,01	0,069	0,005	0,02	°	°
Bottom Ash I	1,15	26,89	0,99	0,16	0,063	0,01	0,01	5,28	1,8
Raw Bottom Ash K	1,77	52,81	0,94	0,22	0,075	0,01	0,01	11,37	3,46
Bottom Ash K	1,35	27,57	0,85	0,24	0,53	0,01	0,01	9,71	3,82
Raw Bottom Ash L	1,04	48,89	0,55	0,02	0,007	0,01	0,01	11,78	2,51
Bottom Ash L	2	26,51	1,27	0,32	0,066	0,01	0,01	12,58	7,64
Raw Bottom Ash M	1,03	26,95	0,59	0,15	0,003	0,01	0,01	11,14	0,47
Bottom Ash M	0,69	24,27	0,69	0,26	0,065	0,01	0,01	7,73	5,92
Raw Bottom Ash N	0,48	23,39	0,54	0,05	0,77	0,004	0,01	5,18	1,6
Bottom Ash N	0,37	34,44	0,86	0,12	0,4	°	0,01	11,6	1,82

° no test result

Lysimeter Tests

Lysimeter tests (Table 4) were performed on fresh raw bottom ash in order to be able to characterize the solution behaviour versus time and to determine approximate values of the actual percolate water contamination. The following results were obtained:

- The largest fraction of cations of the lysimeter percolate waters is made up of sodium, potassium, calcium and, in the second place, by strontium. The principal anions are chloride, sulfate and, in the second place, fluoride.
- The high sodium, potassium and chloride concentrations are attributable to the readily soluble mineral phases halite ($NaCl$) and sylvite (KCl).
- The drop in calcium and strontium concentrations is due to the precipitation of carbonate.
- The reduction of the heavy metals lead, arsenic and zinc can be attributed to a precipitation of hydroxides. Likewise, precipitation as carbonate is possible.
- The steady increase in copper concentration allows the conclusion to be drawn that soluble organocopper complexes are formed.
- The solution of anhydrite and bassanite is responsible for the sulfate concentration.

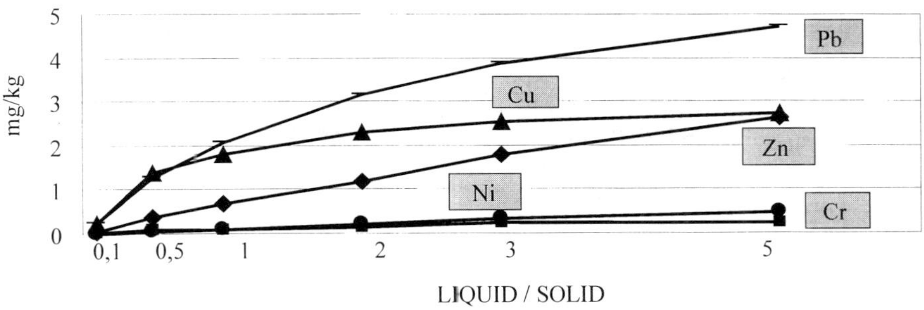

Figure 4 Column test: Plant O: Elution data of raw bottom ashes

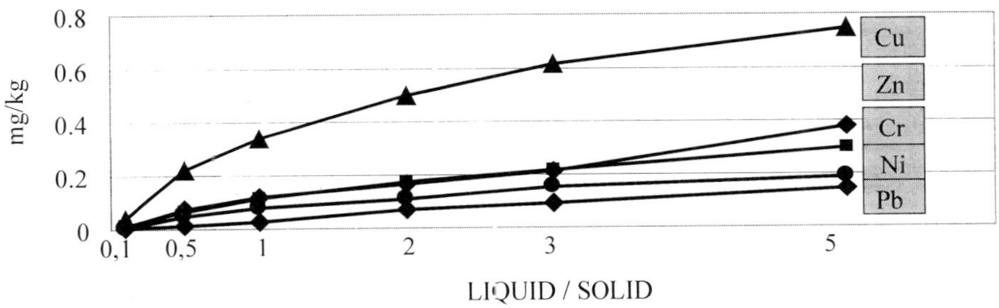

Figure 5 Column Test: Plant O: Elution Data of Bottom Ashes

Table 4 Constituents of the lysimeter percolate waters

Sample		L2	L3	L4	L5	L6
Parameter	unit	6 days	12 days	19 days	27 days	34 days
Sodium	mg/l	2430	2300	2500	3290	2610
Potassium	mg/l	1270	1300	1390	1470	1690
Calcium	mg/l	780	509	421	287	130
Strontium	mg/l	11.6	10.6	9.46	7.8	6.03
Rubidium	mg/l	2.82	3.07	3.18	3.15	3.85
Lead	µg/l	2090	222	161	203	293
Arsenic	µg/l	217	59	97	84	30
Zinc	µg/l	199	72	79	107	73
Copper	µg/l	184	271	395	476	543
Iron	µg/l	124	112	N.A.	N.A.	55
Chromium	µg/l	100	N.A.	94	86	89
Manganese	µg/l	71	142	140	125	64
Nickel	µg/l	40	38	39	35	74
Cadmium	µg/l	< 0,5	< 0,5	< 0,5	N.A.	N.A.
Chloride	mg/l	3760	4280	4500	4310	4310
Sulfate	mg/l	1610	1680	1570	1350	1350
Fluoride	mg/l	N.A.	N.A.	N.A.	16.4	16.4
pH		12.4	8.7	8.1	6.7	7.2
el. Conductivity	mS/cm	14.4	14.6	15.1	14.4	16.3

N.A. = not analysed

STRUCTURAL ENGINEERING PROPERTIES

Structural engineering investigations were made on MSWI bottom ashes stored for three months and processed, which had been produced in 15 municipal solid waste incineration plants operated on different process technologies, in order to be able to characterize the structural engineering properties for utilization in road construction Table 5.

Based on the results of investigations, the following summarizing conclusion can be drawn:

The crucial limits laid down in TL HMVA-StB 95 [8] are met by all investigated MSWI bottom ashes investigated, with the exception of the parameters „metal content“ and „freeze/thaw resistance“.

Table 5 Structural engineering properties of MSWI bottom ashes

Parameter	Unit	A	B	C	D	E	F	G	H	I	J
Water Content	wt.%	7.5	N.A.	N.A.	12.6	6.3	N.A.	6.8	6.1	9.7	7.2
Bulk Density	g/cm³	2.697	2.665	2.71	2.65	2.69	2.72	2.658	2.662	2.534	2.68
uncompacted bulk Density	g/cm³	1.228	N.A.	1.2	1.21	1.16	N.A.	1.234	1.215	1.049	1.27
Proctor-Density	g/cm³	1.608	1.566	N.A.	1.64	1.91	N.A.	1.915	1.772	1.628	1.99
Op. Water Content	wt.%	19.4	18.6	N.A.	15.7	19	N.A.	16.3	14.8	18.7	18.2
Freeze/Thaw	wt.%	9.2	6.4	12.9	8.9	6.7	11.3	3.6	4.4	2.9	3.5
Impact Test	wt.%	35.1	37.3	35.7	38.9	37.6	36	38.1	38.4	40.6	37.5

*TL = Technical Supply Terms for MSWI Bottom Ashes [8]
N.A. = not analysed

The high contribution of metallic specimens is attributable to poor processing of the raw bottom ashes. The low resistance to freeze-thaw-changes can be attributed to the high contribution of low resistant mineralogical attributes such as high porosity, specific mineral composition and relatively weak carbonate bond between the bottom ash constituents.

The impact destruction values ($SZ_{8/12}$) of the Impact Test determined as a measure of resistance to mechanical loading are within the limits laid down in TL HMVA-StB 95 [8]. However, a comparison with the impact destruction values of natural mineral substances, i. e. basalt ($SZ_{8/12}$: 9 - 20 wt.%) makes evident that similar stabilities cannot be achieved by MSWI bottom ashes. This is attributable to the relatively weak carbonate bond between the bottom ash constituents and to the specific composition of mineral substances. Thus, a high content of stable silicates and oxides exerts a positive influence on the stability properties whereas high fractions of salts and sulfates reduce the stability.

SUMMARY

The MSWI bottom ashes consist mainly of silicates, oxides and carbonates. Secondary phases occuring are amorphous glass phases, several iron minerals and feldspars. Accessory and trace substances detected include metals, alloys as well as a great variety of sulfates.

After discharging from the combustion chamber via a quench tank the primarily formed raw bottom ashes do not reach thermodynamical stability. Under ambient conditions they undergo with respect to their mineral composition further mineralogical processes. In the first three months of storing mineralogical alteration processes take place and stable bottom ashes were formed.

A possible criterion of distinction between fresh MSWI raw bottom ash from incineration and aged MSWI bottom ash may be the absence of the portlandite fraction and the ratio of anhydrite/bassanite measured in the MSWI bottom ash.

The mineralogical composition of MSWI bottom ashes and the contents of the principal elements deviate from the contents found in natural rocks and industrial by-products like blast furnace slags or electro oven slags.

The statistical interpretation of the result of the German DEV S4-test shows: The largest fraction of cations in the solution is made up by calcium, sodium and potassium. The anions are sulfate and chloride. The largest standard deviations are refered to the elements calcium, sodium, potassium, chloride and sulfate. The arithmetic means of the concentrations of the heavy metals, chloride, sulfate and cyanide are clearly below the stringent limits set by LAGA. The test elucidates the positive effect of three months storing on the environmental compatibility of the MSWI bottom ashes.

The availability test is used to determine the maximum leachability of each of the components of the bottom ashes. The test results are: The highest availabilities are associated with calcium and potassium. For the leaching of the heavy metals the sequence zinc, lead, copper, nickel and chromium is recognizable.

The column test used for the determination of the short- and medium-term leaching behaviour. The test results are (max. L/S = 5): A comparison of the elution behaviour based on the results of the column test shows, that the elution curves of raw bottom ashes and bottom ashes are similar, but the absolutely measured elution rates of the raw bottom ashes are much higher.

The low resistance against freeze-thaw changes and mechanical stress can be attributed to the high contribution of low resistant mineralogical attributes such as high porosity, specific mineral composition and relatively weak carbonate bond between the bottom ash constituants.

REFERENCES

1. Gesetz zur Förderung der Kreislaufwirtschaft und Sicherung der umweltverträglichen Beseitigung von Abfällen (Kreislaufwirtschafts- und Abfallgesetzt), Bundesgesetzblatt, Jahrgang 1994, Teil 1
2. DIN 38 414, Teil 4: Deutsche Einheitsverfahren zur Wasser-, Abwasser- und Schlamm-untersuchung: Schlamm und Sedimente (Gruppe S), Bestimmung der Eluierbarkeit mit Wasser (S4), Beuth Verlag, Berlin 1984.
3. VEHLOW, J: Bewertung des Umweltverhaltens von Reststoffen der Müllverbrennung. KfK-Nachrichten, 25, 4 1993, p. 173-181.

4. Mitteilung der Länderarbeitsgemeinschaft Abfall (LAGA), Merkblatt über die Entsorgung von Abfällen aus Verbrennungsanlagen für Siedlungsabfälle, Erich Schmitt Verlag, Neuburg, Ausgabe 1994.

5. NEN 7341: Determination of leaching characteristics of inorganic components from granular (waste) materials. Netherlands Standardization Institute (NNI), Delft, (1993).

6. NEN 7345: Determination of the release of inorganic constituents from construction materials and stabilized waste products. Netherland Standardization Institute (NNI), Delft, (1993).

7. VEHLOW, J.: Reststoffbehandlung – Schadstoffsenke Thermische Abfallbehandlung. Beitrag zur FDBR-Konferenz: Die thermische Abfallverwertung der Zukunft – Mit 100 Jahren Erfahrung ins nächste Jahrhundert. Düsseldorf, 28.September (1995).

8. Technische Lieferbedingungen für Hausmüllverbrennungsasche im Straßenbau (TL HMVA-StB). Forschungsgesellschaft für Straßen- und Verkehrswesen, Köln, 1995.

VOLUME STABILITY OF HYDRAULIC BOUND MSWI ASH

E M Westiner

L Löcherer

Th Wörner

Technical University of Munich

Germany

ABSTRACT. Mineral residues from the incineration of municipal solid waste (MSWI Ash) are used as building material in road construction. By consolidation of MSWI Ash using hydraulic binder several environmentally critical aspects of MSWI Ash can be improved. Although MSWI Ash in Germany is already be used in hydraulic bound sub-bases, it is important to investigate the potential physicochemical processes that may dissolve the structural components of hydraulic bound MSWI Ash. At the Technical University of Munich, Germany a time reducing procedure was developed to examine the volume stability of hydraulic bound MSWI Ash. The volume stability of hydraulic bound MSWI Ash depends on the composition (chemistry, grading) and time of storage of MSWI Ash and the chemistry of hydraulic binder. MSWI Ash from 20 selected incineration plants was examined taking into account the different ways of municipal solid waste, incineration processing and upgrading technologies. This was done to be able to set up a specification allowing the use of MSWI Ash in hydraulic bound sub-bases in Germany without causing any decrease in road quality.

Keywords: MSWI Ash, Hydraulic bound, Volume stability.

Dr E M Westiner, is head of the task group mineral aggregates of the Institute for Building Construction and Materials, Technical University of Munich. He specialises in testing natural, artificial and recycled aggregates for use in concrete, asphalt and unbound mixtures.

Mr L Löcherer, is a research student at the Technical University of Munich. His research interests include the use of recycled aggregates and industrial by-products in road construction.

Dr Th Wörner, is head of the task group asphalt and sealing of the Institute for Building Construction and Materials, Technical University of Munich. He has published widely and serves on many Technical Committees in Germany.

INTRODUCTION

Municipal solid waste ash (MSWI Ash) arises through upgrading of residues produced by incineration of communal waste in incineration plants [1-3]. MSWI Ash is a very heterogeneous and predominantly mineral mixture of leachable incineration products, the actual ash, fragments of glass or ceramics, mineral (concrete, brick, gravel, natural stones) and metallic parts, and also non combusted residues [4].

In Germany unbound MSWI Ash is used in road construction predominantly for subgrade, capping layer and sub-bases [5-8]. Observations in laboratories and experience on testing roads also show the fundamental suitability of MSWI Ash for hydraulic bound sub-bases (hydraulic bound MSWI Ash) in addition to the possible effort from MSWI Ash in unbound sub-bases [9-14]. As MSWI Ash can contain substances which affect the environment, consideration of hydraulic character of MSWI Ash could be profitable referring to water economy. Hereby the possible discharge of harmful substances like heavy metals into the environment could be reduced [15].

The developed testing procedures and the fixed standards for conventional mixtures of construction materials using natural stone, gravel and sand are based on empirical results and can not be transferred without restrictions to hydraulic bound MSWI Ash [16]. At present, a testing procedure for the rating of volume stability is not available [17]. To exclude an enlargement of volume in hydraulic bound sub-bases, the only requirement is to store MWSI Ash at least for three month before use. The reasons, which are given in the technical literature for insufficient volume stability of hydraulic bound MSWI Ash, are reactions of glass particles under alkaline conditions, formation of ettringite and reactions under participation of aluminium and hydration of free calcium oxide [18-21]. With regard to the volume stability the content of ash is of great interest, as substances responsible for such reactions appear mainly in ash aggregate [4].

This paper describes the work of a study carried out at the Technical University of Munich, aimed at developing a testing procedure to investigate the volume stability of hydraulic bound MSWI Ash. The *steam test* should above all be usable within the framework of suitability tests. By using the test it should be possible to estimate if hydraulic bound MSWI Ash has sufficient volume stability in sub-bases and if an extension of storing time or any other measures, like adding conventional minerals, lead to a better volume stability. To find out as soon as possible if the chosen mixture shows a sufficient volume stability, the testing procedure should also be practicable for hydraulic bound MSWI Ash with a storing time less than three months.

MIX PROPORTIONS

MSWI Ash

20 MSWI Ashes from selected plants were chosen for examination taking the regional municipal solid waste collection and handling technology, incineration processing technology and upgrading technology into consideration. Unbound MSWI Ash was examined. The results of these tests are given in Table 1. Typical ranges [2] and requirements [1] on MSWI

Ash are given for comparison. The examined MSWI Ashes fulfil by far the constructional requirements of TL HMVA-StB 95 [1]. However the results show that the MSWI Ashes must be classified as critical with regard to their resistance to freezing and thawing (frost resistance). Table 2 shows that MSWI Ashes fulfil the requirements of the sulphate content and also, under consideration of the free calcium oxide contents, the requirement on the total calcium oxide content [3]. Based on these results, an examination of volume stability of the discussed MSWI Ashes would not be necessary.

Table 1 Test results for examined unbound MSWI Ashes

Testing Feature		Test Results Ranges	[2] Ranges	[1] Requirements
material composition				
	ash	34.7 – 90.7	30 – 80	---
	glass/ceramics	4.9 – 62.4	20 – 70	---
	metals	0.9 – 5.0	0 – 20	max. 5.0
	other materials	0.0 – 8.2	0 – 20	---
	non combusted	0.0 - 0.3	0 – 10	max. 0.5
maximum density [kg/m³]		2340- 2850	2200 – 2700	---
flakiness index [%]		8.0 – 12.9	0 – 20	max. 50
aggregate < 0,063 mm [%]		2.5 – 6.7	0 – 10	max. 7.0
frost resistance [%]		1.5 – 17.4	0.5 – 8.5	max. 3.0
fragmentation resistance (impact test) [%]		31.0 – 39.0	35 – 44	max. 40
max. dry density [kg/m³]		1610 - 1790	500 – 1900	---
opt. water content (w_{opt}) [%]		10.5 – 15.0	11 – 18	---

Table 2 Chemical analysis

	Hydraulic Binder			MSWI Ash		
	A	B	C	Test Results	Ranges [18]	Requirements [3]
SO_3	2.2	2.3	2.9	1.3 – 1.7	0.6 – 1.2	max. 2.0
CaO_{total}	63.4	47.7	57.7	18.1 – 23.0	11 – 19	max. 4.0
CaO_{free}	0.3	0.2	0.4	0.0 – 0.6	---	max. 1.5
Al_2O_3	5.3	8.8	5.8	9.6 - 8.8	8 – 14	---
C_3A	8.6	3.5	4.3			
$Na_2O_{equ.}$	0.51	0.61	0.51			

Hydraulic Binder

The three hydraulic binders used (A: Portland cement CEM I 32,5 R; B: blastfurnance slag cement CEM III 32,5-NW/HS; C: road binder HAT 35 according to DIN 18506) have a low alkali content. As Table 2 shows the Na_2O equivalent lies between 0.51 and 0.61 %. A possible reaction of alkaline and silicic acid in the binder with the glassy parts of the MSWI Ashes could thereby be decreased. The content of C_3A ranges from 3.5 to 8.6 %. Consequently, in the presence of sulphate containing components, these reactions, which increase the volume, could not be excluded. Free calcium oxide (CaO_{free}) and free magnesium oxide (MgO_{free}) are included in all three hydraulic binders in only low amounts.

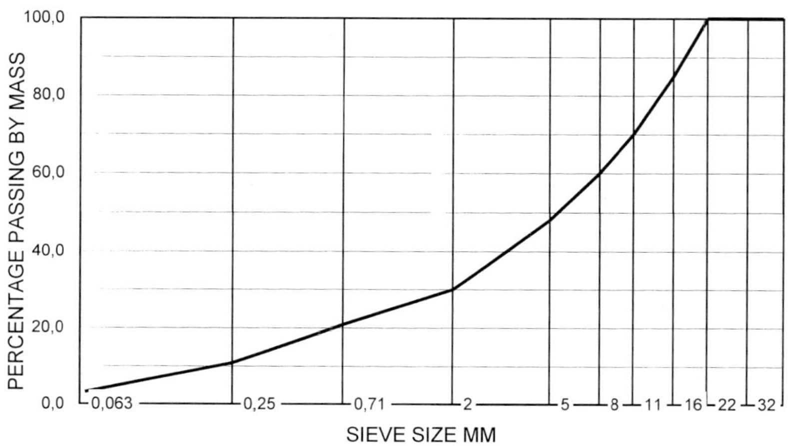

Figure 1 Grading curve of examined MSWI Ash

Mix Proportions

The suitability tests corresponded to the requirements of TP HGT-StB 94 [23]. The tests were made with a uniform, Fuller-like composed grain size (Figure 1). Within the framework of the suitability tests, the water contents required for maximum dry density and the necessary binder contents were determined for the material variants (three MSWI Ashes with three binders, 20 MSWI Ashes only with blastfurnace slag cement). The compressive strength (uniaxial compressive strength after a storage period of 28 days) and the frost resistance (extension of material after 12 cycles of freezing and thawing) were determined for three different binder contents.

As a result of the suitability tests, the optimal water content is between 11 and 16 %, independent of the ash content and of binder type. The bulk density is between 1570 and 1860 kg/m³. The evaluation of the results only shows that a larger amount of binder increase compressive strength. In agreement with the literature [9] no evidence could be found that the frost resistance depends on the kind as well as the content of Ash of the MSWI Ash, or the binder type. The requirement of strength (design strength min. 7 N/mm^2) complies with a binder content of min. 7 % (Figure 2a), whereas the frost resistance (design change of length max. 1 ‰) complies with a content of at least 9 % (Figure 2b). The evaluation of the suitability tests confirms the conclusions of literature data, that to fulfil the requirements of compressive strength and frost resistance higher binder contents in comparison to the use of natural aggregates are necessary [10-11].

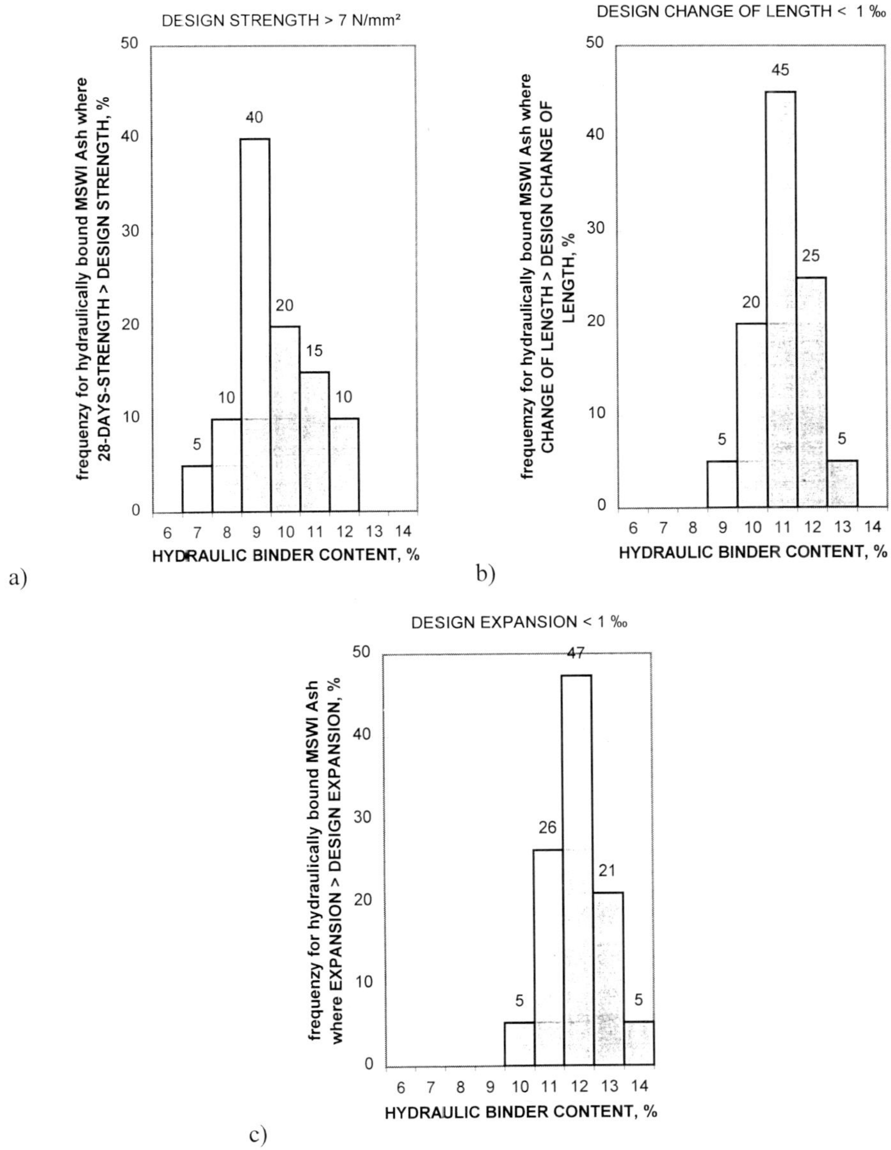

Figure 2 Effect of hydraulic binder content on 28 day compressive strength, freeze thaw resistance and volume stability

VOLUME STABILITY OF HYDRAULIC BOUND MSWI ASH

Testing Procedure

At the Technical University of Munich a time reducing procedure called the *steam test* was developed which allows the determination of the volume stability of hydraulic bound MSWI Ash by measuring the expansion of a cylindrical test specimen.

The test specimen for the steaming test were made and stored according to the standards of TP HGT-StB 94 [17]. The test specimens were cylindrical with a slenderness ratio of 0.83 (diameter d=100 mm and height h=120 mm). They were compacted under vibro compression on a vibrating table with a weight of 12 kg. After storage of 28 days in a climatic chamber (T 20°C, > 95 % relative humidity) small metal plates, which allow to determinate the expansion, were fixed on top and bottom of the test specimen. In the steam test, the test specimen is placed on a grid in a glass jar (d=150 mm, h=320 mm) which is filled in part with water (1 litre) and possesses a cooled lid (Figure 3). The glass jar stands on a heated plate. The test specimen undergoes 21 cycles of steam for 16 hours followed by storage under dry conditions for 8 hours in the glass jar. The determination of the expansion corresponds to freeze-thaw testing, whereby the expansion of the test specimen is refers to its height at the beginning of the test. The expansion of a sample results in a mean value for three single tests.

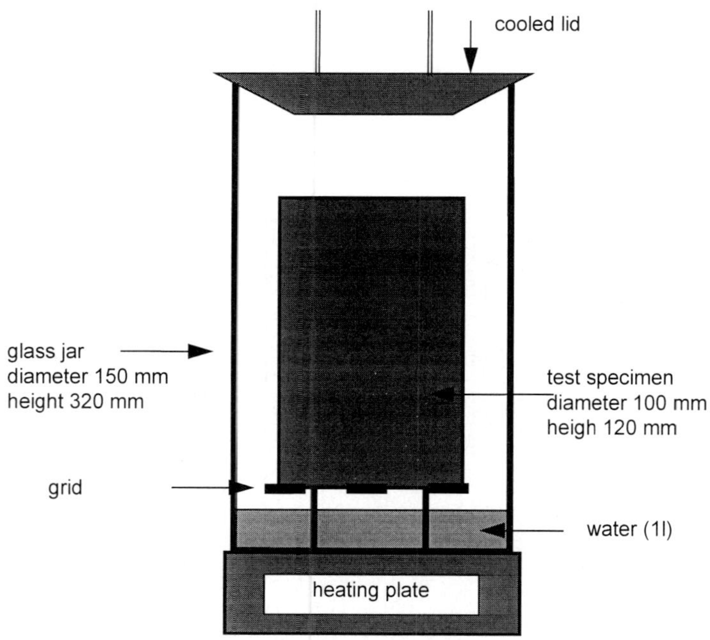

Figure 3 Testing equipment to examine the volume stability (steam test)

During the test the temperatures in the test specimen increases from 20°C (storing at dry conditions) to 70°C (at the end of time of steaming). This range of temperature can also be observed in waste disposal sites for MSWI Ashes [24]. The glass jar enables an observation of the test specimen during the test. Cooling of the lid decreases the loss of steam. Thereby the emission of unpleasant odour is clearly reduced. The range of the expansion, found by the examination, reaches from 0.1 to 14 ‰.

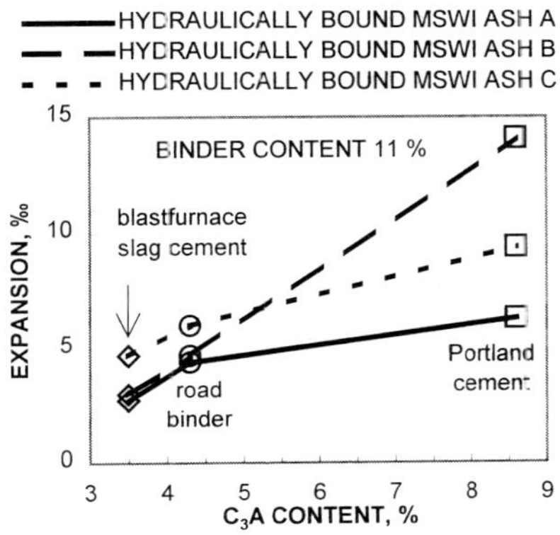

a)

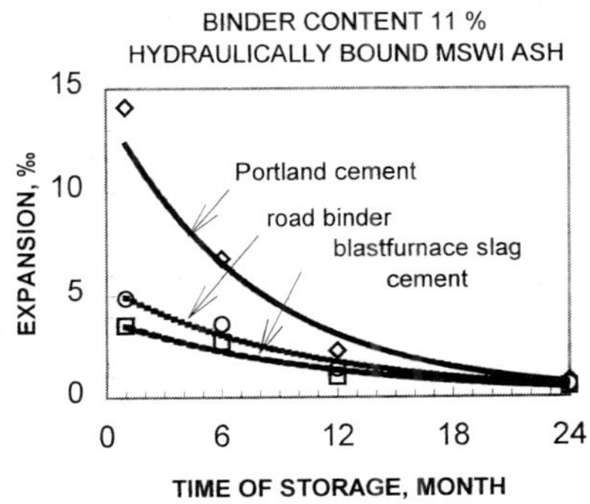

b)

Figures 4 Effect of hydraulic binder on hydraulic bound MSWI Ash

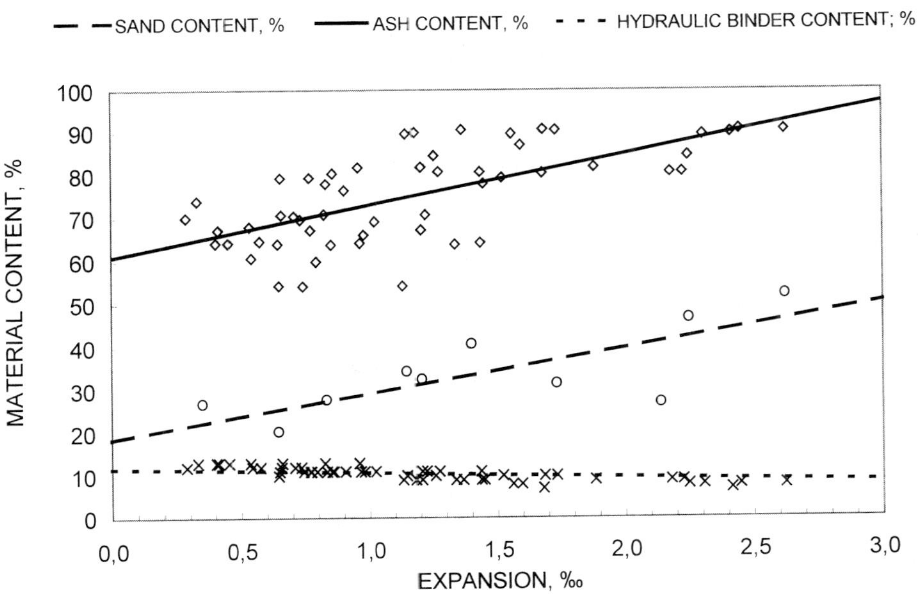

Figure 5 Results of the examination of the volume stability of 20 hydraulic bound MSWI Ashes

Test Results

MSWI Ash

An increase of ash content in hydraulic bound MSWI Ash causes an increase of expansion in the sample during the steam test (Figure 5). Since temperatures in the combustion chamber of incineration plants have increased in the past years as well as the sorting of communal waste, the ash content of the MSWI Ash has risen. According to this a disadvantageous volume stability of the MSWI Ash could be supposed [22]. An increasing amount of sand and/or ash lead to larger extensions of the hydraulic bound MSWI Ash (Figure 5). To improve the volume stability of hydraulic bound MSWI Ash, the processing facilities should be designed for a reduction of the MSWI sand part or its replacement by natural sand. Moreover, the measured extensions depend strongly on the deposit age of the inserted MSWI Ashes (Figure 4a).

The longer the MSWI Ash has been stored, the lower is the volume stability of hydraulic bound MSWI Ash due to the ash content and the particle size distribution. MSWI Ash should therefore, over the 3 months indicated in the regulation, be deposited before use in hydraulic bound sub-bases.

Hydraulic Binder

At higher binder contents, smaller expansions result (Figure 5). With the same composition of the hydraulic bound mixtures higher binder contents result with the steam test than for compressive strength and frost resistance (Figure 2). As well as the binder content, the C_3A content of the hydraulic binder influences decisively the dimension of the expansion during the steam test (Figure 4b).

To reach a successful volume stability a larger content of hydraulic binder is necessary shown by the results of compressive strength and frost resistance. Otherwise appropriate selection of hydraulic binder can reduce the necessary binder content of the hydraulic bound MSWI Ash.

Physical and Chemical Processes

In view of the results, parallel chemical reactions during the steam test can be assumed. Ettringite formation, which could be detected in only small amounts in MSWI Ash, will be enlarged by a (1) ettringite or (2) mono sulphate formation, which increase the volume of the hydraulic bound MSWI Ash. This occurs as a result of reaction of calcium aluminates and calcium sulphates in the hydraulic binder in the presence of water. Also (3) aluminium reacts with calcium hydroxide, which is set free during hydration, or (4) C_3A, included in binders respectively in MSWI Ashes, is being hydrated. This leads to the formation of C_3AH.

These reactions in hydraulic bound MSWI Ash appear also at room temperature. Nevertheless the higher temperature (up to 70°C) of the steam test leads to an acceleration of reaction time and thereby to a larger expansion by a factor of 5 to 10, depending on the C_3A

contents of the binder. During the steam test the porous sample (porosity is between 30 and 35 %) can be penetrated by steam. Thus more water can be provide for the reaction at higher temperatures.

(1) $3CaO{\cdot}A_2O_3 + 3CaSO_4{\cdot}2H_2O + 30H_2O \rightarrow 3CaO{\cdot}Al_2O_3{\cdot}3CaSO_4{\cdot}32H_2O$

(2) $3CaO{\cdot}A_2O_3{\cdot}3CaSO_4{\cdot}32H_2O \rightarrow 3CaO\ Al_2O_3{\cdot}CaSO_4{\cdot}12H_2O + 2(CaSO_4{\cdot}2H_2O) + 16H_2O$

(3) $2Al + 3Ca(OH)_2 + 6H_2O \rightarrow 3\ CaO{\cdot}Al_2O_3{\cdot}6H_2O + 3H_2$

(4) $3CaO{\cdot}Al_2O_3 + 6H_2O \rightarrow 3\ CaO{\cdot}Al_2O_3{\cdot}6H_2O$

From the separate assessment of the MSWI Ash or the hydraulic binder, no statement can be made with regard to the volume stability of the unbound MSWI Ash. On one hand the unbound and hydraulic bound MSWI Ashes contain nearly the same volume extending mineral phases. On the other hand the volume enlargements in the case of hydraulic bound MSWI Ash are considerably greater. This is the result of adding a binder, which contains calcium aluminates and calcium sulphates and the hydration progressing in connection with water. Therefore no connection is demonstrable between the volume stability from unbound and hydraulic bound MSWI Ash.

PRACTICAL ISSUES FOR USE

The 28 days compressive strength and freeze thaw resistance have proved to be a criterion of the mix design for hydraulic bound mixtures. The results of literature evaluation and the present studies indicate that both testing features allow no sufficient assessment of hydraulic bound MSWI Ash. No well-defined relation was found between compressive strength or frost resistance and MSWI Ash or binder (type and content). However, such a relationship could be proved in spite of the inhomogenity of the MSWI Ash for the steam test. Therefore, an enlargement of the suitability test appears as reasonable for hydraulic bound MSWI Ash through the steam test for the assessment of volume stability.

In order to guarantee sufficient volume stability of MSWI Ash a reduction of fines (sand content) of MSWI Ash should be aimed at. In addition, an expansion of storage time affects volume stability favourably.

MSWI Ash with small contents of CaO, Al_2O_3 and SO_3 effects positively the volume stability of hydraulic bound MSWI Ash. A determination of these characteristic values for the calculation of the reactivity of hydraulic bound MSWI Ash is useful. However, there is no point in ascertaining free CaO and MgO since virtually no free CaO and/or MgO is present at MSWI Ashes.

Even if a hydraulic binder with low effective alkali content and high sulphate resistance is used, as recommended in literature, the measured expansions in the steam test pass still up to 5 ‰. If possible hydraulic binder with low contents of C_3A (best below 5 %) should be selected. The binder content in the construction material mixture can be held low by selection of a suitable binder for the MSWI Ash being used.

Testing of the material mixture with the steam test is decisive for the estimation of a sufficient volume stability. A sufficient volume stability is then available if virtually no expansions occurs with the steam test. Until a large experience background and a correlation of the results of the steam test with practice behaviour of material mixture is found, a sufficient volume stability independent from the deposit age of the MSWI ash can be assumed, if the measured extensions of the steam test are smaller than 1 ‰. This procedure for the definition of the permissible expansion corresponds to the procedure for the assessment of the frost resistance. Therefore a sufficient volume stability complies with a binder content of at least 10 %.

CONCLUSIONS

Investigations show, that it is not sufficient to determine compressive strength and freeze thaw resistance in the scope of suitability tests for hydraulic bound MSWI Ash. Rather the volume stability is a very important aspect for opinions on such construction material mixtures.

During the steam test a sample (d=100 mm, h=120 mm) is located in a glass jar, partly filled with water. The sample undergoes 21 cycles in steam for 16 hours and in dry conditions for 8 hours. The expansion determined during the steam tests is a standard value for volume stability of hydraulic bound MSWI Ash. Reactions during the steam test, enlarging the volume, pass off between MSWI Ash, hydraulic binder and water. It could be demonstrated that the steam test for hydraulic bound MSWI Ash enabled differentiation between the quality of MSWI Ash as well as the binder used (kind and contents).

ACKNOWLEDGEMENTS

The authors would like to acknowledge the financial support provided for the project by the Bundesministerium für Verkehr (German department of traffic).

REFERENCES

1. Technischen Lieferbedingungen Für Hausmüllverbrennungsasche Im Straßenbau (Tl Hmva-Stb 95) Ausgabe 1995. Forschungsgesellschaft für Straßen- und Verkehrswesen, Köln, 1995.

2. Merkblatt Über Die Verwendung Von Industriellen Nebenprodukten Im Straßenbau, Teil Müllverbrennungsasche (Mv-Asche), Ausgabe 1986. Forschungsgesellschaft für Straßen- und Verkehrswesen, Köln, 1986.

3. Merkblatt Für Die Verfestigung Von Müllverbrennungsasche Mit Hydraulischen Bindemitteln, Ausgabe 1992. Forschungsgesellschaft für Straßen- und Verkehrswesen, Köln, 1993.

4. PFRANG-STOLZ, H., KÖSL, H., SCHNEIDER, J. Charakterisierung von MV-Schlacken. Beihefte zu Müll und Abfall 31 (1994), S. 38-43.

5. KRASS, K., RADENBERG, M., ANFALL. Aufbereitung und Verwertung von industriellen Nebenprodukten und Recycling-Baustoffen. Straße und Autobahn (1993), Heft 12, S. 720-724.

6. HIERSCHE, E.-U., WÖRNER, TH. Verwendung von Müllverbrennungsaschen im Straßenbau. VDI-Berichte Nr. 753 (1989), S. 167-178.

7. GÄHL, N., HERZ, G., RAUTER, A., TOUSSAINT, A. Erfahrungen mit der Verwendung von Müllverbrennungsaschen als Dammschüttmaterial. Straße und Autobahn 38 (1987) Heft 5, S. 186-189.

8. KLUGE, G. Erfahrungen mit der Verwendung von Müllverbrennungsaschen im Straßenbau. IRC 1986, S. 375-383.

9. RADENBERG, M. Ermittlung und gegebenenfalls Verbesserung der bautechnischen Eigenschaften von MV-Asche. Schriftenreihe des Instituts für Straßenwesen und Eisenbahnbau der Ruhr-Universität Bochum, Band 6, Bochum, 1994.

10. SCHMIDT, M. Verwertung von Müllverbrennungsrückständen zur Herstellung zementgebundener Baustoffe. Beton 38 (1988) Heft 6, S. 238-245.

11. SCHUBENZ, D. Müllverbrennungsaschen als Baustoff für Tragschichten mit hydraulischen Bindemitteln. Straße und Autobahn (1990), Heft 9, S. 394-400.

12. VOGEL, P. Verwertung von Müllverbrennungsaschen mit hydraulischen Bindemitteln im Straßenbau. Straßen und Tiefbau 45 (1991) Heft 9, S. 16-22.

13. VOGEL, P. Entwicklung bei der Verfestigung von Müllverbrennungsaschen. Straße und Autobahn Heft 4 (1992) S. 235-244.

14. Zusätzliche Technische Vorschriften und Richtlinien für die Ausführung von Tragschichten im Straßenbau (ZTVT-StB 95), Ausgabe 1995. Forschungsgesellschaft für Straßen- und Verkehrswesen, Köln, 1995.

15. SCHELLENBERG, P. Entwicklung eines Verfahrens zur Reduzierung leichtlöslicher Salze in Müllverbrennungsaschen. Schriftenreihe des Instituts für Straßenwesen und Eisenbahnbau der Ruhr-Universität Bochum, Band 10, Bochum, 1995.

16. WÖRNER, TH., WESTINER, E. Anforderungen und Kriterien für den Einsatz von Müllverbrennungsaschen im Straßenbau. Prüfamt für Bituminöse Baustoffe und Kunststoffe der TU München, Forschungsbericht Nr. 3173, 1994.

17. WESTINER, E. Untersuchungen zur Raumbeständigkeit von hydraulisch gebundener Hausmüllverbrennungsasche. Berichte aus dem Baustoffinstitut, Heft 1/1998, 190 S., 1998.

18. SCHMIDT, M. Stoffliche und konstruktive Eigenschaften hydraulisch gebundener Tragschichten. Schriftenreihe der Zementindustrie, Heft 51/1991.

19. REICHELT, J. Mineralogische Aspekte bautechnischer und umweltrelevanter Eigenschaften von Müllverbrennungsschlacken. Veröffentlichungen des Institutes für Straßen- und Eisenbahnwesens der Universität Karlsruhe (TH), Heft 47, 1996.

20. DORNER, H., WESTINER, E. Heißhydratation von Hausmüllverbrennungsaschen – Ergebnisse baustofftechnischer Untersuchungen. Berichte des Bayerischen Forschungsverbundes für Abfallforschung und Reststoffverwertung (BayFORREST). Berichtsheft 7 zum 5. Statusseminar München., Oktober1997.

21. GOHLKE, O., WESTINER, E. Bautechnische Aspekte der Wäsche von Müllverbrennungsschlacken. Müll und Abfall, Heft 3/99, S. 74-77, 1999.

22. KLEPPMANN, F. Zentrales Verwertungskonzept von MV-Schlacken einschließlich Schlacke-Monodeponie am Beispiel ZAR Würzburg. VDI-B Handbuch zur Tagung "Schackeaufbereitung, -verwertung und -entsorgung", München, 1995.

23. Technische Prüfvorschriften für hydraulisch gebundene Tragschichten (TP HGT-StB 94), Ausgabe 1994, Forschungsgesellschaft für Straßen- und Verkehrswesen, Köln, 1994.

24. LICHTENSTEIGER, TH. Langzeitverhalten von Schlacken in Deponien. VDI-B Handbuch 43-76-01 zur Tagung "Schackeaufbereitung, -verwertung und -entsorgung" München, 1992.

MSWI PLANTS: SUSTAINABLE QUARRIES

D M York

Ballast Phoenix Ltd.

United Kingdom

ABSRACT. The paper examines progress to date with the processing of Incinerator Bottom Ash (IBA), the mechanical properties of IBA and the potential for incinerator plants to be urban quarries. After appropriate processing, some bottom ash can meet the requirements of BS 882; 1992 - Aggregates from Natural Sources for Concrete. A number of schemes that incorporate IBA in their pavements are given as examples.

The paper also outlines the interim policy for the use of IBA, negotiated with the Environment Agency (EA).

Keywords: Incinerator bottom ash, Mechanical properties, Urban quarries, Interim policy .

Mr D York, is the Managing Director of Ballast Phoenix Ltd, a company that specialises in the processing and marketing of IBA. The company currently has three operational sites in the UK.

INTRODUCTION

The incineration of domestic refuse is a very old practice, with its roots in the garden fire. The commercial plants now operating in the UK have generally been built since 1970. Until recently it has been common practice to landfill the bottom ash arising from the process.

A simple visual inspection of bottom ash will reveal its make up as:

- glass
- porcelain
- brick
- concrete
- flint/stones
- clinker
- ash
- ferrous metal
- non-ferrous metal
- unburnt organic matter

A plant that has operated since 1970 and that has produced 100,000 tonnes per annum of IBA will have landfilled over 20,000 tonnes of non-ferrous metals, over 150,000 tonnes of light iron plus approximately 2.5 million tonnes of potentially usable aggregate.

In today's world, there is a need to conserve natural resources and to re-use, recycle and generally make use of materials which in the past have been discarded. The ability to process IBA to a consistent quality is essential in promoting the use of IBA. It is also essential to have environmentally acceptable uses for IBA that will not impact on human health or cause pollution.

Other European countries have processed and used IBA and continue to do so, generally as an unbound fill material. This paper will detail some higher value applications derived in the UK and made possible by innovation.

PROCESSING OF IBA IN THE UK

At this time in early 2000, there are four IBA processing sites in the UK. Between them they may handle up to 350,000 tonnes per annum (t.p.a). However, the amount of ash being processed will undoubtedly increase in the coming years, probably to somewhere around the 1 million t.p.a. mark.

The processing currently employed is mechanical and involves:

- ferrous and non-ferrous metal extractions;
- screening to size;
- removal of unburnt organic matter;
- natural drying.

Table 1 shows the Mechanical Properties of the stronger quality ash after processing.

Table 1 Mechanical Properties of Processed IBA

PARAMETER		STANDARD	VALUE
10% fines value		BS 812 Part 111 (soaked)	75kN
Flakiness index		BS 812 Part 105	21
Acid soluble sulphate		BS 812 Part 118 1988	0.53%
Specific gravity	Ovendry	BS 812 Part 2 1995	2.19Mg/m³
	SSD		2.35Mg/m³
	Apparent		2.62Mg/m³
Water absorption		BS 812 Part 21995	7.40%
Drying shrinkage		BS 812 1989/ BRE Digest 357, 1991	0.070% (Category A)
pH		BS 1377	9 to 11 (age dependent)
Acid soluble chloride		BS 812	0.04%
Co-efficient of thermal expansion			13.5×10^{-6} per °C

N.B. Not all IBA will have these properties

No chemical processing or washing is currently involved. Whilst the processing may appear simple, there are many difficulties in actually doing the work. The material is generally in a wet or damp condition and does not handle easily.

Also the material undergoes certain natural changes in the first few weeks after creation. These processes include oxydation, carbonation and hydration.

On initial production, the IBA has a pH of around 12. However, after some weeks stood in the open, the material will have a final pH of about 9.

For some European countries that use the IBA as unbound fill, this is a state at which the pollution potential is minimised. Here in the UK, however, the industry has decided to generally pursue bound applications for IBA. This means the IBA, if processed properly, can have a higher value when used as aggregate in a bitumen bound or cement bound mixture. It also means that any pollution potential is greatly reduced due to the binder. Bitumen acts as an encapsulent whilst cement chemically bonds metal ions into the mixture. For these types of applications, a high pH does not pose a problem.

The location of incinerators (and hence IBA processing) is generally in urban areas. The positive use of IBA in such areas has many environmental benefits.

Environmental benefits of IBA usage in urban areas

- Saves transport of IBA to landfill

- Saves import and transport of primary aggregates over great distances on busy roads or rail routes

- Saves quarrying of primary aggregate often in environmentally sensitive areas
- Provides a huge recycling opportunity for local authorities:

Local Authority produces waste
↓
Incineration produces energy and bottom ash
↓
Bottom ash produces aggregate
↓
Aggregate used in asphalt or cement bound material (CBM)
↓
Asphalt or CBM is used on local authority infrastructure

EXAMPLES OF CEMENT BOUND IBA USAGE

During 1999, a number of major civil engineering schemes incorporated IBA into their pavements in a cement bound form.

- Industrial Hardstanding – East London – 60,000m^2
- 4km single carriageway bypass – North London
- Heavy Duty access road to Hypermarket Complex – East Midlands
- Distribution Centre – Midlands – 35,000m^2

Industrial Hardstanding – East London

This scheme was designed for road/rail transfer of containers using a large forklift vehicle with maximum loads of 70 tonnes on the front axle. The construction of the pavement incorporated a 260mm thick layer of CBM with a specified strength of 20N/mm^2 at 28 days.

The aggregates for the scheme were totally recycled. 75% of the aggregate came from site arising , mainly old foundry wastes which had been tipped on the site. The remaining 25% was processed IBA. Typical Test results are shown in Table 2.

4km Single Carriageway Bypass – North London

In this instance, the project required approximately 6,500m^3 of CBM3 [1]. This grade of CBM requires a BS 882 quality aggregate [2]. IBA was processed to this quality and formed 100% of the CBM aggregate. The mixing and laying of the CBM was all undertaken as with natural aggregates and lying operatives in particular were impressed with the ease with which the CBM handled and the good finish that could be obtained.

Table 2 Results of Tests on CBM Produced Using IBA Aggregate.

SCHEME	TYPICAL CUBE DENSITY, kg/m³	COMPRESSIVE STRENGTH, N/mm²			
		7 day	28 day	56 day	90 day
Industrial Hardstanding	2220	17	22	24	25
4km bypass	2090	11	16	18	19
Heavy-duty access road	2340	21	26	29	31
Distribution Centre	2250	11	15	16	17

Routine testing of aggregates and CBM showed full specification compliance. Technical data is provided in Table 2.

Heavy Duty Access Road to Hypermarket Complex – East Midlands

This project required a CBM4 [1] grade roadbase extending to 8,000m². In order to achieve the strength required, a mixture of 75% IBA aggregate and 25% natural sand with ordinary Portland cement and water was used.

Once again, the laying of the CBM was reported to be easier than with conventional aggregate. The strengths actually obtained were well in excess of the specified minimum and are shown in Table 2.

Distribution Centre – Midlands

A blend of 75% IBA and 25% sand was employed as the CBM aggregate for this scheme. Typical results are shown in Table 2.

USE OF IBA IN BITUMINOUS MIXTURES

IBA processed to meet the requirements of BS 882 [2] in terms of 10% fines value and flakiness index with consistent grading and with an acceptable removal of metals can replace virgin aggregate in bituminous mixtures.

A number of local authorities have laid trial areas over the past few years and all have recorded performance equivalent to that of a conventional mixture.

Some authorities now accept IBA as a recycled aggregate for routine use in their supplies. Bituminous mixtures incorporating IBA aggregate do not need increased amounts of binder and operatives laying such mixtures have no difficulties in handling and laying the material.

There is some evidence of improved stiffness of mixtures when a virgin aggregate has been replaced. It is certainly technically acceptable to replace 50% of virgin aggregates in bituminous mixtures, though in theory, this percentage may be considerably increased.

ENVIRONMENTAL ACCEPTABILITY

In order to gain environmental approval for the use of IBA aggregates, a working party was set up by the Energy from Waste Association during 1997 to firstly collect data and then to join in discussion with the Environment Agency and eventually to negotiate a national policy for the use of IBA.

Test Work

It was recognised in the early stages of investigating IBA that unbound use would be least environmentally acceptable (in general) and also be least commercially interesting.

Much leach testing was done on IBA and on cement bound IBA to determine the effects of ordinary Portland Cement on the mobility of potential pollutants. These are mainly in the form of heavy metals. Some organics are present but generally in very low concentrations. Work was also done with bituminous mixtures along similar lines.

Both cement and bitumen are very effective in immobilising the heavy metals present in IBA. However, in order to get a proper 'feed' for IBA it can be characterised as similar chemically to coal fired bottom ash or indeed coal fired fly ash. These materials are in common use and are not generally regarded as a threat to either humans or the environment, except in particular circumstances.

Results of X-Ray Fluorescence Spectrometry for Major Element Analysis of IBA, Coal fired furnace bottom ash, conditioned pfa from coal burning and indeed granite show results all broadly of the same order.

A trace element analysis does show 'spikes' for elements such as lead or copper, where pieces of these metals were in the original samples.

Discussions and negotiations have taken place with the Environment Agency and an interim policy for IBA usage has been agreed. This approves the use of cement or bitumen bound IBA in all areas of the UK except on chalk or limestone formations without reference to the Environment Agency. Proposals for unbound use must receive individual approval from the Agency.

A simple Code of Practice accompanies the policy dealing with such matters as temporary storage of IBA on site prior to mixing and laying.

CONCLUSIONS

The use of processed IBA in bituminous and cement bound mixtures is now established. Technically, IBA performs similarly to some virgin materials. A national policy for usage has now been agreed with the Environment Agency.

The location of incinerators that produce IBA makes then a perfect urban quarry offering an opportunity to minimise transport of ash to landfill and of virgin aggregate into already congested urban areas. Incinerator plants are generally built around twenty or twenty-five year contracts to burn waste and the numbers of such facilities will increase in the coming years.

These long-term commitments will provide a sustainable source of aggregates and provide a wonderful recycling loop where waste produced in an area can be burnt to recover energy and the ash residue used to produce local infrastructure.

REFERENCES

1. HIGHWAYS AGENCY. Specification for Highway Works: 1996. HMSO. Norwich, UK. 1992. Clauses 1035 to 1042 – Cement Bound Materials.

2. BRITISH STANDARDS INTITUTION: BS 882:1992. Aggregates from Natural Sources for Concrete.

UNPROCESSED AND PROCESSED INCINERATOR BOTTOM ASH AS A CEMENT BOUND MATERIAL

K A Paine

R K Dhir

V P A Doran

University of Dundee

United Kingdom

ABSTRACT. The results of a laboratory investigation into the potential of using incinerator bottom ash as a cement bound material (CBM) in pavement sub-bases is presented. Two forms of IBA were investigated. Firstly IBA in a raw (unprocessed) form, to investigate whether the aluminium content would have any detrimental effects, and secondly, processed material in which a large proportion of metals and unburnt material had been removed. The effect of IBA on the compressive strength of CBM was studied using a range of IBA and cement contents. It was shown that processed IBA led to higher strength than unprocessed IBA for the same cement content. However, to achieve the categories of CBM defined in the UK Highway Specification it was necessary to use higher cement contents than in natural aggregate CBMs. For example, up to 4% increase in cement content when replacing all natural aggregate with processed IBA. All natural aggregate and IBA specimens passing the strength requirement for a CBM2 material were tested for leaching of polluting species by means of a tank-leaching test. It was shown that whilst IBA gave rise to higher levels of cadmium, copper and potassium leaching, the service water concentration values were well below UK environmental and drinking water quality limits.

Keywords: Incinerator bottom ash, Pavements, Cement bound material, Sub-bases, Leaching, Mineralogical properties, Compressive strength

Dr Kevin A Paine, is a Research/Teaching Fellow in the Concrete Technology Unit, University of Dundee, Scotland, UK. His research interests concern the use of recycled materials as binder and aggregate components in concrete, and the development of steel fibre reinforced concrete.

Professor Ravindra K Dhir, is Director of the Concrete Technology Unit, University of Dundee. He specialises in binder technology, permeation, durability, and protection of concrete. His interests also include the use of construction and industrial wastes in concrete to meet the challenges of sustainable construction. He has published and travelled widely and serves on many Technical Committees. He is a past chairman of the Concrete Society, Scotland.

Mr Vincent P A Doran, is currently completing a thesis on the potential of using incinerator bottom ash within pavement construction; work which he carried out as a research student at the University of Dundee. He currently works for Oscar Faber Consulting Engineers.

INTRODUCTION

Approximately two million tonnes of the total annual arising of UK municipal solid waste is currently incinerated at modern mass-burn waste-to-energy plants. About one third of the material burnt is deposited as bottom ash from the furnace [1]. Traditionally, much or most of this incinerator bottom ash (IBA) has been deposed of in landfill facilities. However, in recent years the UK Government has adopted the principals of sustainable development, which encompasses the need to conserve natural resources and encourage the use of recycled products and secondary materials in construction. IBA may have some potential as an alternative aggregate. However there are some environmental concerns that its use in construction may give rise to the leaching of polluting chemicals. In addition, IBA may require some processing and treatment prior to its use for engineering and economical reasons.

A number of studies have assessed the potential of utilising IBA as unbound fill or as granular sub-base in pavement construction. However, there have been some concerns regarding the environmental impact of such practices. This is because IBA possesses enriched concentrations of metals and salts that may leach into the watercourse. A recent report issued by CIRIA [2] has recommended that some restrictions will apply when using IBA as an unbound material in road construction. In particular, this is because of concerns with the potential level of cadmium leaching. Leach tests showed cadmium in the leachate in excess of 100 times the UK Drinking Water Standard. In addition, concerns have been expressed elsewhere of the relative mobility of copper from IBA, particularly in the presence of dissolved organic carbon [3].

In addition to the leaching of heavy metals, results of leach tests to assess the environmental impact of IBA have shown that sodium, potassium, chlorides and sulfates are the most heavily leached species. This is because these species are present largely as salts, which start to dissolve as soon as IBA comes into contact with water. Despite their high availability for leaching, sulfates tend to be retained for slightly longer in the field due to their incorporation in gypsum and ettringite, which form due to ageing and weathering [4]. The high leachability of these salts has consequences for both vegetation and underground water. It has been shown, for example, that due to absorption of high amounts of sodium, plants grown in areas subject to IBA encounter a transitory salt shock that leads to loss of germination and reduced growth [5].

As a consequence of the concerns with leaching from unbound granular IBA a test programme was set up to investigate the effect of binding IBA in cement for use as pavement sub-base, and to measure the subsequent strength and leaching properties. The tests were carried out in two stages :

(i) An investigation into the effect of IBA on the strength of CBM, to identify the increase in cement content required to meet the strength classes of the UK Highway Specification [6].

(ii) Leachate tests on cement bound IBA, meeting the strength requirements of the UK Highway Specification, to measure the concentration of selected species contained in the leachate for comparison with the UK drinking water quality standard [2].

The UK Highway specification [6] currently defines five classes of cement bound material (CBM) for pavement construction, of which CBM1 and CBM2 are generally used as sub-base. The limits on CBM1 and CBM2, are minimum seven-day compressive strengths of 4.5

N/mm^2 and 7.0 N/mm^2, respectively. The limit applying to CBM2 was used within this study as a minimum strength requirement due to concerns with other aspects of performance. In addition, the UK Highway Specification denotes that aggregates used must fall within specified grading bands (as shown in Figure 1), and for use in CBM2 have a soaked ten-per-cent fines value (TFV) greater than 50 kN.

MATERIALS

In this study, both unprocessed and processed IBA were compared, to show the benefit of processing in terms of strength and leaching properties.

Processing of the IBA to upgrade it to a more consistent and homogeneous material, and to improve its aggregate characteristics, was carried out by a UK company specialising in the recycling of industrial residues. The process consisted of the following stages: (i) screening and crushing cycles, (ii) removal of the ferrous fraction with magnets, (iii) removal of unburned material via trommels and (iv) eddy current separators to remove non-ferrous metals: aluminium, brass, red copper, tin, zinc, alloys and some precious metals. In addition to improving its quality as an aggregate, processing of IBA permits profitable recycling of metals (particularly ferrous metals). Both the unprocessed and processed IBA were stockpiled in the open air and allowed to drain naturally prior to transportation to the University of Dundee for testing.

The unprocessed material was supplied in an all-in form that was subsequently screened to a size fraction of less than 30mm and graded to meet the particle size distribution requirements for CBM2. Processed IBA was supplied in two size fractions: 30-8 mm and 8-0mm. These fractions were combined to produce a material meeting the grading for CBM2. Characteristics of the ashes in terms of major oxide analysis and mineralogy are shown in Tables 1 and 2, respectively.

The reduction in metallic phases as a consequence of processing is shown in the IBA compositions given in Table 2. Note particularly the reduction in ferrous metal as expressed by the wüstite and magnesoferrite phases. This is also indicated in the reduction in the iron content expressed as oxide in Table 1. Because of this, non-ferrous phases, particularly quartz and albite, dominate the crystalline component of processed IBA. However, it should be noted that both processed and unprocessed IBA is predominately made up of amorphous material (≥ 70%).

Table 1 Major Oxide analysis of Incinerator bottom ashes (% mass)

		LOI	CaO	Al_2O_3	SiO_2	K_2O	Na_2O	Fe_2O	MnO	MgO	Cl	SO_3
U	30-0mm	2.4	17.6	11.5	47.4	1.0	6.6	10.7	0.3	1.5	0.9	1.2
P	30-8mm	2.0	13.1	9.4	59.8	0.9	7.2	5.6	0.1	1.1	0.3	0.4
	8-0mm	3.2	15.1	10.0	48.6	1.1	4.9	7.0	0.2	1.2	0.5	0.8

U = unprocessed material, P = processed material

Table 2 Mineralogy of Incinerator bottom ashes (% mass)

PHASE	UNPROCESSED	PROCESSED	
	30-0mm	30-8mm	8-0mm
Calcite, $CaCO_3$	1.18	0.82	1.91
Quartz, SiO_2	4.05	7.62	7.23
Graphite, C	2.32	2.42	1.73
Margarite, $Ca_2Al_2(Si_2Al_2)O_{10}(OH)_2$	1.00	1.28	0.33
Melilite, $(Ca,Na)_2(MgAl)(Si,Al)_2O_7$	2.60	1.22	2.14
Magnesioferrite, $MgFe_2O_4$	1.62	0.73	0.85
Sanidine, $(K,Na)AlSi_3O_8$	1.50	2.00	1.06
Vishnevite, $(Na,K,Ca)_8(Si_6Al_6)O_{24}(SO_4).2H_2O$	0.08	0.78	0.30
Albite, $NaSi_3AlO_8$	2.10	6.68	2.96
Wüstite, FeO	0.71	0.26	0.22
Rozenite, $FeSO_4.4H_2O$	0.03	0.79	0.00
Amorphous	82.81	75.40	81.27

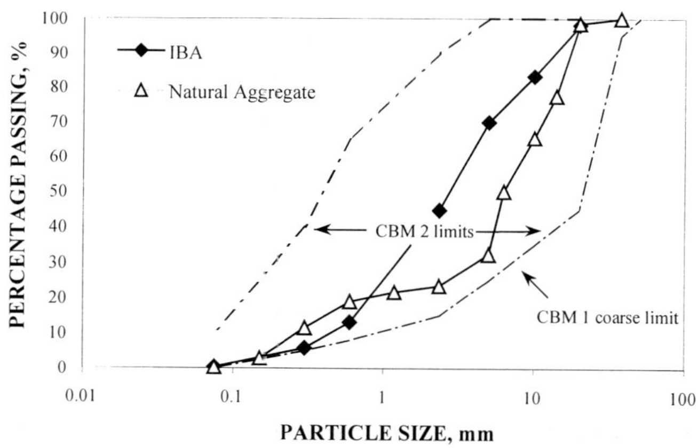

Figure 1 Grading of bottom ashes used in study

Soaked TFV tests carried out in accordance with BS 812: Part 111 [7] on the 14-10mm range of both unprocessed and processed IBA gave results of 45 kN, which are less than the specified minimum value of 50 kN in the UK Highway Specification. However, this value of 50 kN is based on an unrepresentative fraction of the full-grading and has been criticised as being unnecessarily harsh for waste materials [8]. Failure to meet this criterion was, therefore, not seen as critical.

The Portland cement (PC) used was of Grade 42.5N to BS 12 [9]. The natural aggregate was a combination of natural sand and river gravel proportioned to meet the grading requirements of CBM 2 in the UK Highway Specification [6].

TEST PROGRAMME

Strength

To investigate the effect of IBA on the strength of cement bound material, tests were carried out on CBMs containing a range of IBA contents (0%, 40%, 70%, 100%) as natural aggregate replacement. A range of selected cement contents (2 – 10%) were used to determine the cement content required to achieve classes CBM1 and CBM2.

All specimens were cast at a moisture content corresponding to the maximum dry density, known as the optimum moisture content. This was determined in accordance with BS 1924: Part 2 [10] and was found to be approximately 12% for specimens containing 100% IBA. This was approximately twice that measured for the natural aggregate CBMs. The optimum moisture content for the IBA specimens was observed to be independent of cement content over the range tested. This is probably due to the inhomogeneity of IBA.

Specimens were cured in accordance with BS 1924: Part 2 [10], and tested for compressive cube strength in accordance with BS 1881: Part 116 [11] at seven days.

Leaching

A tank-leaching test was chosen to investigate the leaching behaviour of cement bound material since leaching in monolithic wastes is controlled by diffusion [12]. The test is therefore designed such that water flows around rather than through the specimen. Alternative tests based on percolation more readily simulate the conditions for granular waste materials. The test used was a modified version of the CEN Protocol test [13]. Test specimens with dimensions of 100 mm x 100 mm x 25 mm were used, and were immersed in 1.5 litres of local drinking water maintained at 20°C. The leachant was replaced at 8 time intervals: 0.25 days, 1 day, 2.25 days, 4 days, 9 days, 16 days and 64 days. Samples of leachate were filtered through a 45mm sieve, analysed for pH and preserved for analysis by acidifying with nitric acid at pH 2.0 ± 0.2 before storing at 5°C.

Each leachate was analysed for trace metal concentrations of cadmium, copper, potassium and sodium by atomic absorption spectrometry (AAS), and for chlorides and sufates by ion absorption chromatography.

COMPRESSIVE CUBE STRENGTH

The recorded seven-day compressive cube strengths for all specimens are shown in Table 3. It can be observed that the higher the natural aggregate and cement content the greater the cube strength. In addition, the results show that unprocessed IBA produces CBMs of lower strength than processed IBA.

Figure 2 shows that for all combinations of IBA and natural aggregate there is a linear relationship between cube strength and cement content. This is typical of CBM compacted at the moisture content corresponding to maximum dry density. Figure 2 also shows the CBM2 compressive cube strength limit of 7 N/mm^2. From interpolation of the linear relationships shown, the minimum cement content required to achieve CBM2 class can be calculated for each combination of aggregate. For example, the cement content required for natural aggregate CBM is approximately 3%, whilst that of a CBM comprising wholly unprocessed IBA as aggregate is 10%, over three times greater.

Table 3 Seven-day compressive strength of CBMs

IBA CONTENT, %	CEMENT CONTENT, %	COMPRESSIVE STRENGTH, N/mm^2	
		Unprocessed IBA	Processed IBA
0	2	-	3.5
	4	8.5	11.0
	6	14.0	19.0
	8	23.0	25.5
40	4	5.5	6.0
	6	12.5	13.0
	8	18.5	22.0
70	6	-	6.5
	8	6.0	8.0
	10	10.5	12.5
100	6	-	5.5
	8	-	6.5
	10	7.0	11.0

- signifies no recorded compressive strength

The relationship between the IBA content and the minimum cement content to achieve a strength correlating with CBM2 is shown in Figure 3. For relatively low IBA contents the increase in cement content compared with natural aggregate CBMs is small and relatively unaffected by the type of IBA. However, at higher IBA contents the minimum cement content is much greater and depends significantly on the type of IBA. This confirms that processed IBA is a higher quality aggregate than unprocessed IBA.

All combinations of natural aggregate/IBA passing the strength requirements for CBM2, except those comprising 4% cement contents, were then tested for leaching using the CEN Protocol Test [13].

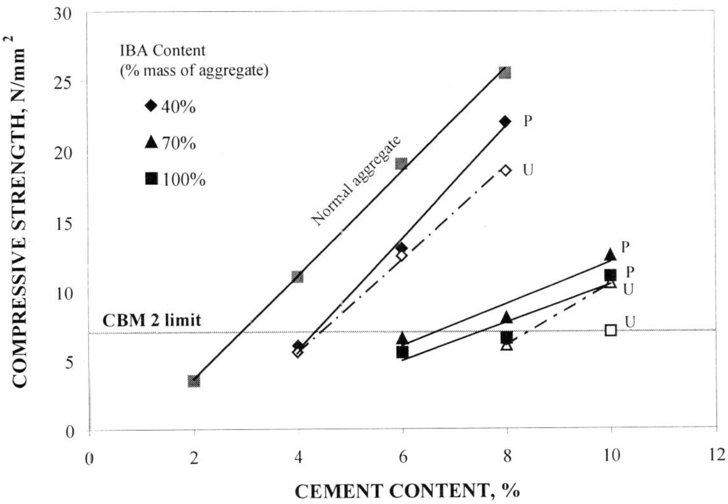

Figure 2 Relationship between cement content and compressive strength

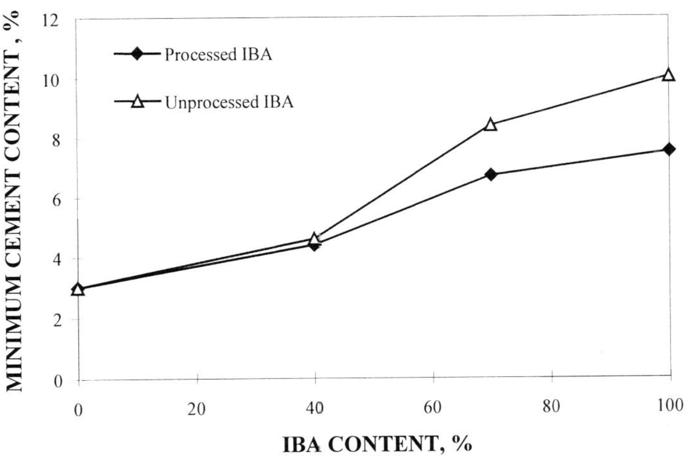

Figure 3 Relationship between minimum cement content required to produce CBM 2 and IBA content (% by mass of aggregate)

LEACHING

The tank-leaching test used produces a cumulative emission of species in terms of mass leached per surface area. This was converted in to a concentration value for service water, expressed in mg/l, so that the results could be compared with the UK Drinking Water Quality limits (shown in Table 4) [2]. For this conversion a factor of 0.3 and an assumption of linear migration of species with time, as recommended by the European Cement Association [13], were used.

Table 4 Regulatory limits for UK environmental water quality for selected elements

Element	UK limits, mg/l
Cd	0.005
Cl	400.0
Cu	3.0
K	12.0
Na	150.0
SO_4	250.0

Tables 5 and 6 show the service water concentration value for each species as percentages of the UK drinking water quality limits, for the unprocessed and processed IBA specimens respectively. The amount of leachate from processed IBA specimens was lower in all cases than the equivalent specimen made from unprocessed IBA.

Of the six species examined only cadmium, copper and potassium were leached at levels comparable with the water quality limits. However, in all cases the service water concentration value was at least one order of magnitude lower than the respective water quality limit.

This relationship between concentration values and the water quality limits is only an indication of the relative environmental impact of IBA use and should not be used as an acceptance value. However, the low results reported do suggest that IBA could be used as an alternative to traditional materials in cement bound material of class CBM2 without giving rise to ground and surface water pollution. A summary of the leaching of each of the six species is given below.

Cadmium

Figure 4 shows the relationship between the service water concentration value for cadmium and IBA content. It can be seen that at low IBA contents there is little cadmium leaching but that this increases sharply as the IBA content approaches 100%, particularly for the unprocessed IBA specimens. At 100% IBA the amount of cadmium leaching from the unprocessed specimens is approximately 3 times greater than that from the processed specimens. This suggests that a significant amount of soluble cadmium is removed during processing. In both cases the service water concentration value is below the water quality limit, but for the unprocessed IBA it is of a similar order of magnitude. Note that there appears to be no significant effect of cement content on the service water concentration value.

Table 5 Leachate results for Unprocessed IBA mixes

IBA CONTENT, %	CEMENT CONTENT, %	LEACHATE SERVICE WATER CONCENTRATION, as % of water quality limit					
		Cd	Cl	Cu	K	Na	SO_4
0	6	0.0	0.0	0.0	3.2	0.1	0.0
	8	0.0	0.0	0.0	3.2	0.1	0.0
40	6	0.2	0.0	3.0	5.2	0.1	0.0
	8	0.0	0.0	3.0	4.8	0.1	0.0
70	10	2.6	0.0	6.3	5.9	0.2	0.0
100	10	16.4	0.0	13.7	7.7	0.5	0.0

Table 6 Leachate results for Processed IBA mixes

IBA CONTENT, %	CEMENT CONTENT, %	LEACHATE SERVICE WATER CONCENTRATION, as % of water quality limit					
		Cd	Cl	Cu	K	Na	SO_4
0	6	0.0	0.0	0.0	3.2	0.1	0.0
	8	0.0	0.0	0.0	3.2	0.1	0.0
40	6	0.0	0.0	0.3	3.8	0.1	0.0
	8	0.0	0.0	1.3	3.8	0.1	0.0
70	8	1.0	0.0	3.0	4.1	0.2	0.0
	10	1.3	0.0	2.0	4.1	0.2	0.0
100	10	5.2	0.0	7.0	5.8	0.3	0.0

Copper

The relationship between copper leaching and IBA content is shown in Figure 5. In contrast to the leaching of cadmium, there is a steady increase in the amount of copper leached as the IBA content increases. When viewed as a percentage of the UK water quality standard, it can be seen from Tables 5 and 6 that the amount of copper leaching is as significant as the amount of cadmium leaching at 100% IBA, and more so at lower IBA contents. This suggests that copper leaching is a greater environmental concern than cadmium, with regard to the use of IBA as a sub-base material. However, the concentration values for all specimens are much lower than the water quality limits.

Potassium

It can be calculated from Tables 5 and 6 that as the IBA content increases from 0 to 100% the amount of potassium leaching increases by about 136% and 80%, for the unprocessed and processed IBA, respectively. The service water concentration value for potassium from IBA specimens is therefore of the same order of magnitude as that from normal aggregate specimens. All results are well below the UK water quality limits.

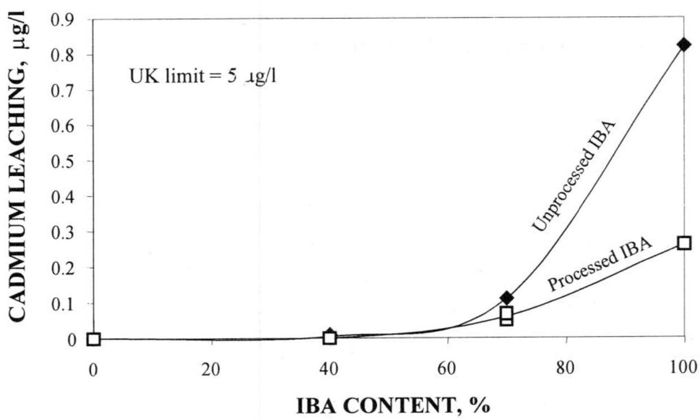

Figure 4 Effect of IBA content on leaching of cadmium

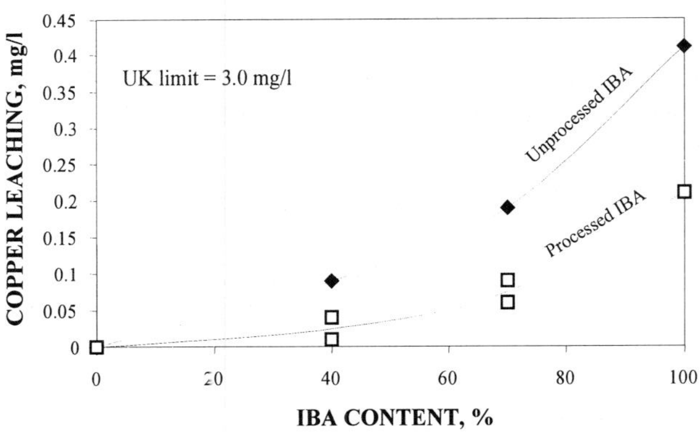

Figure 5 Effect of IBA content on leaching of copper

Sodium

Sodium leaching increases with IBA content but the concentration values are much lower than the water quality limit.

Chlorides and Sulfates

For all specimens tested, adsorption of sulfates from solution was detected. This was greatest for the natural aggregate specimens and the amount of adsorption decreased with an increase in IBA content.

In relation to the water quality limit no significant leaching of chlorides was observed.

CONCLUSIONS

The following conclusions can be drawn from this work:

1. The optimum moisture content for compacting IBA is approximately twice that of natural aggregate, and is independent of cement content over the range tested.

2. CBMs achieving the strength class for sub-base material defined in the UK Highway Specification can be produced using unprocessed and processed IBA as aggregate. However, to achieve the strength for a class CBM2 material it is necessary to use higher cement contents when using IBA. Processed IBA is a better aggregate than unprocessed IBA, and gives higher strength CBM for the same cement content.

3. Monolithic tank-leaching tests were performed on all specimens achieving the CBM2 strength class. Analysis for selected species showed that only cadmium, copper and potassium were leached in concentrations that could be sensibly compared with the UK drinking water quality standard. The leaching of these species was found to increase with IBA content, and were much higher in specimens produced with unprocessed IBA than processed IBA. However, all service water concentration values were at least one magnitude lower than the water quality limit

4. It is probable that IBA can be used as an aggregate in CBM2 class cement bound material without giving rise to ground or surface water pollution.

REFERENCES

1. COVENTRY, S, WOOLVERIDGE, C. AND HILLIER, S. The reclaimed and recycled construction materials handbook. Construction Industry Research and Information Association (CIRIA), Report C513. 1999. 186pp

2. BALDWIN, G, ADDIS, R, CLARK, J AND ROSEVEAR, A. Use of Industrial By-Products in Road Construction: Water Quality Effects. Construction Industry Research and Information Association (CIRIA), Report 167, 1997. 151pp

3. MEIMA, J A , VAN ZOMEREN, A AND COMANS, R N J. Complexation of Cu with dissolved organic carbon in municipal solid waste incinerator bottom ash leachates. Environmental Science and Technology. Vol. 33 No. 9. 1999. pp 1424-1429

4. MEIMA, J A AND COMANS, R N J. Geochemical modelling of weathering reactions in municipal solid waste incinerator bottom ash. Environmental Science and Technology. Vol. 31 No. 5. 1997. pp 1269-1276

5. FUCHS, B, TRACK, C, LANG, S AND GIMMLER, H. Salt effects of processed municipal solid waste incinerator bottom ash on vegetation and underground water. Journal of Applied Botany. Vol. 71 No. 5-6. 1997. pp 154-163

6. DEPARTMENT OF THE ENVIRONMENT, TRANSPORT AND THE REGIONS. Specification for Highway Works. Volume One. Road pavements – Concrete and cement bound materials, Series 1000. 1998

7. BRITISH STANDARDS INSTITUTION. BS 812: Part 111, Testing aggregates. Methods for determination of ten per cent fines value (TFV), 1990

8. NUNES, M C M, BRIDGES M G AND DAWSON, A R. Assessment of Secondary Materials for Pavement Construction: Technical and Environmental Aspects. Waste Management. Vol. 16, Nos 1-3, 1996, pp 87-96

9. BRITISH STANDARDS INSTITUTION. BS 12, Specification for Portland cement, 1996

10. BRITISH STANDARDS INSTITUTION. BS 1924: Part 2, Stabilized materials for civil engineering purposes. Methods of test for cement-stabilized and lime-stabilized materials, 1990

11. BRITISH STANDARDS INSTITUTION. BS 1881: Part 116, Method for determination of compressive strength of concrete cubes, 1983

12. KOSSON, D S AND VAN DER SLOOT, H A. Integration of testing protocols for evaluation of contaminant release from monolithic and granular wastes. Waste Materials in construction: Putting Theory into Practice. 1997. pp 201-215

13. EUROPEAN CEMENT ASSOCIATION. Use of non-ferrous slag in cement and the leaching behaviour of concrete in drinking water, BSI Document 95/106424, 1995

CLOSING THE LOOP

J van Beurden

J de Wijs

Feniks Reclying

Netherlands

ABSTRACT. The Dutch Highway 15 (RW 15) constructed between 1989 and 1993. In this embankment approximately 470,000 tonnes of MSWI-ashes have been used as a fill material. During the construction of this embankment several tests and reports were produced to gain practical knowledge about the use of MSWI-ashes in large bulk fill applications

To improve access to the Europoort and Maasvlakte area, major Dutch industrial areas, a new tunnel will be constructed (Calandtunnel).Therefore RW 15 will have a different route and approximately 260,000 tonnes of MSWI-ashes will have to be excavated and used in a different bulk fill project. The excavation of RW 15 will offer an unique opportunity to compare the initial results of investigation with long term results.

Keywords: MSWI-ashes, Recycling, Bulk fill, RW 15, Leaching results, Civil behaviour.

Jos (A.C.G.) van Beurden, MSc (Mechanical Engineering) is manager of both Feniks Recycling, one of the leading companies for upgrading and use of MSWI-ashes in the Netherlands, and Ballast Phoenix, the leading UK company.

Joost (J.J.) de Wijs, MSc (Chemical Engineering) is head of the research and development department of Feniks Recycling.

INTRODUCTION

In the reconstruction of highway RW 15 in 1989, the objective was to gain additional knowledge of the materials MSWI bottom ash and sand bentonite, and the civil engineering and environmental consequences of using MSWI bottom ash as embankment fill and sand bentonite as a top seal in a road embankment.

The test project concerned an embankment using MSWI bottom ash 3 to 4 m thick, laid on a bed of sand and finished on the top with a sand bentonite seal of 0.2 m, with a packet of sand and ground on it of approx. 1 m. Three test sections were included in the embankment. In test section 1, the porosity of the sand bentonite seal was measured. In test sections 2 and 3, the quantity and extent of contamination of the water percolating from the MSWI bottom ash was determined. Within the test project, a study was conducted to characterise the MSWI bottom ash.

Because this embankment now has to be excavated a unique opportunity is available to investigate the following details of the use of MSWI-ashes in bulk fill:

- in situ behaviour of MSWI-ashes during a long period;
- the performance of soil and groundwater protecting measures;
- the performance of the measuring devices;

With these results future design of bulk fill applications for MSWI-ashes can be optimised.

HISTORY

Introduction and targets

At the beginning of 1989, the decision was taken to implement a major test project in a section part of RW 15, in which the use of MSWI bottom ash in road embankments and the insulation provisions required could be examined in detail.

The general objective of the RW 15 test project was:
"To build and expand knowledge of the material MSWI bottom ash and sand bentonite in general and the civil engineering and environmental consequences of the use of MSWI bottom ash as embankment fill and sand bentonite as surface seal in embankment constructions by conducting practical and laboratory research."

The research was split in three main streams:

1. *Laboratory research into civil engineering and environmental properties of the materials*
   In parallel with the in-situ research on RW15, a characterisation study was conducted in which data were collected about the nature and elementary behaviour of both MSWI bottom ash and sand bentonite. A laboratory study was carried out for this purpose, with the following objectives:

   - To gain greater insight into some civil engineering properties and relevant characteristics for the moisture of the MSWI bottom ash and sand bentonite used.
   - To determine the chemical composition and the extractability of contaminants in MSWI bottom ash.

2. *In-situ environmental study*
   The most important objectives of the in-situ environmental study of RW15 were as follows:

   Study of the quality and quantity of the water flowing from the bottom ash core, both before and after the sealing layer is applied.Three test sections were set up for this purpose, with a system for retaining and recording the discharge of water.In the first test section, a plastic film was laid immediately beneath the sand bentonite seal in order to be able to investigate how much water filtered through the sand bentonite layer. In the other two test sections, a plastic film was laid under the bottom ash core, with drainage pipes and receiving tanks. During the installation phase and in the following years the quantity of water discharged and the concentrations of contaminants in it were measured.

3. *In-situ civil engineering study*
   The in-situ civil engineering study of RW15 was broken down as follows:

   - In-situ study prior to execution, with the objectives of gaining insight into the degree of compaction of the different layers of MSWI bottom ash from which the embankment was constructed, and of determining the composition of the sand bentonite mixture which would meet the specifications.
   - An execution and control study in the final embankment structure, with the objectives of gaining practical experience of laying the sand bentonite seal, and of obtaining insight into aspects of construction and the long-term behaviour of MSWI bottom ash. In addition to the programmed aspects of the study, general know-how was also gathered in the course of the test project, for example with regard to the settlement of MSWI bottom ash or the degree to which sand bentonite was prone to erosion.

The test project was started in July 1989 with a preliminary study into the compactibility of MSWI bottom ash, and the mixing ratio and laying method for the sand bentonite seal. The embankment was built in the period from October 1989 to November 1990. A civil engineering and environmental characterisation study of the properties of MSWI bottom ash and sand bentonite was conducted within the test project (in the period September 1990 to June 1991). Three permanent test sections were set up in the embankment to identify in-situ leaching behaviour during and after laying. The asphalt was laid between February and the end of April 1993, and the road was opened to traffic in July 1993.

Design of Embankment

The following requirements were made on the design of the embankment structure, Figure 1, using MSWI bottom ash, on the basis of the IBC criteria laid down in the IPO/VROM guideline and the properties of MSWI bottom ash:

1. The bottom ash laid on the soil had to be removable.
2. The top and sides of the MSWI bottom ash embankment had to be provided with a non-porous layer.
3. The bottom edge of the MSWI bottom ash had to be permanently a minimum of 0.5 metres above the average highest groundwater level in the new geohydrological situation at the site, or had to be permanently prevented from touching the groundwater level by technical means.

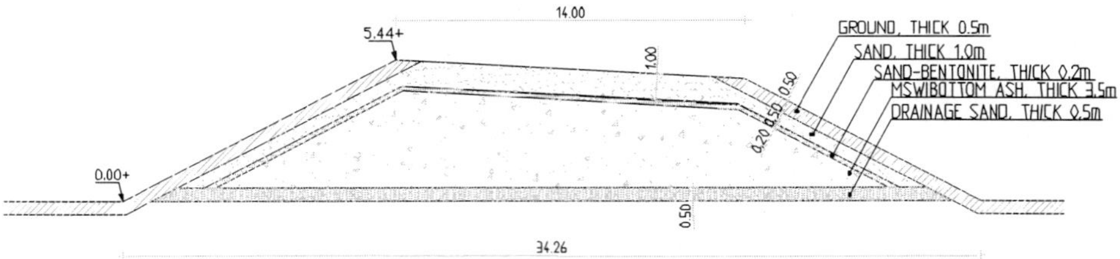

Figure1 Cross-section of embankment, Rijksweg 15

This meant that the design had to take into account the average highest groundwater level in the new situation and settlement of the underground as a consequence of the embankment.

The MSWI bottom ash was laid directly on this sand drainage layer, with a total layer thickness of 3 to 4 m.

The top and sides of the MSWI bottom ash were covered with a sand bentonite layer of 0.2 m. A drainage layer of sand and a covering ground layer were then laid on the sand bentonite layer, with a total layer thickness of no less than 0.8 m.

MSWI Bottom Ash

In the RW15 test project, a total of 470,000 tonnes of MSWI bottom ash was used in the embankments. The origin of the MSWI bottom ash used in the RW15 test project was as follows:

- N.V. Afvalverwerking Rijnmond (AVR bottom ash): 338,000 tonnes.
- GEB Zuid-Holland West, The Hague: 19,000 tonnes.
- ROTEB of Rotterdam (Roteb bottom ash): 113,000 tonnes.

1. The MSWI bottom ash used had relatively good friction and compression characteristics. Within a certain margin (approx. 5%), the level of compaction appeared to have little influence on these characteristics. Partly depending in the underground, slope gradients of up to 1:1 could be used in embankments made with MSWI bottom ash. The settlement of MSWI bottom ash will be limited to approx. 2% of the thickness of the layer. In addition, MSWI bottom ash appear to be sufficiently strong for the envisaged processing method (transportation and compaction).

2. The MSWI bottom ash used in the test project has a low absorbance capacity with regard to moisture content. If this capacity is exceeded, there is a fair chance of the material becoming mushy. With the moderate porosity found, puddles may form in extreme rainfall which may possibly lead to such a development. The influence of this on the environmental impact of MSWI bottom ash was not investigated.

The environmental study was intended to gain an insight into the environmental impact (particularly soil pollution) of large-scale use of MSWI bottom ash as embankment fill in a body in general and in the Rijksweg 15 body in particular. The environmental study was made up of a number of parts:

- Determination of individual properties of the bottom ash material relevant to moisture;
- Determination of the composition of the bottom ash material and the extractability of micro- and macro-contaminants from it;
- Installation of a system for catching and recording the discharge of water and to record precipitation;
- Placement of water gauges to check the groundwater level and determine the spread profile;
- Determination of the quantity of water discharged and the concentration of micro- and macro-contaminants in it during the construction phase and in the following years;
- Determination of the porosity of the sand bentonite seal.

Three test sections were set up at a selected part of Rijksweg 15 for the in-situ environmental study. The measuring programme in these sections covered the period from 28 August 1990 (laying down of MSWI bottom ash) to 11 May 1993.

There were three phases in the assessment of the test sections:

- Open phase: during the first 50 days;
- Closed phase, with sand bentonite sealing and finishing layers: after approx. 50 days;
- Closed phase, as above including asphalting: after approx. 950 days.

Chemical Analysis

The percolate samples were analysed for the heavy metals arsenic, barium, cadmium, copper, molybdenum, nickel, lead, tin, zinc and mercury. From July 1991 onwards, the analysis of mercury was omitted. No concentration of mercury was found above the detection limit of 0.1 μg/l. The percolate samples were also analysed for fluoride, chloride, sulphate, sulphide (or pure sulphur) and phosphate.

The analysis results of the percolate samples were compared with the test values from the Leidraad Bodembescherming [Soil Protection Guidelines]. A formal test was not intended, but an impression was given of the absolute concentrations of contaminants in the percolate. After all, if contaminants attach themselves to particles of soil and are diluted with (rain) water, the initial concentrations in the percolate can undergo major changes (reductions) before the percolate passes into the soil and the groundwater.

The curve of the concentrations measured in the percolate samples results from a number of processes which cannot be distinguished from one another: displacement of "clean" moisture from the drainage sand layer, desorption of components from the drainage sand layer and bottom ash leaching.

The curve in the initial phase may, in addition to leaching of substances from the MSWI bottom ash, be a consequence of components being washed out of the drainage sand layer (displacement by "salt" percolate).

A number of components may be assigned to three groups on the basis of the concentration curve over time [Lit. 5]:

1. The concentration dropped (sharply) after the initial period: cadmium, copper, lead and zinc;
2. The concentration remained reasonably constant during the entire measuring period: nickel, barium, sulphate, fluoride and chloride;
3. The concentration tended to rise during the measuring period: chrome, arsenic and phosphate.

Civil Engineering Aspects

The construction executed was tested according to requirements in the phases of execution and use with reference to strength, stability and deformation.

Strength

No problems occurred with regard to crushing of the MSWI bottom ash as a consequence of processing (transportation and compaction) during the execution of the RW15 test project. However, lumps formed repeatedly during execution. The contractor found some crushing as a result of rolling. However, this did not cause any problems.

Stability of MSWI bottom ash embankment and sand bentonite sealing layer

It appeared from the laboratory research that with an MSWI bottom ash embankment a slope gradient of 1:1 is still stable. therefore, the ground under the MSWI bottom ash will often decide the stability. For the sand bentonite cover layer, no stability problems are to be expected at a slope gradient of 2:3. Consequently, no such problems with either the MSWI bottom ash or the sand bentonite layer occurred during execution. It did prove to be difficult to lay and compact the sand bentonite cover layer at this steep gradient. An extra layer thickness of sand bentonite of 0.05 to 0.10 m was laid in order to be able to guarantee the minimum layer thickness of 0.20 m.

Settlement of MSWI bottom ash

On the basis of the results of the laboratory research, primary settlement to a maximum of 2% of the layer thickness laid is to be expected for an embankment of 4.5 metres of MSWI bottom ash. During the in-situ study, the MSWI bottom ash laid was found in measurements to have settled by 0 to 1.5%. It was concluded from this that virtually no additional deformation occurs due to the use of MSWI bottom ash.

Settlement of the underground

Settlement of the underground varied from +0.60 m to -0.40 m compared to the calculated settlement. In combination with a poorly functioning drainage system, this led to a steep rise in the water level at the site of the environmental test sections and to a distortion of the results of the environmental study. At some points, the MSWI bottom ash itself seemed to be in contact with the groundwater. In addition, it appeared that as a consequence of the uneven settlement the planned cover of 1 metre of sand/ground above the sand bentonite layer was insufficient to accommodate the drainage system. A layer thickness of 2 metres would have been more appropriate.

Processing of the MSWI bottom ash

The requirements set for the compaction of MSWI bottom ash were attained in the work. The characterisation study, however, showed that the requirements with regard to strength, stability and settlement could have been satisfied even with slightly lower levels of compaction. Measures were taken during the processing of MSWI bottom ash to limit the formation of dust during transportation and processing and also to protect people and equipment. Because the work was executed in a dry period, no extra measures were taken to prevent rainwater from causing erosion and flowing away during the open phase (laying the bottom ash).

PRESENT RESEARCH

Target

The research will give an answer to the question whether the present civil and environmental design parameters offer the expected construction to fulfill the requirements of the Building Materials Decree (Bouwstoffenbesluit).

The following questions can be raised:

- what is the main mechanism for aging of material;
- can these mechanisms be controlled;
- do the present laboratory testing offer a reliable prediction of leaching behaviour;

Policy

Since July 1999 the Netherlands has a new law for the use of materials in construction (Bouwstoffenbesluit, building material decree). In this decree rules are made for the leaching of anorganic and composition of organic components in all materials te be used in civil works, both primary and secundary materials. If the materials offer a higher leaching protection measures have to be taken to prevent pollution of soil and groundwater.

Ageing of MSWI-ashes

Literature shows that MSWI-ashes undergo a sequence of ageing reactions, which all influence the leaching of certain components. The following processes can be mentioned:

- absorption;
- precipitation;
- complexation;

These processes depend on the surface properties of the MSWI-bottom-ash particles and the conditions within the liquid in the solid phase, i.e. pH, redox potential, ion strength and the presence of complex formers. The different circumstances which may occur in the embankment body and which can influence ageing reactions are of importance to this study. These are, among other things, the initial composition, the ambient temperature, the level of saturation with the presence of water and air.

Because the test sections are different from the rest of the structure due to the presence of film, the redox potential will be different and there will be no emissions into the groundwater and the soil, the study of the quality of the MSWI-bottom-ash will concentrate on two cross-sections of the structure. Groundwater measurements have shown that some parts of the MSWI bottom-ash were to be found in the groundwater. The study will concentrate on one cross-section, which was always dry, and one cross-section, which was sometimes below groundwater.

In addition, this research will study the quality of the soil, groundwater, drainage sand, filter gravel sand bentonite and the film. A brief description of the analyses to be performed for each component follows below.

MSWI Bottom-Ash

In the excavation, attention is to be paid to the presence of clumping, crystal formation and metal corrosion. During excavation, part samples will then be taken which have to give a picture of the present quality in relation to re-usability and quality 10 years ago, determining whether a variation in quality is observable in relation to the position and moisture content within the embankment. These results could be used to forecast quality changes in the future. The study is to address both environmental and civil engineering aspects.

Sand Bentonite

The study will determine the current quality of the sand bentonite, whether there has been negative interaction between the sand bentonite and the MSWI bottom-ash, and determine any differences in quality in areas with different moisture levels.
The study will focus on dry density, level of compaction, moisture content, porosity and proctor density.

Ground And Groundwater

The aim is to discover whether contaminants from the MSWI-bottom-ash have led to a significant increase within the surrounding ground, and whether differences can be found between wet and dry areas. The soil will be sampled in layers. This, too, will primarily be an environmental study.

In view of the speed of the groundwater, 10 m/year, mobile contaminants will be able to move by a maximum of 100 metres. A number of gauges will be placed over this distance, starting from the dry and damp parts of the embankment. The groundwater will be analysed for mobile and critical components.

Results

Upon completion of the study, answers should have been found to the following research questions:

- Are the applied protection measures (IBC-rules) adequate;
- What is the effect of MSWI-ashes on the quality of ground water;
- To what extent has clumping occurred and what effect has this had on stability, porosity and ageing;
- Have deviations and phenomena occurred in the embankment;
- Is the civil engineering and environmental quality suitable for re-use;
- To what extent can MSWI bottom-ash be re-used in works after excavation;
- To what extent has ageing occurred;
- A picture must have been obtained of the moisture levels in and porosity of the embankment;
- What was the actual leaching from the MSWI-bottom-ash, and to what extent does this correspond to the expected laboratory results;
- Is optimisation of the structure possible in order to improve the quality of MSWI-bottom-ash faster;
- To what extent did the field measurements performed correspond to the values found;
- Are the design criteria adequate;
- Are the current risk regulations adequate.

According to our expectations, the results of the study will be available in mid-2000.

Closing The Loop

As stated, rules are laid down in the Bouwstoffenbesluit [Building Materials Decree] for the environmental quality of materials. One of the main criteria is the possibility to excavate the material and use it in another project. However, requirements are also made on the manner in which quality is determined. These refer to sampling, pre-treatment of samples, research methods and analysis. In the case of materials from ongoing production, a certificate is preferred, otherwise, batch inspections have to be carried out. One of the requirements of batch inspection is that the samples have to be selected at random and these have to cover the whole batch. In other words, all the research conducted is not sufficient to allow the MSWI-bottom-ash to be used in new civil engineering works.

It is expected that the excavated MSWI-bottom-ash will be used in two projects (through the company Feniks Recycling). The first of these is a noise-protection scheme at a Vinex location in the municipality of Vleuten-De Meern on the RW12 national highway. Around 600,000 tonnes are to be used in the noise-protection dam, of which 100,000 tonnes will come from RW15. 15,000 tonnes have already been used at the present moment. No difference was found during laying between this MSWI-bottom-ash and fresh bottom-ash. The MSWI-bottom-ash is to be insulated with a combined seal consisting of a bentonite mat with film. A natural layer of approximately 1 metre will be laid over this.

The second project will possibly be a decontamination location in Amsterdam. The soil contamination here is not to be removed, but completely sealed off by placing deep walls consisting of cement bentonite, topped off by a Trisoplast layer with an HDPE film. The Trisoplast, sand bentonite with polymer, will require good compaction in order to achieve the acoustic baffle. MSWI bottom-ash is an outstanding material for this purpose. In total, 450,000 tonnes of MSWI bottom-ash will be used, of which 50,000 tonnes may come from RW15.

CONCLUSION

With the re-use of MSWI bottom-ash from RW15 in the Vleuten-De Meern noise-protection dam and at Diemerzeedijk, it can be said that Feniks Recycling has now closed the re-use cycle completely.

As for the results of the different investigations they are not yet available fur publishing but will be ready in the course of 2000. Regarding workability of recovered MSWI-ashes no problems have occurred so far.

SWELLING PRESSURE TEST OF CEMENT-TREATED MSWI BOTTOM ASH IN TRIAXIAL CELL

C Crignon

G Pecqueur

Ecole des Mines de Douai

F Skoczylas

Ecole Centrale de Lille

France

ABSTRACT. Currently, use of MSWI bottom ash in road construction seems to be the way to go. In such application, bottom ashes are often treated with binder to improve their mechanical properties. Nevertheless, this material could cause a chemical reaction resulting in expansion, leading to cracking and road destruction. In order to investigate this, an assessment of the swelling potential of this material was undertaken. This is one of the aims of the European contract 'Mashroad'. To reach this goal, a swelling pressure test under unsaturated conditions was developed. This paper introduces the the experimental device which was used to measure swelling pressure and air permeability. Results of such tests allows the evaluation of both the efficiency of different hydraulic treatment to prevent bottom ash expansion, and air flow conditions which lead to the maximum deformation.

Keywords: MSWI bttom ah, Expansion, Swelling pressure, Air permeability, Compressibility, Triaxial cell

Caroline Crignon is in the last year of her Ph.D. studies in the Département Génie Civil of the Ecole des Mines de Douai, France.

Frédéric Skoczylas is a Professor at the Ecole Centrale de Lille, France.

Guillaume Pecqueur is a research scientist in the Département Génie Civil of the Ecole des Mines de Douai, France.

INTRODUCTION

Many reasons were supposed to induce MSWI bottom ash expansion (formation of aluminium hydroxide, ettringite, magnesium oxide hydration). Nevertheless, investigations show that metal aluminium is the main expansion source [1]. In fact, metal is oxidised and forms a gel. As binder treatments influence the magnitude of deformation, we showed in a previous study [2] [3] that compounds formed are the same whatever the treatment. The main parameter of the resulting expansion seems to be the kinetic of binder hardening and the variation pH due to binder hydration. It appears that clinker treatment reduces the deformation. However, in order to design the road structure, the engineer need a swelling pressure value. Moreover as metal oxidation depends on oxygen, we can suppose the acceleration of the swelling phenomenon by air flowing within bottom ash specimens. The materials ability to be crossed by fluid is given by permeability. The air permeability throughout the swelling test can decrease due to the filling up of the porosity by gel. Its increase is also possible, as a consequence of fracture formation. Moreover, carbonation of the bottom ash, as a result of air flow, leads to a fall in pH, a variation in metal aluminium reactivity and a change in binder hydration. Thus, air permeability seems te be a property which can give some indications on the bottom ash structure evolution. For this reason a modified triaxial cell was developed which allows swelling pressure and air permeability to be measured.

After a presentation of the materials used and of the experimental device, we introduce the poromechanical theory applied to the bottom ash, and first obtained results. Then, air permeability measurement is presented.

MATERIALS

The MSWI bottom ash used comes from Lagny sur Marne where the company Yprema is implanted. It is a coarse material (0-31.5 mm), matured for one month. The results presented in this paper were made on bottom ash both with and without treatment. The two binder treatments used were:

- an artificial Portland cement with low C_3A content,
- a lime activated slag.

The cement content was about 3 % of the total weight. Samples were compacted under Modified Proctor Optimum (OPM) conditions. The main characteristics are given table 1.

Table 1 Properties of the materials used.

	RAW BOTTOM ASH	CLINKER TREATED BOTTOM ASH	ACTIVATED SLAG BOTTOM ASH
W_{OPM} (%)	12.3	13	14.2
γd_{OPM}	1.84	1.79	1.76
u_a-u_w (kPa)	40	40	40

After preparation, samples were laid within a sheath to be protected from the confining fluid. Following this the bottom ash before the air pressure was applied. In fact, if we proceed directly to this last step without saturation, water continuity is broken down, and the sample can undergo desiccation. Once saturation is reached, with the application of water pressure at the bottom of the sample, a few days later, air pressure at both ends can be applied to obtain the wanted suction value. It takes a few days for equilibrium to be achieved, at which point no more water is released from the sample.

EXPERIMENTAL DEVICE

The test is based on the air pressure technique, which allows the unsaturated conditions (air and water pressure) to be controlled. If u_w and u_a are respectively the water and air pressure, the suction (negative pressure) is equal to $(u_a - u_w)$. Modified Wykeham Farrance triaxial cells were used, where the two fluid phases are separated with a system of semipermeable porous stones which are described in detail in the following section. Figure 1 shows the experimental device.

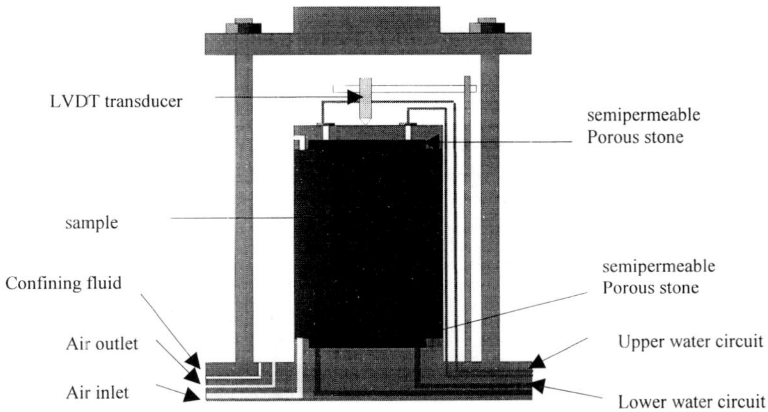

Figure 1 Modified triaxial cell

The air inlet and outlet are placed at both ends of the sample. Air flow is induced by a slight pressure gradient which is controlled by precision manometers. The whole experimental device is presented in Figure 2. To prevent the passage of air into the water circuit, porous stones semipermeable to air are also located at both end of the sample. Water flows in each porous stone at atmospheric pressure so as to maintain steady water content, and finally steady negative pressure.

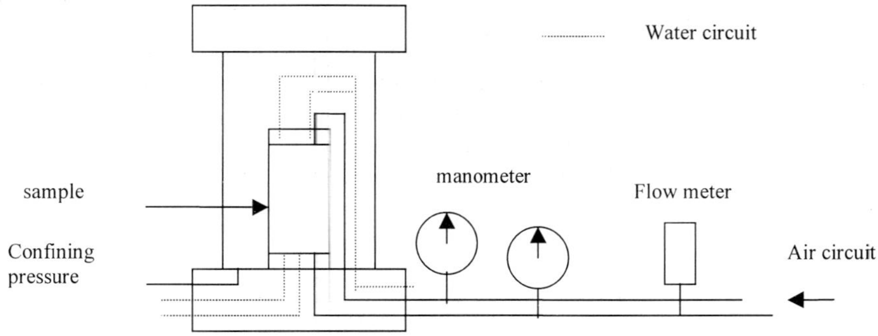

Figure 2 Whole experimental device

In our test, swelling pressure is defined as the hydraulic confining pressure to be applied to cancel sample expansion as soon as its volumetric stabilisation is achieved. During free swelling, axial deformation is measured. The measuring device consists of a LVDT transducer. As we assume the hypothesis of isotropic deformation, we deduce volumetric deformation from the axial displacement. Therefore, both information on the volumetric expansion and the consequent stress are available.

As seen more precisely in the following part, flow rate measurement leads to the air permeability determination. A schematic diagram of the device and the boundary conditions are given in Figure 3.

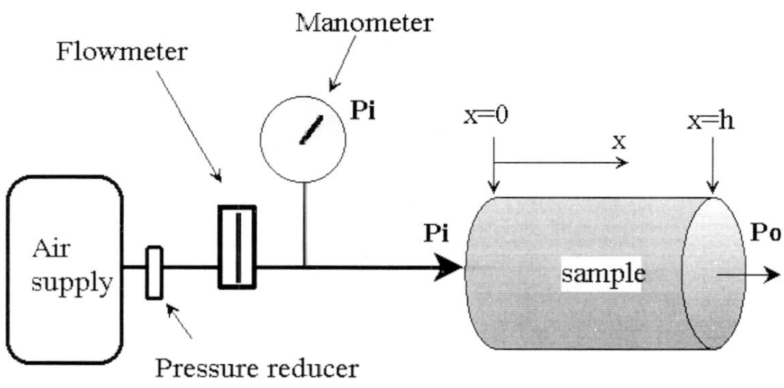

Figure 3 Air permeability measurement

POROMECHANICAL BEHAVIOUR

Theory

According to Biot's theory [4] and from a general point of view, bottom ash has to be considered as a porous medium as this material is a superposition of two continua : a rock (or solid) matrix and a connected interstitial space. The isolated pores, which are not connected, can be seen as a part of the matrix (Figure 4). The macroscopic approach to such materials was first developed by Biot, using the thermodynamic principles of open continuous media. Biot [4] showed that, in isothermal transformation, the state variables that describe the state of the material are the skeleton (matrix and connected porosity) strain and the variation in the fluid mass which fills up the interstitial space. In France, Coussy [5] extended Biot's studies to elastoplastic behaviour, putting forward certain hypotheses which can be experimentally tested [6].

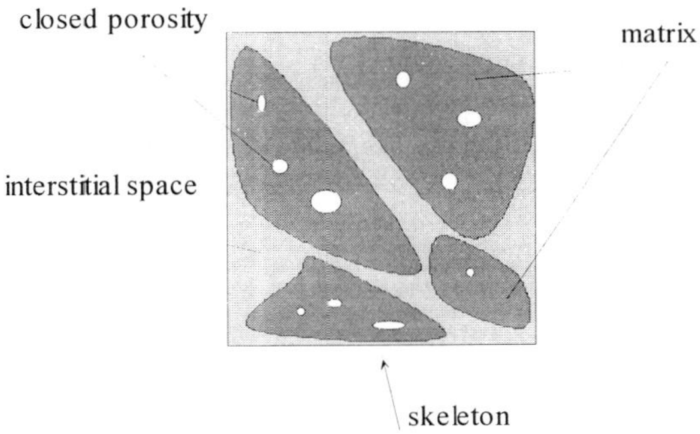

Figure 4 Schematic diagram of a porous medium

For an isotropic poro-elastic material, saturated with a single fluid, it can be shown [5] - in the case of infinitesimal strains - that the behaviour law is written :

$$\sigma_{ij} - \sigma^o_{ij} = (K_b - \frac{2G_b}{3})tr(\varepsilon).\delta_{ij} + 2G_b\varepsilon_{ij} - \alpha(P - P_o).\delta_{ij} \qquad (1)$$

σ_{ij} and P are the total stress state and the interstitial pressure, σ^o_{ij} and P_o are their initial values. ε_{ij} is the strain skeleton, and K_b and G_b are, respectively, the drained isotropic incompressibility and shearing modulus of the skeleton. α is the Biot coefficient.

In experimental terms, hydrostatic tests, are most useful to characterise the material behaviour and in particular the drained isotropic compressibility test ($P=P_o$, $\sigma_{ij}=\Delta P_c\ \delta_{ij}$). ΔP_c is the increase in confining pressure. In that case equation (1) leads to :

$$\Delta P_c = K_b \varepsilon_v \qquad (2)$$

ε_v is the volumetric strain of the skeleton.

Application to the bottom ash - Results

As previously mentioned, a swelling test, performed on a bottom ash sample, consists in applying a slight constant confining pressure and a continual air flowing through the interstitial space. The material is thus partially saturated with air and water. The experimental device allows a constant suction pressure (u_a- u_w) to take place into the sample until the end of the swelling process. This leads to a volumetric variation of the sample noted ε_v^f (Figure 5).

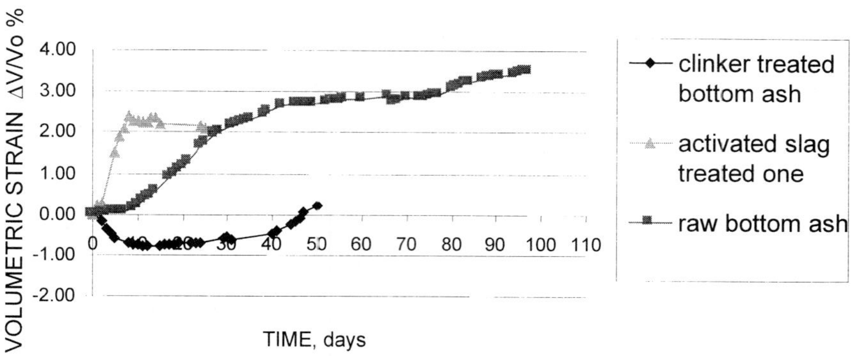

Figure 5 Free swelling curves

The swelling pressure P_s can thus be defined as the confining pressure variation which cancels ε_v^f. This is similar to a compressibility test, ε_v^f and P_s can be linked by:

$$P_s = C_b\, \varepsilon_v^f \qquad (3)$$

Similarity can be observed between Equations (2) and (3); however Equation (2) is written for an isotropic poro-elastic behaviour under drained conditions. As the suction pressure remains constant during the compressibility test, drained conditions can be admitted. Isotropic behaviour is assumed as a first approach and could be the purpose of a further study.

Several preliminary compressibility tests, with loading-unloading cycles, have been carried out on bottom ash samples (under initial state -ie- before the swelling test). The results of one of these tests are shown in Figure 6; the confining pressure is plotted versus volumetric strain.

The material exhibits a quasi linear behaviour which is not reversible - as shown by unloading. ε_v^p is the permanent volumetric strain which can be assumed to be a plastic strain. Thus, under drained conditions, equation (2) becomes:

$$\Delta P_c = K_b(\varepsilon_v - \varepsilon_v^p) \qquad (4)$$

This result shows that C_b (from Equation 3) is not equal to K_b and, therefore, is not the real incompressibility modulus. C_b can be considered as an apparent or a secant modulus. As the behaviour is linear, this apparent modulus is constant – ie - does not depend on the confining pressure. From this point, the question is whether swelling will alter the apparent modulus or not. If not, C_b could be used to evaluate the swelling pressure at any stage of the swelling process. On the other hand, if the final volumetric strain ε_v^f is estimated by extrapolation starting from initial results of the swelling test, the swelling pressure can immediately be calculated by Equation (3). This would thus allow an appreciable saving of time if the main scope of such a study was to compare the efficiency of different preparations or compositions of the material. This is quite interesting as we showed in Figure 5 that raw bottom ash expansion lasts more than 100 days.

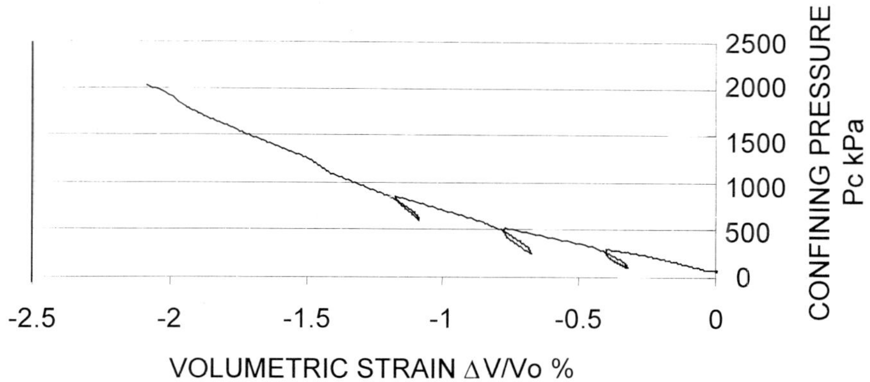

Figure 6 Loading-unloading cycles on raw bottom ash

During a swelling test, mechanical cycles are performed. These cycles consist in applying, at various stages of the swelling process, a slight increase in the confining pressure, to measure the volumetric strain due to the stress variation and then to return to the initial confining pressure as shown in Figure 7.

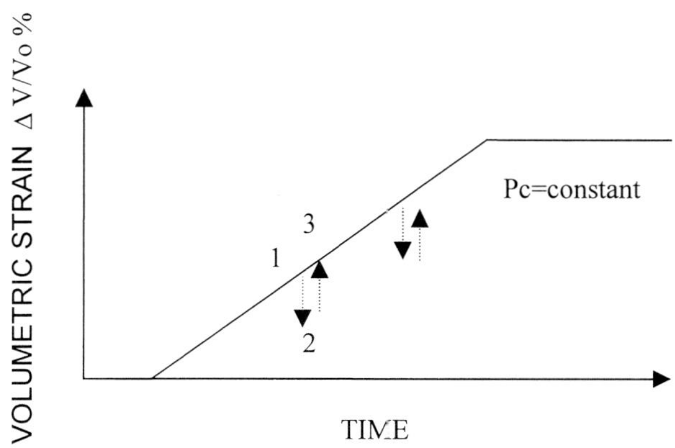

Figure 7 Compression during swelling test

With 1 – 2 loading path, 2 – 3 unloading path.

Figure 8 shows the results obtained on raw bottom ash. These results show that C_b can be assumed to be constant –ie- is not significantly modified when swelling occurs (Table 2).

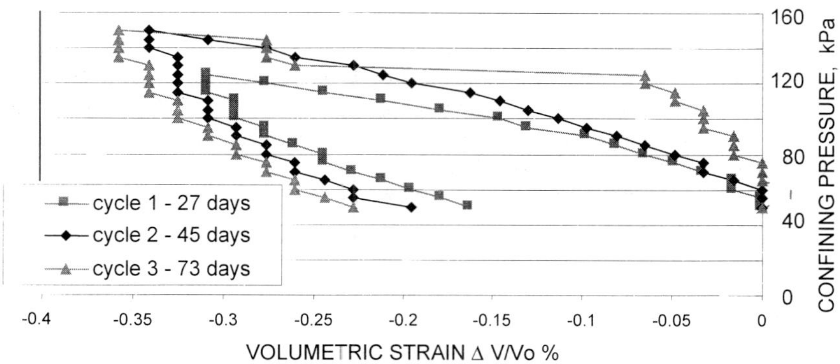

Figure 8 Compression cycles during swelling of raw material

Table 2 Cb values during swelling test of raw bottom ash

	TIME, days	Cb, MPa
Loading 1	27	23.94
Loading 2	45	28.25
Loading 3	73	19.91

PERMABILITY MEASUREMENT

This part concerns the hydraulic behaviour of the material –ie- air permeability- while a swelling test is being performed. As the material air permeability is supposed to vary with swelling, this parameter is continuously measured during that test. Figure 3 presents a schematic diagram of the experimental device designed for such a test. In this study, it is assumed that the fluid flow obeys Darcy's law. To apply this law, the flow must be laminar. One can assert that a laminar flow takes place if the Reynolds number is lower than 10^3 [5]. This number can be evaluated by $R = \frac{\rho V_c l_c}{\mu}$ where ρ is the specific mass of the air, μ its dynamic viscosity and V_c its relative velocity with respect to the skeleton. l_c is a characteristic length which is of the same order as $\sqrt{\mathrm{tr}\overline{\overline{K}}}$ where $\overline{\overline{K}}$ is the permeability tensor. With an expected value of $10^{-13}\mathrm{m}^2$, we obtain an extremely low value of R (less than 10^{-3}). That proves the flow will be laminar.

As the material is partially saturated with air and water, we use the generalised Darcy's law [7]. In such a case, for a given level of water saturation S_w, it can be written that:

$$\vec{V} = -\frac{K(S_w)}{\mu}\overrightarrow{\text{gradP}} \tag{5}$$

In this equation, permeability is considered to be an isotropic property and $K(S_w)$ is thus the effective permeability of the medium to air. As the saturation S_w is kept constant during the test, any variation of $K(S_w)$ will be due to the expansion effect of swelling compounds. The boundary conditions to which the sample of length h is subjected are shown in Figure 3. For x=0 $P(x)=P_i$ injection pressure (Pi=40 kPa) and for x=h $P(x)=P_o$ draining pressure (P_o=30 kPa, h=150mm). In a steady injection state, the pressure variation in the sample is given by the well known expression :

$$P(x) = \sqrt{P_i^2(1-\frac{x}{h}) + P_o^2\frac{x}{h}} \tag{6}$$

The use of Darcy's law and of equation (6) gives the following :

$$K(S_w) = \frac{\mu Q}{S}\frac{2hP_i}{(P_i^2 - P_o^2)} \tag{7}$$

where Q is the measured volumetric flow and S the sample surface.

Only two initial permeability values are currently available. These permeability were measured with two different samples : a raw materiel and a clinker treated one (table 3).

Table 3 Air permeability values

	RAW BOTTOM ASH	CLINKER TREATED BOTTOM ASH
Q (m^3/s)	1.2 X10^{-5}	10^{-5}
K (m^2)	1.4 X 10^{-13}	9.6 * 10^{-14}

CONCLUSIONS AND PERSPECTIVES

The experimental device, presented in this article, allowed the swelling pressure and the air permeability of MSWI bottom ash to be measured. Possible interactions between the swelling phenomenon and variation of air permeability have not been assessed yet. That is the main scope of further tests which are in progress.

In our study, all of these characteristics are determined in well-defined conditions. In fact, negative pressure is kept constant and equal to the OPM value. For incoming tests, it appears quite interesting to assess swelling pressure for different conditions (ie negative pressure, confining pressure, compaction...), so as to show the worst conditions which lead to the maximum deformation and swelling pressure.

ACKNOWLEDGEMENTS

The authors would like to thank the financial support of the European Community for the project MASHROAD (contract BRST CT97-5150). We would like also to thank Yprema, Calcia (Italcementi group) and Eurovia for there financial and technical supports.

REFERENCES

1. ALKEMADE M.M.C., EYMAEL M.M.Th., MULDER E., DE WIJS W. How to prevent expansion of MSWI bottom ash in road construction ? Environmental aspects of construction with waste materials, JJJM. Goumans, HA. Van der Sloot and ThG. Aalbers (Editors), 1994, p863-876.

2. CRIGNON C., PECQUEUR G., GARCIA DIAZ E., GERMANEAU B., SIWAK J.M., Study of cement-treated MSWI bottom ash expansion, Waste Stabilization and Environment 99, Lyon, Villeurbanne, 13-16 avril 1999.

3. CRIGNON C., PECQUEUR G., GARCIA DIAZ E., SIWAK J.M., Influence du traitement aux liants hydrauliques sur la stabilité dimensionnelle des MIOM, 2ème Congrès Universitaire de Génie Civil, AUGC, Des Matériaux à l'Ouvrage, Poitiers, 6 et 7 mai 1999.

4. BIOT M. A., General theory of three dimensional consolidation, Journal of Applied Physics, 1941, Vol. 12, pp. 155-164.

5. COUSSY O., Mécanique des milieux poreux, Technip Editions, Paris – 1991, 437 pages.

6. KERBOUCHE R., SHAO J.F., SKOCZYLAS F., HENRY J.P., On the poroplastic behavior of porous rocks, European Journal of Mechanics, A/Solids, 1995, Vol 14, N° 4, pp. 577-587.

7. DANA E. , SKOCZYLAS F., Gas relative permeability and pore structure of sandstone, International Journal of Rock Mechanics and Mining Sciences, 1999, Vol. 36, pp. 613-625.

PART THREE:

USE OF INCINERATOR BOTTOM ASH IN CONCRETE CONSTRUCTION

THE USE OF INCINERATOR ASH FOR MAKING LIGHTWEIGHT BRICKS

P M Abrahams
V G Nanikashwili
Kharkov State Academy of Municipal Management
H E Brown
N V Brown
Brown Consultants
Ukraine

ABSTRACT. The problem of waste pollution in both industrially developed and developing countries is getting top priority with increasingly frequent discussions both in government and non-government sectors, as national economies suffer seriously from these unwanted losses. Very often these discussions are irrelevant and do not help to predict actual ways to take advantage of these wastes. In this work the authors will investigate the possibility of using incinerated ashes as raw material for the manufacturing of lightweight bricks. The possibility of using incinerated ashes for brick making depends on mixture ratio and the curing process

Keywords: Waste, Incinerated ashes, lightweight bricks, Mixing ratio, Strength, Curing period

Mr Patton Mwasha Abrahams, is a research fellow at Kharkov State Academy of Municipal Management in Ukraine. He is a specialist in expansive clay soils. His interests also include the use of wastes in construction industry to meet the challenges of sustainable construction

Dr V G Nanikashwili, is an associate professor in the Department of Town Planing at Kharkov State Academy of Municipal Management in Ukraine. His research interests include the problems of designing concrete strength, and use of alternative building materials

Mr Hazael Elisha Brown, is a research assistant and consultant in waste management. He is also managing director of Brown consultants.

Mrs Natalie V Brown, is director of Brown consultants in Kharkov, Ukraine. Her area of interest includes the recycling of industrial wastes.

INTRODUCTION

Conservation of material and environment is one of the most acute problem facing everyone of us. Today more than a billion people in today's urbanizing world live and work among the worlds most degraded environment where garbage lies to rotting outside their makeshift shelters while more and more urban resident lack resources to meet their basic need for sustainable shelter giving ways to slums and squatters settlement. Today, scientists throughout the world are investigating the most economical ways of taking advantages of accumulating wastes, the most recent method of using incinerated ashes in construction industries has shown promising positive effects especially as fill in highway works and building materials. Building material industry plays an important part in the sustainable social development in every corner of this world as its output and the standards of its products have direct bearing upon the amount and the quality of construction works. Bricks have been used for hundreds of years as the most essential building material but at the moment in many part of this world we are noticing the acute shortage of raw materials for manufacturing bricks. A considerable reduction in the cost of raw material may be achieved by utilizing and recycling industrial, municipal and special wastes as source of raw material for the manufacturing of lightweight bricks. In this work the authors will investigate effective ways of using incinerated ashes from Kharkov waste incarnating factory for the manufacturing of lightweight bricks. Kharkov municipal waste incarnating factory has daily production capacity of sixty ton of ashes.

PROBLEMS WITH ACCUMULATING WASTES.

Land pollution is one of the result of industrial development. Annually in Ukraine and in other former Soviet States, more than 40 million tons of wastes are generated contaminating large parts of land which could be used for farming or other economical activities. According to Ministry of Metallurgy in the former Soviet Union about 500 million tons of slag's has been accumulated with projected annual increase to 50 million tons. The cost of maintaining landfill annually can reach up to 30 million rubles. The other major sources of industrial wastes, like electric plants where more than one billion cubic meter of slag has been accumulated, contaminate about 18 thousand hectares of valuable land. The problem of wastes accumulation is continuing to be the center of political and economical debate in almost in every country, according to Anold [1] in U.K alone there 200,000 ha of contaminated land which will cost 20 billion pounds to clean up the process estimated to take 200 years at the current rate of process. The possibility of using incinerated ashes in construction industry could be the answer to ever accumulating wastes. In many countries the use of wastes as source of energy has gained high popularity. Figure 1 shows the role waste plays as source of energy in comparisons with other sources of energy according to Alexis Neumann [2].

DESCRIPTION OF WASTES

According to Althers and Flynn [3], wastes are primarily divided into liquid and solid wastes and secondarily into municipal industrial, special wastes and radioactive wastes. The municipal liquid wastes in primarily domestic sewage. Municipal solid wastes is that refuse from residential and commercial sources that is collected in garbage cans and trash bins.

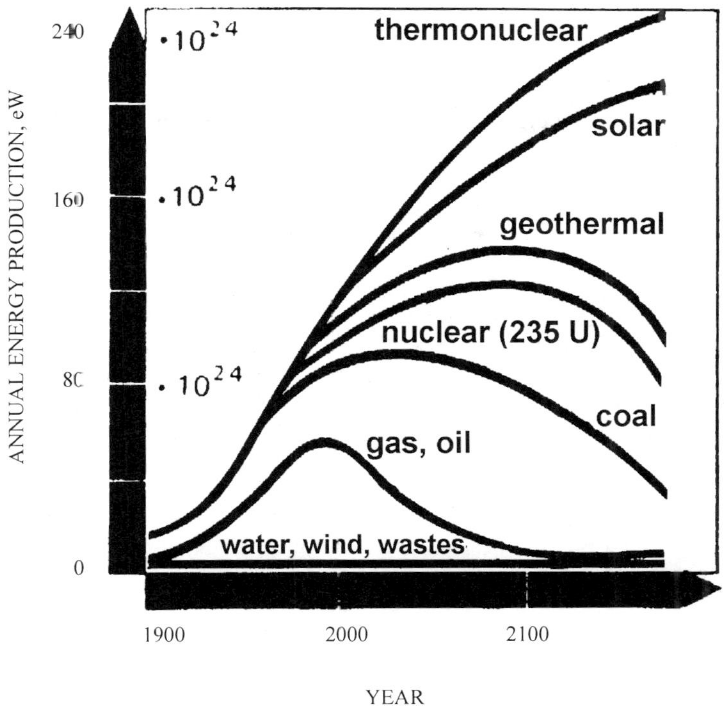

Figure 1 The Position of waste playing as source of energy in comparison with other sources of energy

Industrial wastes include aqueous organic and inorganic sludge , slurries and solids but the main type of industrial wastes can be pesticides, batteries, chemical manufacturing, electroplating, explosive textiles, pharmaceutical, mining and metallurgy, coal and petroleum wastes. There also special wastes generated by power plants such as air pollution control equipment, cooling tower, boiler and metal cleansing wastes. Considering the different types of wastes, proper selection of wastes should be considered for the reuse, especially for the construction of living houses. The availability, durability and strength should not be the final criteria for the selection of different wastes for utilizing in construction industrial especially in construction of living houses. Figure 2 shows the methods of separation of wastes for further processing.

Figure 3 illustrates the processing procedure for incinerated ashes.

EXPERIMENTAL DETAILS

Materials

The chemical properties of the ashes from Kharkov Municipal incinerator was analyzed. The pH of water content of this ash was detected as alkaline. The chemical properties have almost

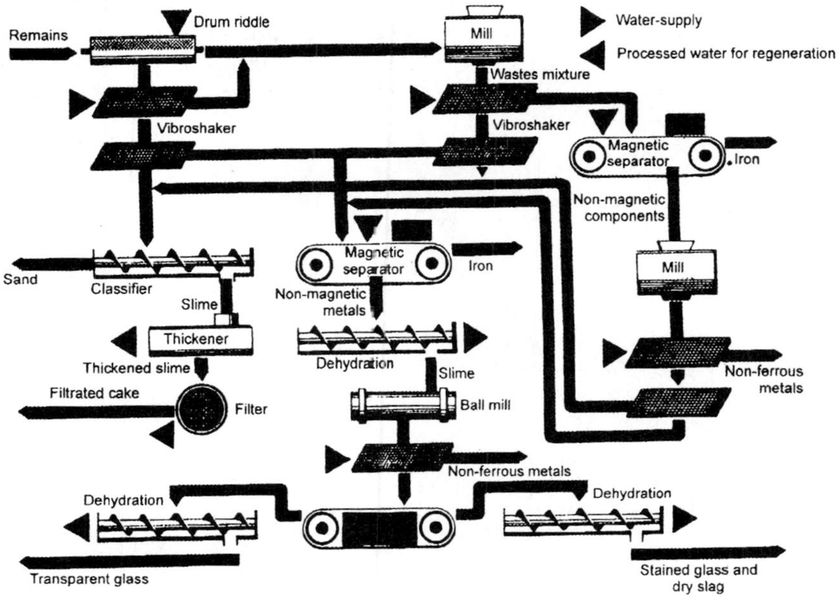

Figure 1 Processing of wastes

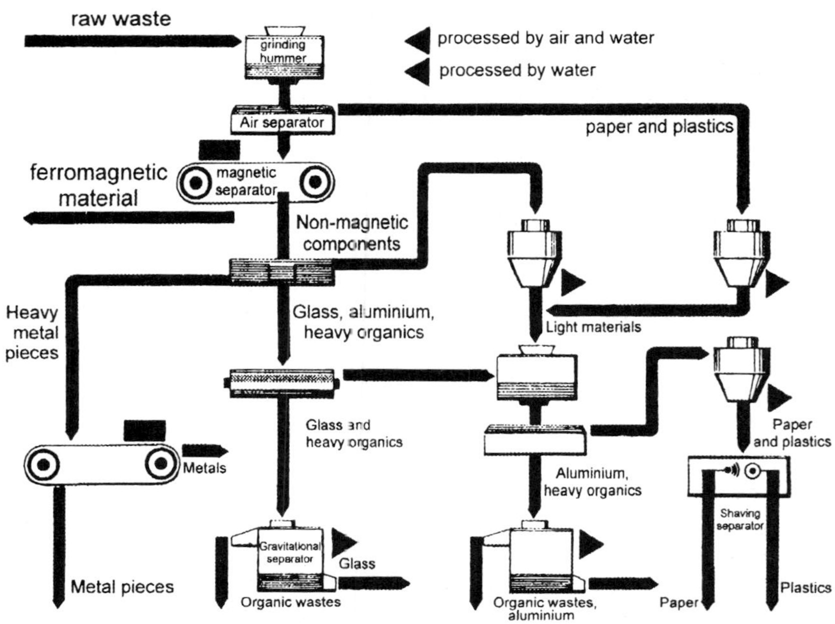

Figure 2 Processing after incineration of wastes

the same properties as the raw materials for manufacturing of Portland cement. Due to low burning temperature almost 500°C, there is no possibility of formation of clinker minerals. In fact $CA(OH)_2$ occupies almost 50%. Ash has an average bulk density ranging from 0.7 T/m^3 to 0.8 T/m^3. For reactivating inorganic acid was used for example H_3PO_4. The second component used was amorphous consisting of a mixture of Mono and Diammoniphosphates, i.e.

$NH_4H_2PO_4$ and $(NH_4)_2HPO_4$

Mix Proportions

On mixing of the above mentioned components with calcium hydroxide the following reactions take place:

$$Ca(OH)_2 + 2NH_4(H_2PO_4) \rightarrow Ca(H_2PO_4)_2 + 2NH_3 + H_2O$$

$$Ca(OH)_2 + (NH_4)_2HPO_4 \rightarrow CaHPO_4 + 2NH_3 + 2H_2O$$

Where $Ca(H_2PO_4)_2$ is soluble in water

Where $CaHPO_4$ is partially soluble in water

In order to speed up the reaction amorphous material should be in fine form. This should be done by grinding in a ball mill up to specific surface area of 1000-1500 cm^2/gm. After grinding the incinerated ash is added. At this stage two processes are expected:

1) Grinding of the ash
2) Mixing the processed ashes with amorphous.

In order to determine the optimum amount of ash and amorphous material, a test is performed. It was found that a mixing ratio of 7:3 by weight worked well. The mix of ash and amorphous material gave a fat cream like consistency when water was added.

Curing

After the preparation of the mix of ash with amorphous material, this mixture was poured in to special curing forms. After 15 to 20 minutes the mixture showed signs of expansion with the release of a specific type of ammonia like smell. Therefore, the casting mould for manufacturing of lightweight bricks of sizes 250 x 120 x 65mm using the above mixture should leave an allowable space of 10-20mm for expected expansion of the mix. After expansion process the heaved part of the brick was leveled and the brick allowed to dry in the casting mould for 3 days, and then demoulded to dry for a further 25 days. The drying process can be accelerated if the mixer of ashes and amorphous material is flooded with warm water ranging from 50 60°C with this process it is possible to demould the brick after a single day. It was also noted that after two or three times of using the metal casting forms, the forms were highly corroded. Because of this it is recommended that plastic casting forms made from fluoroplastic be used instead.

Physical mechanical properties

The density of this material is up to $2T/M^3$ and the compressive strength 7-8 N/mm^2. Water resistance was based on degree of softness of which was found to be 0.87. The colour of the ash and its product was grey in colour.

CONCLUSION

This study has shown that incinerated ashes can be used for the manufacture lightweight bricks for building.

REFERENCES

1. ANOLD. Progress in environmental science, Volume 1 number1 1999.

2. ALEXIS NEUMANN. Werstafft mit zuk uuft perspektiv . Volkswirts chafting werkstaffe Urania –Vertage ,berlin 1988.

3. ALTHER, G.R AND FLYNN, B.N. The role of bentonite on sealing application Pages 144-155 Detroit USA

THE LONG-TERM PERFORMANCE OF UNTREATED BOTTOM ASH, AND THE PRODUCTION AND USE OF A LIGHTWEIGHT SYNTHETIC AGGREGATE FROM THE BOTTOM ASH FINE FRACTION

P J Wainwright

J R Barton

D J F Cresswell

University of Leeds

United Kingdom

ABSTRACT. As the incineration of domestic refuse increases as a waste management option there will be larger volumes of incinerator residues to dispose of. One method of utilisation is the use of the bottom ash produced by incineration as an aggregate in concrete. This work investigates the use of '*untreated*' bottom ash in concrete and the use of a sintered synthetic aggregate made from the fine fraction of bottom ash. The long-term strength and medium-term expansion results are reported for concrete incorporating '*untreated*' ash, as are short-term strength tests for concrete using the synthetic aggregate as a coarse aggregate replacement. The results of the strength tests for '*untreated*' ash and synthetic aggregate are encouraging, although a longer more comprehensive study is necessary to substantiate the findings.

Keywords: MSWI bottom ash, Synthetic aggregate, Concrete, Trefoil kiln, Compressive strength.

Dr P J Wainwright is a senior lecturer in the Civil Engineering Materials Unit with over 20 years experience in concrete research, especially the incorporation of waste materials such as MSW incinerator ash, ground granulated blast furnace slag and recycled demolition waste.

Mr J R Barton is a senior Lecturer in Waste Management he previously worked at Warren Spring Laboratory as head of the materials recovery division. He has 20 years of experience in all aspects of solid waste treatment & recovery

Mr D J F Cresswell is a research assistant dealing with the use of waste materials in construction. He has three years experience of working with the Trefoil kiln to thermally treat wastes and produce synthetic aggregate at laboratory and pilot scale.

INTRODUCTION

As the reliance on landfill as the primary waste management option declines in the UK more Municipal Solid Waste incinerators (MSWI) will be needed. Depending on the levels of minimisation, reuse, and recycling the number of new incinerators is estimated to be between 28 and 165 (200 000 t/a) [1] in the coming decades which will produce a large and consistent supply of MSWI Bottom Ash. This in turn will help to encourage the use of bottom ash (BA) as a construction material, which is seen as the preferred method of utilisation to avoid the problems and costs associated with disposal, and provide an alternative aggregate source. Concern has been expressed on the use of BA in construction due to the metals contamination and the associated expansion when incorporated in concrete [2]. Methods of sintering and vitrification of BA to decrease the availability of toxic elements are well documented [3,4 5,6,7] but energy costs are high and use is limited.

The levels of metal contamination within BA mean that without vitrification utilisation requires some form of encapsulation or covering to prevent the leaching of contaminants into the groundwater. At present the main use of BA is in bound applications to reduce the potential for contaminant release [8,9,10], such applications include: asphaltic concrete, concrete blocks and cementitious sub-bases in roads. Improvements in BA quality can be bought about by washing and or ageing [11], this decreases the mobility of metals by oxidation and carbonation reducing the pH and producing more stable mineral phases.

The highest levels of contamination appear to be in the fine fraction of BA [6,12]. The removal of the fines and further processing may allow for the use of BA in greater number of, and higher quality, applications. The separation of the fine fraction (under 2mm), that makes up 30 –50 % of BA, to help reduce the risk of contamination is seen as undesirable because this fraction would still need to be disposed of [13].

The separation of the fines from the BA would reduce the availability of metals and any potential deleterious effects upon the concrete in which it is incorporated. Studies have shown that over one third of the metals and 40% of the un-burnt carbon are present in the size range below 2 mm and the great majority of sulphates and chlorides leach from the fraction below 0.44mm [2].

Evidence from continental Europe (See [14] for example) suggests that BA is a potentially suitable material for some construction purposes. This has been borne out by recent developments in the UK where ash is being used from 3 incinerators and industry is drawing up guidelines on its use [15].

Work has been carried out previously by Wainwright [16-19] on producing a quality aggregate from sintering BA in a rotary kiln. The resultant aggregate was used to replace natural coarse aggregate in a number of concrete specimens, which were then tested over a 5-year period for structural performance & long term degradation. The results were satisfactory although the tests did not extend to emission monitoring during production of the aggregate or leach testing during its use in concrete. Since this initial research work no progress has been made to commercialise the idea, a major deterrent being the high capital cost and design deficiencies associated with traditional rotary kiln of the type used in this earlier study.

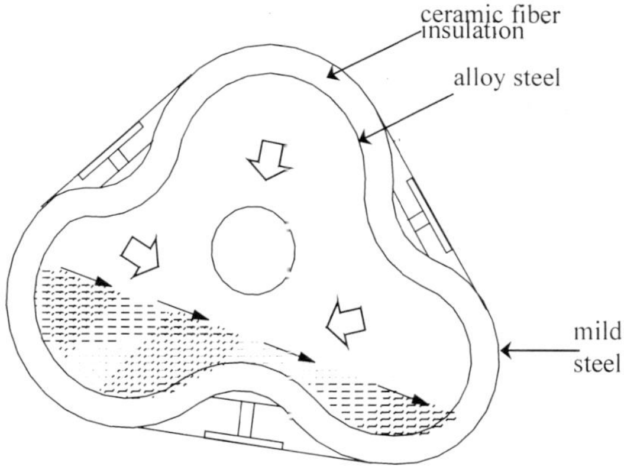

Figure 1 Cross section through the Trefoil rotary kiln

The work reported in this paper involves the use of an innovative rotary kiln, which is much more thermally efficient than a traditional kiln. The traditional rotary kiln is a thick steel drum lined with heavy manually laid fire bricks which have a limited life and which need regular maintenance entailing total shut down which can involve weeks of lost production. They can be extremely thermally inefficient and very slow to respond to firing adjustments. Their weight stresses motors, gear boxes, bearings and transmissions. It is difficult for them to fire material either consistently or quickly, a deficiency which can only be partly compensated for by making them very long in proportion to their diameter. By contrast the Trefoil kiln; (see Fig 1) which is potentially the only real advance in kiln design for the last 100 years [20], compresses a ceramic fibre insulation between a thin steel alloy hot face and a tensioned thin mild steel cold face to form a cartridge with a triple-lobed cross section. The resulting kiln has the following characteristics :

- remarkable thermal efficiency (an inner kiln temperature of 1200° C and an outer temperature of 60° C).

- a much smaller and lighter structure which does not need costly maintenance or relining and which is easier to manufacture, transport, install, remove and replace.

- much faster heat-up and cool-down periods (50 times quicker than conventional kilns),

- vastly improved responsiveness (operating temperatures can be increased or decreased in minutes)

- higher load factors.

The research presented in this paper covers two different methods of utilising BA in concrete. The first assesses the long-term effects of using, what is termed *untreated* BA, as an aggregate replacement in concrete. The term *untreated* refers to ash that has only been screened for the removal ferrous and non-ferrous metals. The second approach considers the production of a sintered synthetic coarse aggregate from BA using the Trefoil kiln. The later work forms the main focus of the ongoing research in this area.

EXPERIMENTAL WORK

Untreated Bottom Ash Utilisation in Concrete

Specimens of concrete were cast using varying amounts of BA as a replacement of both coarse and fine natural aggregates. The BA originated from two UK incinerators, Sheffield and Selchp. The preparation of the bottom ash consisted of grading, and magnetic separation to remove ferrous metals. The BA used for the direct replacement of aggregate complied with BS 882 grading requirements.

Substitution of the aggregates was on a weight basis and the free water / cement ratio was kept constant at 0.53. Table 1 shows the concrete mix ratios used. Long-term compressive strengths of the specimens were conducted, as were shorter-term expansion / shrinkage tests. Compressive strength was conducted in accordance with BS1881 [21] on 100mm specimens. Expansion tests were carried out by measuring the change in length of 40 x 40 x 200mm prisms. Four prisms were used for each concrete mix, the first two were stored at 20 °C and 65 % relative humidity (referred to as 'dry'), and the second two were kept in water at 35 °C ('wet'). Unfortunately because of a shortage of material it was only possible to perform these expansion tests on mixes made with the Selchp ash.

Table 1 Mix ratios (by wt.) used for BA as a replacement for natural aggregates

MIX	NATURAL CA	COARSE BA	NATURAL FA	FINE BA	OPC	W:C	WORKABILITY SLUMP (mm)
Control	4.00	0.00	2.00	0.00	1.00	0.53	25
Selchp C50/F50	2.00	2.00	1.00	1.00	1.00	0.53	0
Selchp C50	2.00	2.00	2.00	0.00	1.00	0.53	85
Sheff C50	2.00	2.00	2.00	0.00	1.00	0.53	0

Thermal Processing for Synthetic Aggregate Production From Bottom Ash

The bottom ash

The aim of the current work is to take the finer fraction of BA and process it into a coarse aggregate for use in concrete.

The bottom ash used to produce the aggregate was obtained from an ash recycling plant taking ashes from one incinerator in the UK. A sample of the 0 – 6mm size fraction of the ash was taken and this was further processed to take out the material < 600 μm which was the fraction used in this current investigation.

Pelletisation of the bottom ash

To enable the ash to be sintered into an aggregate it is necessary to pelletise it before input into the kiln. The 600μm maximum size limit mentioned above has been derived from experience gained over a number of years in producing pellets from a variety of wastes. Materials with particle size distributions over 600μm are possible to pelletise but are more difficult and larger proportions (30 –40 %) of clay binder are needed.

In previous work on pelletisation [16-19] a disc (or pan) pelletiser was used to produce pellets with moisture contents ranging from 5 – 10 % and with 15 % clay content as the binder. To enable the pelletisation to work the ash had to be ground to a fine powder whereby the majority of the material passed 150 μm. Commercially however size reduction to this is degree is not considered to be practical which is why this more recent study has concentrated on utilising ashes that have not undergone any size reduction and have a with coarser grading.

As an alternative to the pelletising disc, the method used to produce pellets in this recent work was extrusion of a 'plastic' mix made from the BA fines and a clay binder. The extrusions were fed directly into a rotating drum in which they were rolled into balls. The drum also acted as a dryer which produced a surface dry green pellet which was much easier to handle than when wet. The surface dried green pellets were then fed into the Trefoil kiln for firing at temperatures of approximately 1200°C.

In this work aggregates were produced from two different fractions of the same original < 600 μm sample, these were nominally < 600 μm >300 μm, and < 600 μm (see Table 2). A clay binder was used to enable the ash to be extruded and to provide sufficient green strength to the pellets produced. The clay content (see Table 2) of the two mixes differs, and is considered the minimum amount to balance green pellet production rates and green pellet strength. Once the green pellets had been produced they were left to air dry before being fired.

Table 2 Mixes used for green pellet production

	BOTTOM ASH		
	< 600 μm	< 300 μm	Clay
Mix 1 (Agg 01)	82 %	----	18 %
Mix 2 (Agg 02)	4.6 %	85.7 %	9.7 %

Aggregate firing

Firing consisted of two stages: the first is known as the 'burnout' stage where the carbon content of the ash/clay mix is allowed to combust (burnout) at a temperature below that where the inert material of the pellet begins to sinter. From previous experience it was known that this stage can take between 5 and 40 minutes and occurs at between 700 and 800 °C, depending on the amount and type of carbon (organic material) present and the internal structure of the pellet.

According to BS 3797 [22] the carbon content of a synthetic aggregate produced from pulverised fuel ash needs to be below 4% and it will be assumed that a similar figure is applicable to the material being produced here. To ensure that this 4% level is reached before the sintering of the ash begins the carbon content has to be assessed during the burnout period.

To assess the burnout time needed the first batch of pellets was sampled at time intervals up to 20 minutes. A loss on ignition test at 775 °C was then performed on each sample to assess the residual carbon content at different burnout times, from this a minimum burnout time is obtained. At the same time a visual assessment was made of the appearance of the inside of each pellet looking in particular at the size and colour of the grey centre. From the experience gained it was then possible in all subsequent tests to assess the carbon content from a visual inspection of the inside of the pellets. Using this procedure the burnout time was estimated at approximately 5 minutes.

Once pellet burnout had been achieved the kiln temperature was raised rapidly (4-5 minutes) to firing temperature which from previous experience was known to be approximately 1200 (+/- 20) °C. In the tests reported here the two aggregates were fired at slightly different temperatures which were: 1220 °C for Agg 01, and 1180 °C for Agg 02. This difference in temperature is thought to be related to the different PSD's of the bottom ashes used to make the pellets (see Table 2). The average residence time for each batch of pellets was 2 minutes.

Once the aggregates had been fired tests were performed to measure water absorption and relative densities in accordance with to BS 812 [23].

Casting of Concrete from the Synthetic Aggregate

A total of 5 concrete mixes were made to assess the performance of the synthetic aggregate in concrete. (see Table3). Two control mixes were made, one with all natural aggregates and one containing Lytag as the coarse aggregate. Two mixes were made with the synthetic coarse aggregate replacing the natural coarse aggregate (mixes 1,2) and one in which the natural fines was part replaced by the 5mm - 600μm fraction of the original sample of the bottom ash (mixes 3). All mixes were made with nominally the same volume proportions. For each mix a number of 100mm cubes were made for measurement of compressive strength at a loding rate of 3.0 kN/second.

RESULTS AND DISCUSSION

Long Term Use of Untreated BA Incorporated in Concrete

The workabilities of all 4 mixes made to a constant water / cement ratio of 0.53 have been shown earlier in Table 1. As might be expected the partial replacement of the coarse and / or fine aggregate with bottom ash leads to a reduction in workability (compared to the control) with the exception of mix 3 in which 50% of the coarse aggregate has been replaced with Selchp bottom ash. This observed increase in workability from 25 to 85mm slump is somewhat surprising and no apparent reason could be found for this unusual trend.

Table 3 Concrete mix ratio's (by weight) for sintered aggregate

AGGREGATE AND MIX CODE	NATURAL CA	SYNTHETIC CA	NATURAL FA	RECYCLED FA	OPC	W:C	WORKABILITY SLUMP (mm)
BA Aggregate 1 (A 01) mix 1	--	2.52	2.23	--	1.00	0.53	95
BA Aggregate 2 (A 02) mix2	--	2.37	2.23	--	1.00	0.53	135
Fines Replacement (FRV) mix 3	3.64	--	1.12	1.01	1.00	0.53	10
Control (Con) mix4	3.64	--	2.23	0.00	1.00	0.53	20
Lytag mix5		2.80	2.23		1.00	0.53	10

The compressive strength development of all mixes up to an age of 3 years is shown in Fig 2. Replacement of 50% of the natural aggregate with either Sheffield ash or Selchp ash resulted in strength reductions of 38% and 45% respectively at 3 years compared to the control. Such strength reductions are to be expected due to the poorer quality of the bottom ash but what is of more importance is the fact that none of the mixes show any reduction in strength with age. This is a good indication that there are no detrimental reactions taking place between the bottom ash and the cement which would lead to long term degradation of the concrete. The Sheffield ash showed a consistently higher strength of approximately 5 N/mm^2 compared to Selchp although had the concretes been made to a constant workability the situation may have been reversed. The mix in which both the natural coarse and fines have been partially replaced by the Selchp bottom ash achieved the lowest strength of all but the difference between this mix and the one containing natural fines and Selchp coarse was less than 5 N/mm^2. The reason this difference is so small is probably related to the fact that the mixes were made to nominally the same free water / cement ratio. As with the other mixes this mix also showed no reduction in strength with age.

The shrinkage / expansion results [Figure 3] up to an age of 6 months show similar trends for all 3 mixes which again is probably due to the fact that they have been made to the same water / cement ratio. The larger increase in expansion between 3 months and 6 months in the case of the mix containing bottom ash fines may suggest the beginning of some of expansive reaction but this is not reflected in the strength results discussed earlier.

Synthetic Aggregate Production from Bottom Ash

Aggregate properties

The properties of the aggregate produced are shown in Table 4. The fired synthetic aggregate has a loose bulk density just below the maximum limit of 1000 kg/m^3 required for it to be classified as a lightweight Aggregate. The higher water absorption of Agg. 01 is due its more open textured surface probably as a result of the different PSD's of the bottom ashes used in its production and the higher firing temperate (1220 °C cf. 1180 °C for Agg. 02).

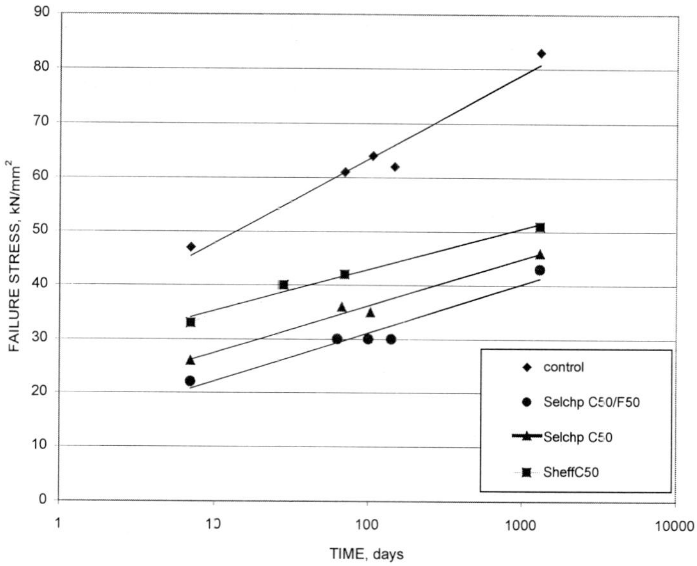

Figure 2 Long term compressive strength tests of ba incorporated as a direct replacement of coarse and fine aggregate in concrete

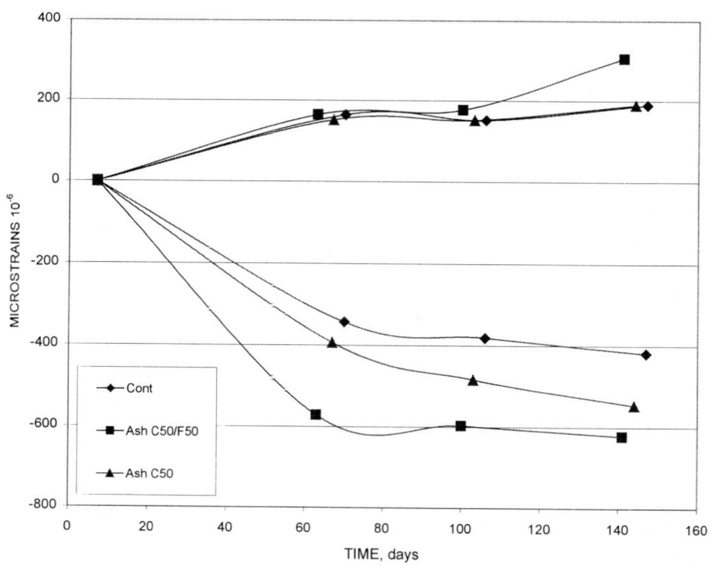

Figure 3 Shrinkage / expansion results of concretes containing bottom ash as a aggregate replacement

Table 4 Aggregate properties

AGGREGATE	LOOSE BULK DENSITY kg/m^3		RELATIVE DENSITY (SSD)	24HR WATER ABS. (% dry mass)
	Dry	SSD		
Agg 01	920	1036	1.88	12.7
Agg 02	985	1034	1.85	4.5
Lytag	780	876	1.76	12.3
Nat. Gravel Aggregate	1510	1520	2.61	0.8

Testing of the sintered BA aggregate concrete

The workabilities of all five mixes made to nominally the same water / cement ratio of 0.53 have been shown earlier in Table3. All mixes with the exception of the control mix (mix 4) have similar densities and it is reasonable therefore to make comparisons between the Lytag control and the remainder of the mixes containing bottom ash aggregates. As can be seen using the synthetic aggregate as a coarse aggregate replacement instead of Lytag leads to a significant increase in workability: from 10mm to 95mm in the case of mix1 to 135mm in the case of mix2. This is thought to be largely due to the more spherical rounded shape and smoother surface texture of the bottom ash aggregates compared with Lytag. Partial replacement of the natural fines with bottom ash appears to have little effect on the workability; compare mix 3 with mix 5.

At the time of writing this paper only strength results up to the age of 28 days were available and these are presented in Fig 4. The mixes made with the bottom ash, synthetic aggregate and the Lytag show a consistently lower strength than the control mix made with all natural materials. The differences in strength though are not very large, the Lytag mix (mix5) is less than 5 N/mm^2 weaker than the control mix at all ages and mix 2 (made with synthetic aggregate 02) is only marginally weaker than Lytag. Of the two synthetic aggregates made in this study type 01 produced a slightly lower strength concrete than type 02. This may be an indication that the aggregate itself is weaker a fact which may also be reflected in its higher water absorption (12.7% cf. 4.5% see Table 4). Clearly a much more comprehensive and longer term test programme is required to provide long term performance data on the synthetic aggregates but the results to date are encouraging. The trends, in particular those relating to the relative differences between the control mix and the synthetic aggregate mixes are similar to those found in the earlier work [12] in which 6 years data are reported. It must also be remembered that if the mixes had been made to a constant workability it would have been possible to reduce the water / cement of mixes 1 & 2, which would have increased their strengths.

A partial replacement of the natural fines with the untreated bottom ash (mix 3) leads to a much more significant drop in strength than when using the synthetic coarse aggregate (mixes 1 & 2). Mix 3 is more than 50% weaker than the control mix at all ages and its strength at 28 days is only about 20 N/mm^2. This would clearly mean that this concrete could not be used as a structural material but provided there is no long term degradation in strength it could for low grade applications. Again a longer term and more comprehensive test programme is required to verify its properties.

Conclusions and Future Work

The results reported here form only part of the initial phases of a much larger study that is being carried out at Leeds into the utilisation of waste materials in construction. As the work is at such an early stage it is difficult to draw too many firm conclusions from the data presented here, but the results, particularly those relating to the synthetic coarse aggregate, look encouraging. Some of the ongoing / future work includes:

- Influence of mix composition and firing parameters on synthetic aggregate properties.
- Long term properties of concretes made with synthetic aggregate produced from different waste streams
- Leaching characteristics of aggregates / concretes produced waste.

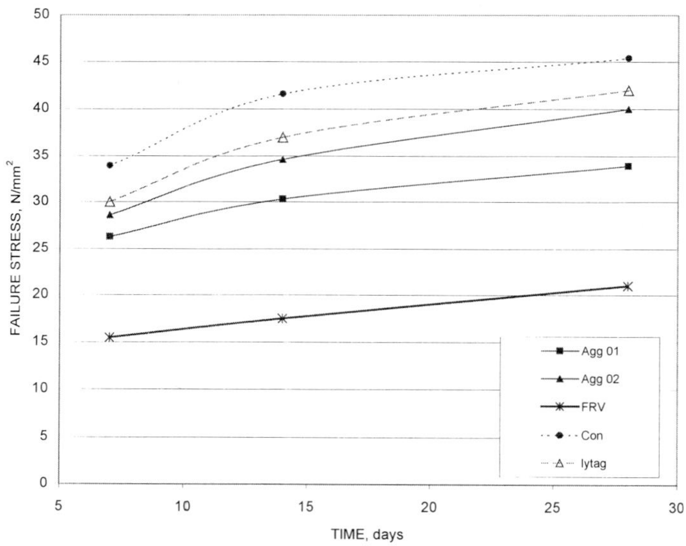

Figure 4 Compressive strength tests of concrete containing ba as a fine aggregate replacement and sintered to a coarse aggregate.

REFERENCES

1 DETR 1999 A Way With Waste. A Draft Strategy for England and Wales Page 24-5

2 ALKEMADE, M.M.C. How to Prevent Expansion of MSWI Bottom Ash in Road Construction. in GOUMANNS J.J.JM. etal (eds) Environmental Aspects of Construction with Waste Materials. 1994. Elsevier Amsterdam. pp863-877

3 ABE, S. Construction Materials Manufacturing by the Technology of Melting. in GOUMANNS, J.J.J. et al (eds). Waste Materials in Construction: Putting Theory into Practice. 1997. Elsevier. Amsterdam. pp25-30.

4 NISHIGAKI, M. Producing Permeable Bricks and Paving Slabs from Molten Slag. in GOUMANNS, J.J.J. et al (eds). Waste Materials in Construction: Putting Theory into Practice. 1997. Elsevier. Amsterdam. pp 31-40.

5 OSADA, M. Direct Melting Process for MSW Recycling in GOUMANNS, J.J.J. et al (eds). Waste Materials in Construction: Putting Theory into Practice. 1997. Elsevier. Amsterdam. pp 73-78.

6 SELINGER, A. & SCHMIDT, V. The ABB Dry Ash Concept: InREC™ in GOUMANNS, J.J.J. et al (eds). Waste Materials in Construction: Putting Theory into Practice. 1997. Elsevier. Amsterdam. pp 79-84

7 SELINGER, A. et al. Investigation of Sintering Processes in Bottom Ash to Promote their Reuse in Civil Construction. Parts 1 & 2. in GOUMANNS, J.J.J. et al (eds). Waste Materials in Construction: Putting Theory into Practice. 1997. Elsevier. Amsterdam. pp 41 – 58.

8 PFRANG-STOTZ, G. & REICHELT, J. Municipal Solid Waste Incinerator Bottom Ashes as granular Base Materials in Road Construction. in GOUMANNS, J.J.J. et al (eds). Waste Materials in Construction: Putting Theory into Practice. 1997. Elsevier. Amsterdam. pp85-90

9 JANSEGERS, E. The Use of MSWI Bottom Ash in Hollow Construction Materials in GOUMANNS, J.J.J. et al (eds). Waste Materials in Construction: Putting Theory into Practice. 1997. Elsevier. Amsterdam. pp 431-436.

10 PHIL, K.A. The Application of Incinerator Bottom Ash in Road Construction. in GOUMANNS, J.J.J. et al (eds). Waste Materials in Construction: Putting Theory into Practice. 1997. Elsevier. Amsterdam. pp 541-550.

11 STEKETEE, J.J. Quality Improvements of MSW Bottom Ash by Enhanced ageing Washing and Combination Processes. in GOUMANNS, J.J.J. et al (eds). Waste Materials in Construction: Putting Theory into Practice. 1997. Elsevier. Amsterdam. pp 13-24.

12 STEGEMANN, J.A AND SCHNEIDER J. Leaching potential of Municipal Waste Incinerators Bottom Ash as a Function of Particle Size Distribution. in GOUMANNS J.J.JM. etal (eds). Waste Materials in Construction – Proc. Int. Conf. On Env. Implications of Construction with Waste. 1991. Elsevier. Amsterdam. pp 135 – 44.

13 LAMERS, F.J.M. & BORN, J.G.P. Upgrading Techniques for the Quality Improvement of Municipal Waste Incinerator Residues. in GOUMANNS J.J.JM. etal (eds) Environmental Aspects of Construction with Waste Materials. 1994. Elsevier Amsterdam. pp 645-654.

14 BORN J.G.P. & VEELENTURF R.A.L. 1997 MSWI Residues in The Netherlands, Putting Policy into Practice. in GOUMANNS, J.J.J. et al (eds). Waste Materials in Construction: Putting Theory into Practice. 1997. Elsevier. Amsterdam. pp841-850.

15 ENDS Progress on Recycling Incinerator Ash. ENDS Report 290 March 1999. p16.

16 WAINWRIGHT, P J & BONI, S P K. Artificial Aggregate from Domestic Refuse. Concrete, 1981. 5(5), pp 25-29.

17 WAINWRIGHT, P J & BONI, S P K. Some Properties of Concrete Containing Sintered Domestic Refuse as a Coarse Aggregate. Magazine of Concrete Research, 1983. 35 (123), pp 75-85.

18 WAINWRIGHT, P.J., HADZINAKOS, I & ROBERY, P. A Review of the Methods of Utilisation of Incinerator Residues as a Construction Material. Proceedings of the International Conference on Low Cost & Energy Saving Construction Materials. Rio de Janeiro, Brazil, 1984

19 WAINWRIGHT, P J & ROBERY, P. Production & Properties of Sintered Incinerator Residues as Aggregates for Concrete. in GOUMANNS J.J.JM. etal (eds). Waste Materials in Construction – Proc. Int. Conf. On Env. Implications of Construction with Waste. 1991. Elsevier. Amsterdam. pp 425-432.

20 L.F.PARKS & R.G.SHERWEN The Use of New Materials in an Improved Design of Rotary Kiln. Materials & Design Volume VII (5). 1986.

21 BS 3797 Lightweight Aggregate for use in Masonry Units and Structural Concrete

22. BS 812 Methods for the Testing of Aggregates for use in Concrete

ASH REUTILIZATION RESEARCH AT THE UNIVERSITY OF CONNECTICUT

G C Frantz

K R Demars

University of Connecticut

United States of America

ABSTRACT. University of Connecticut researchers studied the reuse of bottom and fly ashes from burning municipal solid waste in refuse derived fuel (RDF) and mass burn (MB) incinerators. Ash was used in portland cement concrete, masonry, bituminous asphalt concrete, and as structural fill. Chemical and environmental aspects of raw ash and "recycled" ash were studied. Large amounts of ferrous particles remained even after pre- and post-processing of the municipal solid waste and ash. Leachate tests showed unacceptable concentrations of heavy metals in the fly ash; bottom ash concentrations were below "hazardous" limits. The ash contained large amounts of soluble salts. Compared with a control concrete, RDF fly ash produced acceptable concrete compression strengths at up to 30 percent cement replacement rates, but MB fly ash produced much lower strengths. Both fly ashes reduced concrete durability. Very little heavy metal leached from crushed portland cement concrete or masonry. Bottom ash had lower compacted density and specific gravity than typical well-graded coarse-grained natural aggregate but had equivalent or higher strength properties. Without any post-processing, RDF bottom ash produced an aggregate of marginal quality for Class 1 bituminous concrete. MB ash, with some post-processing and blending with other aggregates, performed as well as a control aggregate for Class 2 bituminous concrete. Required asphalt contents were 20 to 30 percent higher when using municipal solid waste ash as aggregate.

Keywords: Ash, Asphalt, Bottom ash, Chemical properties, Concrete, Fly ash, Masonry, Mass burn, Municipal solid waste, Pavement, Physical properties, Pozzolans, Refuse derived fuel, Structural fill, Wastes

Professor G C Frantz of the University of Connecticut has conducted research in concrete materials (corrosion inhibitors, durability, repair materials, MSW ash) and in structural concrete (prestressed concrete bridge beams, high strength concrete beams).

Professor K R Demars of the University of Connecticut has performed research on the geotechnical properties and behavior of natural and recycled aggregates, such as incinerator ash.

INTRODUCTION

Cities and towns are rapidly running out of landfill space to dispose of Municipal Solid Waste (MSW). Towns collect recyclables to reduce the metal, glass, plastics, batteries, and paper content from the waste stream. Incineration of MSW, with 60 – 75 percent reduction in waste volume, is one alternative many cities are using. In 1990, the State of Connecticut incinerated 59 percent and landfilled 41 percent of its total MSW. By 1997 with a more aggressive recycling program in place, these figures were 64 percent incinerated, 12 percent landfilled, and 24 percent recycled. The State has the highest incineration rate in the Nation (national averages: 8 percent incineration 61 percent landfilled, 32 percent recycled). The State's incineration plants either burn as-received MSW as mass burning (MB) or burn processed refuse-derived-fuel (RDF), where ferrous and glass materials are removed and the remaining MSW is shredded prior to burning. Heat generated during RDF burning is used to make electricity. Disposal of the residue fly and bottom ashes, which are usually contaminated with heavy metals, creates additional problems. Most ash is currently disposed of in landfills, but space is becoming scarcer.

The ash from an incinerator may be used in its "natural" state as pure bottom ash, pure fly ash or a combined ash. In general, the fine silt-sized fly ash is most contaminated and will not pass toxicity tests unless co-mingled with cleaner bottom ash. An ash may also be processed, such as washed to remove salts or fines, magnetized to remove ferrous particles, or sieved to alter particle size among other treatments. An ash may also be "amended" to improve its properties by blending with other materials such as cement, bitumen, clay, or coal fly ash.

This paper is a brief summary of University of Connecticut research on the possible reuse of MB and RDF ashes in portland cement concrete, masonry, bituminous asphalt concrete, and as structural fill. Chemical and environmental aspects of the raw and "recycled" ashes were studied.

GENERAL CHARACTERIZATION OF THE ASH

Ash was obtained at several times during the research. Most of the laboratory testing was done using ash described in this section. Fly ash and combined fly/bottom ash were obtained from one burning run at a MB plant. Fly ash and bottom ash were obtained from three separate burning runs at an RDF plant. Each RDF ash sample was taken over one day during a series of 15, one-day long plant performance tests to study how operating conditions effected ash volume, steam production, emissions, etc. (Table1). Notice Run RDF7 had poor burning conditions (high CO level) but low emissions (low outlet SO_2 and HCl). Plant operating conditions for the MB sample were not available. Both plants added lime during scrubber operations to reduce emissions.

As-received bottom ash and combined ash samples were full of large particles, such as cans, wire, unburned cloth, unburned grass (especially in mass-burn ash), paper, ceramics and stones.

Ash was tested using EPTOX [1] and/or TCLP [2] tests to measure the potential for leaching heavy metals, not the total metals content, under conditions similar to a landfill. EPTOX maintains an acidic test solution; TCLP allows pH to vary. The fly ashes generally had leachable cadmium and lead higher than the toxicity limits (Table 2). Bottom ash generally had lower levels of soluble metals than fly ash did and with only lead being above the toxicity limit (Table 3). Although the concentration for leachable lead may have exceeded the toxicity limit for a particular sample of the bottom ash, the average concentration for four or more samples of bottom ash was always within allowable limits.

Table 1 Operating conditions for RDF runs

	RDF7	RDF8	RDF9
Overall operating conditions (based on loading rate, emissions, etc)	Poor	Normal	Above Normal
Furnace conditions			
Refuse loading, kg/hr	33,350	28,670	30,030
Furnace temperature, °C	1006	1015	1033
At inlet to scrubbers (after burning)			
Inlet CO, ppm	347	90	93
Inlet SO_2, ppm	188	190	181
After scrubbers and fabric filters			
Outlet SO_2, ppm	17	130	194
Outlet HCl, ppm	6	40	102

Table 2 EPTOX results for fly ashes (ppb)

LEACHATE SAMPLE	BARIUM	CADMIUM	CHROMIUM	LEAD
RDF7	1,139	2,475	9	1,890
RDF8	543	1,840	9	11,300
RDF9	775	4,482	< 6	12,700
MB	3,910	4	70	47,300
Detection Limit	2	3	6	59
Toxicity Limit	100,000	1,000	5,000	5,000

Table 3 TCLP results for bottom ashes (ppb)

LEACHATE SAMPLE	BARIUM	CADMIUM	CHROMIUM	LEAD
RDF*	300 (260-360)	150 (61-360)	200 (39-400)	1,800 (410-4,680)
MB*	200 (170-270)	250 (200-400)	190 (ND-400)	3,060 (650-9,000)
Detection Limit	2	3	6	59
Toxicity Limit	100,000	1,000	5,000	5,000

* average of 4 samples (range), ND means not detected

ASH USE IN PORTLAND CEMENT CONCRETE AND MASONRY

Fly Ash as Cement Replacement

References 2 – 6 contain complete details of using ash in concrete and masonry. Fly ash was used as collected. Compared with coal fly ash, which is spherical and smooth surfaced, MSW fly ash particles were generally larger, not spherical, with coarser texture, had high (12 percent) loss-on-ignition, and absorbed up to 20 percent moisture if exposed to air for a sufficient time. The MSW fly ashes were not cementitious when mixed only with water, but they had pozzolanic properties.

Fly ash was used to replace 7.5 to 60 percent of the cement in concrete mixes. At 7.5 and 15 percent replacement, RDF ash produced compressive strengths better than both the control concrete and concrete with coal fly ash at ages beyond 7 days (Figure 1). Even at 30 percent replacement, RDF ash produced strengths comparable to control concrete at 1 to 2 months of age. MB fly ash produced significantly lower strengths, especially at higher replacement rates. MSW fly ash significantly extended setting time, probably due to its lead and zinc content. High dosages of set-accelerating admixtures were required to reduce setting time.

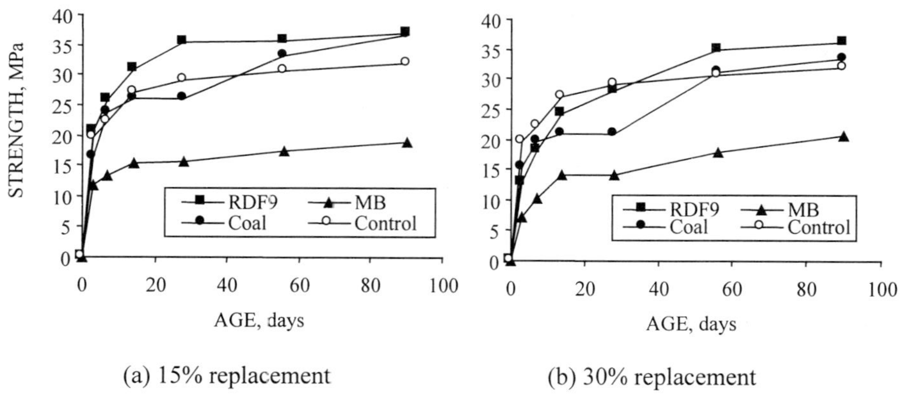

Figure 1 Effect of fly ash on concrete compression strength

MSW fly ash greatly reduced the ability of air-entrained concrete to withstand freeze-thaw cycling (lower relative dynamic modulus) (Figure 2). They also required significantly increased amounts of air-entraining admixture. Deicing salts caused much more surface damage (higher surface rating) for concrete with RDF fly ash. MSW fly ashes had total chloride contents exceeding 0.5 percent, indicating potential problems for reinforcement corrosion.

Although MSW fly ashes had high leachable levels of some heavy metals, very little heavy metal (well below toxicity limits) leached from concrete containing these fly ashes (Table 4).

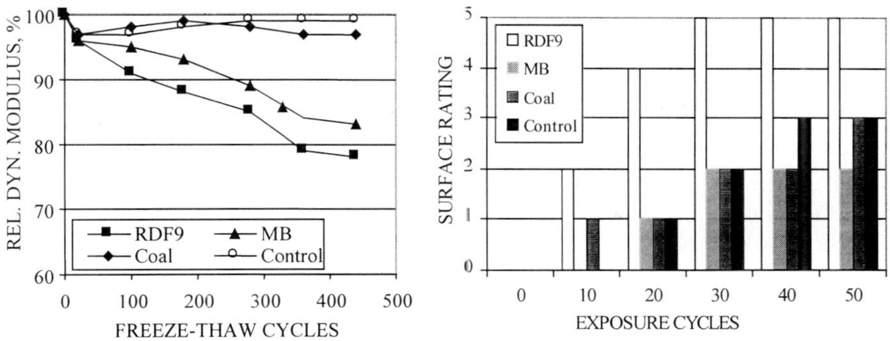

(a) Freeze-Thaw Resistance

(b) Deicer Scaling Resistance

Figure 2 Effect of fly ash on concrete durability, 15% replacement

Table 4 EPTOX results for concrete and masonry (ppb)

LEACHATE SAMPLE	BARIUM	CADMIUM	CHROMIUM	LEAD
Concrete				
Control	950	10	45	88
RDF (15% fly ash)	1,353	4	74	93
RDF (bottom ash aggregate)	734	7	34	85
MB (15% fly ash)	838	3	122	80
MB (fly/bottom ash aggregate)	374	9	22	95
Masonry				
Control	895	< 3	< 6	< 59
RDF (bottom ash aggregate)	475	< 3	14	< 59
MB (fly/bottom ash aggregate)	336	3	47	< 59
Detection Limit	2	3	6	59
Toxicity Limit	100,000	1,000	5,000	5,000

Bottom Ash or Combined Ash as Concrete Aggregate

Bottom ash and combined ash samples were dried and sieved through either 4.8-mm (for RDF) or 9.5-mm sieves (for MB) to remove as much metal and glass as possible, leaving a useable "fine aggregate" ash about 30 percent of the original ash dry weight. But, even after sieving, ashes contained some glass, wire, etc., which, if seen, were removed. The remaining ash had loss on ignition values of 5 percent. Generally, the ashes had too many fines for ASTM limits for concrete sand but did satisfy limits for lightweight fine aggregates. Bulk specific gravities were about 1.8, more typical of lightweight aggregate than of concrete sand.

Processed MSW ash was used to replace 50 or 100 percent of the fine aggregate in concrete and compared with normal weight and all-lightweight aggregate concrete. Demolding of specimens was delayed until 2 days of age because the ash concrete was still soft 1 day after mixing. During three hours after casting, the ash concrete expanded about 1 or 6 percent for the MB and RDF ashes, respectively. While curing in saturated limewater, the ash concrete produced relatively large volumes of gas, identified as hydrogen, that cracked the concrete.

Expansion and cracking of the RDF ash concrete significantly reduced its compression strength (Figure 3). MB ash concrete was much stronger than RDF, but weaker than the normal or lightweight aggregate controls. Prewashing the MB ash in limewater produced higher strengths but had little effect with the RDF ash. Replacing only 50 percent of the fine aggregate with ash produced higher strengths for the MB ash but not for the RDF ash.

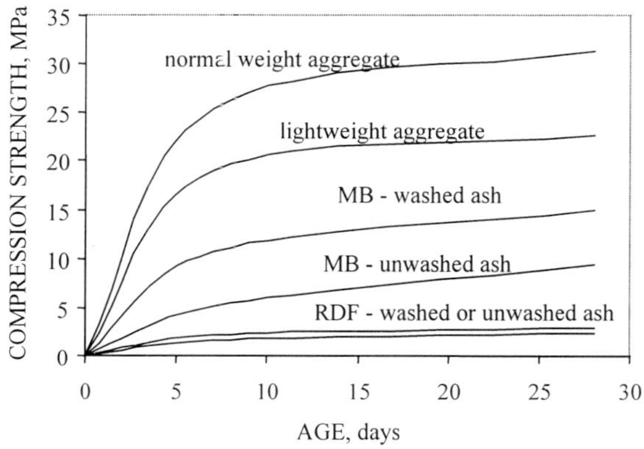

Figure 3 Concrete strength with MSW aggregate

Simulated masonry specimens were also made. Specimens were molded under pressure and cured at elevated temperature, as for masonry block. Specimens with 100 percent MB ash and specimens with up to 50 percent RDF ash had compression strength comparable to the control, which used a 50:50 blend of sand and lightweight aggregate. At each percentage replacement, the MB specimens were stronger than the RDF specimens, sometimes more than double. Expansion of the ash masonry was less than 1.6 percent, probably due to the drier and more porous mix compared to concrete. Ash masonry had a distinct, unpleasant odour. Some specimens cracked due to growth of white particulate material, probably the result of reaction between aluminum in the ash and cement. Very little heavy metal leached from either concrete or masonry containing the ash as aggregate (Table 4).

ASH USE AS STRUCTURAL FILL

References 8 – 10 contain complete details of the geotechnical testing of the ash. Bottom ash was evaluated as replacement for natural soil aggregates such as sand and gravel in earth fill applications. To be cost competitive, the bottom ash was subjected to minimal post-processing using only a 4-inch sieve to remove oversize particles. The RDF incinerator also used magnetic separation prior to combustion.

Bottom Ash Properties and Behavior

Both MB and RDF bottom ashes were evaluated for physical and chemical properties and engineering behavior. Multiple samples of each ash type were subjected to laboratory tests to determine properties variation. Gradation analyses for ten MB ash samples are shown in Figure 4; the ash was similar to a well-graded gravel (GW by Unified Soil Classification). Gradations for RDF and MB ashes were about the same.

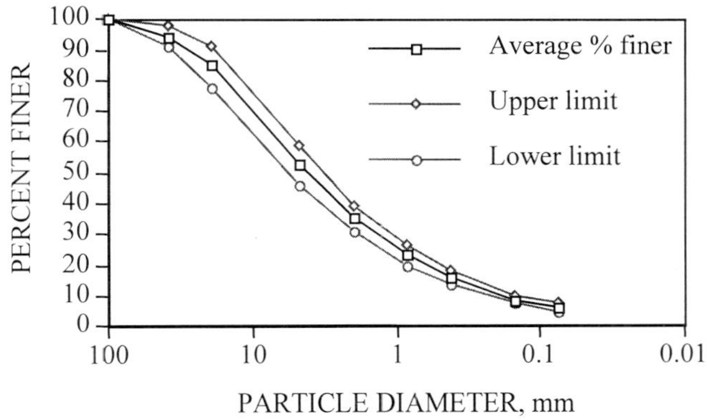

Figure 4 Grain size data for ten MB samples

Ferrous metal and organic matter contents were measured in MB and RDF bottom ash samples and varied considerably with particle size (Figure 5, for RDF bottom ash). These results show the magnetic separation procedure used prior to combustion failed to remove 100 percent of the ferrous metal in the RDF. Additionally, the ferrous metal content was inversely proportional to the organic matter content. MB bottom ash showed a similar trend to those shown in Figure 5, with slight differences. The average ferrous metal contents for MB and RDF bottom ash samples were 19.3 and 15.0 percent, respectively. The average organic matter contents for MB and RDF samples were 2.3 and 2.0 percent, respectively.

The larger ash particles may be described as clinkers of fused smaller particles comprising metal, glass, ceramic, concrete, soil, etc. These clinkers were fragile and had an irregular geometry. During modified Proctor compaction, as needed for a structural fill, the particles tended to crush (Figure 6). Particle crushing was necessary to attain the high densities needed for a strong, stable structural fill. Moisture-density, specific gravity, permeability and strength properties for MB and RDF bottom ash are summarized in Table 5. While there was some variation in the range of measured properties, these materials exhibited a behavior similar to granular soils.

Structural Fill Demonstration Project

MB bottom ash and gravel were used to construct three road section embankments at a landfill. Each test section was 2 lanes wide, 30 m long, and 3 m thick. One section was constructed of 100 percent MB ash, a second section of 50 percent ash and 50 percent gravel, and a third section of 100 percent gravel. Standard road construction procedures were used. Each embankment section was instrumented with settlement platforms, slope stakes and groundwater lysimeters to compare the performance of each section. Traffic consisted of a limited number of heavy trucks carrying ash to the landfill.

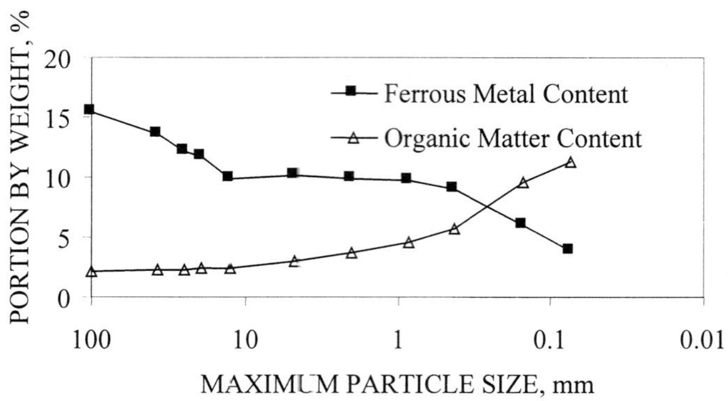

Figure 5 Ferrous metal and organic content in RDF bottom ash

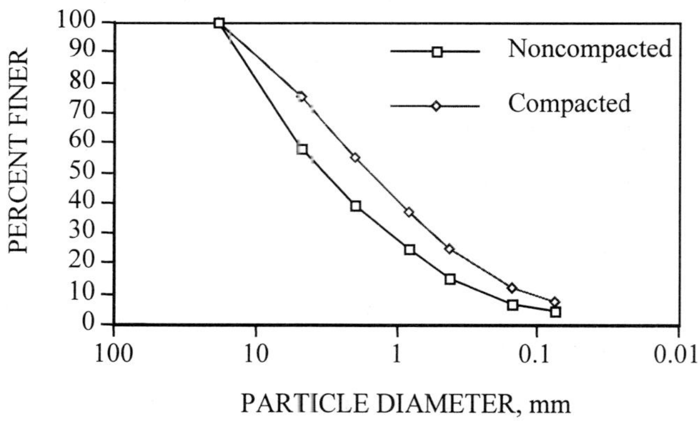

Figure 6 Effect of dynamic compaction on bottom ash gradation

Table 5 Summary of physical and geotechnical properties

PHYSICAL CHARACTERISTIC	MB BOTTOM ASH	RDF BOTTOM ASH
Mod. Proctor dry density (kN/m^3)	18.7 (18.0 - 19.9)	16.7 (16.2 – 17.1)
Optimum moisture content (%)	11.5 (10.0 – 12.5)	14.0 (13.5 – 14.9)
Specific gravity	2.71	2.54
Permeability (cm/s)	5.9E-5 to 1.3E-4	3.9E-5 to 5.1E-4
Friction angle (degrees)	45 (42 - 52)	44 (41 – 47)

The magnitudes of both the embankment and underlying fill settlements are critical to roadway performance. Large differential settlements can impede traffic flow. The settlement platform assembly measured compression of both the embankment fill and the underlying landfill materials. Data for the embankment compression is shown in Figure 7 for the bottom ash, mixed material, and gravel sections.

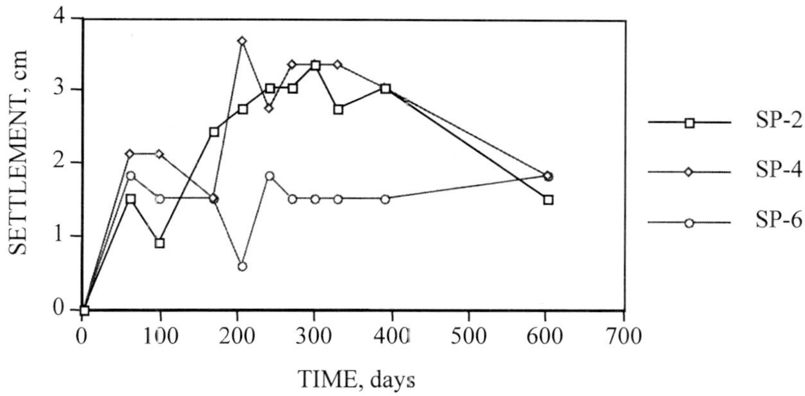

Figure 7 Compression of embankment subsections from end of construction

All three fill materials compressed rapidly following the end of construction. The rate of compression then decreased, and the subsections maintained a relatively constant thickness. The bottom ash layer and the mixed material layer each compressed more than 3 cm, double the gravel control section amount of approximately 1.5 cm. At day 600, the bottom ash and mixed material subsections had recovered some of their prior settlement. The overall compression of the bottom ash and mixed ash sections reduced from 3.2 cm at 400 days to 1.5 cm and 2.4 cm, respectively, at 600 days. This reduction in compression may have been be due to expansion of the ash or it could have been because a new survey crew replaced the original crew following the 400-day measurement.

The difference in compression between the bottom ash and the gravel may have resulted from two sources. Compression in the bottom ash can occur from the presence of large quantities of thin walled metal containers and hollow metallic particles. High chloride levels in leachate within the bottom ash subsection may have caused rapid deterioration of the particle's metallic walls. The weak, deteriorated particles can create voids that collapse causing compression of the fill. The gravel may undergo less compression than the bottom ash because it does not degrade in the presence of chlorides. Settlement in each of the fill materials may have also occurred as a result of particles rearranging in response to repeated heavy loadings from trucks. Another possible source of settlement for the bottom ash is post-construction dewatering. The bottom ash section was constructed at a moisture content above optimum. Loading and seepage may have caused water to migrate out of the fill resulting in compression. Settlements within the embankment fill were small compared to those occurring in the underlying MSW, which formed the subbase for the test sections.

ASH USE IN BITUMINOUS CONCRETE

References 8 and 11 contain details of the research using ash in bituminous concrete.

RDF Bottom Ash as Aggregate

RDF bottom ash was evaluated for use as aggregate. After removing all particles larger than 25 mm and visible unburned paper, the usable material was about 78 percent of the original bottom ash. Most remaining organic matter was below the 1.2 mm sieve and was less dense and more absorbent than the coarser material. This post-processed ash satisfied Connecticut Department of Transportation gradation requirements for Class 1 asphalt mixes. As expected, the ash was not very tough and had a 44.7 percent loss (40 percent maximum acceptable for pavement use, 50 percent for base course use) on the Los Angeles abrasion test.

Table 6 summarizes the properties of trial mixes to determine the maximum useable ash content for an acceptable asphalt mix. Trap rock, a strong basaltic rock, was used as a control aggregate and a blending aggregate. Mix 1, all ash, and Mix 2, blend of ash and trap rock of all sizes, had very high air void contents and very low Marshall stability values, probably due to unburned organic matter in the finer ash sizes. Improvement resulted when the finer portions of the ash were removed and replaced with trap rock fines. Mixes 3 and 4 used ash for the coarser sizes and trap rock for the finer sizes. Mix 4 had a stability value suitable for Class 1 mix, although its void content was slightly higher than allowed by specifications.

Additional performance properties were evaluated for Mix 4. About 20 percent additional asphalt was needed for Mix 4, compared to control. The indirect tension strength of the ash mix was 20 percent lower than the control. Mix 4 may be susceptible to water damage since it performed poorly in the stripping test (test result of 0.70, 0.75 considered minimum acceptable). In leachate tests (Table 7) lead was the only metal to exceed the toxicity limit.

Table 6 Properties of trial mixes with RDF ash

ITEM	CLASS 1 CONTROL -TRAP	MIX 1 – 100% ASH	MIX 2 - 40% ASH 0% TRAP	MIX 4 - 32% ASH (TRAP BELOW 2.4 mm)	MIX 4 – 32% ASH (TRAP BELOW 4.8 mm)
Asphalt content, %	4.9	7.0	7.0	6.5	6.2
Stability, lb, 60°C	2200	650	600	1050	1150
Unit weight, kg/m^3	2,540	1,640	1,660	2,240	2,280
Flow, 0.01 in.	11.0	11.5	13.0	11.0	12.0
Air void, %	5.0	23.5	21.5	13.4	7.2
VMA, %	16.3	28.5	27.5	19.5	18.6

Note: stability and flow values are shown in Customary units since these units are used in these standardized tests; 1 lb (force) = 4.48 N, 1 in. = 25.4 mm

Table 7 TCLP tests of asphalt mixes, ppb

LEACHATE SAMPLE	BARIUM	CADMIUM	CHROMIUM	LEAD
Class 1 control mix	247	3	11	26
RDF ash Mix 4	204	19	26	10,500
Class 2 control mix	56	2	< 6	< 59
MB ash mix	280	60	40	2,050
Detection Limit	2	3	6	59
Toxicity Limit	100,000	1,000	5,000	5,000

MB Ash in Test Pavement

Asphalt pavement test sections with MB ash as aggregate were constructed as part of the structural fill demonstration project described in a previous section. Each test pavement was 7.2 m wide, 30 m long, over a 350 mm base of gravel, and was either (1) two 62.5 mm thick layers of Class 2 (for low volume traffic) mix, (2) one layer of Class 2 mix over a layer using ash, or (3) two layers using ash. "Ash" mixes actually consisted of 50 percent ash with the remaining aggregate being trap rock and sand. A polyethylene membrane under each base course allowed collecting leachate that passed through the pavement. The test pavements and test embankments were at different locations.

Ash was screened to remove particles larger than 19 mm. Metal tearing and paper clogging the screens hampered the screening operations. Mixes were processed through a drum mix plant.

Abrasion test values for the ash were too high for pavement courses but were acceptable for unstabilized base courses. As with the RDF ash mixes, the MB ash mixes required higher asphalt contents than the control mix (8 vs. 6 percent) for satisfactory stability. TCLP tests showed no metals above the toxicity limits for these ash mixes (Table 7).

Pavement performance was assessed with skid resistance tests and analysis of water collected after passing through the pavement layer. The skid resistance of both the Class 2 and the ash pavements were comparable, but after six months there was a significant decrease in skid resistance for both mixes. Performance may be different on a more heavily traveled highway. There was no significant difference in the leachate collected beneath any of the three test pavements and no fixed pattern of change with time. Furthermore, since the site was a landfill, it was very difficult to assess if any cross contamination may have resulted.

CONCLUSIONS

1. It is doubtful if these MSW fly ashes have much potential use in structural quality concrete. Concrete strength and durability properties varied greatly depending on the type of MSW fly ash used.

2. With minimal post-processing, MSW ashes are not suitable as fine aggregate for structural quality concrete because of expansion due to gas generation that causes cracking and low strength. Further post-processing may alleviate this problem. Performance as fine aggregate varied with type of MSW ash.

3. MSW ash performed better in masonry than in concrete due to the lower strength requirements and the more porous texture of masonry. However, the ash masonry had a distinct, unpleasant odor and had formation of white particulate material.

4. RDF bottom ash, as it leaves the incineration plant, produced an aggregate of marginal quality for use in a Class 1 bituminous concrete mix. A blend of MB ash, with some post-processing, and normal aggregate performed about as well as a control aggregate Class 2 mix. Post-processing must remove a significant amount of material as oversize and finer particles.

5. Asphalt mixes with RDF or MB ash required 20 to 33 percent higher asphalt content and had lower strength and increased stripping potential compared to non-ash mixes.

6. MB and RDF bottom ash, with or without post-processing, exhibited properties and behavior similar to coarse-grained natural aggregates.

7. High amounts of heavy metals, especially lead and cadmium, and salts were leachable from the as-received ash, especially in the fly ash.

8. Crushed concrete or masonry containing MSW ashes had levels of leachable heavy metals well below toxicity limits. Bituminous concrete containing ash generally had low levels of leachable heavy metals, except of some samples where lead exceeded toxicity limits.

9. These RDF and MB ashes may have application in a landfill setting for access road construction aggregates, structural fill, and daily cover over MSW. Post-processing of the ash may make it suitable for use in bituminous concrete for low volume traffic and in masonry.

ACKNOWLEDGEMENTS

These investigations were supported by grants from the Connecticut Department of Environmental Protection to the Environmental Research Institute of the University of Connecticut, from the Connecticut Department of Transportation, and from the Connecticut Resources Recovery Authority. The assistance of colleagues, Drs. Norman W. Garrick, George E. Hoag, Richard P. Long, and Jack E. Stephens, whose works are also discussed in this paper, is gratefully acknowledged. Thanks are also given to the many undergraduate and graduate students and to the laboratory personnel of ERI who worked on these projects.

REFERENCES

1. U.S. ENVIRONMENTAL PROTECTION AGENCY. 1982. Extraction Procedure (EP) Toxicity Test Method and Structural Integrity Test, Test Methods for Evaluating Solid Wastes. Physical/Chemical Methods, Second Edition, SW-846, Cincinnati, OH.

2. U.S. ENVIRONMENTAL PROTECTION AGENCY. 1979. Methods for Chemical Analysis of Water and Wastes, EPA-600/4-79-010, Cincinnati, OH.

3. HAMERNIK, J. D. AND FRANTZ, G. C. 1991. Physical and Chemical Properties of Municipal Solid Waste Fly Ash, ACI Materials Journal, American Concrete Institute, Vol. 88, No. 3, pp. 294-301.

4. HAMERNIK, J. D. AND FRANTZ, G. C. 1991. Strength of Concrete Containing Municipal Solid Waste Fly Ash, ACI Materials Journal, American Concrete Institute, Vol. 88, No. 5, pp. 508-517.

5. TRIANO, J. R. AND FRANTZ, G. C. 1992. Durability of MSW Fly Ash Concrete, Journal of Materials in Civil Engineering, American Society of Civil Engineers, Vol. 4, No. 4, pp. 368-384

6. ABDUL RASHID, R. AND FRANTZ, G. C. 1992. MSW Incinerator Ash as Aggregate in Concrete and Masonry, Journal of Materials in Civil Engineering, American Society of Civil Engineers. Vol. 4. No. 4, pp. 353-368.

7. RIVARD-LENTZ, D.J., SWEENEY, L.R., AND DEMARS, K.R. 1997. Incinerator Bottom Ash as a Soil Substitute: Physical and Chemical Behavior, Testing Soil Mixed with Waste or Recycled Materials, ASTM STP 1275, Wasemiller and Hoddinott, Eds, American Society for Testing and Materials.

8. DEMARS, K.R., GARRICK, N.W., LONG, R.P., LENTZ, D., MADDALI, R., STEPHENS, J.E. AND RECCHIA, C. 1993. Municipal Waste Combustor Bottom Ash Road Paving and Structural Fill Demonstration Project, Proceedings 6th International Conference on Ash Management and Utilization, Arlington, VA, November.

9. LENTZ, D.J. 1994. Performance of Incinerator Bottom Ash as Structural Fill, Master of Science Thesis, University of Connecticut, Storrs, CT.

10. SWEENEY, L.R., 1994. Physical/Chemical Behavior of Incinerator Bottom Ash and Stabilization with Bentonite, Master of Science Thesis, University of Connecticut, Storrs, CT.

11. GARRICK, N. W. AND CHAN, K.L., 1993. Evaluation of Domestic Incinerator Ash for Use as Aggregate in Asphalt Concrete, Transportation Research Record, No. 1418, Performance Related Testing and Evaluation of Characteristics of Aggregates and New Geomaterials, National Academy Press, Washington, D.D., pp. 30 – 34.

MSW COMBUSTOR ASH UTILIZATION IN MARINE AND TERRESTRIAL SYSTEMS

F J Roethel

V T Breslin

State Universitt of New York

United States of America

ABSTRACT. Since 1985, scientists at the State University of New York have been evaluating the physical, chemical and environmental properties of municipal solid waste (MSW) combustor residues along with the feasibility of utilizing this material as an aggregate substitute in concrete products. Laboratory investigations revealed that combustor ash could be stabilized using Portland cement. The stabilized ash products possessed improved physical and environmental properties. In 1987, the research moved into the field with the construction of the first artificial reef within the coastal waters of the United States constructed from blocks of stabilized combustor residue. Tests have also shown that construction quality cement blocks can be manufactured from combustor ash. As a second demonstration program, 14,000 construction quality cement blocks were produced using both combined and bottom ash in the construction of a boat house on the campus of the University. Both of these multi-year field studies have provided data indicating that the ash/concrete products maintain their structural integrity and leach constituents in such a manner as not to adversely impact the environment. A third application focused on a laboratory study to assess the use of stabilized combustor ash in the construction of a revetment on the south shore of Long Island.

Keywords: Artificial reefs, MSW combustor ash, Structural integrity, Leaching behavior, Environmental issues.

Frank J Roethel, is a Research Professor in the Waste Reduction and Management Institute, Marine Sciences Research Center at the State University of New York at Stony Brook. His research interests includes assessing the structural integrity and environmental acceptability of waste materials, especially combustion by-products.

Vincent T Breslin, is an Assistant Professor in the Waste Reduction and Management Institute, Marine Sciences Research Center at the State University of New York at Stony Brook. His research examines the release of metals from treated waste products and more fundamental studies examining the biogeochemical behavior of metals in the water column and sediments in freshwater, marine and estuarine environments.

INTRODUCTION

In 1983 the New York State Legislature, in an effort to protect Long Island's groundwater aquifers, enacted legislation that in effect closed all of this regions landfills on December 18, 1990. Many municipalities incorporated combustion of their municipal solid waste (MSW) as an integral part of their redefined waste management strategy. The 2.7 million residents of Long Island generate nearly 3.5 million tons of MSW per year. Assuming the State's recycling goal of 50% is achieved, there exists the need to manage, primarily through combustion, nearly 2 million tons of garbage and trash annually. Combustion of MSW results in the production of ash and for Long Island communities this residual must either be utilized or trucked to distant landfills.

Combustion of municipal solid waste produces particulate residues that are often rich in lead, cadmium, copper and zinc [1]. Results of laboratory leaching studies show that metals of environmental concern may be released from the particulate residues, resulting in the need for the residues to be disposed of in an environmentally acceptable manner. Research has shown that particulate combustor residues can be combined with cement and stabilized to form a solid block. The stabilization process has been shown to effectively prevent the release of metals from stabilized combustor residue blocks, and may provide an alternative method for the disposal or reuse of the by-products of combustion [1].

A research program was initiated at the Waste Reduction and Management Institute (WRMI) at the Marine Sciences Research Center (MSRC) at the State University of New York at Stony Brook. This program examined the feasibility of utilizing stabilized combustor residues in a variety of applications in lieu of landfilling this combustion by-product. Results of laboratory studies showed that stabilized combustor residues possessed structural properties that would allow for their use in a wide variety of applications. The stabilized residues were subjected to several regulatory extraction protocols and in no instance did the metal concentrations in the leachate exceed regulatory limits for toxicity [2]. Bioassays using a variety of marine organisms (phytoplankton, zooplankton, fish eggs and larvae) did not reveal any adverse impacts to communities exposed to elutriates prepared using the stabilized residue blocks. The success of the laboratory studies led to the development of field demonstration programs designed to evaluate the potential for manufacturing and utilizing stabilized combustor residues in secondary products. This paper provides an overview describing the results to date of programs designed to determine the suitability of using MSW combustor residues to manufacture blocks for artificial reef construction, the development of ready-mix ash concrete mixtures for use in manufacturing shore protection devices, and using combustor residues as a substitute for natural aggregate in the manufacture of construction quality cement blocks.

The successful use of MSW combustor residue secondary products requires that the products meet or exceed existing engineering and environmental standards prior to use. Engineering criteria for materials used for artificial reef development [3], shore protection devices [4] and construction blocks [5] are shown in Table 1.

Both the recommended criteria and achieved values are obtained using unconfined compressive strength test according to ASTM C39 testing protocols. These strength criteria are well established and provide a minimum strength required for the successful performance of a product used for each application. The environmental criteria are not as well established. Several regulatory leaching protocols such as the E.P. Toxicity test and the TCLP leaching protocol were not designed to evaluate the leachability of metals from a product destined for marine use. Seawater or freshwater tank leaching studies or modified ASTM leaching techniques using seawater or synthetic acid rain are more useful techniques in determining the leachability of metals from stabilized products for marine or terrestrial use. Since one test may not be sufficient, several leaching protocols were employed to evaluate the leachability of metals from MSW combustor residue products.

Table 1 Engineering criteria for residue block secondary products.

SECONDARY PRODUCT	COMPRESSIVE STRENGTH CRITERIA	COMPRESSIVE STRENGTH ACHIEVED
Artificial Reef Blocks	300 psi [3], 2.1 N/mm^2	800-1200 psi, 5.5-8.3 N/mm^2
Shore Protection Devices	3000 psi [4], 20.7 N/mm^2	3100-4200 psi, 21.4-29.0 N/mm^2
Construction Blocks	1000 psi [5], 6.9 N/mm^2	1620-2570 psi, 11.2-17.7 N/mm^2

BLOCK MAKING TECHNOLOGY

One of the objectives of the research is to manufacture MSW combustor residue products using existing technology. In the case of block making, for either reef or construction quality blocks, the technology is well established. Typically, the mix components are fed in weighed amounts from hoppers into a mixer. Water is then added to the mix to adjust the moisture content. The well mixed materials are then transported to the feeder of a Besser Vibrapak block making machine. A block mold is positioned on a steel pallet in the block machine while material is fed into the mold box by vibration. The material is then quickly vibrated and compacted into the mold consolidating the material into blocks. After compaction, the mold lifts, and the pallet holding the blocks emerges from the machine as a new pallet is positioned beneath the mold box for the next molding cycle. At full capacity, a block making machine can produce more than 1500 blocks per hour. The pallets holding the blocks are stacked and transported to curing chambers where the blocks are steam cured for 12 hours. Cured blocks are unracked, depalletized and stacked by a cubing machine as cubes of interlocking blocks on carrying pallets.

DEMONSTRATION PROJECTS

MSW Combustor Ash Reef Construction and Monitoring Activities

During the fall of 1986, blocks destined for reef construction were manufactured in Alpena, Michigan. The block mix consisted of 85% combined combustor residue and 15% Portland (Type II) cement, with a total moisture content of 10.2%. The well mixed materials were hopper fed into the Besser Vibrapak block making machine where stabilized incineration residue blocks were produced. Approximately sixty stabilized reef blocks were manufactured for physical and chemical evaluation. On April 27, 1987 approximately thirty stabilized MSW combustor residue blocks and thirty standard cement blocks were submerged in 10 meters of water in Conscience Bay, Long Island Sound, New York. The blocks were arranged underwater to produce two separate structures, one constructed using stabilized combustor residue blocks and one constructed using cement blocks, approximately 3 meters apart.

Periodically, divers have returned to the reef site to study the interactions of the stabilized ash blocks with the marine environment. The 28 day net unconfined compressive strength of the stabilized residue blocks prior to seawater submersion was 1120 psi (7.7 N/mm^2). This value far exceeds the minimum unconfined strength value of 300 psi (2.1 N/mm^2) recommended for marine disposal (Table 1). Results of unconfined compression testing of combustor residue blocks retrieved from Conscience Bay showed that the blocks retained approximately 90% of their initial strength after five years of seawater submersion.

Analysis of the elemental composition of the stabilized ash blocks submerged in Conscience Bay was performed to detect any release of metals from the blocks to the surrounding seawater. Results show that even after prolonged seawater exposure these metals of environmental concern were effectively retained within the cementitous matrix [6]. Analysis of the dioxin and furan content of blocks prior to and following submersion in the sea show that these organic compounds are also effectively retained within the ash reef blocks [7]. The reef structures were rapidly colonized by a diverse community of marine organisms and provided suitable habitat for a wide variety of finfish and crustaceans.

Metal analysis of digested tissues of organisms removed from both reef structures showed that metals of environmental concern were not elevated in tissues of organisms removed from the ash reef [6]. In addition, similar results were obtained using mussels (*Mytilus edulis*) as an indicator organism. An identical program was initiated in 1988 to assess the potential for dioxins and furans to bioaccumulate with the tissues of both the fouling community and filter feeding organisms associated with the ash reef. After extensive testing, no uptake of these organic compounds were observed [7].

To date, no adverse environmental impacts have been measured at the Conscience Bay reef site due to the presence of the stabilized ash blocks

Construction Block Manufacturing

The favorable physical and chemical properties of the stabilized MSW combustion residue blocks used for artificial reef construction led to an investigation to determine if incineration residues could be used in lieu of natural aggregate in the manufacture of concrete masonry blocks. During September 1988, block making activities were again carried out at the research facilities of the Besser Company at the Alpena Community College in Alpena, Michigan. The primary goal of this research was to produce construction quality blocks, using conventional block making equipment, from combustor residues.

Construction Block Mix Designs

Combined combustor residues were collected during August 1988 from the Baltimore RESCO facility in Baltimore, Maryland. The combined residues were processed at the Alpena facility to remove both the ferrous and non-ferrous metals and improve the particle size distribution of the ash. Five residue mixes were tested to determine their suitability for fabricating construction quality blocks (Table 2).

Table 2 Compressive strengths of Alpena 88 construction blocks. [a]

SAMPLE BATCH	MOISTURE [b] (%)	SAND (%)	UNIT WEIGHT (kg)	1 DAY COMPRESSIVE STRENGTH		28 DAY COMPRESSIVE STRENGTH	
				(psi)	(N/mm⁵)	(psi)	(N/mm⁵)
A	8.3	33	17.16	1010 ± 330	7.0 ± 2.3	2190 ± 39	15.1 ± 0.3
B	8.7	33	17.96	1330 ± 255	9.2 ± 1.8	2570 ± 42	17.7 ± 0.3
C	6.4	33	16.59	829 ± 15	5.7 ± 0.1	1620 ± 148	11.2 ± 1.0
D	5.6	50	17.90	1680 ± 12	11.6 ± 0.1	2570 ± 98	17.7 ± 0.7
E	8.7	25	17.30	1080 ± 18	7.4 ± 0.1	1990 ± 29	13.7 ± 0.2

[a] Construction blocks (8" x 8" x 16" ie 203 mm x 203 mm x 406 mm) were fabricated on 8/24/88.
[b] Moisture contents were measured after heating the sample to 110°C for 24 hours.

Three sand contents (25%, 33% and 50%) for the mixes were investigated. Fifteen percent by weight Portland cement Type IP was added to each of the combined residue and sand mixtures. Type IP Portland cement is a pozzolanic cement containing between 15-40% coal fly ash. The moisture content of the 33% sand mixes (mix A, B and C) was varied to determine an optimum moisture content for block making. In addition, Acme-Hardesty superplasticizer was added to the mixes to allow the mixes to flow more easily.

The unconfined compressive strengths of the construction blocks were determined after 1 and 28 days (Table 2). According to Underwriters Laboratories specification 618, concrete masonry units produced using combustor residues as an aggregate require an average unconfined compressive strength of 1000 psi (6.9 N/mm^2) and a minimum strength of 800 psi (5.5 N/mm^2) (Table 1). The 28 day unconfined compressive strengths of the residue blocks easily exceeded the average strength criteria.

The unconfined compressive strengths of the 33% sand mixes were shown to increase with increasing moisture content (Table 2). As the moisture content of the mix increased from 6.4% for mix C to 8.7% for mix B, the one day unconfined compressive strengths likewise increased from 829 psi (5.7 N/mm^2) to 1330 psi (9.2 N/mm^2), respectively. Compressive strengths of the blocks also increased as the sand content of the mix increased from 33% to 50%.

During October 1988, thirty two construction grade MSW combustor residue blocks were shipped to Underwriters Laboratories, Inc., Northbrook, Illinois to undergo fire resistance testing. The Underwriters Laboratories standard UL 263, Fire Tests of Building Construction and Materials is intended to evaluate the length of time a wall assembly will contain a fire, retain structural integrity, and test the wall assembly's resistance to a hose stream. Results of this test showed that the passage of flame to the unexposed surface of the wall assembly constructed using the combustor residue blocks did not occur during the 2 hour fire exposure. In addition, transmission of heat through the wall assembly did not raise the temperature of the unexposed surface at any one point more that 325°F (163°C) above the initial temperature and the average temperature did not exceed 250°F (121°C) above the ambient temperature during the initial 2 hour fire exposure. Finally, the wall assembly withstood the hose stream test without developing openings or allowing penetration of the water stream.

The physical properties of blocks produced using combustor residues as a substitute for natural aggregate meet or exceed existing standards for use as a construction material.

Further physical evaluations of the construction quality incineration residue blocks were undertaken at the Department of Civil Engineering at the State University of New York at Buffalo [8]. These investigations provided additional support for a full-scale demonstration program.

To more fully evaluate the potential for the use of residue blocks as a building material, a boathouse was constructed at the Marine Sciences Research Center at SUNY Stony Brook using masonry blocks fabricated using MSW combustor residues. Over 14,000 incineration residue blocks were manufactured using bottom and combined MSW combustor residues at Barrasso and Sons, Inc. Islip Terrace, Long Island for use in the boathouse construction. Unconfined compressive testing of the blocks show strengths exceeding 1200 psi (8.3 N/mm^2) after 7 days curing. With the construction now completed, a comprehensive monitoring program is being initiated to further evaluate the environmental acceptability of residue blocks as a construction material. Air quality within the demonstration building was monitored for a period of two years prior to occupation. In addition, rainwater running off the exterior blocks was collected and analyzed for the presence and concentration of trace metals. Finally, modifications in the chemistry of the soils surrounding the building were evaluated. To date, no adverse environmental impacts have been identified [9].

Shore Protection Devices

Concrete, due to its physical properties and its adaptability to various designs and shapes, is widely used in coastal and waterfront construction. Applications include docks, piers and wharves at port facilities, bridges, outfall structures and erosion control structures. The durability, strength and economy of concrete make it ideal for the marine environment. Although concrete is already one of the most popular materials for coastal construction, its unique properties will result in increased usage in the future [4].

Marine uses of concrete will increase even if conservative estimates for sea level rise are assumed. The rise in sea level would necessitate the need for gradual shoring or reconstruction of port and waterfront facilities and stepped up efforts in erosion control [10]. As one would expect, the greatest level of construction activity in response to rising sea level would most likely occur in the highly-developed and populated coastal urban centers [11]. Thus the areas requiring the largest volume of construction materials suitable for marine applications would coincide with the areas experiencing the greatest difficulty disposing of MSW combustion residues. Development of secondary materials-using combustor residues for marine construction would provide a needed resource that may alleviate the need to dispose of particulate combustion residues. The Waste Reduction and Management Institute conducted a research program funded by the Center for Hazardous Waste Management, SUNY at Buffalo designed and evaluate MSW combustion by-products in the fabrication of shore protection devices.

The objective of the investigation was to develop ready-mix material using combustor residues that meet design criteria established by the Army Corps of Engineers (COE) for erosion control concrete armor units. The ready-mix material has the ability to be poured into molds and cured allowing a greater variety of products to be manufactured. The concrete armor units require the highest standards in terms of strength and durability since they are most often used in high energy wave environments. The first phase of this program showed that combined combustor residues can be combined with cement and additives to form a product which exceeds the COE strength criteria for use in the manufacture of tetrapods and pilings. Working with an industrial partner, Hazcon, Inc. Brookshire, Texas we have evaluated the effects of adding their patented additive Chloranan to the mix. Chloranan alters the morphology of the bonding crystals that are formed when the cement cures resulting in a denser material with enhanced strengths.

A series of mix designs were developed incorporating Chloranan at a constant cement to admixture ratio of 10:1. Table 3 shows the compressive test results for a variety of residue/cement/Chloranan/water mix designs. Three of the mix designs tested exceeded the recommended compressive strength criteria of 3000 psi (20.7 N/mm^2) (Table 1). The optimum mix design shown had an unconfined compressive strength of 4200 psi (29.0 N/mm^2) using a cement addition well within the range used in commercial applications. The chemical properties of the mixes which exceed the COE physical requirements were evaluated in greater detail to assess the environmental acceptability. Subsequent testing indicated that no adverse marine impacts were associated with the ash/concrete mix design associated with this marine application.

Table 3 Compressive strengths of mix designs incorporating Chloranan.

MIX DESIGN	CEMENT (%)	WATER/CEMENT RATIO	COMPRESSIVE STRENGTH (psi)	(N/mm^2)
I 1	33	0.40	4020	27.7
I 2	28	0.70	2880	19.9
I 3	23	0.69	1900	13.1
I 4	23	0.60	2110	14.6
I 5	23	0.51	2860	19.7
I 6	20	0.72	1260	8.7
I 7	17	1.00	690	4.8
I 8	23	0.49	4200	29.0
I 9	20	0.50	3190	22.0
I 10	17	0.50	2680	18.5

The chloranan to cement ratio was held constant at 1:10.
The compressive strengths were determined after 28 days curing.

SUMMARY

Long Island communities are currently limited with respect to their options for managing incineration residues. Landfilling is constrained by both new landfill siting difficulties and the Long Island Landfill law which has mandated the closure of Long Island landfills during 1990. Currently, the majority of the ash produced on Long Island is exported, most often out of state. A shift in state regulations or a cancellation of these out of state contracts could result in the disruption of solid waste management for many Long Island communities.

At the present time it is estimated that approximately 250,000 tons per year of ash generated on Long Island is available for potential use as an aggregate substitute material [12]. Table 4

Table 4 Current aggregate use and potential for residue reuse in lieu of natural aggregate use on Long Island [12]

APPLICATION	ESTIMATED QUANTITY OF AGGREGATE USE	ESTIMATED ASH MARKET CAPACITY
Portland Cement Concrete	1,800,000 tpy[a]	85,000 tpy[b]
Artificial Reefs	7,000,000 tons[a]	4,900,000 tons[b]

[a] Quantities reported in tons per year(tpy) or tons. Portland cement concrete figures are based on annually recurring consumption based on a 1986 survey and estimate quantity. The artificial reef market is a fixed quantity market based on an estimate of reef capacity using 20% of available permitted reef space.

[b] Portland cement concrete estimate based on the potential market and a 25% blend of ash with other aggregate materials and a 25% application rate (i.e., percentage of construction application using ash). Artificial reef estimate based on the total potential market and a 70% blend and a 100% application rate.

shows the estimated quantity of aggregate use on Long Island and the potential quantities of combustor residues that could be substituted for natural aggregate in secondary products.

On the basis of these estimates, the reuse of combustor residues in secondary products can be expected to reduce, but not eliminate, existing ash disposal problems. Factors such as environmental suitability of the residues, engineering restrictions and the length of the construction season could further reduce the use of the residues in secondary products.

The economic benefits associated with residue reuse in secondary products are primarily derived from the avoidance of disposal costs rather than the market value of the residue. Potential savings that could be realized if residues were used as an aggregate substitute in secondary products are estimated to range from $50 to more than $80 per ton. Such savings could amount to a total annual cost savings for Long Island communities ranging from $20 million to $40 million [12]

REFERENCES

1. ROETHEL, F J, BRESLIN, V T, SCHAEPERKOTTER, V, PARK, K AND GREGG, R. The Fixation of Incineration Residues; Phase II. Marine Sciences Research Center 87-3, State University of New York at Stony Brook, Stony Brook, New York, 221pp, 1987.

2. PARK, K. Leaching Behavior of Incineration Residues from Municipal Solid Wastes. M.S. Thesis, SUNY at Stony Brook, Stony Brook, New York, 90pp, 1987.

3. WOODHEAD, P M J, PARKER, J H, CARLETON, H R AND DUEDALL, I W. Coal Waste Artificial Reef Program, Phase 4B; Interim Report. EPRI Report CS-3726, Electric Power Research Institute, Palo Alto, California. 1984.

4. MOFFAT AND NICHOLS. Construction materials for Coastal structures. U.S. Army Corps of Engineers, Coastal Engineering Research Center, Springfield, Virginia 427 pp, 1983

5. UNDERWRITERS LABORATORIES, INC. Standard for Safety 618. Standard for Concrete Masonry Units, UL 618, Eight Edition, 8pp, 1979

6. BRESLIN, V T, ROETHEL, F J AND SCHAEPERKOETTER, V P. Physical and chemical interactions of stabilized incineration residue with the marine environment. Marine Pollution Bulletin, 19(IIB), pp. 628-632, 1988.

7. ROETHEL, F J, BRESLIN, V T, AND ALDOUS, K. Mobility of dioxins and furans associated with stabilized incineration residues in the marine environment. 1990.

8. BERG, E R AND NEAL, J A. CMU Mix Designs using Municipal Waste Incineration Ash; Durability Test Results of Sample Units. SUNY at Buffalo, 100 pp, 1990.

9. STEINBERG, M. The Environmental Effects Associated with a Boathouse Constructed Using Blocks of Stabilized Municipal Solid Waste Combustor Ash. M. S. Thesis, SUNY at Stony Brook, Stony Brook, NY, 150 pp, 1995.

10. NATIONAL RESEARCH COUNCIL. Responding to changes in sea level, engineering implications. Marine Board, Commission on Engineering and Technical Systems, National Research Council, National Academy Press. Washington, D.C. 148 pp. 1987

11. TITUS, J G. Greenhouse Effect, Sea Level Rise and Coastal Zone Management. Coastal Zone Management Journal, 14(3), pp. 147-171, 1986

12. KOPPELMAN, L E AND TTANENBAUM, E G. A summary of the findings, conclusions and recommendations of Phase I of the Long Island Ash Management Study. Long Island Regional Planning Board, Hauppague, New York. March, 1990.

PRECAST CONCRETE PRODUCTS PRODUCED USING INCINERATOR BOTTOM ASH

T D Dyer

R K Dhir

T C Collier

University of Dundee

United Kingdom

ABSTRACT. This study investigates the use of municipal waste incinerator bottom ash (MWIBA) as an aggregate component of precast concrete masonry units (CMU). CMUs were manufactured at a commercial plant with MWIBA aggregate content ranging from 0-100% and cement content from 140-200 kg/m³. Physical and chemical properties were assessed according to BS 6073 and MWIBA CMUs were found to conform in all respects apart from chloride content. Increased cement content had little effect on properties. Durability of CMUs was investigated using a 24-hour wetting and drying cycle and the ASTM 666C 8hr freeze thaw cycle. No significant deterioration has been observed in any MWIBA CMUs during the wetting and drying cycles. In the freeze thaw exposure CMUs with up to 30% MWIBA performed far better than those with greater than 30% MWIBA. Increased cement content was found to enhance performance. Leaching potential of CMUs was investigated using the CEN protocol for concrete in contact with drinking water. Releases of heavy metals were well below EU limits for drinking water. CMUs with MWIBA aggregate are a practical proposition for appropriate applications

Keywords: Municipal waste incinerator bottom ash, Concrete masonry units, Physical and chemical properties, Durability, Leaching, Heavy metals, Appropriate application

Dr T D Dyer, is research/teaching fellow in the Concrete Technology Unit (CTU) at the University of Dundee. His research interests include

Professor R K Dhir, is director of the Concrete Technology Unit, University of Dundee, Scotland, UK. He specialises in Binder technology, permeation, durability and protection of concrete. His interests also include the use of construction and industrial wastes in concrete to meet the challenges of sustainable construction. He has published and travelled widely and serves on many technical committees. He is past chairman of the Concrete Society, Scotland.

Mr T Collier, is a research student at the University of Dundee. He is currently researching the use of MWIBA in precast concrete products

INTRODUCTION

The environment act of 1990 [1] sets out to put into practise the principals of sustainable development. In the context of this study, it seeks to conserve natural resources whilst encouraging the recycling of materials and minimising environmental impacts. A hierarchy of preferred waste disposal routes was established which in order of preference are reuse, recycling, incineration with heat recovery, incineration and landfill. Incineration reduces the waste mass by about 70%, this leaves 30% to be disposed of to landfill [2]. The incinerator residue can be further divided into bottom ash, which is deposited in the furnace and fly ash, which is removed from the flue gases by filters and precipitators. This paper explores the use of this bottom ash fraction as an aggregate in precast concrete masonry units.

A number of recycled waste materials are already firmly established as sources of raw materials used in the building and construction industries [3]. So widespread is the use of PFA, GGBS, silica fume and other wastes that they have British and European standards to control their production and use. These materials are so important that raw materials going into the parent processes are selected to give a consistent quality of waste and the processes themselves are designed with the quality of waste as a consideration. This results in a waste product, which is homogenous and has controlled properties. Municipal waste is by definition not homogeneous and will contain undesirable and potentially hazardous components. This project sets out to identify potential uses for this waste where the product has fitness for purpose and the environmental risk can be minimised.

Berg and Neal [4] and other workers have investigated trial mixes for concrete masonry units (CMUs) in the laboratory using municipal waste incinerator bottom ash (MWIBA) as an aggregate. This study was designed to investigate CMUs produced in an industrial plant using MWIBA as an aggregate. Three aspects of CMUs performance were investigated

1. Physical and chemical characteristics i.e. their ability to satisfy the criteria of BS 6073 [5];
2. durability or long term performance, which may dictate environments where these CMUs may be used;
3. hazardous substances contained within CMUs and their environmental availability

Materials and Test Samples

The MWIBA used in the production of the masonry units was derived from a modern, mass burn waste incinerator. The MWIBA had been processed to remove metals (ferrous and non ferrous) and had been screened to produce an 8mm down grading similar to the sand normally used on the production site. Material greater than 8mm is used for road sub base. The MWIBA was not washed.

Test samples were full size blocks (440mm x 220mm x 99mm) produced at a precast concrete plant. The process is a static hydraulic/vibrated press method where the product is demoulded after a few seconds and which requires a concrete with virtually zero slump. This requires a mix with a low water/cement ratio. Two series of blocks were cast, one with 100% MWIBA as aggregate and different cement contents (140,170 and 200 kg) the other with a standard cement content of 140 kg but with varied MWIBA content (0, 30, 50 and 100% by mass of aggregate). A standard production block was tested as a control. The objective was to produce a 3.5 N/m² block that conformed to BS 6073 [5].

PHYSICAL AND CHEMICAL PROPERTIES

Blocks were assessed for their physical properties to conform to BS 6073 [5], for durability in service and potential environmental risk. The properties measured included dimensional checks, compressive strength, flexural strength, drying shrinkage, chloride content and sulphate content.

Test Methods

Crushing strength of blocks were measured on 5 full size samples of each mix type using a suitable crushing machine with platens of 120mm x 460mm and a capacity of ~ 450 kN. Samples were mortar capped and soaked for 16 hours prior to crushing.

Flexural strength was measured to BS EN 1521 [6] on 3 whole samples of each mix type. A four point loading rig was used with supports at 400 mm centres and the load applied at the two third points.

Drying shrinkage measurements were carried out on prisms 200mm x 100mm x 50mm cut with a diamond tipped saw from whole blocks. Stainless steel reference studs were inserted in the centres of the 50 mm x 100mm faces and fixed with epoxy resin. Three prisms of each mix type were assessed. Prisms were soaked in water at 20°C for 7 days. Their length between studs was compared to a standard bar, using a suitable dial gauge rig (accurate to 0.002mm). Prisms were then placed in a controlled drying environment (20°C. RH 95%). At 13 and 15 days their length was again compared to the standard bar and the mean change in length recorded. The drying shrinkage is expressed as the % change in the original length.

Chloride and sulphate concentrations were determined using X-ray fluorescence techniques. Samples of blocks of each mix type were ground to a fine powder and pressed into discs. These were analysed using a Phillips x-ray flourescence spectrometer.

The wet density of whole blocks was measured using a balance immersion method. Blocks were then dried to constant weight, weighed and measured to determine dry density (BS EN 992) [7]. Water absorption was estimated from the difference in wet and dry densities.

Results

Table 1 Results of chemical and physical tests

	BLOCK REF	CEMENT CONTENT Kg	MWIBA % OF TOTAL AGGREGATE	DENSITY WET kg/m³	DENSITY DRY kg/m³	WATER ABSORPTION kg/m³	COMPRESSIVE STRENGTH N/mm²	FLEXURAL STRENGTH N/mm²	DRYING SHRINKAGE %	CHLORIDE % BY WT	SULPHATE % BY WT
Standard BS 6073							>2.8	>0.65	<0.06	<0.13	
production	7	140	5	1711	1583	128	5.84	1.52	0.038	0.13	0.19
trials	4	200	100	1856	1701	155	4.21	0.86	0.037	1.14	1.78
	3	170	100	1852	1682	169	3.85	0.76	0.048	0.67	1.72
	2	140	100	1838	1651	187	3.96	0.76	0.036	0.74	1.82
	5	140	50	1906	1742	164	3.93	0.91	0.061	0.4	0.61
	6	140	30	1729	1588	141	4.82	0.82	0.043	0.33	0.28
	1	140	0	2032	1916	115	9.41	1.95	0.038	0.2	0.20

Compressive and Flexural Strength

It can be seen from Table 1 that all trial CMUs satisfied the minimum strength requirement of BS 6073 [5]. It was found that increased MWIBA content produced a reduction in strength with a minimum at 50% (Figure 1). However increased cement content produced little strength gain in 100% MWIBA CMUs. This is in contrast to production CMUs, where the same increase in cement content would raise the design strength from 3.5 to 7.0 N/mm². Similar relationships can be seen for both compressive and flexural strength (Figures 1 & 2).

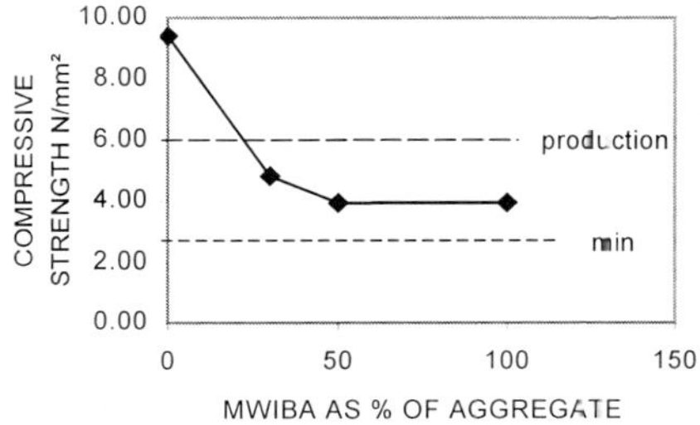

Figure 1 Effect of MWIBA content on compressive strength

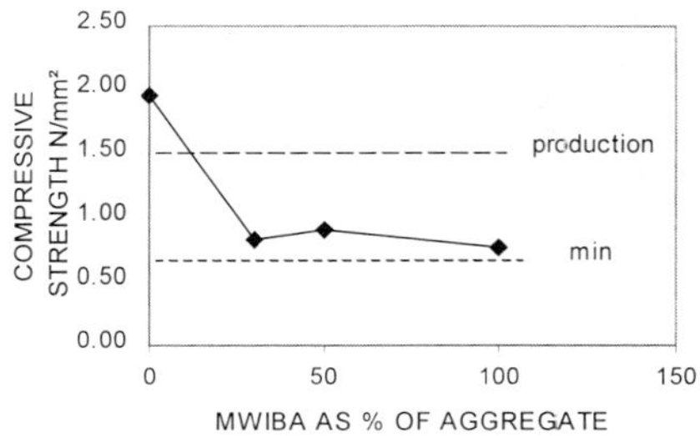

Figure 2 Effect of MWIBA content on flexural strength

Drying Shrinkage

A maximum drying shrinkage of 0.06% is allowed for a CMU to comply with BS 6073. All the trial CMUs achieved this although the blocks with 50% MWIBA aggregate were at the limit. The relationship between shrinkage and MWIBA content is non-linear as shown in Figure 3, with shrinkage rising as MWIBA increases to 50% and falling back as it approaches 100%.

Chloride content

BS 6073 [5] allows a maximum chloride content of 0.13% by weight. All MWIBA CMUs tested exceeded this limit, and even the production CMU was found to comply only marginally. Chloride content was found to increase linearly with MWIBA content up to a maximum of 0.75% at 100% MWIBA. This can be seen in Figure 4.

Sulphate content

Above 30% MWIBA content, sulphate content was found to increase with MWIBA content as can be seen from Figure 4.

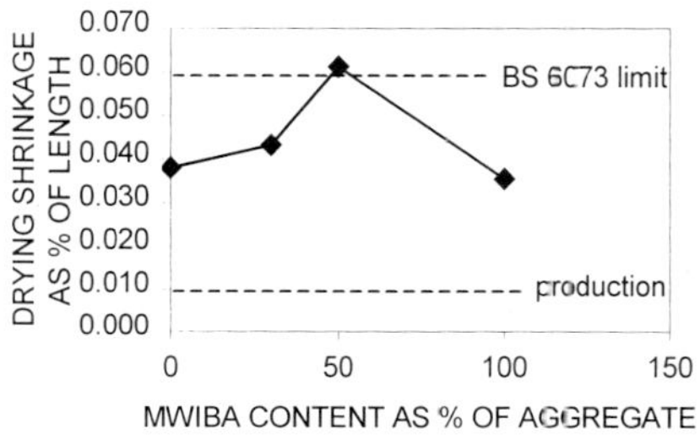

Figure 3 Effect of MWIBA content on drying shrinkage of CMUs

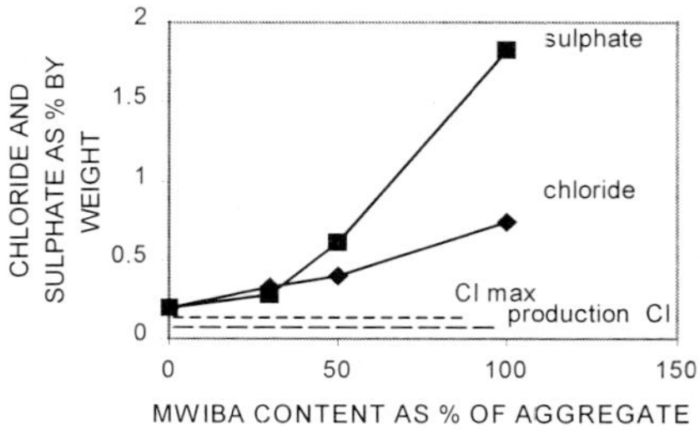

Figure 4 Effect of MWIBA content on chloride and sulphate content

Water absorption

Water absorption of CMUs was found to increase with increased MWIBA content (Figure 5) and decrease with increased cement content (Figure 6). The production CMU had a lower absorption than the MWIBA CMUs but higher than the CMU with no MWIBA which is perhaps explained by the use of expanded clay lightweight aggregate in the production block.

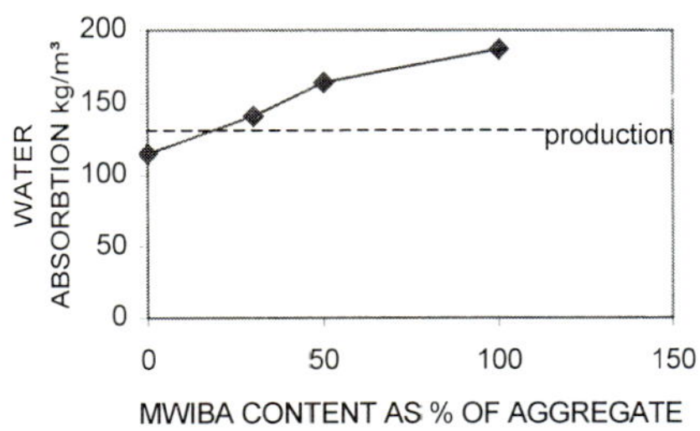

Figure 5 Effect of MWIBA content on water absorption

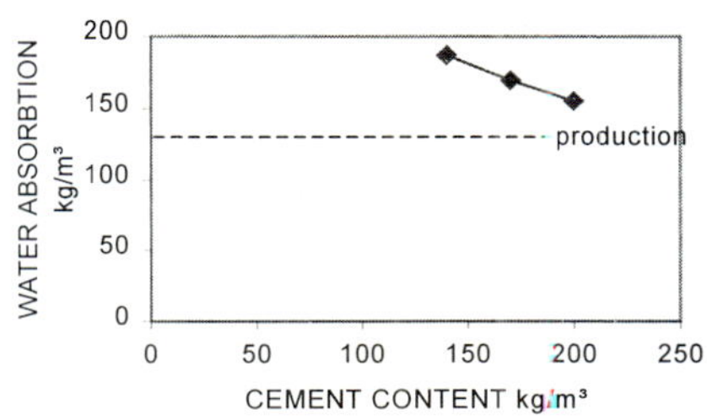

Figure 6 Effect of cement content on water absorption

DURABILITY TESTS

Three aspects of durability in service were investigated. These were; freeze-thaw resistance, sulphate attack and the effect of cyclical wetting and drying. Sulphate exposure is not reported here.

Test Methods

Freeze-thaw effects on masonry units were tested in accordance with ASTM C 666-92 [8]. 300mm x 75mm x 75mm prisms were sawn from whole blocks and stainless steel reference studs fixed into the centre of each 75mm x 75mm face with epoxy resin. Three prisms of each mix type were tested. The prisms were soaked in water for sixteen hours and then initial values of weight, length (with respect to a standard) and dynamic modulus of elasticity taken. They were then placed in steel containers, which gave a clearance of 1-3mm on each face. The containers were topped up with water and subjected to an eight-hour cycle of alternative freezing and thawing (-17°C to +4°C). At intervals of 36 cycles the prisms were removed from the containers and the measurements repeated.

Dynamic modulus of elasticity was measured by using the Erudite resonant frequency apparatus according to BS 1881 part 77 [9]

The weathering test was carried out on 100mm cores drilled perpendicular to the 440mm x 220mm face of whole blocks. Six cores were extracted from each block. The cores were placed in a tank, which was filled and drained automatically in a timed sequence. During the dry period an air blower assisted the drying of the specimens. The tank was programmed for a 24-hour cycle with a 4 hour wet period. Cores were weighed weekly to monitor weight loss and wet density was measured every two months. Initially, one core of each set was mortar capped and crushed to BS 1881 [10]. Further cores were crushed at two monthly intervals.

Durability Test Results

Weathering

The wetting and drying cycle has produced no significant change in wet weight of any of the CMU cores during the six months the test has run so far, however the 100% MWIBA aggregate CMUs appear to have suffered a reduction in wet density of the order of 2.5%. This implies that since the weight has remained constant the cores have expanded slightly. All cores have suffered only minor surface damage.

Freeze thaw

The results of the freeze thaw tests show a distinct increase in the performance of CMUs with increased cement content or lower MWIBA content. Figure 7 shows that an increase of cement content from 140 to 200 kg/m³ in a 100% MWIBA CMU, produces a much higher relative modulus of elasticity (RME) after both 36 and 72 cycles. Figure 8 shows that higher cement content produces much lower linear expansion. These two sets of results indicate that increased cement content produces a more durable 100% MWIBA aggregate block.

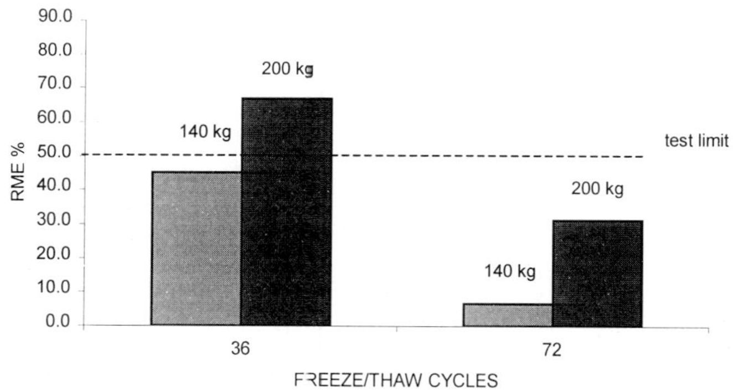

Figure 7 Effect of cement content on RME during freeze thaw exposure

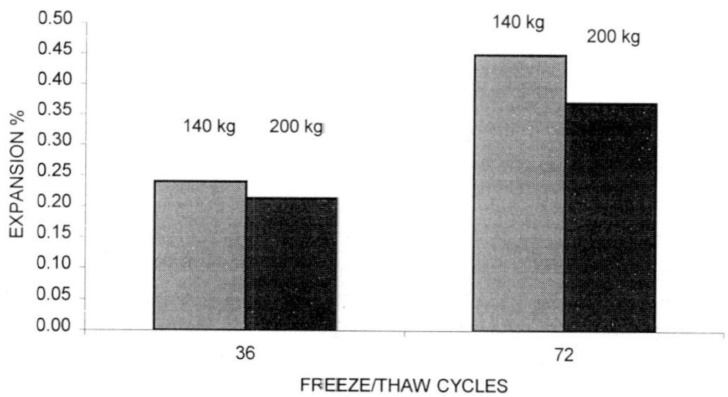

Figure 8 Effect of cement content on linear expansion during freeze thaw exposure

Figures 9 and 10 illustrate the relative behaviour of CMUs with increasing MWIBA content. It can be seen that performance decreases rapidly with MWIBA content above 30%. This is true after both 36 and 72 cycles where above 30% MWIBA aggregate content, RMD falls rapidly and linear expansion increases. It is interesting to note the poor performance of the production CMU in freeze/thaw exposure. This may be due to the vulnerability of expanded clay aggregates used to reduce unit weight. The results indicate that durability is much improved if the MWIBA content is kept down to 30%.

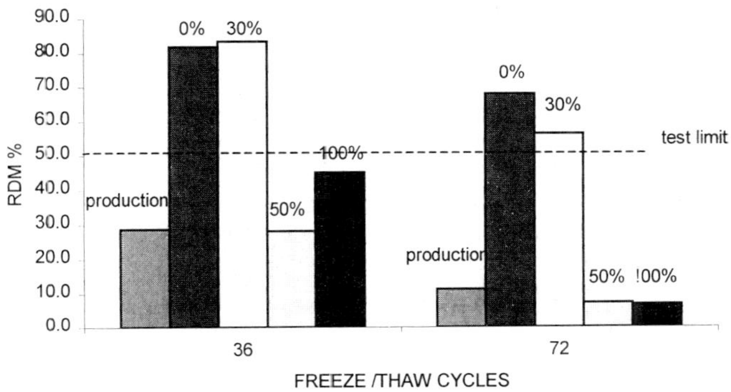

Figure 9 Effect of MWIBA content on RME during freeze thaw exposure

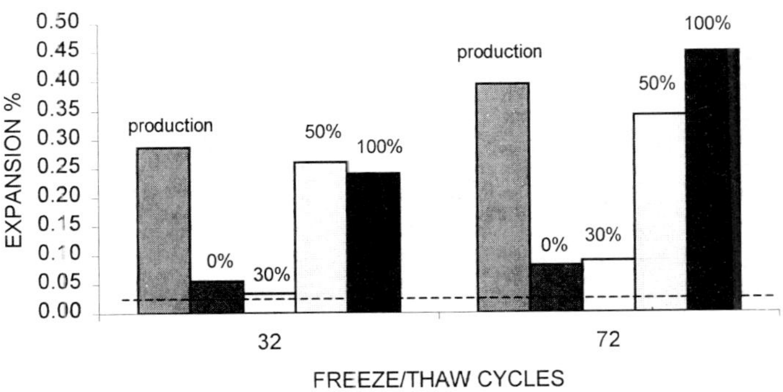

Figure 10 Effect of MWIBA content on linear expansion during freeze thaw exposure

LEACHING TESTS

Test Method

Samples from the CMUs were subjected to the CEN protocol leaching test [11]. This test is designed to investigate the leaching behaviour of concrete in contact with drinking water and is similar to the Dutch Tank Test. The test specimens are 100mm x 100mm x 25mm prisms cut from full CMUs. Each specimen is immersed in local drinking water, arranged such that all surfaces are equally available for leaching. The volume of water is in the proportion of 5cm³ for each cm² of specimen surface area (in this case 300 cm² requires 1500 cm³ of water). The water is replaced 8 times. The replacement times measured from the commencement of the test are: ¼ day, 1 day, 2¼ days, 4 days, 9 days, 16 days, 36 days and 64 days. Samples of each leachate are filtered over a 0.45μm membrane, analysed for pH and redox potential and then preserved for analysis by acidifying with nitric acid to pH 2.0 +/-0.2 and stored at 5°C.

Analysis of leachate

Trace metals concentrations; Fe, Pb, Cu, Zn, and Cd, were determined by Atomic Absorption Spectrometry (AAS) calibrated using standard solutions. Major components K and Na were also determined using AAS. Cl- and SO4- ion concentrations were determined by ion absorption chromatography. Si, Al and Ca were determined by X-ray florescence (XRF). Samples were prepared by evaporating 1 cm³ of solution onto a copper disc.

Results of Leach Tests

The leachate service concentrations (LSC) of species are derived according to the CEN protocol [11] from the cumulative results of the 64-day leaching sequence. The leaching factor in the formula has been taken as the higher value given (3), since none of the species analyzed appeared to be fully leached. This can be seen in Figure 11, which shows the leaching behaviour of aluminium from CMUs with MWIBA contents from 0 – 100%.

The results (Table 2) show that the LSCs of two species, potassium and aluminium, exceed the EU maximum for drinking water. Two species, silica and iron, are seen to be absorbed from solution by all the CMUs. The leaching of lead is seen to be greatest from the production CMU which is in contrast to the results of most other species. It is also notable that the absorbance of iron from solution is greatest in the case of the production block (Figure 12).

Table 2 Service concentrations derived from leach tests on CMUs using the CEN protocol

MIX				LEACHATE SERVICE CONCENTRATION (LSC)											
	Mix	Cement kg	MWIBA %	Cl- ppm	SO4- ppm	Na ppm	K ppm	Ca ppm	Si ppb	Al ppb	Fe ppb	Pb ppb	Zn ppb	Cu ppb	Cd ppb
EU limit max conc				400	250	150	12	250		200	200	50	5000	3000	5
Production	t7	140	5	1.46	0.4[illegible]	4.32	11.02	0.61	-25	34	-4.8	7.5	8.8	5.1	0
Trials	t4	200	100	16.5	16.4	17.6	14.81	3.18	-94	915	-4	3	8.5	13.2	0
	t3	170	100	16.8	17.6	14.2	10.65	2.9	-166	991	-4.2	3.7	10.9	11.7	0
	t2	140	100	13	9.57	14.6	9	3.96	-88	1134	-2.2	3.1	20	4.4	0
	t5	140	50	10.7	12.4	11.6	12.16	1.28	-135	344	-3.8	5.4	9.6	18.1	0
	t6	140	30	9.62	8.14	10.5	14.14	-0.09	-23	143	-4.1	6.8	15.9	8.9	0
	t1	140	0	0.72	2.81	4.96	4.22	4.67	-237	-19	-1.4	0	14.4	8.5	0

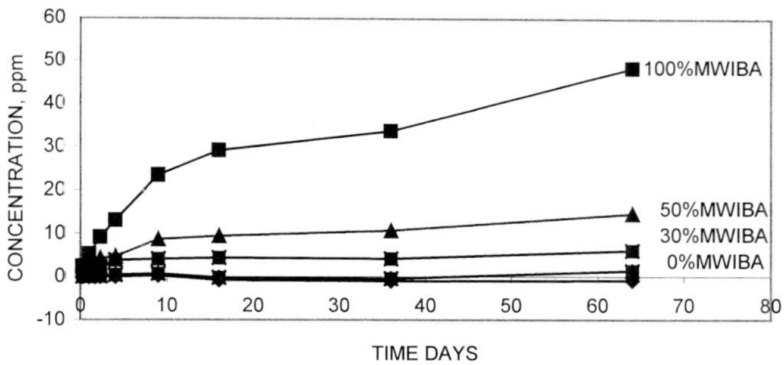

Figure 11 Leaching behaviour of Al for CMUs with increasing MWIBA contents

Aluminium

Figures 11 There was a £7.00 bank charge on the above payment. I would be most grateful if you could please pay this as soon as possible.

12 and 13 show the leaching behaviour of aluminium from the various CMU mixes. It can be seen that Al concentration increases rapidly with increased MWIBA content and at contents over 30% the EU maximum is exceeded. The concentration for the production block is well below the limit. Figure 13 shows that Al concentration falls with increased cement content in 100% MWIBA CMUs, but not sufficiently to reduce it below the EU maximum.

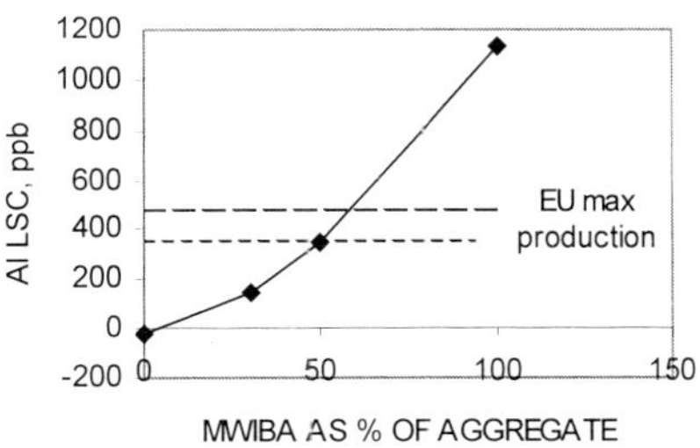

Figure 12 Effect of MWIBA content on Al LSC

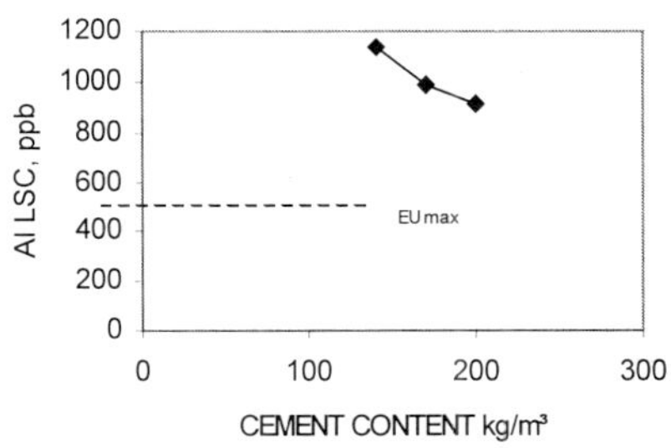

Figure 13 Effect of cement content on Al LSC

Lead

Table 3 shows that the lead LSCs determined were well below the EU maximum. In the case of lead the production block produced the highest LSC with a marked decrease in LSC as the MWIBA content was increased (Figure 14). Cement content appeared to have little effect on lead LSC (Figure 15).

Iron

All CMUs absorbed iron from solution although the effect was greatest in the production block. The amount of iron taken up by a CMU was found to be inversely proportional to its MWIBA content but directly proportional to its cement content (Figures 14 and 15).

The release of lead can be seen to correlate with the take up of iron, perhaps indicating some sort of ion exchange mechanism [9].

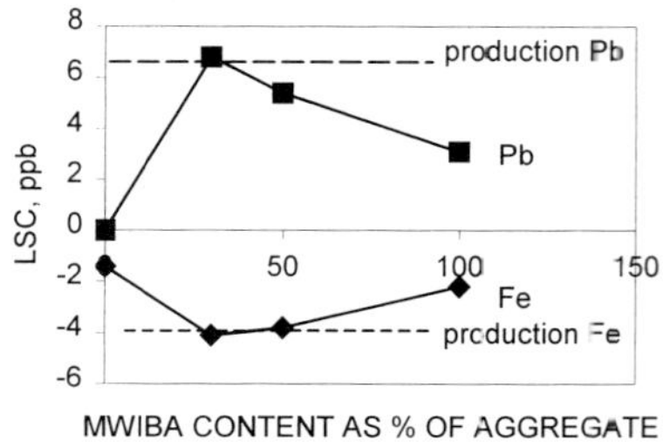

Figure 14 Effect of MWIBA content on Fe and Pb LSCs

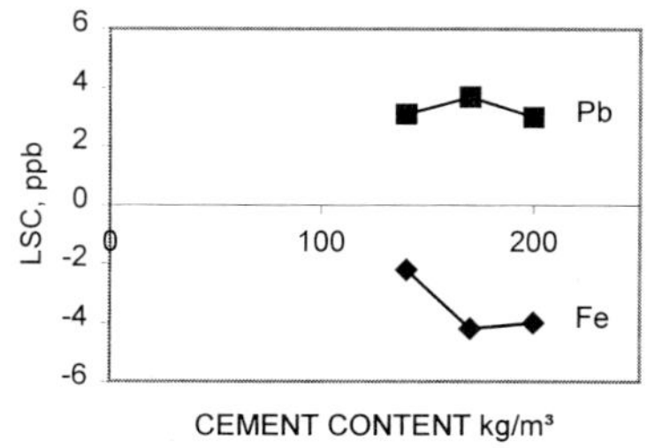

Figure15 Effect of cement content on Fe and Pb LSCs

Zinc and copper

Zn and Cu LSCs are both well below the EU maximum (Table 2) with the production CMU yielding the lowest values. Figures 16 & 17 show an interrelationship between Zn and CU LSCs when MWIBA and cement contents are increased. No definite trend is produced when MWIBA content is increased but the LSCs of the species mirror one another. Increased cement content produces an increased Cu LSC but a reduced Zn LSC.

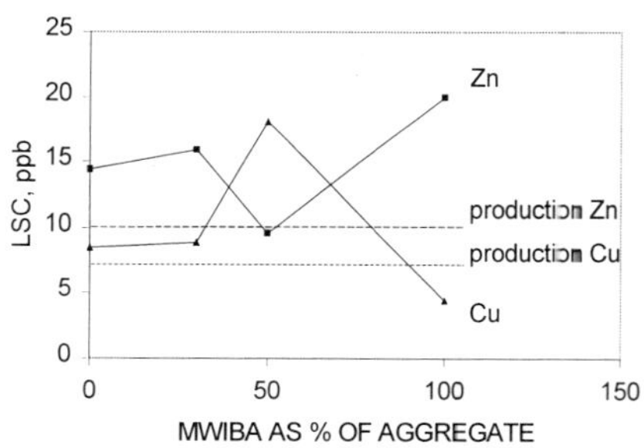

Figure 16 Effect of MWIBA content on Zn and Cu LSCs

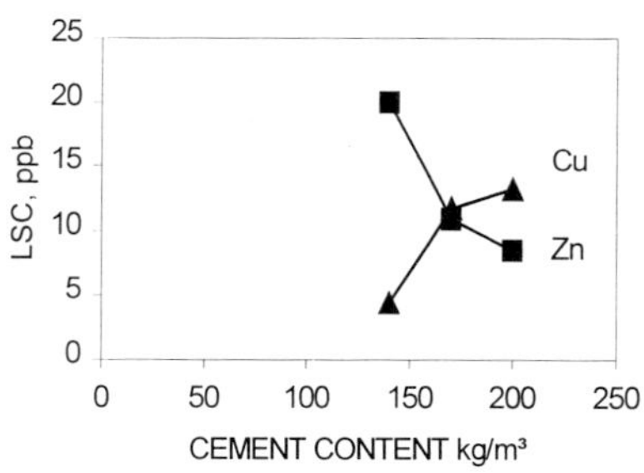

Figure 17 Effect of cement content on Zn and Cu LSCs

Cadmium

Cd levels in all the leachates were below the limits of detection.

Potassium

Potassium LSCs were close to or above the EU maximum for many of the CMUs tested and indeed the production block was only 8% below the limit.

K LSC was found to decrease with increased MWIBA content but to increase significantly with increased cement content (Figures 18 & 19)

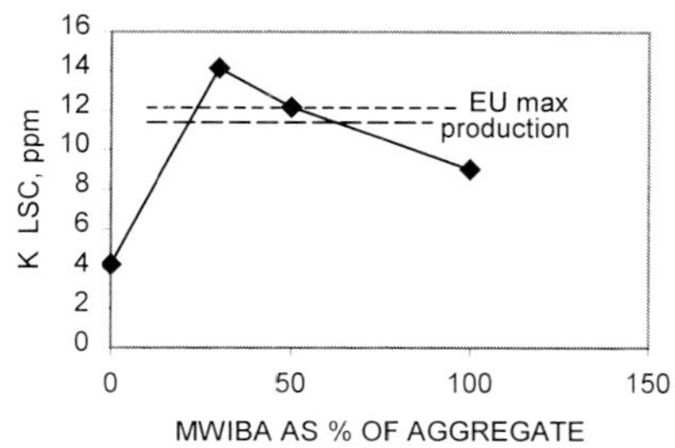

Figure 18 Effect of MWIBA content on K LSC

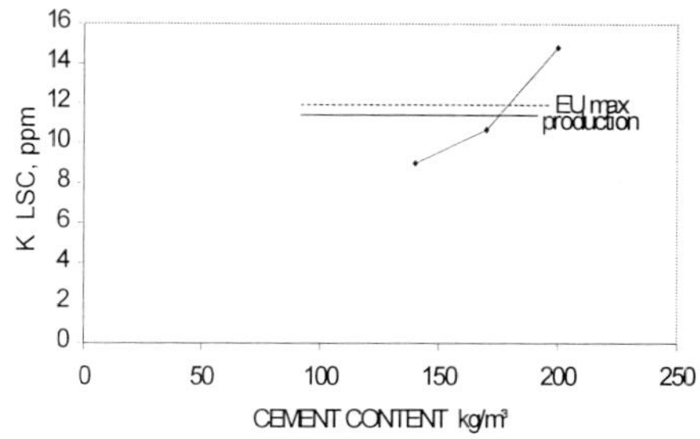

Figure 19 Effect of cement content on K LSC

Other Major Elements and Ions

The SLCs of Na, Si, Ca, Cl and SO_4 were all much lower than the EU maximums (Table 2), typically by a factor of 10 or more.

Chloride, Sulphate and Sodium

Cl, SO4 and Na SLCs increase significantly with higher MWIBA content and 100% MWIBA CMUs have SLCs of the order 10 times higher than the production block (Figure 20). Whilst these are well within the environmental limits they may be significant in engineering terms with regard to the durability of concrete or steelwork. The Cl, SO_4 and Na SLCs appear also to increase with increased cement content (Figure 21).

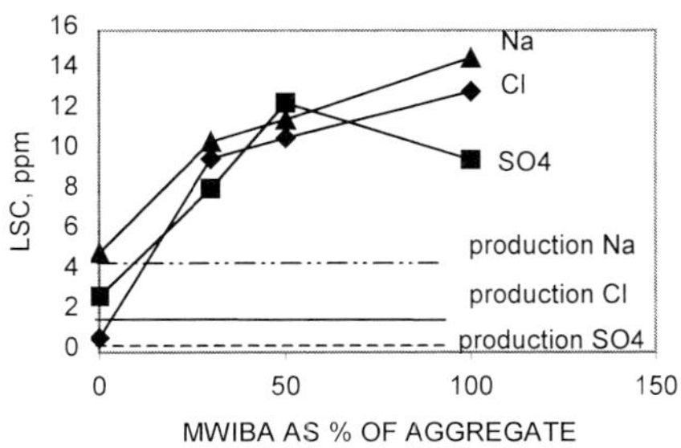

Figure 20 Effect of MWIBA content on Na, Cl and SO_4 LSCs

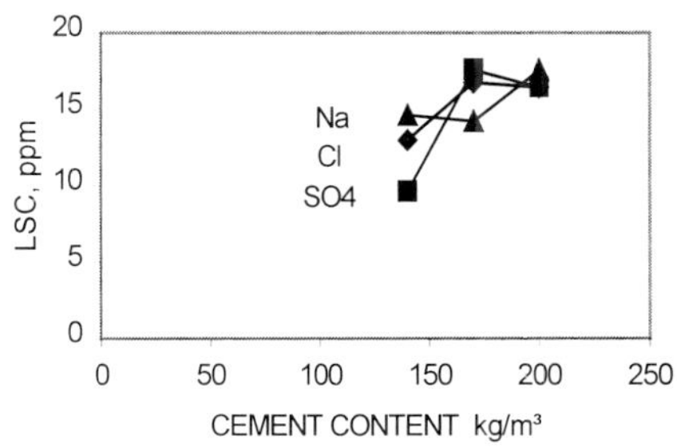

Figure 21 Effect of cement content on Na,Cl and SO4 LSCs

CONCLUSIONS

The trial CMUs performed well in the in the physical tests and satisfied most aspects of BS 6073 however the chloride content of all MWIBA mixes exceeded the limit and indeed the production block was only just within the limit.

The effect of increasing the cement content on strength was minimal perhaps reflecting the low strength of MWIBA as an aggregate. The density of the MWIBA CMUs was similar to that of the production block giving a unit of similar weight. The water absorbency of the MWIBA CMU was greater than that of the production unit, as was its water retention.

The freeze thaw durability tests demonstrated that the trial MWIBA blocks had a performance similar to the production blocks, although it was shown that the CMU with 30% MWIBA performed much better than blocks with a higher content.

The leach tests carried out according to the CEN protocol for drinking water showed that the leachate service concentrations of most of the species analysed for were well within EU guidelines for drinking water. Two species exceeded the prescribed limits for drinking water, potassium and aluminium. Potassium was marginally in excess of the limit in CMUs with 30% MWIBA content but concentration reduced below the limit with increased MWIBA (Figure 18). In contrast Al was just under the limit at 30% MWIBA content but was well over it at 100% MWIBA content (Figure 12).

The leach test results must be taken in context. They apply only for the species analysed and are subject to the interpretation of the CEN protocol. Six of the species (Al, Na, K, Pb, CL and SO_4) had concentrations exceeding the prescribed limits in leachates at various stages in the leaching sequence. Further, these are environmental rather than engineering criteria and levels of chloride, sulphate and other species may be sufficient to have a detrimental effect on other building components such as plaster, mortar, brick ties, steel etc.

PRACTICAL ISSUES

This study shows that it is feasible to manufacture CMUs using various proportions of MWIBA as an aggregate. The grading of the material is similar to that of sand and many of the test results indicate that ~ 30% MWIBA may be an optimum content. Applications are possible where exposure to leaching and weathering are minimised. These may include internal walls and beam and block flooring although interactions with other components may have to be considered. Whilst use in a dry application will result in minimum environmental releases of harmful substances, in the long term future demolition of the structure will produce contaminated building waste. A more versatile product may be obtained if the ash were washed to remove soluble salts such as sulphates and chlorides but this would have cost implications in a very competitive market.

REFERENCES

1. DEPARTMENT OF THE ENVIRONMENT. This Common Inheritence – Britains Environmental Strategy, Cmnd 1200, HMSO, London, 1990.

2. HESTORE R E AND HARRISON R H. Waste incineration and the environment, Royal Society of Chemistry, Bath 1994.

3. COLLINS R AND SHERWOOD PT. Use of waste and recycled materials as aggregates, HMSO, London, 1995.

4. BERG E R and NEAL A J. Concrete masonry unit mix designs using municipal solid waste bottom ash. ACI Materials Journal, Title no. 95-M45, July/August 1998, pp 470-479.

5. BRITISH STANDARDS INSTITUTION. BS 6073 part 1, Precast masonry units: part 1: Specification for precast masonry units, 1981.

6. BRITISH STANDARDS INSTITUTION.BS EN 1521, Determination of flexural strength of lightweight concrete with open structure, 1997.

7. BRITISH STANDARDS INSTITUTION. BS EN 992, Determination of the dry density of lightweight aggregate concrete with open structure, 1995.

8. AMERICAN SOCIETY FOR TESTING AND MATERIALS. C 666: Procedure A, Standard test method for resistance of concrete to rapid freezing and thawing, ASTM C666, 1988.

9. BRITISH STANDARDS INSTITUTION. BS 1881: part 77, Method for the determination of dynamic modulus of elasticity of concrete, 1977.

10. BRITISH STANDARDS INSTITUTION. BS 1881: part 120, Method for determination of the compressive strength of concrete cores, 1983.

11. CEMBUREAU. Use of non -ferrous slag in cement and the leaching behaviour of concrete in drinking water, 1995.

PART FOUR:

USE OF INCINERATOR FLY ASH IN CONSTRUCTION

RESONANT SHOCK COMPACTION: ASH-BASED CONSTRUCTION BLOCKS

R C Amme

University of Denver

R Pressey

K Weir

D Frey

Resonant Shock Compaction, LLC

United States of America

ABSTRACT. The Resonant Shock Compaction process (RSC) has been used commercially for ten years to manufacture large refractory objects for the steel industry. In laboratory studies, this process has recently been applied successfully to soils and to coal combustion by-products containing various contaminants (carbon, heavy metals, sodium) and appropriate binders (e.g., phosphates, cement, clay) for the manufacture of construction blocks, pavers, and building panels. Blocks may also be crushed and screened to produce lightweight aggregate. The granular mixture is placed into a mold which is subjected to vigorous vibration and shock to produce solid, dense products. We report investigations of mixes of fly ash and bottom ash from utilities and from incinerated sewage sludge with sand, steel slag, cement and other binders in various proportions. Product strengths and leaching resistance are presented, along with data relating to durability.

Keywords: Vibratory compaction, Granular materials, Leaching, Heavy metals, Ash

Dr Robert C Amme is Manager of the Environmental Materials Laboratory and Professor of Physics at the University of Denver, Denver, Colorado. His specialties are atomic and molecular physics and granular materials. He is a Fellow of The American Physical Society.

Robert Pressey is a Managing Director of Resonant Shock Compaction, LLC, Denver, Colorado. He has had 30 years in technical applications and business development, as well as 30 years of materials physics and chemistry.

Keith Wier is a Managing Director of Resonant Shock Compaction, LLC, Denver, Colorado. He is a Co-inventor of the RSC process and has had 30 years in technical product development and sales.

David Frey is the Chief Technical Officer of Resonant Shock Compaction, LLC, Denver, Colorado. He is a Co-inventor of the RSC process and has had 25 years of materials production management experience.

VIBRATIONAL COMPACTION AND THE RSC PROCESS

General Description

Vibratory compaction has been used in the powder metallurgy industry to produce composite materials made of dissimilar materials that are not otherwise readily formed. Granular materials (powders) are combined with lubricants and chemical binders, then vibrated to achieve maximum-density (minimum porosity) shaped parts; these "green" parts are then fired at sintering temperatures. Granular material may be pressed into shaped parts, but inter-particle bridging forces limit the maximum achievable size and density; thin parts may be easily pressed but very large and expensive presses are required to overcome these internal bridging forces when making large products.

The physicochemical principles of vibratory compaction and chemical additives are discussed in *Vibratory Compacting, Principles and Methods, Part 1*, a translation of the USSR Institute of Physical Chemistry of the Academy of Sciences of the USSR. Other sections of this treatise address mechanical packing of spherical particles (binary, ternary, and quaternary packing), experimental vibratory methods and vibratory compaction of nuclear fuels: *Vibratory Compacting, Principles and Methods, Vol. 2, Perspectives in Powder Metallurgy*, edited by Henry H. Hausner, Kempton H. Roll and Peter K. Johnson, 1967, Plenum Press (New York).

Resonant Shock Compaction

Resonant shock compaction (RSC) is a unique vibratory process in which the mold containing the granular material is placed on a shake table driven by a synchronized pair of motor-vibrators. The mold is opposed from above by a pneumatically controlled ram or die. The pneumatic pressure is in turn regulated by a programmed computer. In most circumstances, the pressure profile is selected so that the mold actually leaves the shake table and is slammed back onto the surface, resulting in very high accelerations. This (patented) scheme (Figure 1) produces shock waves passing through the material and ultimately produces a higher density than that otherwise obtained. The process is optimized when the velocities of the mold and of the shake table are out of phase, i.e., when the impacts occur with the highest relative velocity.

The RSC process has been used to compact coal combustion products (CCPs) and other materials such as sand, clay, aggregate, recycled asphalt and recycled concrete rubble into durable and high compressive-strength material similar to masonry and concrete. Little liquid (usually in the form of water) is needed. Reinforcing bars, wire mesh or other physical strengtheners can be added as well, as can various types of binder fluids.

First used commercially in 1985, the RSC technology is being employed to manufacture refractory well blocks, slide gates, ladle bottoms and tundish dams that meet stringent technical requirements of the steel-making and refractory industries. These are engineered products of some 400 different shapes and sizes and weighing up to about 5,000 pounds each.

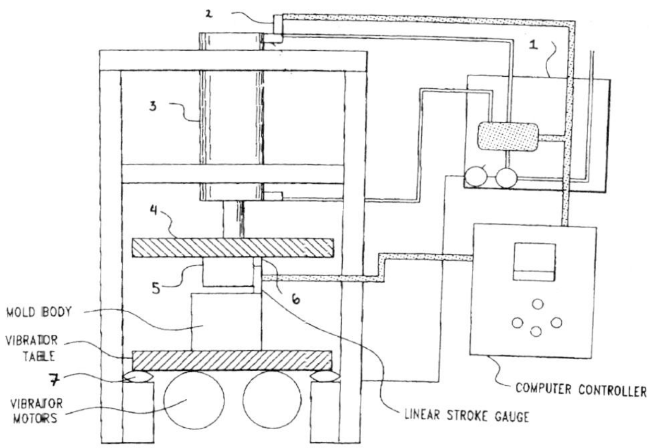

Figure 1 Schematic of resonant shock compaction apparatus (see also K. Wier, U.S. Pat. No. 5,863,476 (1999). 1: pneumatic control system; 2: pressure transducer; 3: pneumatic ram; 4: (rectangular) steel top plate; 5: ram; 6: accelerometer; 7: pneumatic shock absorbers.

The RSC compaction machine employed in the present study is approximately ten feet high and fits within a space less than six feet on a side. Additional space is required to load and unload molds, but the entire machine and associated operations can be placed in a relatively compact area. Figure 2 shows a representative plant layout which includes a feed hopper, material loading into the mold, the RSC machine and an ejection station to remove the compacted products from the molds. The product is typically cured for 24 hours before shipping. Alternatively, the RSC machine can be truck-mounted for on-site production.

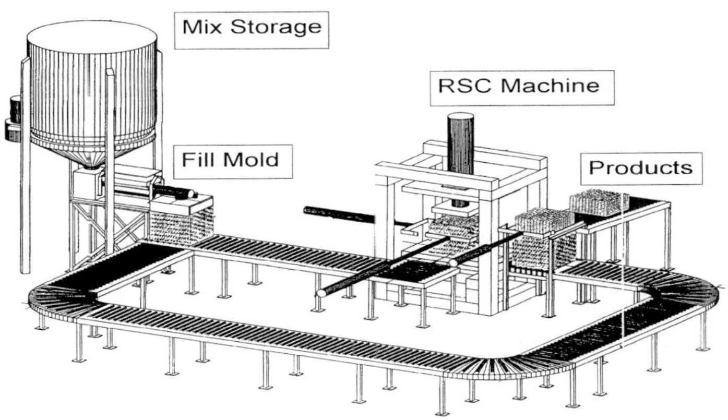

Figure 2 Resonant shock compaction (Represenative plant layout)

RSC technology has also been tested in a number of environmental remediation applications for the treatment and disposal of hazardous and radioactive wastes[1]. A new patent was issued in January 1999 for this application.

For the studies reported here, the compactor is driven by two 15 hp Invicta™ motor-vibrators operating at a frequency of 20 Hz. Running times depend on mold size and the material to be compacted, but are in the vicinity of three to fifty seconds (i.e., less than 1,000 impacts). The impact force is determined by both the accelerations of the ram and of the table, and by the mass suspended from the ram. The latter remains in contact with the top surface of the material in the mold. A typical acceleration profile for the ram is presented in Figure 3. For the first several oscillations, the acceleration is sinusoidal; when the (underdamped) system permits the mold to leave the table, the acceleration becomes larger and sharply peaked. Most of the data reported here are for a test mold with inside dimensions measuring 30.5 cm by 10 cm by 23 cm deep; typically the mold is filled to a depth of about 16 cm, and the final volume is roughly 50% of the starting "loose" volume.

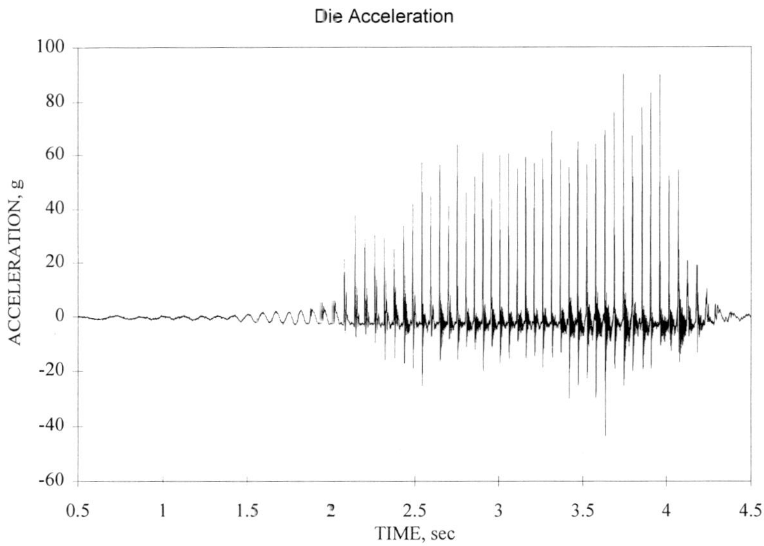

Figure 3 Accelerometer trace in units of g for the die (upper ram), showing spikes resulting from mold impact with table

Preliminary results on compaction of large 3-ft. by 4-ft. by 0.5-ft. (91 cm by 122 cm by 15 cm) panels of coal ash/cement blends have also been obtained[2]. To handle products of this size, hydraulic mold ejection apparatus and fork lift as well as a large, temperature- and humidity-controlled curing chamber are employed. Lacking in this preliminary work was an adequate high-volume mixer; it was necessary to mix three separate batches of ca. 90 kg each and combine them in the mold manually. The goal of this effort is to assess the viability of using RSC technology to compact mixes of fly ash, bottom ash and binders into higher-valued products including construction blocks and panels, highway sound barriers and aggregate. Lightweight aggregate is made by crushing and screening compacted, cured blocks. To date, ashes from four different utility power stations have been compacted into the large test blocks, as well as the smaller version.

Analysis of RSC Products

Six test methods are currently employed to investigate the properties of the specimens fabricated with the smaller mold:

(a) freeze/thaw cycling (ASTM C-666 for concrete);
(b) resonant frequency testing (longitudinal, flexural and torsional) (ASTM C-215 for concrete);
(c) compressive strength (ASTM C39-94 concrete);
(d) absorption ratio (ASTM C216-97);
(e) acoustic emission accompanying compression or bending;
(f) surface hardness test hammer (ASTM C805-94 for concrete).

For specimens from the large mold, some of these methods have been applied to 10-cm cores removed from the center and corners of the panels.

GRANULAR MATERIALS TREATED BY RSC

Soils (Sands/Clays, rubble)

The significant success of manufacturing very large refractory parts under license agreement for over ten years has given ample evidence that superior products can be produced using the Resonant Shock Compaction process. The natural extension of this application led to the current team effort to investigate both the process and products manufactured from a range of other granular materials. (Since the current compactor is not optimized to study several important operating parameters, a more research-friendly apparatus is being planned.) The first material we investigated was a sample of representative soil from the USDOE/Hanford site near Richland, Washington. The properties of the compacted product led to a series of tests of surrogate soils consisting primarily of sand and red clay (plus a binder) similar to that found at the DOE/Rocky Flats site near Denver. Under U.S. Department of Energy sponsorship, these specimens were examined for strength and resistance to leaching (of Cd, Cr, Pb) as a function of curing temperature up to 1000°C [1]. Soil sample surrogates were prepared with the addition of 320 ppm cadmium and 420 ppm chromium and either a (cementitious) Class C fly ash or phosphate binder (Figure 4). Compressive strengths of 2000 psi (14 N/mm^2) were obtained, along with acoustic resonant frequencies from which dynamic elastic moduli could be derived. With this binder, the chromium could be readily tied up, but for cadmium high temperature curing was necessary in order to pass the TCLP. A search for improved binders not requiring elevated temperatures led us to a joint experimental effort with investigators at the Argonne National Laboratory (ANL), who had developed "Ceramicrete," a non-thermal chemically bonded phosphate ceramic [3]. Soil samples were spiked with 5000 ppm each of cadmium, chromium and lead, stabilized with the (proprietary) ANL Ceramicrete and sodium sulfide, then compacted either with ANL's laboratory uniaxial press or with RSC. RSC processing of the soil samples achieved ~45% volume reduction [3],

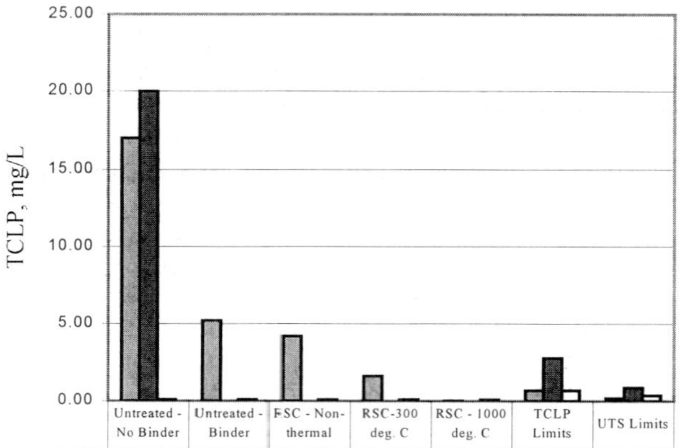

	Untreated - No Binder	Untreated - Binder	FSC - Non-thermal	RSC-300 deg. C	RSC - 1000 deg. C	TCLP Limits	UTS Limits
Cd	17.00	5.20	4.20	1.60	0.02	0.690	0.190
Cr	20.00	0.01	0.01	0.01	0.01	2.770	0.860
Pb	0.10	0.10	0.10	0.10	0.10	0.690	0.370

Figure 4 TCLPs: Compacted and stabilized sand/clay (320 ppm Cd and 420 ppm Cr added)[1]

exhibited 14 N/mm^2 compressive strength, and passed the Toxicity Characteristics Leaching Procedure (TCLP) tests (Figure 5). These results indicated that with 15% Ceramicrete binder, the leach rate for all three heavy metals was below the TCLP minimum limits, without heat treatment.

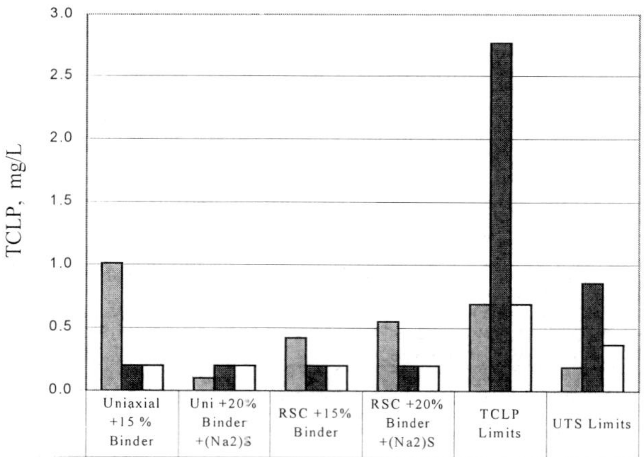

	Uniaxial +15 % Binder	Uni +20% Binder +(Na2)S	RSC +15% Binder	RSC +20% Binder +(Na2)S	TCLP Limits	UTS Limits
Cd	1.0	0.1	0.4	0.550	0.690	0.190
Cr	0.2	0.2	0.2	0.200	2.770	0.860
Pb	0.2	0.2	0.2	0.200	0.690	0.370

Figure 5 TCLPs: Compacted and stabilized soil uniaxial press and resonant shock compaction (5000 ppm Cd, Cr, and Pb added)[4]

Further Studies on Soils and Ashes with Argonne National Laboratory

In a double blind experiment [4] performed by Peters et al., at Argonne, a comparison was made of three compaction techniques employing both soil and ashes spiked with Cd, Cr, Pb, Ni, and Cs, and using Ceramicrete as binder. In these tests, RSC (referred to by them as "Harmonic") compaction for the small mold was compared with results from their uniaxial press and with a proprietary soil/ash compression technique. Compressive strengths for both soil and ash samples were greatest (2100 psi or 14.5 N/mm^2) for the RSC specimens (Figure 6); TCLP for Cd was significantly better than for the other two techniques for both soil and ash specimens (Figures 7, 8) while Cr and Pb were all comparable (Figures 7, 8). TCLPs for ash specimens containing Ni were best for RSC (Figure 9).

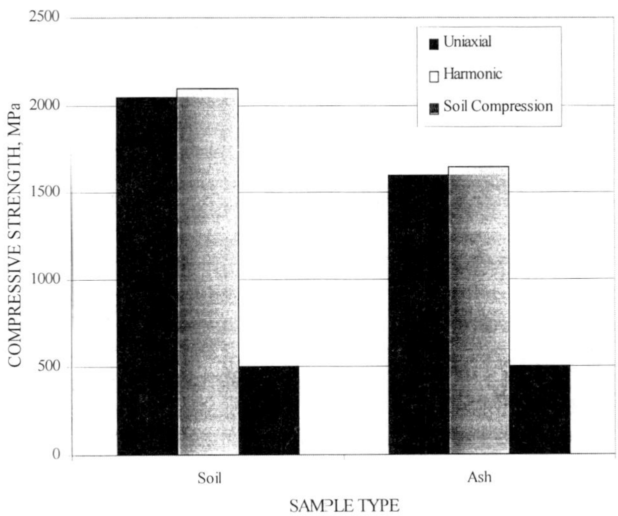

Figure 6 Comparison of compressive strengths using various compaction techniques

In separate tests using the same mold, soil containing concrete and asphalt rubble were compacted with 5% cement as the binder. The resulting specimens showed high strength and good integrity. At several U.S. sites in need of cleanup, very large volumes of such material exist or will be generated, much of which is classified as low-level radiation waste (LLW) as well as low-level mixed waste (LLMW), which contains heavy metals in addition. The RSC process, coupled with suitable binders, appears to be a strong candidate for stabilization and volume reduction of much of these wastes [5].

Utility Ashes: Coal Combustion By-Products (CCPs)

Research Shock Compaction fabrication of building materials (blocks, panels and aggregates) using high loading (over 80%) of coal combustion products appears to be feasible from both an engineering and economic perspective [6]. Specimens prepared with the 31 cm mold

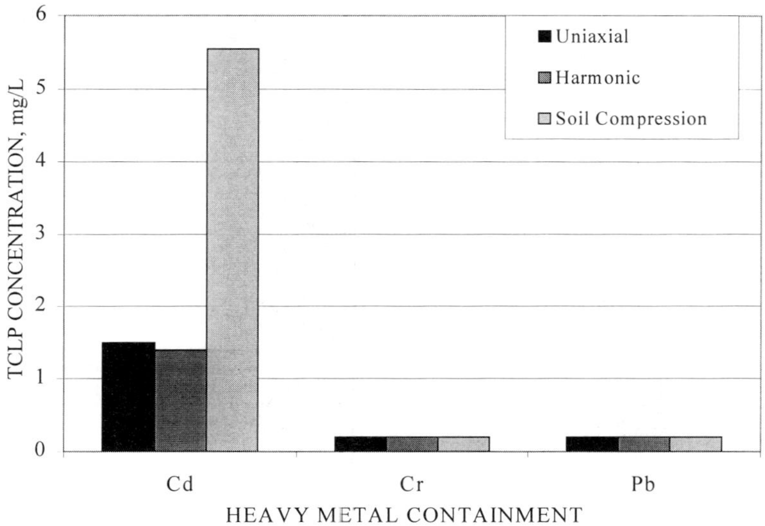

Figure 7 Comparison of TCLP results using various compaction techniques -- Ash samples

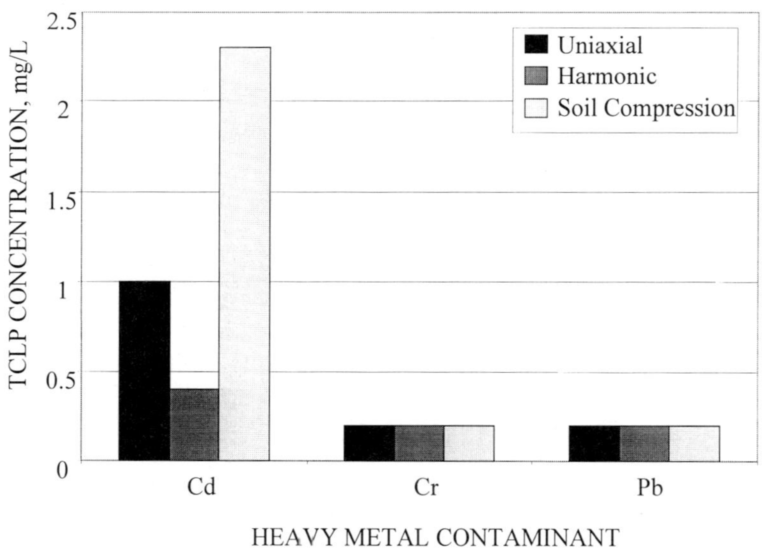

figure 8 comparison of TCLP results using various compaction techniques -- Soil samples

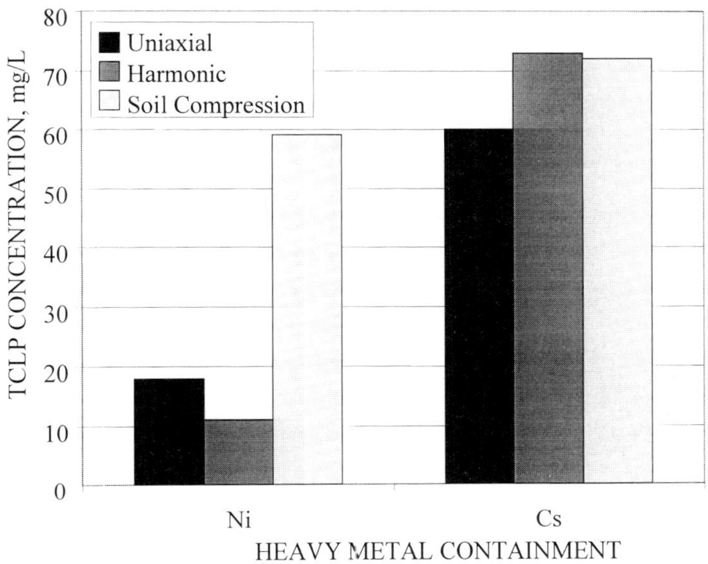

Figure 9 Comparison of TCLP results using various compaction techniques -- Ash samples

described above have exhibited compressive strengths up to 5,000 psi (35 N/mm^2). Product quality is not greatly affected by contaminants often appearing in the ash; unburned carbon or dry sodium compounds added to flue gases to enhance SO_2 removal are examples. 1) Usable specimens have been formed from mixes containing 60% to 98% fly ash plus bottom ash.

Class F fly ash containing approximately 9% unburned carbon and sodium sesquicarbonate mixed with bottom ash and binders (cement, sand and clay) in the ratio 20/60/20 produced specimens with compressive strengths greater than 3700 psi (25 N/mm^2). Class C fly ash plus bottom ash plus sand in the ratio 48/50/2 resulted in strengths of about 23 N/mm^2. Samples of these products placed in a freeze-thaw chamber exceeded the ASTM C-666 (300 cycles) as well as the ASTM C-67 (50 cycles) standards with minimal deterioration, as seen in Figure 10 for eight different series of tests employing a range of ash composition provided to the laboratory by Public Service Company of Colorado. Some of these specimens contained "F" ash with sufficient carbon to render them unsuitable as a concrete additive. For these tests, the decline in the resonant frequency fr (flexural mode) for each specimen was measured with a C.N.S. Electronics Ltd. Erudite MkII system (BS 1881/ASTM C215). Failure is defined as $(fr/fro)^2 \leq 0.8$, where fro is the resonant frequency prior to cycling. This ratio is equivalent to that of the dynamic Young's Moduli.

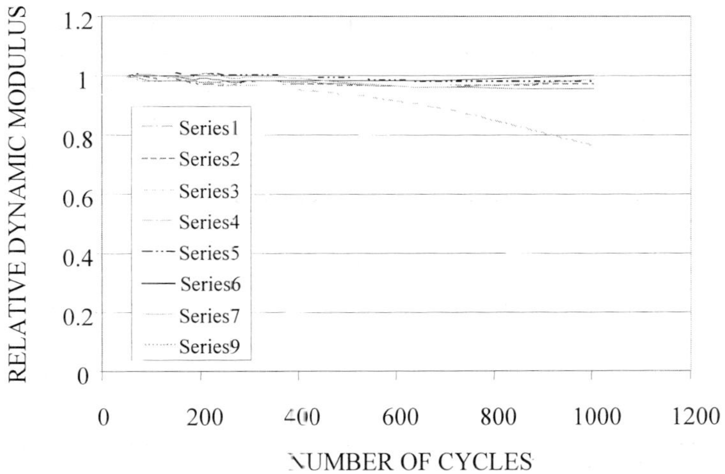

Figure 10 Freeze-thaw durability of RSC-fabricated specimens of Public Service Company and ash mixtures (ASTM C666 is employed).

RSC BINDING OF HAZARDOUS METALS IN SEWAGE INCINERATOR ASH

The objective of these studies was to determine whether the RSC process, when combined with appropriate binders, can immobilize potentially hazardous metals and prevent leaching of these metals into the environment. Applications include volume reduction and stabilization of incinerator ash mixed with other by-products, for safe and economical disposal, and alternatively the production of blocks and panels that may be used in the construction of buildings, retaining walls and other structures.

Again, the procedure was to measure the metals leaching characteristics of the untreated incinerator ash and coal ash, as determined by the Toxicity Characteristic Leaching Procedure, form a test block specimen in the RSC machine, and then measure the TCLP of this block, along with resonant frequencies and compressive strengths.

The incinerator ash and coal ash samples discussed here were provided by a Japanese firm. They exhibit very low TCLP leachable metals and thus these results do not provide a clear demonstration of the benefits of an integrated RSC compaction and volume reduction with chemical binders. (See Table 1.) However, low metals content is preferable when using these materials to manufacture building materials.

In the two following figures (Figures 11 and 12), the compressive strengths of the RSC compacted ash blocks are presented as 1) a function of the added cement binder content to blocks containing fixed amounts of ash, and 2) as a function of the ash content with fixed amounts of cement.

Table 1 TCLP Test results for sewage sludge ash and additional coal fly ash

METALS	EPA STANDARD MAXIMUM ALLOWABLE LIMIT	COAL FLY ASH	SEWAGE ASH ASH SAMPLE	SEWAGE ASH RSC BLOCK 9-24-99-1 10% CEMENT	SEWAGE ASH RSC BLOCK 9-24-99-2 5% CEMENT 5% SLAG
Arsenic, As	5.0	0.81	0.34	0.05	0.01
Barium, Ba	100	0.91	0.20	0.43	0.35
Cadmium, Cd	1.0	<0.005	<0.005	<0.005	<0.005
Chromium, Cr	5.0	0.31	<0.005	0.007	0.052
Lead, Pb	5.0	<0.05	0.05	<0.05	0.05
Mercury, Hg	0.2	<0.0002	<0.0002	<0.0002	<0.0002
Selenium, Se	1.0	0.01	0.19	0.05	0.05
Silver, Ag	5.0	<0.005	<0.005	<0.005	<0.005
Total Hg		0.09	0.05	0.04	0.04

TCLP Procedure: Per EPA Reference SW-846, Method 1311. TCLP results are reported in milligrams per liter (mg/L), on an extract basis.

Total Mercury Procedure: The sample was prepared and analyzed to Bituminous Coal Research Report #2, Double Gold Amalgamation Coal Vapor Atomic Absorption technique for Mercury analysis. Results are reported in micrograms per gram (ug/g), on an as-received basis.

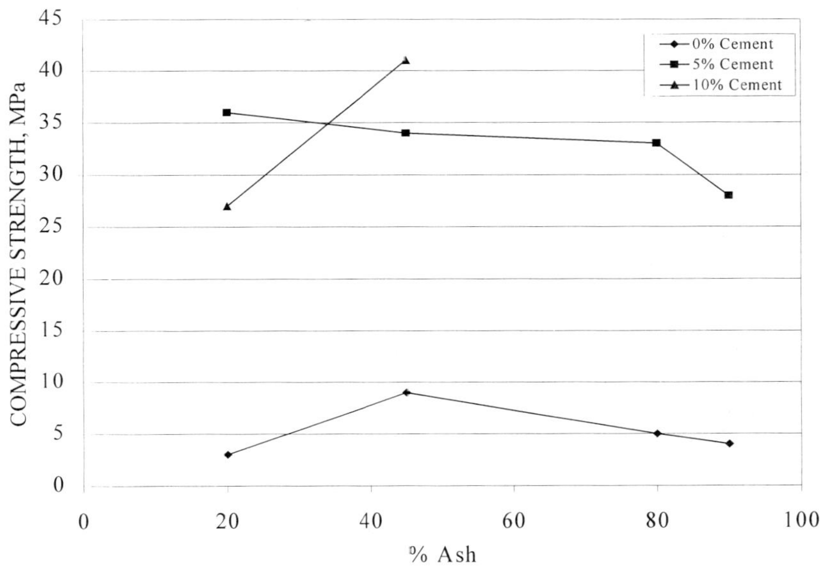

Figure 11 Incinerator ash
compressive strength vs. %ash (and %cement binder)

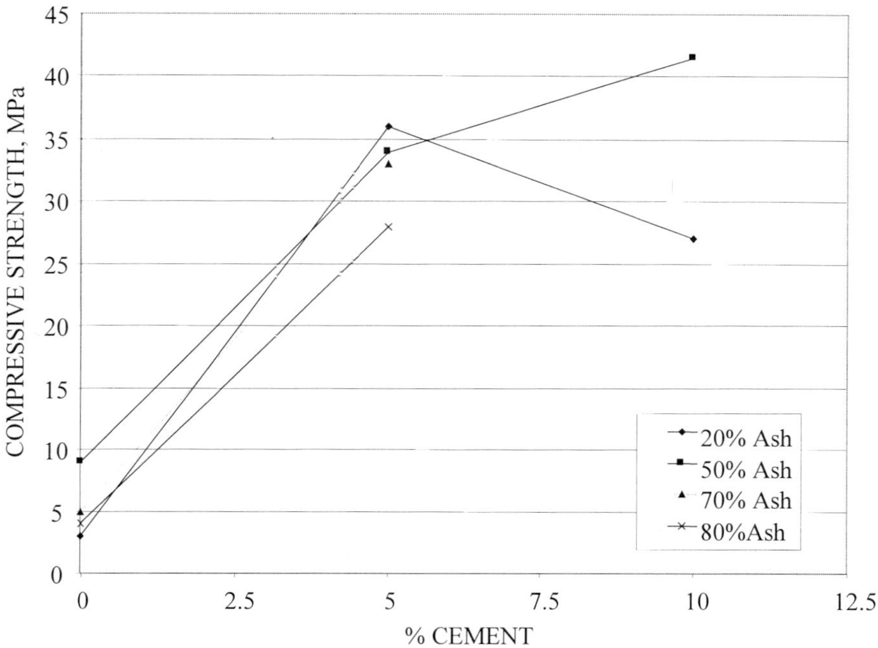

Figure 12 Incinerator ash strength vs. %cement binder (and %ash)

Commercial use of ash in the RSC manufacture of construction materials must satisfy both technical, environmental and economic criteria; low cost wastes are combined to manufacture high value-added materials that are environmentally sound. The goal is to maximize the waste ash content, minimize addition of costly binders and stabilize contaminants. RSC is a dry granular packing process in which the very fine ash is combined with larger granular waste sand (for strength), clay (for "green strength") and a binder (cement, steel slag, phosphates). Test results suggest that RSC products may be able to treat 60-100% waste materials, depending upon the waste composition and product design criteria.

CONCLUSIONS AND FUTURE DIRECTIONS

There is growing evidence supporting our postulate that the incremental improvement in densification obtained with the RSC process, coupled with suitable binder(s), leads to superior products in terms of strength and durability as well as in leach reduction. The same physiochemical processes at work in tying up the hazardous metals are also expected to be operative in the presence of radionuclides. The RSC process appears to be fast, efficient, and economical. It is evident, however, that in treating such wastes, considerable effort will be required to fully characterize the outcomes for any particular granular substance and/or waste stream.

Future directions include the construction of a more versatile research compactor with greater controls over the important research parameters such as acceleration, vibrational frequency and amplitude, and real-time measurement of compaction rates. Porosity/permeability studies will be incorporated.

Since municipal incinerator ashes are expected to have much higher concentrations of heavy metals in several states of oxidation, we plan to make extensive comparative measurements on leaching of various chemical species. Curing will be performed at both room and elevated temperatures. These studies are to be coupled with freeze/thaw testing of the type reported here.

REFERENCES

1. WIER, K. and AMME, R.C. Measurements of physical/chemical properties of compacted soils utilizing the harmonic (shock) compaction process. University of Denver final report to the CAMI-ETAP Dept. of Energy Program, 1 March 1997.

2. FREY, D.J., PRESSEY, R.E., WIER, K.E., and AMME, R.C. Large block testing program - Phase I Report: Resonant shock compaction of coal ash to produce construction blocks, University of Denver Environmental Materials Laboratory Report, 30 March 1999.

3. WAGH, A.S., SINGH, D., JOHNSON, D.O., PETERS, R.W., RAO, A. and JEONG, S.-Y. Phosphate bonding with harmonic compaction for stabilizing low-level mixed wastes. Final report on TTP CH2-6-MW-41 submitted to Mixed Waste Focus Area, U.S. Department of Energy, 1998.

4. PETERS, R.W., WAGH, A.S. and JOHNSON, D.O. Comparison of compaction techniques in terms of structural integrity and leach resistance, presented at 10th International Symposium on Gas, Oil, and Environmental Biotechnology and Site Remediation Technologies, Orlando, Florida, 8 December 1997.

5. AMME, R.C., WIER, K.E., FREY,D.J. PRESSEY, R.E. and PETERS, R.W. Resonant Shock Compaction of low-level waste: solidification, stabilization, and volume reduction. Spectrum '98 Supplement, Inter. Conference on Nuclear and Hazardous Waste Management, (Amer. Nuclear Society) Proceedings of Sept. 1998 Conference in Denver, Colorado.

6. GOSS, D.C., AMME, R.C., FREY, D.J., PRESSEY, R.E. and WIER, K.E. Fabricating high ash-content building materials using resonant shock compaction, 13th Inter. Symposium on Management and Use of Coal Combustion Products, Amer. Coal Ash Association, Jan. 1999, Orlando, Florida.

UTILIZATION OF MUNICIPAL SOLID WASTE FLY ASH AS INNOVATIVE CIVIL ENGINEERING MATERIALS

H-K Cheong

J-H Tay

K-Y Show

Nanyang Technological University

Singapore

ABSTRACT. One of the ways in which highly urbanized cities cope with increasing amounts of municipal solid waste (MSW) generated by the population is incineration. As land is a very precious commodity and landfills are fast dwindling in most highly urbanised cities such as Singapore, the disposal of the ash generated from MSW incineration poses increasingly difficult problems to the municipalities. A viable solution to the disposal problems would be the reuse of MSW ash for civil engineering applications. Research studies have been conducted at the Nanyang Technological University, Singapore, to look into the potential reuse of MSW fly ash. This paper presents a summary of the studies in utilising MSW fly ash as innovative civil engineering materials. Properties and suitability of the products derived from MSW fly ash as an admixture in stabilization of soft soil, non-conventional concrete aggregates, brick-making material, and cement replacement material will be discussed.

Keywords: Incinerator fly ash, Reuse, Concrete, Properties, Strength.

Dr Hee-Kiat Cheong, is a Professor and Dean of the School of Civil & Structural Engineering, Nanyang Technological University, Singapore. His research interests include use of wastes in cement/concrete, behaviour of concrete under multiaxial stress conditions, structural performance of concrete structures, behaviour of upgraded RC beams, and blast effects in soil and on piles.

Dr Joo-Hwa Tay, is a Professor and Head of the Division of Environmental & Water Resources Engineering, and the Director of the Environmental Engineering Research Centre, School of Civil & Structural Engineering, Nanyang Technological University, Singapore. His research interests focus on the hazardous and industrial waste management, waste minimization, recycling and reuse.

Dr Kuan-Yeow Show, is an Assistant Professor in the Division of Environmental & Water Resources Engineering, School of Civil & Structural Engineering, Nanyang Technological University, Singapore. His research interests include concrete materials and waste management, with special emphasis on wastes recycling & reuse.

INTRODUCTION

The disposal of municipal solid waste (MSW) is becoming an increasing concern in many modern urban municipalities because of the increasing volume of solid waste generated, the spiraling costs of operating landfills, and scarcity of landfill sites. In the United States alone, it is estimated that a total of 192,700,000 tons of solid waste will be generated annually by the end of this century [1]. An option increasingly being considered to reduce the volume of waste is the incineration of the solid waste. Incineration reduces the solid waste by as much as 80% by weight. Currently, the residues from incineration are disposed of by landfilling.

For the rapidly developing island city of Singapore which has a land area of about 650 square km, more than 2.7 million tonnes of MSW is generated each year. About 40% of the waste, mainly metal, paper, construction debris and horticultural waste such as tree trimmings, is recycled, while the remaining 20% of the waste is disposed of by landfilling and 40% by incineration. After incineration, about 17% by weight of the original material remains as bottom ash and 3% remains as fly ash; thus, some 33,000 tonnes fly ash is produced annually which is also disposed of by landfilling. As land has become a very precious commodity and landfills are fast dwindling in the highly urbanised island City-State, the authorities face difficult disposal problems with the increasing amount of MSW ash. A viable solution to the disposal problems would be the reuse of MSW ash for civil engineering applications. Reuse of these residues will help to alleviate the disposal costs, preserve landfill capacity, conserve dwindling resources of natural raw materials and mitigate potential environmental impacts. Incinerator residue is a potentially useful material for construction-related applications. They have been used, for example, in experimental bricks, in asphalt mixes, in concrete, and as a fill material [2-6].

In view of the anticipated MSW ash disposal problems and the depleting natural resources, studies have been conducted at the Nanyang Technological University, Singapore, to look into the potential reuse of MSW fly ash. This paper presents a summary of the studies conducted in the University in utilising MSW fly ash as innovative civil engineering materials. Properties of the products derived from MSW fly ash as an admixture in stabilization of soft soil, brick-making material, and cement replacement material are presented. The ash processing technologies as well as suitability of the products for construction applications are discussed.

INCINERATOR FLY ASH AS A CEMENT REPLACEMENT MATERIAL

This section reports on the exploratory work conducted in determining the suitability of using the incinerator fly ash to partially replace cement in structural-grade concrete. The MSW ash is greyish white in colour and is in a very fine powder form. Under the microscope, it appears to be finer than cement. In Table 1, a summary of the results of the physical tests done on the ash is given, and in Table 2, a breakdown of the main chemical elements and compounds of interest is set out. From the chemical analysis, it is clear that the ash has a very high lime content whereas the proportions of silica and alumina are very much lower. The high lime content is almost certainly due to intake of the compound from the lime column through which the gaseous products of incineration pass before the ash is collected. The ash also has very low percentages of elements such as sodium, potassium, magnesium and lead, which might be detrimental when present in concrete.

Table 1 Physical properties of MSW ash

PROPERTY	RESULTS
Effective size (micron)	2 - 50
Specific gravity	2.067
Moisture content	0.5%
Pozzolanic activity index	102%

Table 2 Chemical composition of MSW ash

ELEMENT	CONCENTRATION (% by weight)
Silicon	0.25
Iron	0.73
Calcium	29.00
Potassium	2.75
Sodium	1.30
Magnesium	1.03
Zinc	0.75
Lead	0.01
Aluminium	0.18
Copper	0.12
COMPOUNDS	% CONTENT
Silica (SiO_2)	0.54
Alumina (Al_2O_3)	0.33
Lime (CaO)	40.57
Ferrous oxide (Fe_2O_3)	1.04
Loss-on-ignition	19.5%

The ash has a significant loss-on-ignition value (19.5%), which is evidence of a high carbon content, possibly from the incomplete combustion of the solid waste. Moisture content (0.46%) is low; so is the specific gravity (2.07). As a measure of the ash's reactivity with cement, the pozzolanic activity index test was conducted. The index obtained (102%) suggests that the ash does contribute to the strength of a cement-ash mix in some way, most likely by hydration of the lime in the ash. The strength gain, however, is modest.

Three mixes of concrete containing various proportions of ash were cast and tested. The ash was mixed in as a weight-for-weight replacement of cement, the replacement levels being 10%, 20% and 30% designated by mix C10, C20, and C30, respectively. A normal concrete mix, containing only ordinary Portland cement as the cementitious material, was used as the basis for obtaining the proportions for the three batches of concrete containing the ash, and also served as the control mix for comparison purpose.

In Table 3 and Figure 1 are presented the results of the tests on concrete containing the MSW ash. The control mix was designed to give a slump value of between 60 and 100 mm. In terms of workability, the ash seems to have different effects at different replacement levels. At a replacement level of 10%, a slight reduction in the workability was encountered, as indicated by slump and compacting factor values. However, with higher proportions of ash, a slight improvement of workability was observed. The small differences in relation to the control suggest that at such replacement levels, workability is not significantly affected.

Table 3 Properties of MSW ash concrete

MIX	SLUMP (mm)	COMPACTING FACTOR	INDIRECT TENSILE STRENGTH (N/mm^2)	FLEXURAL STRENGTH (N/mm^2)
C0	70	0.96	1.06	2.15
C10	40	0.93	1.73	3.78
C20	55	0.95	1.31	2.67
C30	75	0.99	1.11	1.85

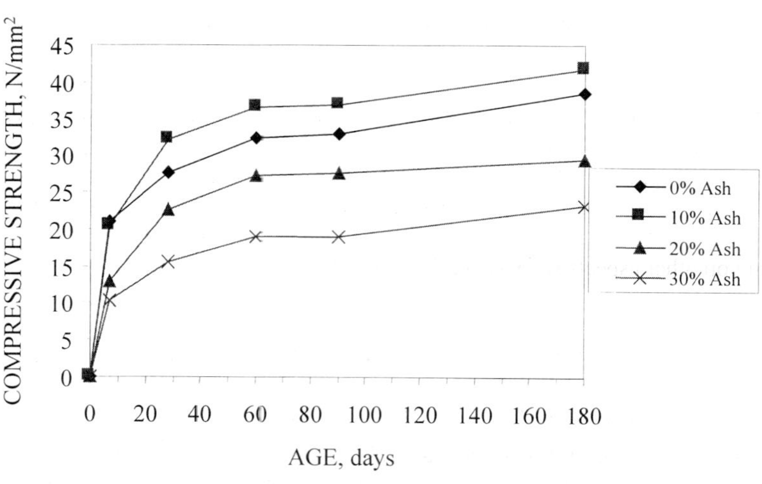

Figure 1 Compressive strengths of various mixes

At an early age, the concrete containing the ash had lower compressive strengths than the control concrete. This was particularly so for the C20 and C30 mixes, which registered respective strengths of 61% an 49% of the control. The C10 mix, however, appears to be similar to the control. At 28 days, the strength of the C10 mix exceeded that of the control and the difference of about 12% - 15% remained at subsequent ages. The strengths of the C20 and C30 mixes remained inferior to that of the control. It is apparent, however, that there was a higher rate of strength gain between 7 and 28 days for the C10 and C20 mixes, whereas strength gain for the C30 mix was comparable with that of the control. Hence, it may be concluded that there is some form of reaction, and strength gain, associated with the MSW ash during this period. In terms of strength development with age after 28 days, all mixes appear to behave in a similar manner, with a less rapid gain in compressive strength after 28 days. It is significant that there is no indication of any pozzolanic activity, as there was no closing of the gap in strength between the MSW ash concrete and the control as these mixes matured.

Clearly, the lower strengths of the C20 and C30 concrete are the result of the lower cement content and of a higher water/cement ratio. However, even in the C30 case, a concrete of reasonable strength is obtained when only about 200 kg/m^3 of cement was used at a water/cementitious material ratio of about 0.65. This suggests that the ash does contribute to the strength and that it is feasible to use the MSW ash as a cement replacement material, as a means of disposing the ash. The reason for the strength contribution probably lies in the high lime content (41%) in the ash, though what mechanisms are involved is still being investigated. The absence of any delayed strength increase due to pozzolanic activity ties in well with the low silicate and aluminate contents of the ash.

In terms of indirect tensile and flexural strength, results again show that concrete of reasonably good strength can be obtained despite a significant replacement of the cement by MSW ash. Thus, the C10 mix exhibited the highest tensile strength, and though this fell off as the percentage of ash increased, strengths comparable to that of the control could still be obtained.

From the tests reported in this section, it may be concluded that the ash generally has composition and physical characteristics that render it suitable for use in concrete as a partial replacement of cement. At this stage of the work, the results suggest that the incinerated MSW ash has promise as a partial replacement of cement in concrete of structural grade. More tests are needed on concrete with other mix proportions, on the variability of the ash itself, and on other aspects such as durability and long term behavior.

INCINERATOR FLY ASH AS A BRICK-MAKING MATERIAL

Experimental work on the feasibility of using the incinerator fly ash mixed with water sludge to partially replace clay in brick making is presented in this section. The water sludge with about 70% water content was collected from the sludge sedimentation tank of a water treatment plant. Various mixes of the bricks using different combinations of incinerator fly ash and water sludge are shown in Table 4. The mixes were designed based on the finding of a preliminary study that the maximum replacement of sludge and fly ash should not exceed 25% by weight. Beyond that, bonding of the mixture is poor and extrusion of the bricks results in "dog-earing" of the products. A set of three bricks were made for each mix combination. Mix A1 represents the control bricks made with 100% normal clay.

Table 4 Mix proportions of clay, fly ash and water sludge (% by weight)

MIX	CLAY	FLY ASH	SLUDGE
A1	100	0	0
A2	80	15	5
A3	80	10	10
A4	80	5	15
A5	75	10	15
A6	75	5	20

After mixing, the mixture was ground and crushed into fine pieces by a crushing machine to particle sizes of less than 20 micron. The crushed mixture was fed into an extruder where it was compacted at a pressure of 0.7 bar and extruded to form brick samples of dimensions 70 x 40 x 30 mm. The bricks were dried and fired in a kiln at one of 3 temperatures, viz., 1050°C, 1100°C and 1150°C.

It can be seen from Figure 2 that the loss-on-ignition values of the bricks with replacement of fly ash and sludge were much higher than that of the control bricks (A1). This was mainly attributed to the fact that sludge and fly ash, being themselves high in organic contents (38.2% and 35.7% respectively), would result in high loss-on-ignition in the bricks after firing at high temperatures. It should be noted that due to experimental errors, the result obtained for sample A6 fired at 1050°C was ignored. The water absorption of the bricks decreases as the firing temperature increases as shown in Figure 3. This could be due to the fact that more of the compounds of various metals are fused together upon firing leaving less void spaces in the bricks. The water absorption increases with the proportion of fly ash replacement, and hence, durability of the bricks would be affected as increased void spaces reduce brick strength.

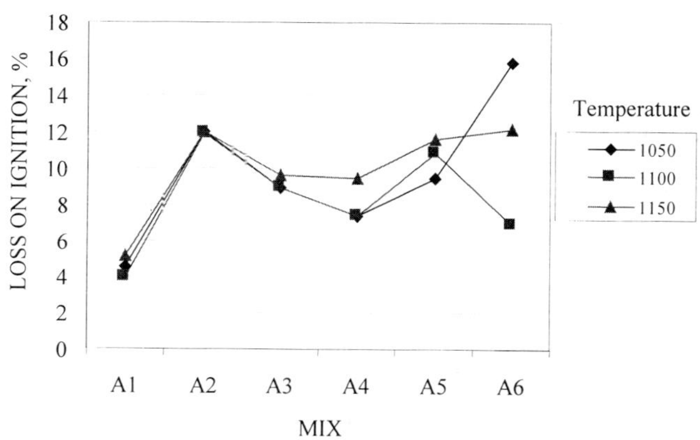

Figure 2 Loss-on-ignition of bricks

Compressive strength results of the bricks, as presented in Figure 4, show that the optimum firing temperatures in terms of strength are 1100°C for mixes A2 and A3, and 1150°C for mixes A4, A5 and A6. The highest compressive strength achieved among the bricks is 51.5 N/mm^2 for mix A6 fired at 1150°C. The same mix (A6) also produce the bricks with the next highest strength of 43.7 N/mm^2 fired at 1100°C. For those bricks fired at a lower temperature of 1050°C, their compressive strengths are relatively lower, with the highest strength of 35.5 N/mm^2 being achieved by mix A4. It should be noted that the strength generally increases as the amount of fly ash decreases. Significant drops in the strength were observed for mix A5 containing 10% fly ash, as compared to mixes A4 and A6 containing lower amount of fly ash (5%).

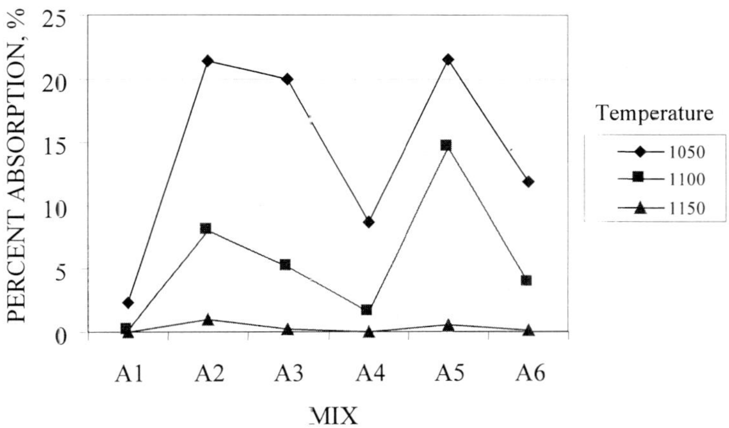

Figure 3 Water absorption of bricks

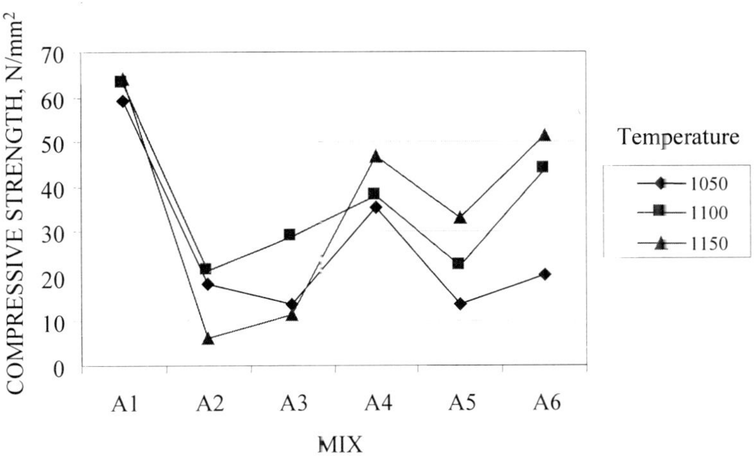

Figure 4 Compressive strength of bricks

Based on the local specifications for load-bearing bricks, one of the criteria of acceptance is that the bricks should have an average compressive strength of 20 N/mm^2. From the compressive strength results, bricks of mixes A1 and A4 fired at all three temperatures, those of mixes A5 and A6 fired at 1100°C and 1150°C, and bricks of A2 and A3 fired at 1100°C satisfy the strength requirement.

INCINERATOR FLY ASH AS AN ADMIXTURE IN SOFT SOIL STABILIZATION

In this section, the results of a laboratory study to assess the feasibility of utilizing MSW fly ash to improve the engineering properties of soft soils are presented. A common practice is to use cement in the stabilization of soft clays. The tests in NTU were carried out to investigate whether the addition of the fly ash would further improve the properties of cement stabilized soft clay. The soft clay was stabilized with different amounts by weight of cement and fly ash: 20% cement, 20% cement + 20% fly ash, 30% cement, 30% cement + 20% fly ash. The laboratory soil stabilization technique adopted essentially followed the procedures of Broms [7].

The clay samples were obtained from the bottom of an open excavation using a backhoe and transported in plastic bags to the laboratory for testing. After mixing of cement grout with/without the MSW fly ash, the stabilized soil was compacted into two types of cylinders which had been cut from standard polyvinyl chloride (PVC) pipes: 30 mm diameter by 72 mm length, and 83 mm diameter by 30 mm height. The samples were then wrapped in a transparent plastic wrapper, and then by aluminium foil. Further change in water content of the samples was prevented by coating with a layer of wax. The samples were then left to cure in a moist room at room temperature. The stabilized samples were divided into 6 batches and cured for a duration of 1, 7, 30, 60, 90 or 120 days before testing, so that the effects of curing time on the various engineering properties could be studied.

The ratio of gain in strength by the stabilized samples over the untreated remoulded clay over time for all four mixtures is shown in Figure 5. There was a substantial increase in strength of the stabilized samples when compared to that of the untreated remoulded clay, which had an undrained shear strength of 7.23 kPa. At the end of 120 days, all the stabilized samples increased in strength by more than 75 times over the untreated clay. It was also apparent that the cement-only mixes constantly showed slightly higher strengths than their counterparts with partial replacements of the MSW fly ash throughout the test period. It was observed that the 20% cement-only mix yielded a strength of 615 kPa, about 8% higher than the mix with partial replacement. Similarly, the 30% cement-only mix also showed a strength that was about 6% higher than the 30%-20% partially replaced mix. This is a positive phenomenon in that although 20% of the cement had been replaced by the ash, it had resulted in a less than 20% reduction in strength in both cases.

It appeared that the presence of MSW fly ash played a more significant role in the initial curing stage, notably for the first 30 days, for both proportions of cement mixes. In both cases, the rate of increase of shear strength was lower when MSW fly ash was present. After 30 days, the gradients seemed to be more or less the same. The only exception seemed to be the 30%-20% mix, which will be further discussed in the following section. Table 5 summaries the strength obtained for all four mixes over 120 days.

It should be noted that due to experimental errors, the result obtained for the 20%-20% mix after 120 days of curing was ignored. An additional test was conducted at the end of 150 days and its result was adopted instead. The value for this mix at 120 days was obtained by interpolation from the graph shown in Figure 5. It is also worth noting, at this point, that further increase in strength would be expected as curing time increases, as shown in the case of the 20%-20% mix. The extent of the effect could only be confirmed upon further curing and testing, which was beyond the scope of this project.

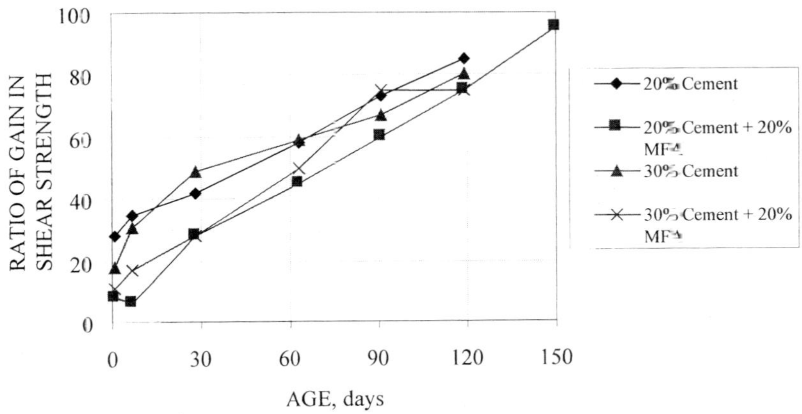

Figure 5 Ratio of increment in strength of stabilized samples over strength of untreated remoulded clay with time

Table 5 Undrained shear strength (kPa) from the unconfined compression test

DAY	20% CEMENT without MSW fly ash	20% CEMENT with MSW fly ash	30% CEMENT without MSW fly ash	30% CEMENT with MSW fly ash	UNTREATED MARINE CLAY
1	43.3	52.3	131.1	77.2	7.23
7	250.0	48.0	221.1	122.3	7.23
30	305.3	202.7	248.0	403.9	7.23
60	289.3	160.5	161.9	360.1	7.23
90	581.2	438.8	490.4	538.2	7.23
120	614.8	(563.0)	581.9	545.2	7.23
150	-	688.7	-	-	7.23

There was an obvious gain in strength over the untreated clay sample by about six to seven times for both mixtures at the start of the curing period. However, at the end of 7 days, the strength of the pure cement sample jumped to 250 kPa, about 35 times of the untreated sample, whereas the sample with MSW fly ash replacement showed little gain. This could be due to the initial slow setting effect due to the presence of MSW fly ash. After curing for 7 days, the mix without MSW fly ash increased at a more gradual rate. The mix with MSW fly ash increased at a faster rate, probably due to an increase in pozzolanic activities. After 30 days, both mixes seemed to gain in strength at a fairly constant rate, exhibiting an approximately parallel relationship. The final strength at the end of 120 days for the cement-only mix remained higher than the mix with MSW fly ash, at 615 kPa and 563 kPa respectively. The failure of closing of the gap in strength probably signifies that a higher percentage of MSW fly ash replacements may be required to induce more pozzolanic reaction.

It can be seen that the 30%-cement mixtures also presented a similar trend as in the 20%-cement mixtures, with the cement-only mix gaining relatively more than the cement-MSW fly ash mix over the testing period. The only difference was that the slow setting effect due to the MSW fly ash was not as significant as in the 20%-20% case. Also, at the end of 120 days, the final gain in strength for the 30%-cement mix was not higher than the 20%-cement mix, as would have been expected with an increase in cement content, these being 81.7 times and 85.7 times respectively. Further investigations should be carried out in the future to verify the results.

A general observation from these comparisons was that the presence of MSW fly ash had clearly served its purpose, as the samples containing it exhibited a trend that featured gains in strength. The only setback, however, was the increase in the setting time. This implies that the time required for a sample partially replaced with MSW fly ash to achieve the same strength as the cement-only mix would be extended. The extent of the effect could only be confirmed upon further curing and testing, which unfortunately, was beyond the scope of this project.

Figure 6 shows the permeability results of all four mixtures obtained indirectly from the Odeometer tests over time. The actual readings derived from the consolidation tests were rather erratic, probably due to the inconsistent initial states of the samples, that arose during sample preparation. At the end of 120 days, the 20%, 20%-20%, 30% and 30%-20% mixes achieved a permeability of 1.34×10^{-9} m/s, 1.91×10^{-9} m/s, 1.79×10^{-9} m/s and 0.9×10^{-9} m/s respectively. Comparing the permeability for the stabilized samples with that of the untreated clay, which registered 1.97×10^{-10} m/s, it is clear that the admixtures had indeed played a part in increasing the drainage property of the marine clay by one order of magnitude.

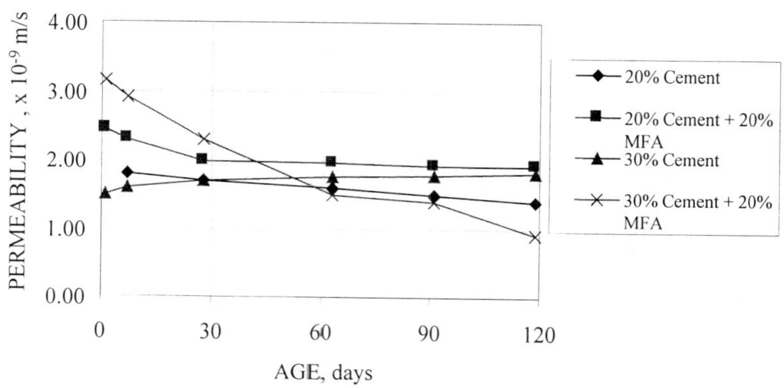

Figure 6 Variations of permeability with time

The cement-only mixes seemed to show little change in permeability over time, while the mixes containing MSW fly ash admixtures showed a falling trend as curing time increased. This was because the addition of the relatively free-draining material (i.e. the MSW fly ash) had created voids right from the first day, thus increasing the drainage paths. Over time, as pozzolanic reactions sped up and the voids were filled up, permeability of the samples fell. When reactions were complete, there would not be further falls in permeability. This probably explains why the cement-only mixes showed relatively constant trends over the 120 days.

It should be noted that permeability of a soil could be directly correlated to compressibility, as an increase in permeability implies a rise in the rate of consolidation for a soil. Hence, since the stabilized samples were more permeable, they were able to achieve a higher rate of consolidation over a much shorter period of time as compared to the untreated marine clay. This is advantageous in the stabilization of soft marine clay as it implies that a large part of foundation settlement would have occurred during construction.

From the experiments conducted, it was sufficient to conclude that MSW fly ash is indeed suitable for use as a partial replacement of cement in soil stabilization. This is largely due to the chemical composition and physical characteristics of the ash, which favour pozzolanic reactions. Where this project is concerned, the chemical compound that had contributed most to such chemical reactions is calcium, which is present in large proportions in the MSW fly ash used.

While the addition of MSW fly ash as a partial replacement of the cement was found to depict the same trends as that obtained by the mixes without MSW fly ash, it remains rather inconclusive as to what is the magnitude of the role that MSW fly ash plays. For contrary to what was reported by Goh and Tay [8], the results obtained from this project seem to indicate that MSW fly ash slows down the age-hardening process of the soil-cement mixtures, instead of speeding it up, after some time. In fact, "parallel relationships" were observed on several occasions between mixes of the same cement content (with and without MSW fly ash), showing that perhaps the percentage of partial replacements of MSW fly ash needs to be increased in order to improve the pozzolanic activities. It may also be possible that there exists an optimum admixture mix ratio that will improve the soil conditions without incurring too much costs in financial terms. Hence, further investigations with different amounts of partial replacement should be carried out to determine the final outcome.

In conclusion, it can be deduced that MSW fly ash, as and admixture in soft clay stabilization, not only resolves the problem of its disposal, but also improves soil properties. Such an application would inevitably reduce cost by cutting down the use of a more costly and scarce admixture, cement. However, one should be aware that due to its metallic chemical composition, leaching may cause environmental problems if not handled appropriately. Further studies should be directed toward the environmental aspects of the potential migration of the contaminants, and should address the implementation of suitable material handling and construction procedures.

REFERENCES

1. CHARACTERIZATION OF MUNICIPAL SOLID WASTE IN THE UNITED STATES 1960-2000, Franklin Associates Ltd., Prairie Village, Kansas,1988.

2. TYRRELL, M. E. AND FIELD, I. L. Structural products made from high-silica fractions of municipal incinerator residues, Proc. 2nd Waste Utilization Symp., Chicago, Ill., 1970, pp. 356-360.

3. WALTER, C. E. Practical refuse recycling, J. Envir. Engrg. Div., ASCE 102(1):139-148, 1976.

4. TAY, J. H. TAM, C. T. AND CHIN, K. K. Utilization of incinerator residue in concrete, J. Conservation and Recycling 5(2/3):107-112, 1982.

5. TAY, J. H. Energy generation and resources recovery from refuse incineration, J. Energy Div., ASCE114(3): 107-117, 1988.

6. SCOENBERGER, R. J. AND FUNGAROLI, A. A. Incinerator residue-fill site investigation, J. Soil Mech. and Found. Div., ASCE 97(10):1431-1443, 1971.

7. BROMS, B. B. Stabilization of soft clay in South-east Asia, Proc. Int. Geotech. Seminar - Case Histories in Soft Clay, Nanyang Tech. Inst., Singapore: 163-198, 1987.

8. GOH, A. T. C. AND TAY, J. H. Municipal solid waste incinerator flyash for geotechnical applications, J. of Geotech. Eng., ASCE 119(5): 811-825, 1993.

SOLIDIFICATION/STABILIZATION OF TEXTILE SLUDGE WITH PORTLAND CEMENT AND BOTTOM ASH

M Cheriaf

J C Rocha

Federal University of Santa Catarina

Brazil

J Péra

National Institute of Applied Sciences

France

ABSTRACT. The use of wastes in the development of building materials can represent a good opportunity in countries like Brazil where this industrial sector employs 1/3 workers and the deficit in dwellings is important (estimated at 12 million). The state of Santa Catarina is the third textile zone in Brazil and more than 35 000 tonnes per month textile sludge are generated and disposed in open landfills. Is was therefore interesting to investigate the feasibility of solidifying these wastes with portland cement and bottom ash from coal power plants.

The different materials were analyzed for their physical and chemical properties. Then, the sludges were solidified using cement and bottom ash in different proportions. The strength and the leaching properties of these products were investigated. A mixture containing 30 % sludges and 70 % cement presented a 28-day strength higher than 30 Mpa; when cement was partially replaced by bottom ash (20 %), the strength reached 20 Mpa. These results prove the possibility of developing blocks for low-cost housing.

Keywords: Textile sludge, Cement, Bottom ash, Building materials, Compressive strength, Leaching, Heavy metals, Organic, Stabilization.

Malik Cheriaf, is a Visitor Researcher in the Department of Civil Engineering, Federal University of Santa Catarina, Florianopolis, Brasil. He specialises in the valorisation of wastes in building materials.

Janaide Cavalcante Rocha, is Assistant Professor in the Department of Civil Engineering, Federal University of Santa Catarina, Florianopolis, Brasil. Her main research interests include the characterisation and the use of wastes in construction as well as leaching properties of such materials.

Professor Jean Péra, is Director of the Inorganic Materials Unit, National Institute of Applied Sciences, Lyon, France. He specialises in the development of new binders for the concrete industry. His interests also include the use of different wastes in cement and concrete. He has published widely and is member of different scientific societies such as ACI, RILEM, Materials Research Society and American Ceramic Society. He serves as expert for the Fifth (EC) Framework Programme - "Promoting competitive and sustainable growth".

INTRODUCTION

The textile industry actually represents a range of industries with operations and processes as diverse as its products. After fabrics are manufactured, they are subjected to several wet processes collectively known as "finishing", and it is in these finishing operations that the major waste effluents are produced (1). Sagar (2), and Kremer et al (3) reviewed the pollutants generated by the textile industry. They divided the pollutants into four general groups, i.e., sizes, detergents, dyes and priority pollutants.

They reported that most of the priority pollutants contained in textile industry effluents are aromatics, halogenated hydrocarbons, and heavy metals. Textile wastes are generally colored, highly alkaline, and high in BOD (Biological Oxygen Demand), suspended solids, and temperature (4). Randall and King (5) reported that wastewaters from textile dyeing and finishing operations may be characterized as high in organic matter, both biodegradable and nonbiodegradable, total solids and color, and very variable in pH. In addition they trend to be high in surfactants and contain potentially significant concentrations of oils, phenols, and heavy metals such as chromium, zinc, and copper.

Generally, releases of textile mill effluents are ultimately to some form of water (rivers, lakes, estuaries, marine environments), whether they are discharged directly or through a wastewater treatment system. Due to their composition, they represent some risk for aquatic systems. This is the problem encountered in the Santa Catarina State in Brazil, the third textile area of the country. Textile mill residues are either landfilled or discharged in lakes and sea.

To solve the problems of treatment of wastewaters from a textile plant, several alternatives should be included (6). The alternatives are the following: in-plant control for waste reduction, treatment for reuse, direct discharge to municipal waste treatment systems. The way adopted in the present study is different: textile sludge is handled using a solidification-stabilization (S/S) process based on cementitious materials. The binder is composed of either ordinary portland cement or a mixture of portland cement and bottom ash. Bottom ash is a coarse-to-fine grain sand – like material collected from the bottom of coal-fired boilers. Like fly ash, it also has beneficial applications (7-8), and specially exhibits some pozzolanic properties (9). It is particularly abundant in the south of Brazil (Santa Catarina, Parana, Rio Grande do Sul), where 3 million tonnes ashes are annually produced: 65 to 85 % fly ash and 15 to 35 % bottom ash. Bottom ash has not yet found any valuable use.

EXPERIMENTAL

Materials

Textile sludge contained 16.7 % solid matter. This solid matter presented a specific gravity of 1.51. It was mainly composed of volatiles and its chemical composition is shown in Table 1. The carbon content was very high (17.9 %) and led to an important loss on ignition (76.4 %).

Table 1 Chemical composition of the different materials (w_t%)

OXIDES	TEXTILE SLUDGE (DRY MATTER)	BOTTOM ASH	ORDINARY PORTLAND CEMENT CP-I-S32
SiO_2	2.9	56.0	19.3
Al_2O_3	10.3	26.7	4.6
Fe_2O_3	0.1	5.8	3.3
MgO	0.8	0.6	5.4
CaO	1.7	0.8	61.4
Na_2O	1.6	0.2	-
K_2O	0.4	2.6	0.6
TiO_2	0.4	1.3	-
P_2O_5	4.8	0.2	-
LOI	76.4	4.6	2.6
Organic carbon	17.9	3.8	-
CO_2 total	107.4	14.0	-
S total	1.5	0.1	2.5
Cl	0.7	< 0.1	-

This dry matter also contained large amounts of Al_2O_3 (10.3 %), P_2O_5 (4.8 %) and sulfur (1.5 %). As shown in Figure 1, not any crystallized phase was identified in the raw dried sludge. The particle size distribution of the dried textile sludge ranged from 1 to 200 μm with 50 % grains smaller than 20 μm.

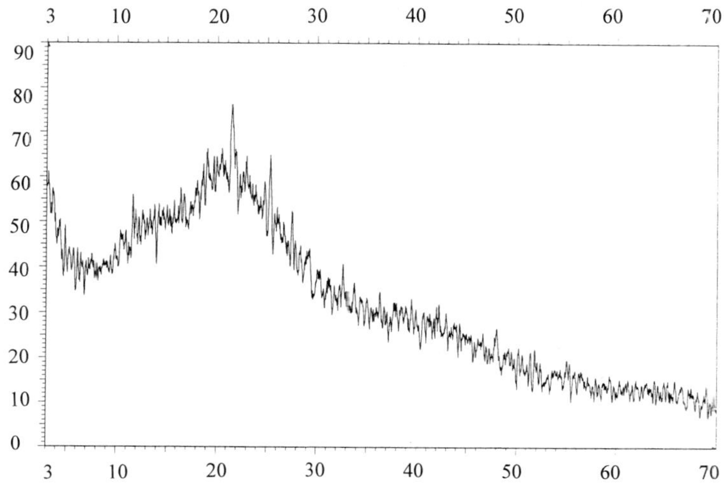

Figure 1 X-Ray Diffractogram of Dried Textile Sludge

In order to destroy all the organics present in the sludge, this latter was heated at 1000°C. Some crystalline phases appeared as illustrated in Figure 2 :

- quartz (SiO_2),
- rutile (TiO_2)
- sillimanite (Al_2SiO_5)
- aluminium phosphate ($AlPO_4$)
- corundum (Al_2O_3)
- akermanite ($Ca_2MgSi_2O_7$)
- calcium phosphate ($Ca_{10}(PO_4)_6CO_3$)

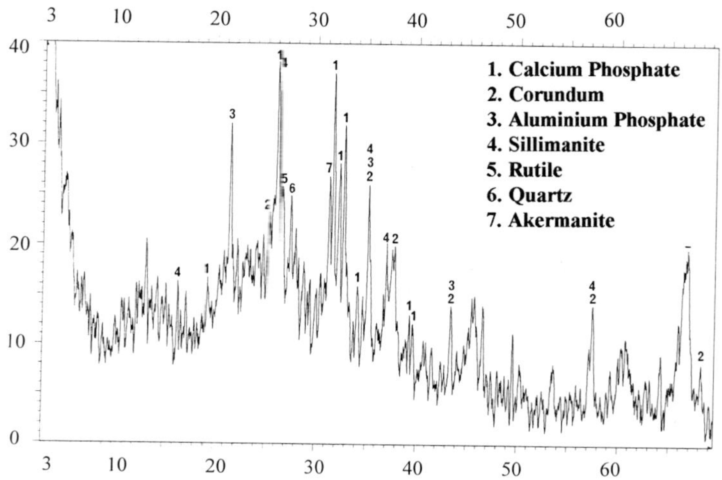

Figure 2 X-ray Diffractogram of Textile Sludge Heated at 1000°C

The textile sludge was also investigated for its minor components by emission-spectroscopy and the results are presented in Table 2. It appears that the heavy metal contents were very low (< 0.1 %).

Table 2 Minor Components of Dried Textile Sludge (w_t %)

Element	Cu	Zn	Mn	Ni	Sr	Cr	Cd	Pb	Ba	V	Co
Content	0.02	0.06	0.01	< 0.01	0.01	< 0.01	0.09	< 0.01	0.01	< 0.01	< 0.01

The chemical composition of bottom ash is also given in Table 1. The calcium content was very low (< 1 %) and the sum ($SiO_2 + Al_2O_3 + Fe_2O_3$) reached 88.5 %, which means that this ash belonged to ASTM Type F ash. The loss on ignition (4.6 %) was mainly due to the presence of free carbon (3.8 %). The X-ray diffractogram of bottom ash (Figure 3) pointed out the presence of a glassy phase (diffused halo maxima at 20-27° 2θ) with two major crystalline phases: quartz and mullite.

The specific gravity of bottom ash was 2.0.. The average diameter of the particule size distribution was 35 μm, with 2 % grains smaller than 1 μm.

The portland cement used in this study was a plain portland cement type CP-I-S32 according to the Brazilian standard (EB.1, 1991).

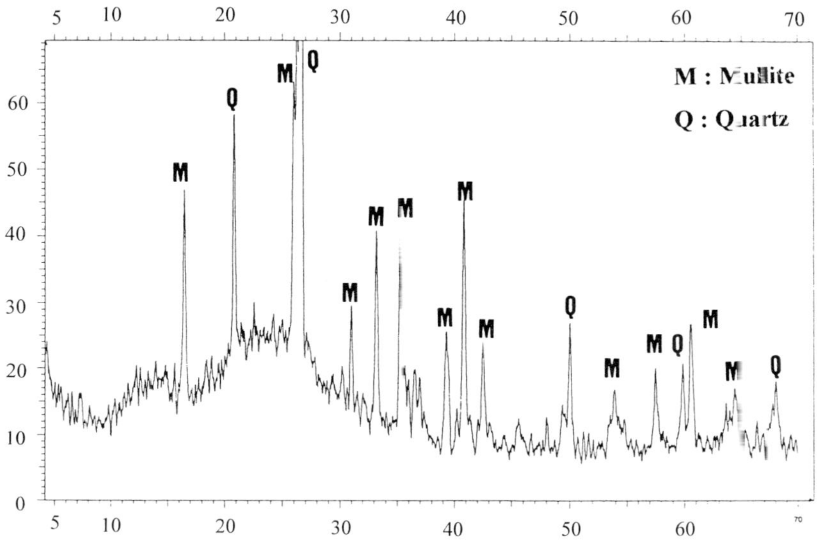

Figure 3 X–Ray Diffractogram of Bottom Ash

Preparation of Samples

Solidification/stabilization (s/s) processes are effective in treating a variety of difficult-to-manage waste materials for reuse or disposal. Some common S/S applications are incinerator ash, wastewater treatment sludge, and low-level waste from nuclear power plants (10-11). Generally, S/S processes are designed and used to accomplish one or more of the following objectives:

- reduce contaminant/polluant mobility or solubility,
- improve the handling and physical characteristics of the waste by producing a solid with no free liquid,
- decrease the exposed surface area across which transfer or loss of contaminants may occur.

The two principal types of inorganic binders used in S/S processes are cement and pozzolanic binders (10-11). They are routinely used to solidify water-based waste liquids, sludges, and filter cakes. This was the way investigated in the present study: the textile sludge was mixed with either plain ordinary portland cement (type CP-I-S32) or a mixture of plain cement and bottom ash. The mixture proportions are given in Table 3. Bottom ash was used because of its pozzolanic properties (9, 12) and its potential absorption of heavy metals like cadmium and zinc (13). The different mixtures were prepared at the same consistency and cast in cylindrical molds (∅ = 50 mm, h = 100 mm). Four moulds were prepared for each mixture. They were demolded after 24 hours and air-cured for 24 hours, before being immersed in water till the day before mechanical testing. Then, they were dried at 60 °C and tested for their strength. The pH of curing water was recorded and this water was analyzed after the samples were immersed for 7 days.

Table 3 Composition of the different mixtures (w_t %)

REFERENCE	CEMENT (%)	BOTTOM ASH (%)	TEXTILE SLUDGE (%)	W/S (*)
1	90	-	10	0.30
2	80	-	20	0.38
3	70	-	30	0.40
4	60	-	40	0.45
5	60	30	10	0.24
6	50	30	20	0.34
7	40	30	30	0.43
8	30	30	40	0.52
9	50	40	10	0.23
10	40	40	20	0.33
11	30	40	30	0.43
12	20	40	40	0.50

(*) = cement + bottom ash + textile sludge.

Testing

The solidified samples were tested for their compressive strength at different ages: 7, 14, 28, and 90 days. Those presenting the best strength were subjected to leaching and solubilization tests according to Brazilian standards NBR 10 004-10 007. The recorded values were compared to the limits given by these standards.

For leaching tests, the samples were crushed to get particles smaller than 9.5 mm, then placed in de-ionized water with a water to solid ratio of 16. The mixture was stirred for 24 hours and the pH was maintained at 4, by adding acetic acid. Then, the suspension was filtered and the collected leachate analyzed by atomic absorption spectroscopy and ionic chromatography. For solubilization tests, a 250 g sample was immersed in 2.5 L de-ionized water and stirred for 5 minutes. Then, the sample remained in water for 7 days. The water was filtered and analyzed.

RESULTS AND DISCUSSION

Compressive strength

The compressive strengths obtained for mixtures containing plain cement and textile sludge are shown in Figure 4; those of mixtures composed of cement, bottom ash, and textile sludge are reported in Figures 5 and 6.

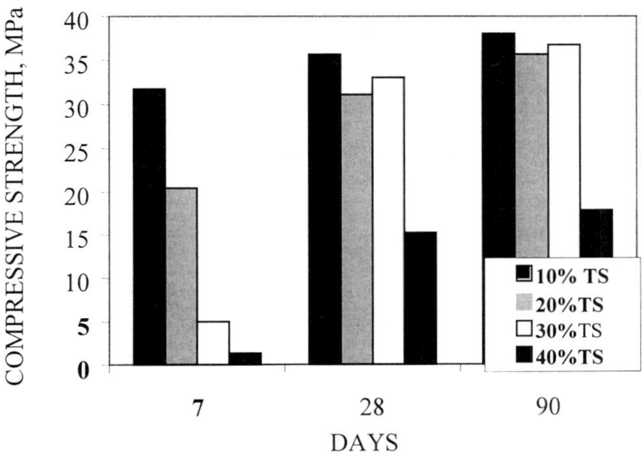

Figure 4 Compressive Strength of Mixtures Containing Plain Cement and Textile Sludge

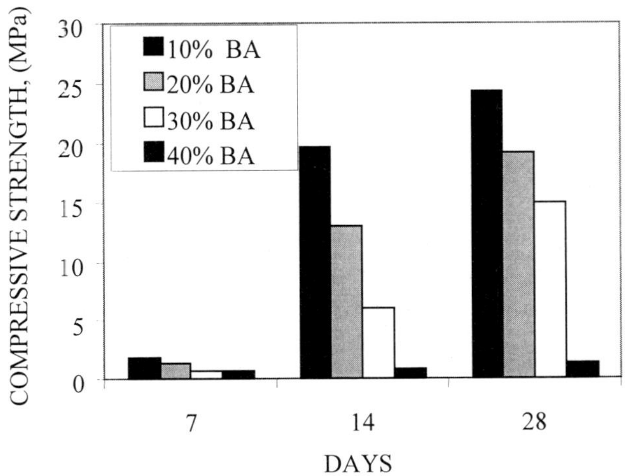

Figure 5 Compressive Strength of Mixtures Containing Cement, Bottom Ash and 30 % Textile Sludge

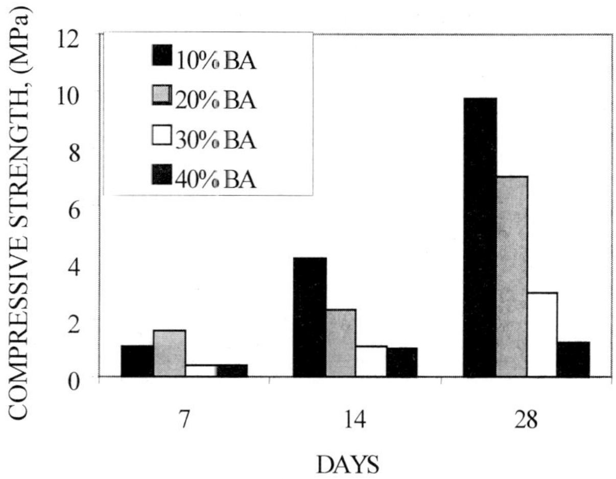

Figure 6 Compressive Strength of Mixtures Containing Cement, Bottom Ash and 40 % Textile Sludge

When plain cement was used alone to stabilize textile sludge (Figure 4), the following conclusions were drawn:

- the addition of textile sludge up to 30 % led to a compressive strength higher than 30 Mpa after 28 days of hydration,
- the presence of 40 % textile sludge drastically decreased the early-age strength but after 90 days of hydration, a strength of 17 Mpa was recorded.

When a pozzolanic binder (cement + bottom ash) was used (Figures 4 and 5), these conclusions became:

- whatever the composition of the binder might be, the presence of textile sludge decreased the early-age strength,
- when the amount of textile sludge was 30 %, the 28-day strength reached 15 Mpa in mixtures containing up to 30 % bottom ash,
- when the amount of textile sludge was 40 %, this strength dropped to 3 Mpa,
- in each case, the good level of cement replacement by bottom ash seemed to be 20 %; a 40 % bottom ash content is too high to allow a valorization of textile sludge,
- the presence of 40 % textile sludge led to very low strength unacceptable for building materials,
- the combination «30 % bottom ash + 40 % cement + 30 % textile sludge» seemed suitable for the manufacturing of concrete blocks for low-cost housing.

The very low strength reached after 7 days of hydration by the mixtures containing bottom ash can be explained by the leaching of calcium ions which occurred during the curing of samples. As shown in Figure 6, the pH reached 12 after 10 days of hydration and then dropped to 10.2, certainly due to the carbonation of calcium hydroxide. These calcium ions present in the curing water were no more available for the pozzolanic reaction and the bottom ash remained unreacted and acted as a filler.

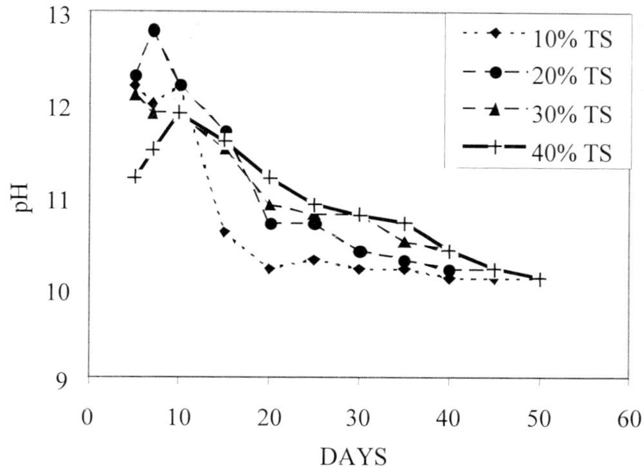

Figure 7 Evolution of the pH of Curing Water for Mixtures Containing Bottom Ash

Leaching and Solubilization

The results of leaching tests carried out on raw materials and samples containing 30 % textile sludge and stabilized either by plain cement or a mixture of cement and bottom ash are given in Table 4.

Table 4 Results of Leaching Tests (mg/l)

ELEMENT	RAW MATERIALS		STABILIZED MATERIALS (30 % TEXTILE SLUDGE)		
	Textile Sludge	Bottom Ash	Plain Cement	10 % Bottom Ash	30 % Bottom Ash
Sodium	94.9	-	12.5	10.2	12.2
Ammonium	134.8	-	3.4	2.2	2.0
Magnesium	16.1	-	0.2	0.2	0.3
Calcium	101.2	-	2331.2	2437.6	3233.9
Chloride	53.5	0.9	< 0.01	< 0.01	< 0.01
Phosphate	0.5	-	< 0.01	< 0.01	< 0.01
Sulfate	5.1	-	13.5	8.6	5.6
Copper	< 0.01	-	< 0.01	< 0.01	< 0.01
Chromium	< 0.01	0.2	< 0.01	< 0.01	< 0.01
Cadmium	< 0.01	-	< 0.01	< 0.01	< 0.01
Iron	0.15	14.9	0.2	0.5	0.3
Aluminium	1.0	-	0.3	0.1	0.5

The results point out a good retention of ammonium, sodium, magnesium, chloride and sulfate ions. There is no problem with heavy metals but calcium ions are leached when the stabilized samples are crushed below 9.5 mm. It seems that bottom ash is not enough reactive to consume all the calcium ions produced by cement hydration.

The results of solubilization tests gave values below the Brazilian specifications for drinking water.

CONCLUSION

From this limited investigation, the following conclusions can be drawn:

1. textile sludge can be solidified and stabilized using either plain cement or a pozzolanic binder composed of cement and bottom ash,
2. when the amount of textile sludge reaches 30 %, the 28-day compressive strength is higher than 30 Mpa when plain cement is used and 15 Mpa when the binder contains 30 % bottom ash,
3. for lower levels of cement replacement by bottom ash (10 and 20 %), the strength is higher than 20 Mpa,
4. the level of stabilization is good and not any pollutant is released at higher values than those recommended by the Brazilian standards,
5. a mixture containing 30 % textile sludge, 30 % bottom ash, and 40 % plain cement can be used to manufacture blocks for low-cost housing,
6. this work represents a new way of valorization for two types of wastes particularly abundant in the Santa Catarina State of Brazil: textile sludge and bottom ash.

ACKNOWLEDGEMENTS

The authors gratefully acknowledge the financial support of the following Brazilian authorities: FINEP, BID and CNPQ.

REFERENCES

1. ABO-ELALA, S.I., EL-GOHARY, F.A., ALI, H.I., and WAHAAB, R. S.A. Treatability Studies of Textile Wastewater. Environmental Technology Letter, 1998, Vol. 9, pp. 101-108.

2. SAGAR, B.F. Textile Wastes. Textile Manufacturing/Knitting World, 1979, Vol. 79, N° 3, pp. 22-24.

3. KREMER, F., BRROMFIELD, B., and FRADLIN, L. Energy and Materials Recovery Options for the Textile Industry. In: Proceedings of the 37th Industrial Waste Conference. Purdue University, Lafayette (USA), 1982, Vol. 37, pp. 157-170.

4. JUNKINS, R. Case History: Pretreatment of Textile Wastewater. In: Proceedings of the 37th Industrial Waste Conference. Purdue University, Lafayette (USA), 1982, Vol. 37, pp. 139-148.

5. RANDALL, C.W. and KING, P.H. Full Scale Physical-Chemical-Biological Treatment of Textile Wastewaters. Progress in Water Technology, 1980, Vol. 12, N° 3, pp. 231-240.

6. BHATTACHARYA, S.K. Treatment of Textile Wastes. In: Handbook of Industrial Waste Treatment. Edited by L.K. Wang and M.H.S. Wang, Marcel Dekker, Inc., New York (USA), 1992, Vol. 1, pp. 307-322.

7. GHAFOORI, N., and CAI, Y. Laboratory-Made Roller Compacted Concretes Containing Dry Bottom Ash – Part I: Mechanical Properties. ACI Materials Journal, 1998, Vol. 95, N° 2, pp. 121-130.

8. GHAFOORI, N., and CAI, Y. Laboratory-Made Roller Compacted Concretes Containing Dry Bottom Ash – Part II: Long-Term Durability. ACI Materials Journal, 1998, Vol. 95, N° 2, pp. 121-130.

9. CHERIAF, M., CAVALCANTE ROCHA, J., and PERA, J. Pozzolanic Properties of Pulverized-Coal Combustion Bottom Ash. Accepted for Publication in Cement and Concrete Research.

10. CONNER, J.R.. Chemical Fixation and Solidification of Hazardous Wastes. Van Nostrand Reinhold, New York (USA), 1990, 691 p.

11. MEANS, J.L. et al. The Application of Solidification/Stabilization to Waste Materials, Lewis Publishers, Boca Raton (USA), 1995, 334 p.

12. CAVALCANTE ROCHA, J, CHERIAF, M., MAIA, M., and PIRES, M. Reaproveitamento das Cinzas Pesados de Jorge Lacerda em Materias de Construçâo Civil: Aspectos Tecnlogico e Ambientais. In: Proceedings of the 2nd Simposio Internacional da Qualidade Ambiental. Projeto de Aterro de Residuos e Areas Degradadas, PUCRS, 1997, pp. 96-105.

13. WENG, C.H., and HUANG, C.P.. Fly ash and Cement Fixation. Journal of Environmental Engineering, ASCE, 1994, Vol. 120, N° 6, pp. 1470-1487.

THE SCIENCE AND POLITICS OF USING STABILISED WASTE MATERIALS IN THE MARINE ENVIRONMENT

K Collins

A Jensen

P Smith

University of Southampton

United Kingdom

ABSTRACT. A wide variety of cement stabilised, finely divided waste materials have been assessed for use in the marine environment. This paper reviews the work that has been undertaken with coal ash, incinerator ash, phosphogypsum and contaminated harbour muds. Such work has largely been undertaken with a view to producing artificial reefs for fishery enhancement where the issue of leaching of toxic compounds is of paramount concern. A common assessment protocol seems to have emerged from this work; monitoring physical integrity of the material, chemical monitoring of the material (to determine leaching rates) and the associated biota (to determine if there is evidence for bioaccumulation above background levels). Monitoring of the colonising organisms (cover, mass, species composition) provides a complimentary indication of the suitability of the material.

The use of waste materials in marine construction is controversial since there is a concern that pollutants may leach into the environment. In addition, there is a strong lobby against the concept of placing any waste material in the sea. International Conventions relevant to the use of waste materials in marine construction are discussed. There is considerable debate between the lobbies advocating a complete ban on the use of any waste and those in favour of using scientific assessment on a case by case basis. One protocol, suggested by the European Artificial Reef Research Network, is described.

Keywords: Coal ash, Incinerator ash, Marine construction, Environmental evaluation, Assessment protocol

Dr Ken Collins, is a Senior Research Fellow in the School of Ocean and Earth Science, University of Southampton. His interests include the environmental evaluation of waste materials for artificial reef construction and marine ecology.

Dr Antony Jensen, is a Lecturer in the School of Ocean and Earth Science, University of Southampton and co-ordinator of the European Artificial Reef Research Network.

Dr Philip Smith, is a Research Fellow in the School of Ocean and Earth Science, University of Southampton studying the behaviour and ecology of reef animals.

INTRODUCTION

Artificial reefs around the world have been built from a considerable range of materials [1], [2]. A generally acceptable material is natural rock, replicating local naturally occurring underwater rocky reefs. However rock may not be available locally and there may be significant environmental impacts resulting from its quarrying on land and subsequent transport. Another widely used material is concrete, particularly in Japan, Italy and Spain. This has the advantage of being readily cast into desired shapes. Again there are environmental costs associated with the extraction of the aggregates used in concrete production. Land based extraction leads to destruction of terrestrial habitats. In the UK, the scarcity of suitable land sites is leading to increasing marine aggregate extraction, with attendant concerns as to the impact on seabed habitats and related fisheries. Other manufactured materials that have been used are steel and plastics, particularly in Japan where considerable government investment has been made in artificial reef programmes.

The majority of reefs around the world are built from waste materials because these are widely available at low cost. In the USA most reefs have been built from large solid waste items such as ships, vehicles, oil platforms, construction rubble, waste concrete products, and tyres. Another potential source of material - finely divided solids such as coal ash, incinerator ash, dredged muds and phosphogypsum, have been stabilised with cement to produce artificial reef building materials. The alternative is to dispose directly into the sea, smothering the sea bed, or into landfill sites. The increasing cost and availability of land disposal is leading to increased research interest in the evaluation of these materials which could be utilised (or recycled) for marine construction [3], [4].

There is a valid concern that placing waste materials in the sea may lead to release of potentially harmful substances into the sea with subsequent incorporation into marine food chains. Much of the scientific research into stabilised materials described below has directly addressed this issue. However there is considerable governmental and non-governmental organisation opposition to placing *any* waste material in the sea. No distinction appears to be made between types of waste material. This argument is based on a anti-dumping philosophy rather than a scientific appreciation of the nature of the material in question. A consequence of blocking any marine use of recyclable materials would be the accumulation of such materials on land where the environmental affects may be negative rather than positive. There are two polarised views of the use of waste materials in marine construction: RECYCLING vs DUMPING

This paper reviews some of the waste materials that have been or could be used for artificial reef construction along with evidence from associated scientific studies. An assessment protocol is suggested, which has arisen from studies considering the environmental acceptability of any potential artificial reef construction material.

STABILISED MATERIALS

Coal Ash

The stabilization of unconsolidated waste coal ash for artificial reef construction was pioneered by the Coal Waste Artificial Reef Program (CWARP) in the USA [5], [6]. Investigations into the use of coal or Pulverised Fuel Ash (PFA) for oyster settlement have been undertaken [7]. Large fishery reefs have been constructed from coal ash concrete in Japan [8], [9] and Taiwan [10]. Japanese workers [9] have developed a physically very strong coal ash concrete, which after 10 years of research and tests, has gained the approval of the Japanese government for public construction works including artificial reefs.

UK

Over half of UK electricity is generated by coal-fired power stations. About 63% of the ash produced is sold, mostly to the construction industry, leaving the rest for disposal to landfill. In the past, two power stations have disposed of ash by marine dumping. In response to a European Community directive to reduce sulphur dioxide emissions to 60% of the 1982 level by the year 2003,retrofitting of Flue Gas Desulphurisation (FGD) plants to major coal-fired power stations was proposed. The favoured option was the use of limestone slurry sprayed though the flue gases producing gypsum. With the adoption of this process, the volume of gypsum produced would far exceed the requirements of the UK construction industry, providing a further disposal problem. A study investigating the possibility of stabilizing PFA and FGD gypsum with cement to produce artificial reefs for fishery enhancement was initiated in 1988, the first of its type in the UK. Whilst other research (above) have provided information on the environmental impact of stabilized coal ash, this study used UK sourced coal ash and gypsum in different mixes.

An experimental structure was deployed in Poole Bay, off the central south coast of England, in June 1989, following a series of laboratory trials (physical, chemical and biological) Three different mixtures of PFA, FGD gypsum, cement and gravel were made up in 40 x 20 x 20cm blocks at a commercial block-making plant [11], [12], [13]. Heavy metals naturally present in coal are concentrated after combustion in the resultant ash. Fifty tonnes of blocks were formed into eight conical units each 1m high by 4m diameter. The eight units replicate three different PFA/gypsum mixtures and one concrete control. The purpose of stabilizing the ash as blocks was to:

- immobilize heavy metals (or other components)
- provide hard substrata for the attachment of organisms.

Chemical studies ([11], [12], [13], [14], [15] during the first four years of immersion have:

- examined the heavy metal concentrations in reef block sections to determine if there is evidence of leaching
- analysed the reef organisms to determine if there is evidence for bioaccumulation of these metals beyond background levels.

No evidence has been detected for excess bioaccumulation of heavy metals by artificial reef biota and their grazers/predators associated with the ash block reefs compared to those from natural and concrete control reefs. No significant difference was found between the development of the epibiota community on the ash blocks and concrete controls [16].

Italy

A collaborative programme, 'CENMARE project', between the Italian Electricity Board (ENEL) and the University of Genoa, [17] aimed to establish the environmental compatibility of coal ash blocks placed in the sea in three phases:

1987-88 Laboratory tests of a physical, mechanical and chemical nature to determine a suitable composition material.

1988-92 Seawater tank trials of small (20cm side) ash cubes alongside concrete controls to monitor the growth of epibiota and their chemical composition.

1992-95 deployment and monitoring of larger (20x20x40cm, 100x100x100cm) ash and concrete blocks in the sea off Loano, north west Italy.

No significant leaching of chemical elements of environmental concern was found. If there is no detectable loss and consequent bioaccumulation of chemicals then there may be effects caused by very low levels of compounds leaching from the material. An alternative approach, based on examining the development of marine growth, has been used to look for such potentially subtle effects based on the argument that if there are any minor effects then these may be revealed by the nature of the species settlement, growth and survival.

Settlement and development of the macrobenthic community was examined on small test blocks recovered at intervals up to 24 months. The biomass of different taxonomic groups was measured as non-decalcified, wet, dry and ash weights. Additionally species richness (number of species per unit area), density (number of individuals per unit area) and covering indices (surface occupied by each taxon) were recorded. SCUBA divers photographed and removed biota from surfaces of blocks deployed in the sea at six monthly intervals. No significant differences in organism settlement were found between the stabilised ash and concrete, though higher friability of the ash material appeared to influence the amount of fouling. Similar work comparing the colonisiation of concrete and coal ash blocks has also been undertaken in the Adriatic Sea [18].

One of the problems associated with cement stabilised coal ash used in the USA, Italy and UK experiments, is the friability of the surface. Whilst the compressive strength of the material is initially low it increases with immersion in seawater [19], [15]. Coal ash is naturally pozzolanic, bonding in seawater, making it a suitable candidate for a marine construction material. The Japanese coal ash formulation [9] has a compressive strength equal to that of normal concrete. An intentionally soft, high coal ash content, mixture has been used by the I.R.Pe.M.-C.N.R. laboratory, Ancona to cultivate piddocks (*Pholas dactylus*), a burrowing mollusc [20].

Israel

Historically Israel has dumped coal fly ash at deep sea disposal site (1500m) in the eastern Mediterranean, some 70 km off the Israeli shore [21], [22]. Sand is in short supply in Israel, following the construction of the Aswan high dam and the high rate of building. The possibility of using coal ash is being considered as a substitute for sand and other conglomerates in concrete. Its use for artificial reef construction is being evaluated [23] using similar approaches described above. Results after 33 months exposure of trial blocks have shown increasing compressive strength (20-35 Mpa), no significant uptake of heavy metals by the colonising organisms and no significant difference in the patterns of colonisation in comparison to concrete controls.

Oil ash

In the USA, cement stabilized oil ash has been investigated as a potential reef construction materials [24], [25], [26],[27], [28]. This is not an ideal material for making concrete [29] and requires the use of coal ash to help bond the mix.

Incinerator ash

Incinerator ash from high temperature furnaces is a more promising material having a range of angular grain sizes and high strength. The first author has attempted to create concrete using the product from a poorly performing local domestic waste incinerator running at about 700 °C which consisted of a fine soft ash and contained fragments of uncombusted paper. Results (unpublished) from this trial were poor in terms of both compressive stength and leaching tests. In the USA the ash from efficient incinerators has been stabilised succesfully with concrete and evaluated as potential reef construction materials [30], [31], [32]. An extensive study in Bermuda [33] examined the potential disposal options; landfill of unconsolidated ash vs use of concrete stabilised block in coastal construction. Whilst the ash from incineration of domestic waste contains high levels of heavy metals and organic compounds these trials have shown the stabilised ash concrete to be both physically and chemically stable. The fate of stabilised incinerator ash blocks, sinking in coastal silts and muds has been considered [34]. Exposure to anaerobic bacteria does not appear to significantly affect the release of heavy metals from the matrix.

There has been considerable opposition to the concept of using incinerator ash in the States of New York and Florida where work to evaluate the environmental acceptability has been carried out, effectively blocking further work in this field. Pressure groups influenced state regulatory agencies in Florida to prevent experimental field trials of stabilised incinerator ash blocks [32].

Furnace slag

Coastal and estuarine sea defences use large quantities of solid materials creating new marine habitats. A wide variety of materials have been used in the construction of Dutch dikes (seawalls) [35]. The extent of growth on different susbstrates is classified:

well covered - concrete, limestone, manganese and steel slags
moderately covered - gneiss, basalt, minestone, phosphorus slag
scarcely covered - asphalt, copper slag

The colonisation and effects on algal growth of a similar range of materials (basalt, granite, steel slags) has been studied on the shores of the Elbe estuary, Hamburg, Germany [36].

Phosphogypsum

Phosphogypsum is a solid by-product of the production of phosphoric acid for fertilizers. The waste phosphogypsum contains elevated levels of radionuclides such as radium-226 and heavy metals which may pose a potential hazard to human health and the environment [37]. Historically in Le Harvre, northern France, waste phosphogypsum slurry has been piped out to sea for disposal [38]. Experiments have been undertaken in Louisiana with cement consolidated phosphogypsum blocks. In aquarium chemical studies no effects solely attributable to the phosphogypsum on copepods, shrimps and fishes were found [39]. A related study with phosphogypsum blocks in estuarine ponds showed no significant effects on the meiofanuna, copepods and shellfish [40], [41].

Harbour dredged muds

In November 1993, mud test blocks composed of 40% muddy dredged harbour sediments along with concrete controls were deployed off the coast of Lavagna, Eastern Ligurian Riverera, Italy [42]. After 12 months of regular monitoring of the organism settlement using the same protocol as for the CENMARE coal ash block study described above [17], no significant difference between the macrobenthos communities which had settled on the two materials, either from a quantitative or qualitative point of view.

REGULATIONS/INTERNATIONAL TREATIES

International legal issues associated with artificial reefs, particularly those in Europe, have been comprehensively reviewed by Pickering [43]. The principal international legislation covering the deposition of waste and other matter in the oceans is the London Convention, 1992 (formerly the London Dumping Convention). Placement of material for the construction of artificial reefs in not covered by the Convention. However aware of the range of materials that have been used for such purposes, the London Convention Scientific Group has recommended that the guidance prepared for the interpretation of the Annexes to the Convention in relation to dumping at sea contains all the considerations that are needed for the assessment of placement of an artificial reef or structure.

The OSPAR (Oslo and Paris Commissions) Convention for the Protection of the Marine Environment of the North-East Atlantic,1992, includes placement of matter such as stabilised or recycled waste materials within its purview and has established a set of technical guidelines for artificial reef deployment [44]. As the oil and gas platforms in the North Sea and Mediterranean reach the end of their useful life there is debate as to whether to remove the larger structures entirely or to continue to use them as artificial habitats. The Oslo and Paris Commission=s working group on sea-based activities (SEBA) has focussed on this question with certain countries, especially Germany, demanding the removal of all oil platforms and opposing the use of any waste material in the marine environment. The Protection of the Marine Environment of the Baltic Sea Area (Helsinki Convention, 1992) and the Barcelona Convention (Protocol for the Prevention of Pollution of the Mediterranean by Dumping from Ships and Aircraft 1976, implementation 1978) have similar characteristics to the Oslo Convention, though the former prohibits any dumping at sea and neither have specific guidelines for the deployment of artificial reefs.

All of the above conventions apply the precautionary principle to varying extents, requiring contracting parties to minimise pollution from sea-based sources. Whilst the proponents of stabilised waste material for marine construction, including artificial reefs, consider that potentially harmful components are immobilised, opponents argue that immobilisation does not render these contaminants inert, since the materials will eventually degrade and disintegrate, releasing pollutants [45].

In the UK, the Ministry of Agriculture, Fisheries and Food (MAFF) is primarily responsible for licensing marine construction including artificial reefs and ensuring adherence to the guidelines laid down by the international treaties. In 1989, the Poole Bay experimental reef [11] was the first waste material construction to be licenced by MAFF.

ASSESSMENT PROTOCOL

The European Artificial Reef Research Network was formed in May 1995, funded (to May 1998) by the European Commission AIR (Agriculture and Fisheries) programme. The formal network brought together 51 scientists from 36 laboratories plus a further 60 associates throughout geographic Europe. It held a conference [46] and a series of workshops to review the state of artificial reef research and suggest future directions. One of the workshops, Artificial Reef Material and Design (Southampton, 8-10 Jan. 1998), recognised a common theme in waste material environmental evaluation and agreed the assessment protocol outlined below [47].

Generally, materials suitable for other marine structures should also be suitable for artificial reefs and not need special tests. The approach to assessing bulk and composite materials can be summarised as:

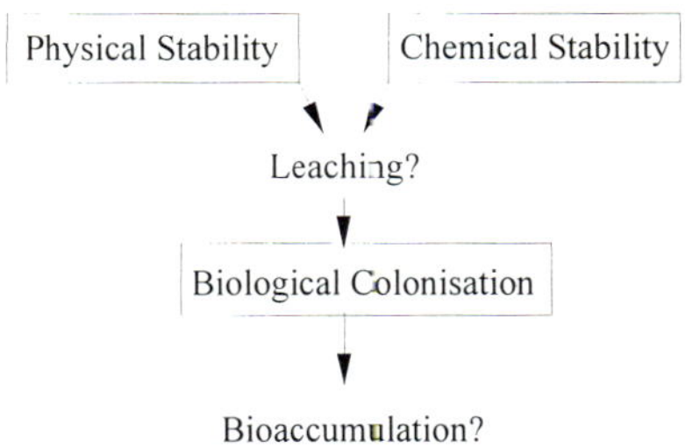

Physical stability

Any structure placed on the seabed should be stable, not be moved or dispersed by storm wave action and not settle into the seabed. Consideration should be given to the placement, maintenance and possible need to remove the structures.
The material should remain intact for a reasonable length of time. For example, car bodies are not suitable because they are rapidly corroded. Cement stabilised materials or any bulk material should be assessed for longevity in seawater using standard methods for marine structures (eg FIB/CEB Federation International Percontrand, European Concrete Association) to determine the performance of the material with extended exposure.

Chemical stability

Any material not normally used in construction in the marine environment should be assessed. Materials should be analysed by standard procedures to determine if they contain any elements or compounds which could affect the marine environment. If materials contain any harmful components do they leach out when the material is immersed in seawater? Samples of the material should be immersed for a period of time and the water analysed for chemical leachates and used for standard ecotoxicological bioassays. Possible leaching from the final composite is dependant on the environmental conditions (temperature, salinity, oxygen concentration) found at the intended destination which should be reflected in the test conditions.

Stabilised materials are designed to >lock up= these compounds and their performance depends critically on their physical stability. After extended exposure solid materials can be sectioned and a cross section chemically analysed to determine if there is any loss of components from the outer margin.

Biological colonisation

The most relevant measure of the suitability of a material is its colonisation (biomass, density and species composition) in comparison to control surfaces. This is potentially a more sensitive test than looking for chemical evidence of leaching.

If there is leaching then the organisms growing on the surface of the reef material may bioaccumulate the leached compounds. These may be in turn taken up by mobile reef associated organisms feeding on the sessile biota. Both should be analysed for the presence of likely leachates along with the same species from a control site.

Protocol

Preliminary phase (1 year)

Laboratory- Design, formulation and physical tests (eg. porosity, compressive strength, density) of the material
Chemical analysis of the material and determination of likely leachates.
Ecotoxicological tests with water exposed to the final composite.

Field- Test samples of the material to check colonisation.

Field study phase (2-5 years)

The preliminary phase laboratory studies should be continued. There is a need to deploy a small experimental structure in the field to provide long term information on the materials performance under realistic conditions. It should be sufficiently replicated to provide statistically valid sample sizes and areas for both biological and chemical studies.

Deployment of small experimental units in the sea alongside controls.
Routine monitoring of biological colonisation using standard methodologies.
Routine chemical analysis of biota for possible leachates.
Chemical analysis of the structure material to determine if there has been loss of key components.
Physical tests of the integrity of the material (e.g. cracking, abrasion, spalling).

Time scale

There is debate as to how long is sufficient to demonstrate the environmental compatibility of the material. Standard environmental assessments have 5 year monitoring phases with a review every 5 years dependant on results. Where possible accelerated test methods could be used. Long term effects should be modelled from the initial results (e.g. to determine if any changes with time are linear or logarithmic). If the material shows promising results after the field study phase and is approved for use in a large construction project then further monitoring (5 year intervals) would be expected.

CONCLUSION

Of the waste products described above stabilised coal ash has received the greatest amount of evaluation as a marine construction material. Studies in USA, Japan, Taiwan, Italy and UK over the past 20 years have demonstrated its suitability for construction of artificial reefs yet there has been no large scale use of this material in the sea in the western world. In contrast, whilst there have been few chemical studies of leaching from tyres, they are used worldwide, by the million, for artificial habitats [48].

There is a need for rigorous scientific evaluation of any material before it is approved for use in the marine environment. The multi-disciplinary approach that has emerged from the coal ash studies provides an excellent model for such work. Equally, attention needs to be paid to disseminating the results of this work to the public and legislators. The blocking of further stabilised incinerator ash studies after promising trials [32] by the pressure groups in Florida provides a salutary lesson to the scientific community. There is a fundamental suspicion that waste material artificial reef research is a step towards sanctioning the dumping of waste in the sea [45]. The alternative view is that use of waste materials in artificial reefs is a way of recycling resources. Waste materials are being produced in increasing quantities in spite of recycling and waste reduction measures. The productive use of materials which can be demonstrated to be environmentally compatible is worth consideration.

REFERENCES

1. SEAMAN, W.J. AND L.M. SPRAGUE, Artificial Habitats for Marine and Freshwater Fisheries. 1991, San Diego: Academic Press. 285.

2. JENSEN, A.C., K.J. COLLINS, AND A.P. LOCKWOOD, Artificial Reefs in European Seas.1999, Dordrecht, Netherlands: Kluwer, ISBN 0-7923-5845-7.

3. COLLINS, K.J., JENSEN, A.C., LOCKWOOD, A.P.M. AND LOCKWOOD, S.J., Coastal structures, waste materials and fishery enhancement. Bulletin of Marine Science, 1994. 55(2-3): p. 1253-1264.

4. COLLINS K.J AND JENSEN, A.C., Acceptable use of waste materials., in European Artificial Reef Research. Proceedings of the first EARRN conference, March 1996 Ancona, Italy., A.C. Jensen, Editor. 1997, Southampton Oceanography Centre. p. 377-390.

5. WOODHEAD, P.M.J. AND M.E. JACOBSON, Epifaunal settlement, the processes of community development and succession over two years on an artificial reef in the New York Bight. Bulletin of Marine Science, 1985. 37(1): p. 364-376.

6. WOODHEAD, P.M.J., PARKER, J.H. AND DUEDALL, I.W., The use of by-products from coal combustion for artificial reef construction., in Artificial Reefs - Marine and Freshwater Applications, F.M. D'Itri, Editor. 1986, Lewis Publications: Chelsea, Michigan, USA. p. 265-292.

7. PRICE, K.S., MUELLER, K., ROSENFELD, J. AND WARREN, T., Stabilized coal ash as a substratum for larval oyster settlement: a pilot field study., in Advances in Chesapeake Bay Research, Proceedings of a Conference. 29-31 March 1988, Baltimore, Maryland. 1988, Chesapeake Research Consortium. p. 128-136.

8. SUZUKI, T., A concept of large artificial ridges using a new hardened product made from coal ash., in Ocean Space Utilization '85, W. Kato, Editor. 1985, Springer Verleg: Tokyo. p. 611-618.

9. SUZUKI, T., Application of high-volume fly ash concrete to marine structures. Chemistry and Ecology, 1995. 10: p. 249-258.

10. KUO, S.-T., T-C HSU AND K-T SHAO, Experiences of Coal Ash artificial reefs in Taiwan. Chemistry and Ecology, 1995. 10: p. 233-247.

11. COLLINS, K.J., JENSEN, A.C. AND LOCKWOOD, A.P.M., Fishery enhancement reef building exercise. Chemistry and Ecology, 1990. 4: p. 179-187.

12. COLLINS, K.J., JENSEN, A.C. AND LOCKWOOD, A.P.M., Artificial Reefs: using coal fired power station wastes constructively for fishery enhancement. Oceanologica Acta, 1991. 11: p. 225-229.

13. COLLINS, K.J., A.C. JENSEN, AND A.P.M. LOCKWOOD, Stability of a coal waste artificial reef. Chemistry & Ecology, 1992. 6(1-4): p. 79-93.

14. COLLINS, K.J., et al., Evaluation of stabilized coal-fired power station waste for artificial reef construction. Bulletin of Marine Science, 1994. 55(2/3): p. 1251-1262.

15. COLLINS, K.J. AND A.C. JENSEN, Stabilized coal ash artificial reef studies. Chemistry and Ecology, 1995. 10(3/4): p. 193-203.

16. JENSEN, A.C., et al., Colonization and fishery potential of a coal-ash artificial reef, Poole Bay, United Kingdom. Bulletin of Marine Science, 1994. 55(2/3): p. 1263-1276.

17. RELINI, G., DINELLI, G. AND SAMPAOLO, A., Stabilised coal ash studies in Italy. Chemistry and Ecology, 1994. 10: p. 217-232.

18. BOMBACE, G., CASTRIOTA, L. AND SPAGNOLO, A., Benthic community on concrete and coal-ash blocks submerged in an artificial reef in the central Ardiatic Sea, in The responses of Marine Organisms to their Environments. Proceedings of the 30th European Marine Biology Symposium, 18-20 September 1995, Southampton, UK., L.E. Hawkins and S. Hutchinson, Editors. 1997. p. 281-290.

19. ROETHEL, F.J.A.O., S.A., Effects of seawater on the mineralogical and chemical composition of coal waste-blocks., in Wastes in the Oceans Vol. 4: Energy wastes in the Ocean, I.W. Duedall, Kester, D.R. and Park, P.K., Editor. 1985, John Wiley & Sons.: New York. p. 691-715.

20. BOMBACE, G., FABI, G. AND FIORENTINI, L., Observations on the settlement and growth of Pholas dactylus (Bivalvia, Pholadidae) on artificial substrata. Biol. Mar. Mediterr., 1995. 2(2): p. 143-150.

21. KRESS, N., Chemical aspects of coal fly ash disposal at sea: Predicting and monitoring environmental impact. Wat. Sci. Tech., 1993. 27: p. 449-455.

22. KRESS, N., AND HERUT, B. 1998., Disposal of coal fly ash at a deep water site in the Eastern Mediterranean off Israel - six years of monitoring. Chemistry and Ecology, 1998. 15: p. 185-198.

23. SPANIER, E., KRESS, N. AND TOM, M., The use of coal ash in marine concrete for artificial reefs in the southeastern Mediterranean., in Proceedings of the Seventh International Conference on Artificial Reefs (7CARAH), Sanremo, 7-11 Oct., G. Relini, Ferrara, G. and Massaro, E., Editor. 1999, Societa Italiana di Biologia Marina: Genoa, Italy. p. 338-346.

24. METZ, S.A.T., J.H., Trace metal considerations in experimental oil ash reefs. Marine Pollution Bulletin, 1988. 19: p. 633-636.

25. NELSON, W.G., NAVRATIL, P.M., SAVERCOOL, D.M. AND VOSE, F.E., Short-term effects of stabilized oil ash reefs on the marine benthos. Marine Pollution Bulletin, 1988. 19: p. 623-627.

26. FREASE, R.A. AND J.G. WINDSOR, JR., Behaviour of selected polycyclic aromatic hydrocarbons associated with stabilized oil and coal ash artificial reef. Mar. Pollut. Bull., 1991. 22(1): p. 15-19.

27. SHIEH, C.S. AND I.W. DUEDALL, Chemical behavior of stabilized oil ash artificial reef at sea. Bulletin of Marine Science, 1994. 55: p. 1295-1302.

28. VOSE, F.E. AND W.G. NELSON, An assessment of the use of stabilized coal and oil ash for construction of artificial fishing reefs: Comparison of fishes observed on small ash and concrete reefs. Marine Pollution Bulletin, 1998. 36(12): p. 980-988.

29. PAYA, J., BORRACHERO, M.V., MONZO, J. AND BONILLA, M., Properties of Portland cement mortars incorporating high amounts of oil-fuel ashes. Waste Management, 1999. 19(1): p. 1-7.

30. BRESLIN, V.T., ROETHEL,F.J. AND SCHAEPERKOETTER,V.P., Physical and chemical interactions of stabilized incineration residue with the marine environment. Marine Pollution Bulletin, 1988. 19(11B): p. 628-632.

31. ROETHEL, F.J. AND V.T. BRESLIN, Behaviour of dioxins, furans and metals associated with stabilised MSW combustor ash in seawater. Chemistry and Ecology, 1994. 10: p. 259-272.

32. SHEIH, C.-S. AND I.W. DUEDALL, Possible use of ash residues for the construction of artificial reefs at sea. Chemistry and Ecology, 1995. 10: p. 295-304.

33. KNAP, A.C., C.B., COOK, S.B., SIMMONS, J.A.K., JONES, R.J. AND MURRAY, A.E., Marine environmental studies to determine the impact of the mass burn incinerator proposed for Tynes Bay, Bermuda..Report to Ministry of Work and Engineering, Bermuda,, . 1991, Bermuda Biological Station for Research Inc.

34. KINNER, N.E., LYONS, W.B., RICE, P.H., DURLING, D.B. AND GRESS, D.L., Geochemistry and metals concentrations in marine sediments exposed to solidified/stabilized municipal solid incinerator ash. Waste Management, 1995. 15(3): p. 221-232.

35. LEEWIS, R., WAARRDENBURG, H.W. AND MEIJER, A.J.M., Active management of an artificial rocky cóast. Hydrobiol. Bull., 1989. 23: p. 91-99.

36. KARBE, L., Auswirkung von in der Elbe im Wasserbau eingestetzter Electroofenschlacke der Hamburger Stalwerke auf aquatische Lebensgemeinschaften., . 1995, Universität Hamburg, D-22765 Hamburg, Germany.

37. VAN DER HEIJDE, H.B., KLIJN, P-J., DUURSMA, K., EISMA, D., DE GROOT, A.J., HAGEL, P. KÖSTER,H.W. AND NOOYEN, J.L., Environmental aspects of phosphate fertiliser production in the Netherlands with particular reference to the disposal of phosphogypsum. The Science of the Total Environment, 1990. 90: p. 203-225.

38. S.P.I., Commission chargée de contrôller l'évoluotion de la pollution en estuaire et en Baie de Seine, . 1990.

39. NIELAND, D.L., WILSON, C.A., FLEEGER, J.W., SUN, B., MALONE, R.F. AND CHEN, S., Preliminary evaluation of the use of phosphogypsum for reef substrate. I. A laboratory study of bioaccumulation of radium and six heavy metals in an aquatic food chain. Chemistry and Ecology, 1998. 14(3-4): p. 305-319.

40. NIELAND, D.L., WILSON, C.A. AND FLEEGER, J.W., Preliminary Evaluation of the Use of Phosphogypsum for Reef Substrate in the Gulf of Mexico. Gulf of Mexico Science, 1998. 16(1): p. 54-63.

41. WILSON, C.A., NIELAND, D.L, FLEEGER, J.W., TORADO, A., MALONE, R.F. AND RUSCH, K.A., Preliminary evaluation of the use of phosphogypsum for reef substrate. II. A study of the effects of phosphogypsum exposure on diversity and biomass of aquatic organisms. Chemistry and Ecology, 1998. 14(3-4): p. 321-340.

42. RELINI, G. Cement stabilised harbour muds. in Third International Ocean Pollution Symposium, Harbor Branch Oceanographic Institution, Florida,7-11 April 1997. 1997.

43. PICKERING, H., Legal framework governing artificial reefs in the EU., in European Artificial Reef Research. Proceedings of the first EARRN conference, March 1996 Ancona, Italy., A.C. Jensen, Editor. 1997, Southampton Oceanography Centre. p. 195-232.

44. OSPAR, OSPAR Guidelines on Artificial Reefs in relation to Living Marine Resources, . 1999, OSPAR Convention for the Protection of the Marine Environment of the North-East Atlantic: Kingston Upon Hull.

45. MACDONALD, J.M., Artificial reef debate: habitat enhancement or waste disposal? Ocean Development and International Law, 1994. 25: p. 87-118.

46. JENSEN, A.C., European Artificial Reef Research, Proceedings of the first EARRN conference, March 1996, Ancona, Italy. 1997: Southampton Oceanography Centre. 449.

47. JENSEN, A.C., Report of the results of EARRN workshop 4: Reef design and materials., 1998, European Artificial Reef Research Network, AIR3-CT94-2144, Report to DGXIV of the European Commission, SUDO/TEC/98/10.

48. COLLINS, K.J., A.C. JENSEN, AND S. ALBERT, A review of waste tyre utilisation in the marine environment. Chem. Ecol., 1995. 10(3-4): p. 205-216.

PROPERTIES OF CALCINED PAPER SLUDGE

J Péra

J Ambroise

National Institute of Applied Sciences

France

ABSTRACT. Processing of waste paper generates fiber residues and sorting wastes (de-inking and sewage sludges). These sludges contain cellulose fibers and mineral fillers such as kaolinite, calcite, and talc.

By a controlled-calcination process in the temperature range of 650 to 800°C, it is possible to transform the sludge into either pozzolanic or hydraulic materials, usable in the cement industry. At 650°C, kaolinite is transformed into metakaolinite, while calcite and talc are preserved. At 800°C, calcite starts to decarbonate and yields quicklime.

This paper deals with the influence of the temperature of calcination on the engineering properties of calcined paper sludge: pozzolanic activity by means of the measurement of lime consumption, development of strength and pore size distribution.

Keywords: Paper sludge, Calcination, Calcium hydroxide, Quicklime, Metakaolin, Pozzolan, Temperature.

Professor Jean Péra, is Director of the Inorganic Materials Unit, National Institute of Applied Sciences, Lyon, France. He specialises in the development of new binders for the concrete industry. His interests also include the use of different wastes in cement and concrete. He has published widely and is member of different scientific societies such as ACI, RILEM, Materials Research Society and American Ceramic Society. He serves as expert for the Fifth (EC) Framework Programme - "Promoting competitive and sustainable growth".

Jean Ambroise, is Assistant Professor in the Inorganic Materials Unit, National Institute of Applied Sciences, Lyon, France. His main research interests include the design and the production of innovative building materials as well as the management of industrial wastes.

INTRODUCTION

The trend toward increased recovery and recycling of wastepaper has caused explosive growth in worldwide deinking capacity. Deinked secondary fiber yields generally are between 65 and 80 %. The fraction of the wastepaper that is not recovered becomes deinking sludge. The quantity of deinking sludge could reach 20 million tonnes by 2005 (1) Deinking sludge typically is dewatered in a screw press to about 50 % solids before it is discharged. On a dry basis, it usually countains approximately equal amounts of organic and inorganic components.

This sludge most often is disposed in landfills. However, in Europe, landfills are becoming increasingly undesirable and prohibitively expensive. Therefore, it is important to find alternate disposal means for deinking sludge. Many pulp mills that integrate a deinking plant as part of their operation are choosing incineration as a way to handle the sludge (2, 3). Through incineration, the fuel content of the sludge is recovered in a lime kiln or power boiler, and the volume of incinerator ash that must be landfilled or otherwise disposed is greatly reduced. Another way of handling deinking sludge consists in land-spreading for agricultural uses (4-7). Some sludge is also recovered for cement manufacturing (8, 9).

In this paper, data are presented that show either pozzolanic or hydraulic materials can be obtained by a controlled calcination of deinking sludge. The sludge contains significant fractions of clay and calcium carbonate. Chemical changes occur when clay and calcium carbonate are heated (10, 11). This happens due to the dehydroxylation of clay and decarbonatation of calcite. Once calcined under 750°C, clay is transformed into a reactive pozzolan (12-14). Over 800°C, calcite starts to decarbonate yielding calcium oxide, larnite (Ca_2SiO_4 or C_2S) and calcium aluminosilicates like gehlenite ($Ca_2Al_2SiO_7$ or C_2AS) and anorthite ($CaAl_2Si_2O_8$ or CAS_2). This material behaves as a hydraulic binder when mixed with water.

POZZOLANIC PROPERTIES OF CALCINED SLUDGE

Starting Material

The chemical analysis of raw sludge is shown in Table 1. The following contents were also measured:

$$CO_2 = 80.4\ \% \ ; \qquad C_{org} = 14.9\ \% \ ; \qquad S_{tot} = 0.3\ \%$$

Assuming that all the organic carbon was present in cellulose ($C_{12}H_{20}H_{10}$), the quantity of organics present in raw sludge was estimated at 33.4 %.

The mineralogy of the inorganic fraction of raw sludge was assessed by means of differential thermal analysis, Fourier transform infra-red spectrometry and X-ray diffraction. The X-ray diffractogram of raw sludge is given in Figure 1.

Table 1 Chemical analysis of raw sludge (% by mass)

ICP- EMISSION, %		ICP - EMISSION- MS, ppm	
SiO_2	16.14	As	5.38
Al_2O_3	9.04	Ba	117.00
Fe_2O_3	0.59	Cd	0.54
MnO	-	Cr	52.00
MgO	1.64	Cu	129.00
CaO	20.3	Ni	23.60
Na_2O	0.17	Pb	32.90
K_2O	0.36	Zn	300.00
TiO_2	0.23	Cl	650.00
P_2O_5	0.34		
LOI	51.13		

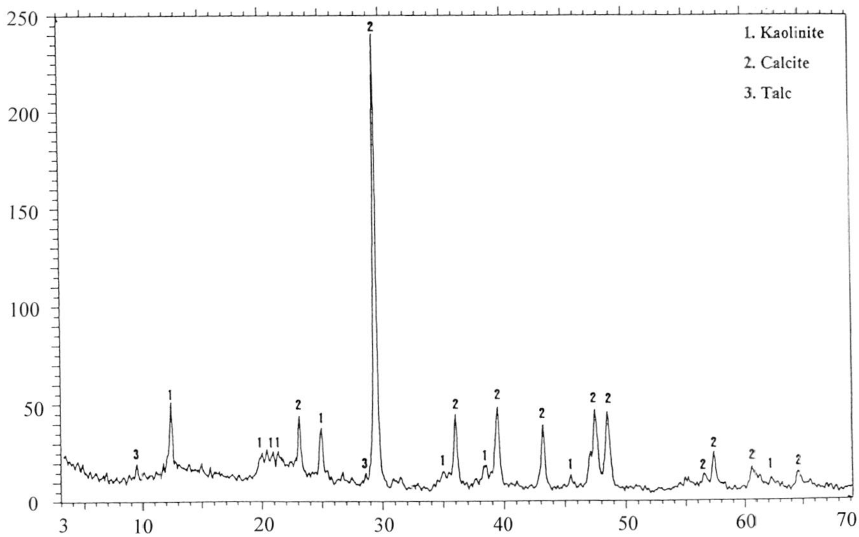

Figure 1 XRD of raw paper sludge

The main phases detected by XRD were: talc, quartz, kaolinite, and calcium carbonate. Based on the chemical composition shown in Table 1, an approach of the quantitative composition of the inorganic fraction of raw sludge gave:

- kaolinite :19.5 %
- talc : 5.0 %
- quartz : 3.5 %
- calcite : 35.5 %
- other : 3,0 %

Calcination of Paper Sludge

The sludge was burnt in a laboratory electrical fixed-bed furnace. The temperature rate was 250°C/hr and the product was maintained at 650°C for 5 hours. Then, it was cooled, with the furnace being closed. The discharging took place as the furnace recovered to ambient conditions. The burnt sludge was desagglomerated before being used for further investigations. As shown in Figure 2, kaolinite was completely dehydroxylated during calcination while talc, quartz, and calcite were preserved.

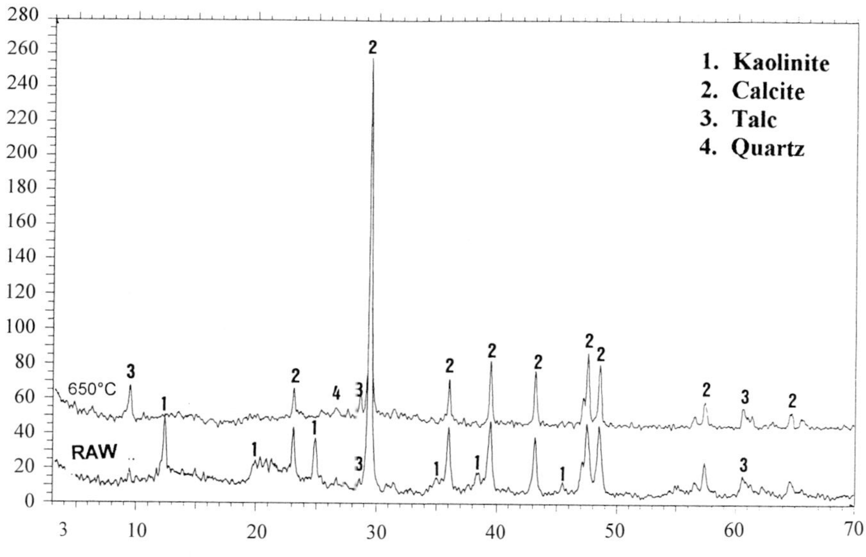

Figure 2 XRD spectra of raw ad calcined sludges

Pozzolanic Activity of Calcined Paper Sludge

The pozzolanic activity of calcined sludge was investigated by measuring the consumption of calcium hydroxide with time on pastes composed of 50 % calcium hydroxide and 50 % calcined sludge, by mass. Pastes were prepapred at equivalent consistency according to the

French Standard NFP 15-402 and cast into cylindrical molds (∅ = 20 mm, h = 40 mm), using a shaking table.

The samples were demolded after 3 days of hydration and immersed in lime-saturated water for 4 or 25 days. Therefore, it was possible to measure the calcium hydroxide consumption after 7 and 28 days of hydration. This measurement was done by differential thermal analysis (DTA) on 600 mg of powder finer than 100 μm. The surface area of the residual calcium hydroxide peak (at 540-560°C) was measured and compared to that of a paste containing 50 % calcium hydroxide and 50 % ground silica. The ratio between these two surface areas gave the lime consumption of calcined paper sludge. The behavior of calcined sludge was compared to that of a pure metakaolin developed by Engelhard Company. The results are presented in Table 2.

Table 2 Calcium hydroxide consumption (% by mass)

POZZOLAN	7 DAYS	28 DAYS
Engelhard Metakaolin	45	94
Paper Sludge Calcined at 650°C	79	87

Table 2 points out that calcined sludge was more reactive than metakaolin at early ages, despite a lower metakaolinite content. This is mainly due to the reaction between metakaolin (AS_2) and calcium carbonate ($C\bar{C}$) contained in the calcined sludge, and calcium hydroxide (CH), which leads to the precipitation of monocarboaluminate hydrate ($C_4A\bar{C}H_{11}$) and C-S-H (10, 13):

$$AS_2 + 5CH + C\bar{C} + 8H \rightarrow C_4A\bar{C}H_{11} + 2C\text{-}S\text{-}H$$

Properties of Mortars Containing Paper Sludge Calcined at 650°C

Standard mortars containing either OPC or a blended cement composed of 80 % OPC and 20 % calcined sludge, were cast at equivalent consistency. The sand to binder ratio was 3: 1, and the water to binder ratio was 1: 2. To get the same workability as the control mortar, a superplasticizer (Glenium 51 from MBT) was added to the mortar containing the blended cement at a binder content of 0.5 %. These mortars were also investigated for their porosity by means of mercury intrusion porosimetry, after 28 days of hydration. The results of these tests are shown in Table 3.

Table 3 Properties of mortars containing paper sludge calcined at 650°C

BINDER	COMPRESSIVE STRENGTH (MPA)		TOTAL POROSITY (%)	AVERAGE PORE DIAMETER (μm)
	7 days	28 days		
100 % OPC	48.9 ± 0.9	58.4 ± 0.5	15.7	0.068
80 % OPC + 20 % Calcined Sludge at 650°C	55.7 ± 0.9	64.2 ± 0.4	15.5	0.048

The replacement of 20 % OPC by calcined paper sludge leads to strength increase (+ 10 % after 28 days of hydration), the reduction of total porosity, and significant decrease in the average pore size (- 30 %). Therefore, when calcined at 650°C, paper sludge can be considered as a very reactive pozzolan with expected benefits on the durability of concrete.

HYDRAULIC PROPERTIES OF CALCINED SLUDGE

Calcination of Paper Sludge

The sludge was calcined at 800°C for 5 hours in the same furnace as previously presented. The X-ray diffractogram shown in Figure 3 pointed out the presence of quicklime (CaO), calcite ($CaCO_3$), talc and calcium aluminium oxide. A vitreous halo (between $2\theta = 22°$ and $2\theta = 35°$) also appeared.

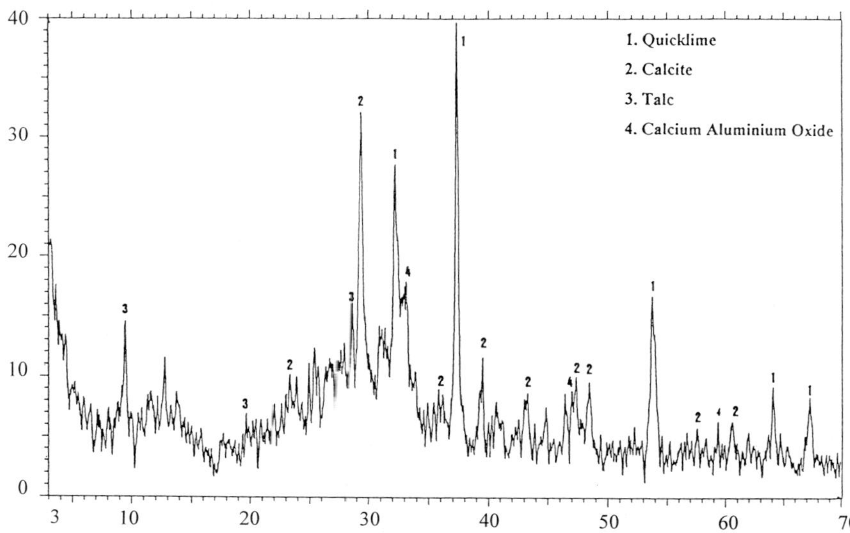

Figure 3 X-Ray diffractogram of sludge calcined at 800°C

Hydraulic Properties of Paper sludge calcined at 800°C

To get a paste sufficiently workable, it was necessary to use a lot of water: 500 g water for 400 g ash, which was a water to solid ratio of 1.25. This was due to the slaking of quicklime occurring in ash:

$$CaO + 2\,H_2O \rightarrow Ca(OH)_2 + H_2O$$

After 7 days of age, this paste hardened and presented a compressive strength of 15.8 Mpa, which meant that paper sludge calcined at 800°C had hydraulic properties and could be considered as self-cementing. After 28 days of hydration, the strength reached 19.7 Mpa.

The sludge calcined at 800°C was also mixed with calcium hydroxide (50 % ash + 50 % calcium hydroxide) and hydrated at a water to solid ratio of 0.78. The lime consumption was as follows:

- 67 % after 7 days of hydration,
- 73 % after 28 days of hydration.

These results prove that the sludge calcined at 800°C exhibited both hydraulic and pozzolanic properties, while that calcined at 650°C was only pozzolanic. The comparison between the two sludges in term of lime consumption is shown in Table 4.

Table 4 Lime consumption of calcined paper sludge (w_t%

TEMPERATURE	7 DAYS	28 DAYS
650°C	79	87
800°C	67	73

When calcined at 650°C, paper sludge developed better pozzolanic properties and consumed more calcium hydroxide (+ 20 % after 28 days), due the preservation of calcite.

Due to an excessive water consumption, sludge calcined at 800°C was not directly used to cast mortars. In a first step, it was slaked at a water to solid ratio of 1.25. Six hours after this operation, the hardened paste was crushed and dried at 80°C for 18 hours. Finally, it was desagglomerated in a hammer-ball and mixed with cement to prepare standard mortars. A blended cement containing 80 % OPC and 20 % slaked ash was compared to plain OPC. It was necessary to add 1 % superplasticizer to get a good workability. The performances measured after 7 and 28 days of hydration are shown in Table 5. The total porosity and the average pore diameter assessed by mercury intrusion porosimetry, are also reported in this Table.

Table 5 Properties of mortars containing paper sludge calcined at 800°C

BINDER	COMPRESSIVE STRENGTH (MPA)		TOTAL POROSITY (%)	AVERAGE PORE DIAMETER (µm)
	7 days	28 days		
100 % OPC	48.9 ± 0.9	58.4 ± 0.5	15.7	0.068
80 % OPC + 20 % Calcined Sludge at 800°C	54.2 ± 1.2	64.3 ± 0.4	16.1	0.060

Comparing Tables 3 and 5 points out that the temperature of calcination, in the range of 650°C to 800°C, has no significant influence on the strength developed by standard mortars containing 20 % calcined paper sludge. Nevertheless, it has an influence on the pore size distribution: the more pozzolanic the sludge (650°C), the smaller the average pore diameter.

CONCLUSION

When calcined at 650°C, paper sludge is a mixture of metakaolin and calcite which is able to consume calcium hydroxide more rapidly than pure metakaolin. It can replace 20 % ordinary portland cement with beneficial effects on compressive strength (+ 10 %) and the size of pores (the average pore diameter decreases by 30 %). It can be classified as a very reactive pozzolan.

When calcined at 800°C, paper sludge is a mixture of calcite, quicklime and metakaolin. It presents both hydraulic and pozzolanic properties. The main problem is linked to the presence of quicklime which considerably increases the water demand and decreases the workability. To be used in concrete, this ash needs to be slaked. In such conditions, it can replace 20 % cement without affecting the strength, but it is less pozzolanic than the sludge calcined at 650°C.

ACKNOWLEDGMENT

The authors gratefully acknowledge support from the European Community, which funded the project under the Brite/Euram Programme (BRE2 CT 94/0626, Project No. PL-8174).

REFERENCES

1. WEBB,L.A.. Host of Options Available for Sludge. Pulp and Paper International, 1996, Vol. 38, N° 11, pp. 44-48.
2. DOERRE, R.. Waste Incineration in the Pulp and Paper Industry. Paper Technology, 1995, Vol. 36, N° 4, pp. 26-30.
3. Mc GRATH, R. Louisiana-Pacific's Samoa Mill May Soon be Using Fuel from Woodwaste. Pulp and Paper, 1987, N° 7, pp. 59-60.
4. MUELLER, J.C., and GAWLEY, J.R.. Gourmet Mushroom Cultivating on Pulp Mill Sludge Proves Feasible. Pulp and Paper, 1985, Vol. 59, N° 7, pp. 140-144.
5. HATCH, C.J., and PEPIN, R.G.. Recycling Mill Wastes for Agricultural Use. TAPPI Journal, 1985, Vol. 68, N° 10, pp. 70-75.
6. COLEMAN, M.J.. Primary/Secondary Sludge Recycled into Valuable Agicultural Fertilizer, Pulp and Paper, 1985, Vol. 59, N° 4, pp. 130-135.
7. PEPIN, R.G.. Paper-Mill Sludge and Ash as Soil Conditioner. Biocycle, 1984, Vol. 25, N° 4, pp. 52-58.

8. ERNSTBRUNNER, L.. Reconstruction of a Cement Plant for Fiber Residue Processing. Ciments, Bétons, Plâtres, Chaux, 1995, N° 815, pp. 244-248.

9. FREDERICK, W.M.J., LISA, K., and LUNDY, J.R.. Energy and Materials Recovery from Recycled Paper Sludge. TAPPI Journal, 1996, Vol. 79, N° 6, pp. 123-131.

10. PERA, J, and AMROUZ, A.. Development of Highly Reactive Metakaolin from Paper Sludge. Journal of Advanced Cement Based Materials, 1998, Vol. 7, N° 2, pp. 49-56.

11. PERA, J., and AMBROISE, J.. Calcined Paper Sludge: An Interesting Admixture for the Cement Industry. In: Proceedings of The Sidney Diamond Symposium-Materials Science of Concrete. Special Volume Published by the American Ceramic Society, Westerville (USA). Edited by M. Cohen, S. Mindess, and J. Skalny, 1998, pp. 467-479.

12. MURAT, M., AMBROISE, J., and PERA, J.. Hydration Reaction and Hardening of Calcined Clays and Related Minerals. Cement and Concrete Research, 1985, Vol. 15, pp. 261-268.

13. AMBROISE, J.. Elaboration de Liants Pouzzolaniques à Moyenne Température et Etude de leurs Propriétés Physico-Chimiques et Mécaniques. Thesis (in French), Lyon (France), 1984, 165 p.

14. AMBROISE, J., MARTIN-CALLE, S., and PERA, J.. Proceedings of the Fourth CANMET/ACI International Conference on "Fly Ash, Silica Fume, Slag and Natural Pozzolans in Concrete". ACI Special Publication SP-132, Edited by V.M. Malhotra, 1992, Vol. 1, pp. 731-748.

SUSTAINABLE WASTE MANAGEMENT - STUDIES ON THE USE OF SEWAGE SLUDGE ASH IN THE CONSTRUCTION INDUSTRY AS CONCRETE MATERIAL

V Pinarli
Uludag University
Turkey

ABSTRACT: A substantial amount of waste materials such as sludge, sludge ash, fly ash, blast furnace slag etc are produced from municipal wastewater treatment plants, coal-fired thermal power plants, iron and steel, copper, ferrochrome, pulp and paper industries etc. In this experimental study carried out in the laboratory, sludge ash obtained from the incineration of municipal wastewater treatment sludges were utilised in concrete production as cement substitute in varying proportions 0, 5, 10, 15 and 20 percent of cement by weight in concrete mixture were substituted by the pulverised sludge ash. Initial and final setting times of the concrete were longer with higher percentages of sludge ash used. With 5 percent replacement of cement by pulverised sludge ash, the compressive strengths at all ages were higher than the control samples. With 10 percent replacement of cement by sludge ash, the compressive strength at different ages were about the same as the reference cubes. The water absorption and permeability of concrete with sludge ash decreased.

Keywords: Wastewater treatment, Sludge ash, Concrete material, Setting time, Compressive strength.

Assoc Prof Dr V Pinarli, is a Deputy Head of Department of Environmental Engineering, Uludag University. He specialises in waste water treatment, disposal and reuse of waste treatment sludges. His research interests also include the use of clay and cement materials for the chemical fixation and solidification of hazardous wastes, and the use of industrial wastes in construction sector as substitute or additive materials.

INTRODUCTION

Sanitary landfill, utilisation on agricultural land, incineration and ocean disposal are widely used methods in the disposal of municipal wastewater treatment sludges. However, all these sludge disposal alternatives have varying degrees of adverse environmental impacts. Currently, incineration of dried sludge is considered as a feasible solution for sludge disposal particularly in large metropolitan cities.

Europe, like the many other parts of the World faces with increasing volume of sludge to be disposed of. Annually more than 6 million tonnes of dry sludge was generated in Europe and this amount was estimated to double by the year 2000. The cost of wastewater treatment and sludge disposal varied between 100 and 300 pound sterling per tonne of dried sludge. Quantities of wastewater treatment sludges generated and their disposal methods used in several European countries are shown in Table 1 [1].

According to a research carried out in 1984, approximately more than 7 million tonnes of municipal wastewater treatment sludges were produced annually in the USA. 25 to 30 percent of these was disposed of by incineration and more than 500.000 tonnes of sludge ash needed to be disposed of was produced [2].

Table 1 How is Europe disposing of their wastewater treatment sludge

COUNTRY	DISPOSAL METHODS				
	Amount of Sludge (tonne * 1000/year)	Agriculture (%)	Landfill (%)	Incineration (%)	Other (%)
Austria	320	13	56	31	0
Belgium	75	31	56	9	4
Denmark	130	37	33	28	2
France	700	50	50	0	0
Germany	2500	25	63	12	0
Greece	15	3	97	0	0
Ireland	24	28	18	0	54
Italy	800	34	55	11	0
Luxembourg	15	81	18	0	1
Netherlands	282	44	53	3	0
Portugal	200	80	13	0	7
Spain	280	10	50	10	30
Switzerland	215	60	30	20	0
England	1075	51	16	5	28
Total	6631	38	43	10	9

In order to incinerate sludge produced from biological wastewater treatment plants, fluidised bed sludge incineration installations were planned by Istanbul Water and Sewerage Administration (ISKI, 1992) for the Greater Istanbul Metropolitan area in Turkey [3]. The incinerator sludge ash generated from these installations required proper disposal.

The magnitude of the sludge disposal problem will be drastically reduced if sludge, be it in the raw, digested, and dewatered or the incinerated form, can be put to large-scale economic use. In recent years, studies have been carried out by various researchers to investigate the use of sludge as nonconventional building and construction material.

It has been reported that the use of sludge mixed with clay-shale material for "Biobrick" production [4]. At the same time it has been reported that sludge mixed with clay could be used in the production of bricks for construction use. The maximum percentage of dried sludge that could be mixed with clay for brick making was 40 % by weight [5,6]. An investigation was also carried out into using dewatered sludge mixed with lime incinerated at high temperature for the production of cementitous materials. The results of this investigation revealed that sludge could be used to produce cementlike material that would satisfy the strength requirements in the ASTM standard C91-89 for masonry cement [7,8]. In the USA California San Diego Region Water Reclamation Agency has devised an innovative wastewater treatment plant that uses sludge mixed with clay to produce a lightweight building aggregate [9,10]. In this plant, screened domestic wastewater is combined with clay, alum coagulant, and polyacrylic acid. The mixture is then coagulated and flocculated. The sludge collected at the settling tank was thickened to 45% solids. Sludge cake was then mixed with clay made into pellets. The pellets were fired at temperatures between 1070-1095 ^{o}C. The resulting product was a strong ceramic-like material that met ASTM standard C330-85 (Standard Specification of Lightweight Aggregate for Structural Concrete) for lightweight concrete aggregate that are suitable for application in thermal insulation and fire protection [11].

In other study, properties of digested, dewatered, dried, and pulverised sewage sludge blended Portland cement mortars were investigated [12]. Sludge was used as replacement for 1, 2, 5, 10, 20, 30, and 40 percent by weight of the cement. The results indicated that initial and final setting times were longer with an increase of sludge. The effects of pulverised sludge on the Le Chatelier expansion and specific surface area of mortar were not significant. For 5 percent replacement of cement by pulverised sludge, the compressive and tensile strengths reduced by 32 and 38 percent, respectively [12].

Subsequently, a comprehensive test programme was developed to examine the potential for using the pulverised sludge ash as a cement substitute in mortar. Sludge ash used as a replacement for 5, 10, 15 and 20 percent by weight of the cement. The results showed that initial and final setting times were longer with an increase of sludge ash. However, with 10 percent replacement of cement by sludge ash, the 28th day compressive and tensile strength for the mortars are similar to conventional mortar [13].

Ash collected after the incineration of sludge was also used in brick making [14]. The maximum percentage of pulverised sludge ash that could be mixed with clay for brickmaking was 50%. Beyond that, bonding of the mixture was poor, and extrusion of the bricks resulted in dog-earing of the products.

Pulverised sludge ash could also be used as a partial replacement for cement in concrete [5]. At 10% replacement of cement by sludge ash, the compressive strength development was similar to conventional concrete. Experimental work by Enviroscience Inc. and other workers has suggested that the ash could be used in small amounts as an essentially inert filler in concrete and other building material components such as bricks, tiles, etc [15,16].

Enviroscience Inc. concluded that the ash could be used as replacement for up to 5 percent by weight of the cement, or 2 percent replacement for fine aggregate in concrete, without affecting its mechanical properties. However, 5 percent replacement specified for cement is much less than the 35 percent volume replacement specified in the ASTM C311-87 test method for mineral admixtures (fly ash, pozzolans) [17]. It has been reported that the use of sewage sludge ash as an admixture in brick appeared to be promising, especially when the ash amended bricks were fired to above ash vitrification temperature [18]. The ash was assessed for the tendency to leach metals during acid extraction, and it was found to be nonhazardous when tested according to the extraction procedure toxicity test (EPTox) [19] and the Toxicity Characteristics Leaching Procedure test (TCLP) [20].

In view of disposal problems and the increasing cost of construction materials such as cement, research studies have been carried out in the laboratory to investigate the feasibility of using sludge ash as building and construction related materials. This paper describes the work of a study aimed at addressing the technical and practical issues associated with the utilisation of sewage sludge ash obtained from the incineration of municipal wastewater treatment sludges in concrete production as cement substitute in varying proportions.

EXPERIMENTAL DETAILS

Incineration of digested and dewatered sludge, yields an odourless, cellular, porous mass of sludge ash with low unit weight. In the present study, dewatered sludge samples were collected from the wastewater treatment plant of Kütahya Municipality over a period of 12 months. Sludge collected was incinerated in the furnace at above 550 + 50 °C. The resulting ash was pulverised using a centrifugal mill. The ash had a mean particle size around 50 μm. Specific surface area for pulverised sludge ash was measured by the Blaine method. Physical and chemical properties of the sludge ash were analysed and the results were given in Table 2 and Table 3 Oxide components of the Portland cement were given for the purpose of comparison in Table 3.

Table 2 Properties of pulverised sludge ash

PROPERTIES	RANGES	AVERAGE
Specific Gravity	2.46-2.99	2.74
Bulk Density, kg/m^3	1150-1580	1365
pH value	4.1-5.8	5.0

A water / cement (W/C) ratio of 0.5 fixed constant was used in the calculation of concrete mixes. For these mixes, the cement dosage was 350 kg/m3. The Portland Cement (PC) used was of Grade 32.5. Maximum grain size of the coarse aggregate was 25.4 mm and fineness modulus of the sand was 3.02. In concrete mixes, 40 percent sand, 35 percent coarse gravel and 65 percent fine gravel were used. In the mixes, sludge ash used as replacement for 5, 10 ,

Table 3 Composition of sludge ash and Portland cement in percent by weight, expressed as oxides

COMPONENT	SLUDGE ASH	PORTLAND CEMENT
SiO_2	28.8	21.7
Al_2O_3	7.8	4.7
Fe_2O_3	3.9	2.2
CaO	21.9	65.7
MgO	4.6	1.3
SO_3	1.4	2.0
$Na_2O_3 + K_2O$	6.8	-
Loss on ignition	24.8	0.7

15 and 20 percent by weight of the cement. A control concrete type without use of sludge ash in the mix was prepared for comparison purposes. Cube molds with internal dimensions of 120x120x120 mm were used. After being removed from their molds, the concrete samples were cured by immersion in water at 20+2 ^{o}C temperature under standard laboratory conditions and tested for compressive strengths after curing periods of 3, 7, 14 and 28 days. Compressive strengths were determined using SEIDNER type LPM 3000M1 model compression machine. Compressive strength values of samples were reported as the average of 6 separate measurements, respectively.

The fresh properties of the concrete mixes were determined by Slump Test Method (TS2871) and Compacting Factor Method (TS 2892). Initial and final setting times of concrete were determined by Penetration Resistance Method (ASTM C403-1980).

24 hour water absorption coefficient of permeability and unit weight of concrete were determined in other parts of the work carried out with the use of sludge ash. Permeability experiments were carried out according to the prescription of TS 3455/81.

PERFORMANCE OF SLUDGE ASH CONCRETE

The performance of fresh and hardened concrete produced with the use of sludge ash was investigated. Initial and final setting times were determined by using the graph of penetration resistance versus time shown in Figure 1. Initial and final setting times were determined as the penetration resistance to reach 3.5 N/mm^2 and 27.6 N/mm^2 values respectively and the results obtained were given in Table 4.

It was seen that the initial and final setting times of concrete were increased as the proportion of sludge ash increased. While the control concrete specimens started to set after 7 hour 9 minutes, this period of time was extended to 11 hours 15 minutes for the concrete with 20 percent sludge ash. Final setting time of the control concrete specimens were determined as 10 hours 27 minutes while the final setting of the concrete with 20 percent sludge ash was completed in 12 hours 51 minutes. A research work carried out previously demonstrated that the initial and final setting times of concrete mix prepared with only dried sludge used as cement replacement material in mortars were longer than the setting times of the concrete produced with sludge ash [12, 13]. Initial and final setting times of control mortar samples

Table 4 Slump, compacting factor and setting times of the control concrete and of the concrete containing sludge ash

SLUDGE ASH (%)	SLUMP (cm)	COMPACTING FACTOR	INITIAL SETTING TIME	FINAL SETTING TIME
0	18	0.94	7 h 9 min	10 h 27 min
5	15	0.94	7 h 34 min	11 h 12 min
10	13	0.93	8 h 34 min	11 h 42 min
15	12	0.90	10 h 6 min	12 h 5 min
20	11	0.85	11 h 15 min	12 h 51 min

and of the mortars containing 20 % sludge were determined as 2 hours 25 minutes, 4 hours 10 minutes and 35 hours, 93 hours 30 minutes, respectively. The effects of materials interfere with the hydration of Portland cement were classified [21]. It was known that organic materials such as oil, grease and phenol even in small quantities increase the initial and final setting times of cement and decrease the compressive strength considerably [22]. As it was seen in Table 3 that ignition loss of sludge ash was found as 24.8 percent. This result has shown that the incineration of sludge in laboratory ash furnace can not be carried out completely. It may be stated that the organic materials remained even after incineration of sludges in laboratory ash-furnace extended setting times and has a negative effect in development of strength by interfering with the hydration reactions of cement. In other related research work, it was also found that the initial and final setting times of the concrete mix and mortars prepared with 10, 20, 30 and 40 percent pulverised sludge ash used as cement replacement material were increased and the compressive strengths were decreased [23].

Experiments carried out to determine the fresh concrete properties were presented in Table 4. As it can be seen in Figure 2 that the workability of concrete measured by slump and compacting factor values decreased as the proportion of sludge ash used increased. However, although finely pulverised sludge ash having a specific surface area of around 600 m^2/kg when used at higher proportions increased the water requirement, since the water/cement (W/C) ratio was kept constant, the workability of concrete was decreased.

Compressive strengths of the concrete samples following 3, 7, 14 and 28 days curing periods were presented in Table 5. The effect of the use of sludge ash on the compressive strengths was seen in Figure 3. On the other hand, the effect of sludge ash in varying proportions in relation to curing periods on the compressive strength development of concrete with sludge ash and of the control concrete samples without use of sludge ash were compared in Table 6. As it can be seen that with 5 percent replacement of cement by sludge ash, the compressive strengths at all ages were higher than the control samples at an average of 7 percent. With 10 percent replacement of cement by sludge ash, the compressive strength at different ages were about the same as the reference cubes. With even 20 percent replacement of cement by sludge ash, the 28th day compressive strength was over 94 percent of the compressive strength of the control samples.

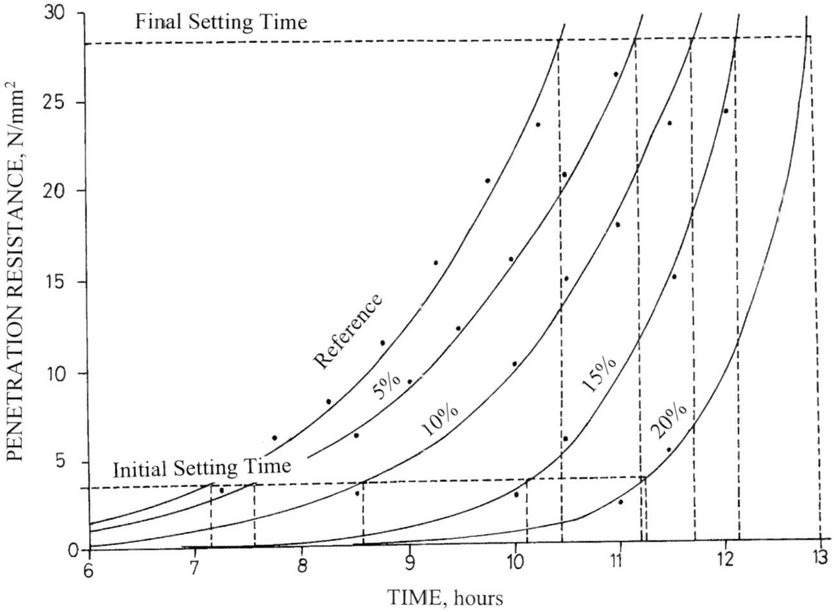

Figure 1 The effect of sludge ash utilisation on the setting times of concrete

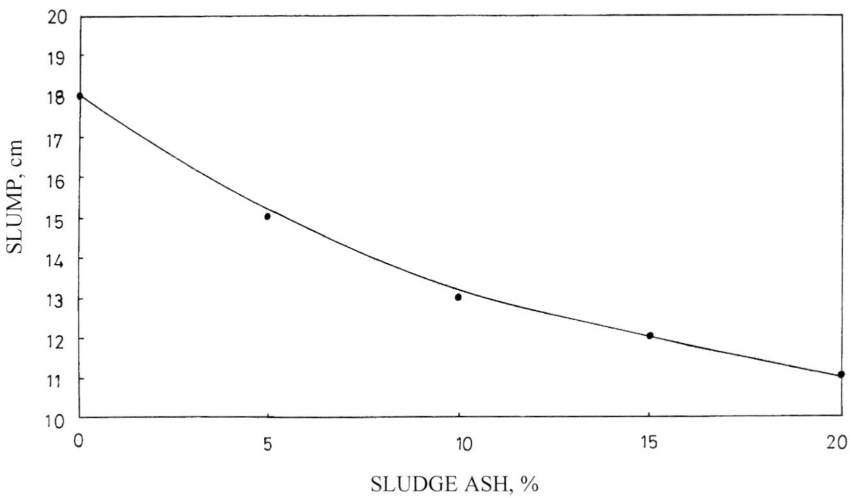

Figure 2. The effect of sludge ash utilisation on the workability of concrete

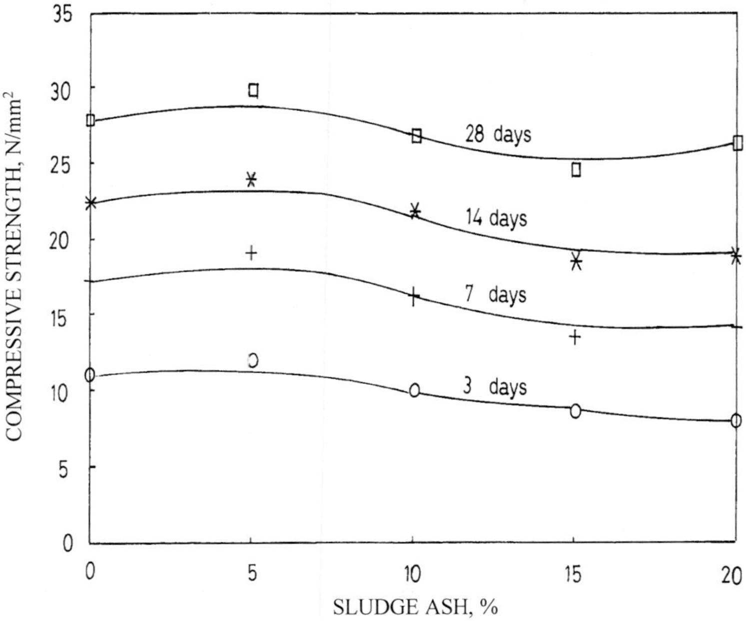

Figure 3 The effect of sludge ash utilisation on the compressive strength of concrete

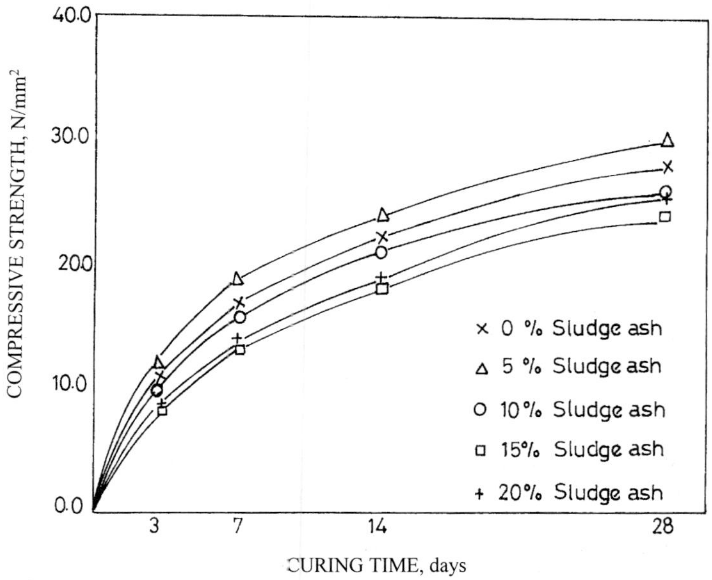

Figure 4 The effect of the utilisation of sludge ash in varying proportions and curing time on the compressive strength of concrete

Table 5 Compressive strength, permeability, water absorption and unit weight of the concrete with sludge ash and control samples

SLUDGE ASH %	COMPRESSIVE STRENGTH (+1.0 N/mm ^{2}) Curing Time, day				PERMEABILITY CONSTANT (cm/sn)	WATER ABSORPTION (%)	UNIT WEIGHT (kg/m ^{3})
	3	7	14	28			
0	10.9	17.1	22.3	27.8	$15x10^{-9}$	6.22	2474
5	11.8	18.8	29.8	29.7	$7x10^{-9}$	5.32	2474
10	9.8	16.2	21.7	26.7	$1x10^{-9}$	5.50	2485
15	8.5	13.5	18.5	24.4	$2x10^{-9}$	5.03	2520
20	7.9	14.2	18.8	26.2	$6x10^{-9}$	5.65	2467

Table 6 Comparison of compressive strengths of control concrete samples and of concrete samples with sludge ash

SLUDGE ASH, (%)	STRENGTH, (%)			
	3 day	7 day	14 day	28 day
0	100	100	100	100
5	108.3	109.9	106.7	106.8
10	89.9	94.7	97.3	96.0
15	77.9	78.9	82.9	87.7
20	72.5	83.0	84.3	94.2

CONCLUSIONS

The effect of utilisation of sludge ash on the properties of concrete depends primarily on physical and chemical characteristics of the sludge ash and on the factors such as the type of cement used, the proportion of sludge replaced cement and the cure conditions applied to the concrete specimens. The following conclusions can be drawn based on the experimental work.

1. Initial and final setting times of the concrete were longer with higher percentages of sludge ash used.

2. When the free water content was fixed constant in the concrete mix. The results showed a reduction in workability with increasing sludge ash content in the mix.

3. The compressive strengths for the concrete at different ages decrease with the increase of sludge ash. However, with 5 percent replacement of cement by sludge ash, the compressive strengths at all ages were higher than the control samples. With 10 percent replacement of cement by sludge ash, the compressive strength at different ages were about the same as the reference cubes. With even 20 percent replacement of cement by sludge ash, the 28th day compressive strength was over 94 percent of the compressive strength of the control samples.

4. The organic matter remaining in the sludge ash due to incomplete combustion upon firing the sludge in the laboratory furnace possibly retards the setting of cement and the strength development by interfering with the normal cement hydration reactions.

5. The water absorption and permeability of the concrete produced with the use of sludge ash decreased.

6. The effect of sludge ash utilisation on the unit weight of the hardened concrete were not significant.

Results obtained from this research study showed that sludge ash generated from the incineration of municipal wastewater treatment sludges can be utilised in concrete production as cement substitute in varying proportions. However, several concrete properties such as long term strength, durability and volume changes etc have to be studied before accepting the sludge ash as a convenient construction material.

REFERENCES

1. DAVIS, R. D. Sludge disposal-Europe's Mountainous Problem, Water Quality International, 1992, 3, pp.22.

2. BHATTY, J. I. AND REID, K. J. Compressive Strength of Municipal Sludge Ash Mortars, ACI Materials Journal, 1989, 86, 4, pp.394-400

3. ISTANBUL WATER AND SEWERAGE ADMINISTRATION (ISKI)- Action Report, 1992.

4. ALLEMAN J E AND BERMAN N A. Constructive Sludge Management Biobricks, Journal of Environmental. Engineering. ASCE, 1984, 110, pp. 301-311.

5. TAY, J. H. Sludge Ash As Filter for Portland Cement Concrete, Journal of Environmental. Engineering, ASCE, 1987, 113, pp.345-351.

6. TAY, J. H. Sludge As Brick Making Material. In : New Directions and Research in Waste Treatment and Residuals Management, Vancouver, Canada, 1985, 2, pp. 661-668.

7. TAY, J. H. AND SHOW, K. Y. The Use of Lime Blended Sludge for Production of Cementitous Material, Water Environment Research, 1992, 64, pp.6-12.

8. ASTM C91-89, Standard Specification for Masonry Cement, American Society Testing Materials, 1989, 52,04-01.

9. ELKIN, B. V., WILSON, G. E. AND GERBERG, R. M. Complete Reclamation of Wastewater and Sludge, Water Science and Technology, 1985, 17, 1453-1454.

10. St. GEORGE M Concrete Aggregate from Wastewater Sludge, Journal of Concrete International, 1986, 8, pp.27-30.

11. ASTM C330-85, Standard Specification for Lightweight Aggregate for Structural Concrete, American Society Testing Materials, 1975.

12. PINARLI, V. AND EMRE, N. K. Constructive Sludge Management- Reutilization of Municipal Sewage Sludge in Portland Cement Mortars, Environmental. Technology, 1994, 15, pp. 833-841.

13. PINARLI, V. AND KAYMAL, G. An Innovative Sludge Disposal Option-Reuse of Sludge Ash by Incorporation in Construction Materials, Environmental Technology, 1994, 15, pp.843-852.

14. TAY, J. H. Bricks Manufactured From Sludge, Journal of Environmental Engineering., ASCE, 1987, 113 (2), pp. 287-243.

15. ENVIROSCIENCE INC. Analysis of Sludge Ash for Use In Asphalt, Concrete, Fertilisers and Other Products, 1982, publication No.12-82-103, Minneapolis.

16. RIGANTI, V., FIUMARA, A. AND OBOBEZ, G. B. The Use of Industrial Sludges as Raw Materials In the Cement Industry, Waste Management Research, 1986, 4 (3), pp.293-302.

17. ASTM C311-87. Standard Test Methods for Sampling and Testing Fly Ash or Natural Pozzolans for Use As a Mineral Admixture In Portland Cement Concrete, American Society Testing Materials, 1989, 04.03,183.

18. TRAUNER, E. J. Sludge Ash Bricks Fired to Above and Below Ash-vitrifying Temperature, Journal of Environmental. Engineering, ASCE, 1993, 119(3), pp.506-519.

19. 40CFR261: Code of Federal Regulations, Title 40, Part 261. U.S. Government Printing Office, Washington, D.C. 1988,

20. 40CFR261: Code of Federal Regulations, Title 40, Part 261. U.S. Government Printing Office, Washington, D.C. 1990,

21. WILDING, C. R., WALTER, A. AND DOUBLE, D. D., A Classification of Inorganic and Organic Admixtures by Conduction Calorimetry, Cement and Concrete Research, 1984, 14, pp.185-194.

22. VIPULANANDAN, C. AND KRISHNAN, S. XSD Analysis and Leachibility of Solidified Phenol-Cement Mixtures, Cement and Concrete Research, 1993, 23, pp.792-802.

23. TAY J H. Properties of Pulverised Sludge Ash Blended Cement, ACI Materials Journal, 1987, 84, pp.358-364.

THE USE OF THE STATISTICAL ENTROPY TO EVALUATE THE UTILISATION OF INCINERATOR ASHES FOR THE PRODUCTION OF CEMENT

H Rechberger
Vienna University of Technology
Austria

ABSTRACT. The paper discusses the utilisation of MSW incinerator residues for the production of cement with regard to materials management. Two objectives of waste management are *the protection of men and the environment* and *resource conservation*. Thus metals have to be recycled or transferred into appropriate final sinks. For both options a concentrating step is indispensable. MSW incinerators are such concentrating processes, as they accumulate metals in their residues. The mixing of high-concentrated residues with low-concentrated clinker in order to produce cement results in dilution of metals. This contradicts the above objectives. Concentrating and dilution processes can be described by materials balances. To evaluate different uses of MSW residues in the cement production the statistical entropy of the resulting distributions of selected metals (Cd, Pb, Cu) is assessed. In order to avoid an increase in the entropy, neither fly ashes nor bottom ashes of MSW incinerators should be utilised in large amounts without pre-treatment. The method can be used to determine the quality of the residues so that they can be used in accordance with the objectives of waste management.

Keywords: Waste management, Materials management, MSW incineration, Entropy, Metals, Cement production, Ash

Dr H Rechberger is a research associate at the Institute for Water Quality and Waste Management of the Vienna University of Technology. His work focuses on the development of methods to evaluate waste treatment systems with regard to the objectives of wastes and materials management. In the recent studies he developed methods and criteria to ensure an environmentally compatible assignment of combustible wastes to different thermal plants.

INTRODUCTION

Fly and bottom ashes from MSW incinerators are not of final storage quality. In contact with the hydrosphere these residues produce emissions exceeding environmental standards. Hence, their management is of special ecological concern. Utilisation of these residues as a secondary resource for cement production is one of several possibilities. Others are direct land filling as well as melting or vitrification processes in order to achieve residues with final storage quality. Since the incineration ashes are products of the waste management system their utilisation or disposal, respectively, has to fulfil the objectives of waste management as good as possible, considering economic frame conditions. This paper investigates the impact of the utilisation of incineration ashes for cement production, exclusively with regard to *substance management*. That this is an important aspect can be justified as follows:

(1) Two objectives of waste management are long-term *protection of men and the environment* and the *conservation of resources*. It can be shown that waste management systems have to concentrate harmful inorganic substances like metals in order to observe these objectives [1]. This can be imagined as follows (cf. Figure 1): Through an optimal concentrating process metals are not relevantly emitted to atmosphere and hydrosphere (protection of men and the environment). Additionally the concentrating results in a high- and a low-concentrated residue which have either at least the quality of an ore or the quality of the average earth's crust. Both residues then can be utilised as secondary materials (resource conservation). These considerations are illustrated in Figure 1 for Cadmium.

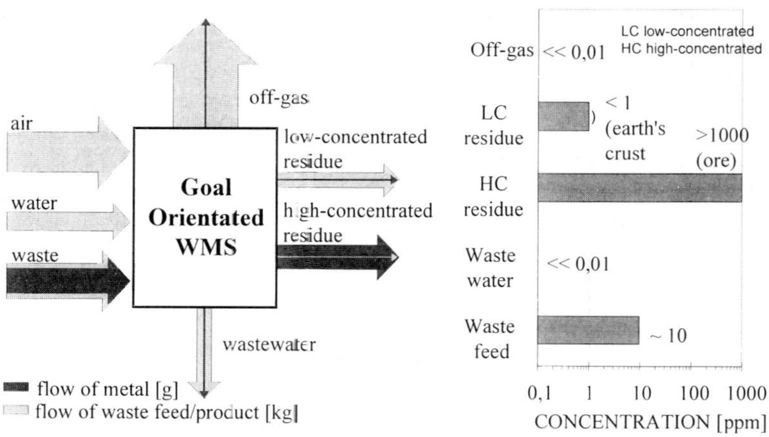

Figure 1 Substance management and resulting concentrations of a goal orientated waste management system (WMS) for heavy metals. The metal (concentrations are given for Cadmium) is concentrated by the system

(2) Investigations show that combustible wastes represent a relevant conveyor belt for some heavy metals. Figure 2 gives the ratio between the metal flow via combustible wastes and the total flow through the economy as it has been estimated for a typical Central European

country [2]. These metals are important with regard to their toxicity. Hence, waste management takes a major role in managing these materials environmentally compatible.

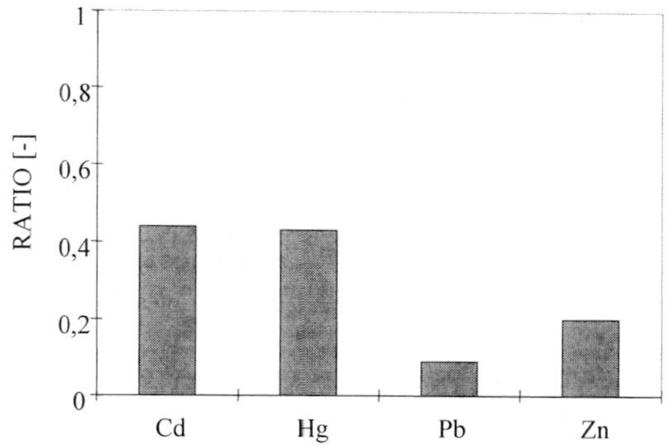

Figure 2 Ratio between the flow over combustible wastes and the total flow through the economy [2]

In this paper substance balances of two opposite scenarios (a) and (b) are investigated for the following reasons and assumptions: (1) The specific data-sets (mass-flows/capita and concentrations) are found for most countries in the European Union. (2) The total amount of MSW of a country is treated in state-of-the-art incinerators. (3) The residues (fly ash or bottom ash, respectively) are not fed into the cement kiln but added to the clinker when milled to cement. Emissions of the mill are not considered as these material flows are very small (<<1%). (4) Additional raw materials are not considered in both scenarios either. It is presumed that the flows of the investigated substances via these materials are small in comparison. In scenario (a) fly ash is mixed with clinker. The bottom ash is landfilled. In scenario (b) fly ash and bottom ash change places. Scenarios (a) and (b) are given in Figure 3.

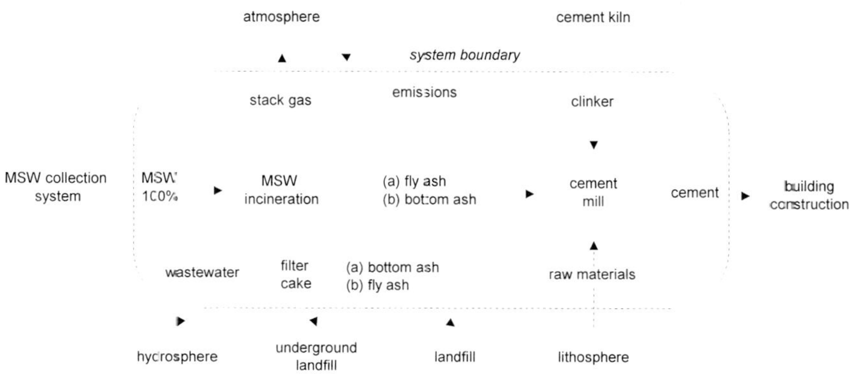

Figure 3 Investigated system: scenarios (a) and (b)

METHOD

The method applied in this paper to evaluate the scenarios is a tool to quantify the statistical entropy of the distribution of substances by a process or system, respectively [1]. The distribution has to be determined by a substance flow analysis (SFA) as described in [3]. The results of an SFA are the mass-flows (m) and the concentrations (c) of the analysed substances (j) for all in- and outputs of the investigated system. In the science of Statistic the entropy is a measure for the variance of e.g. a probability distribution ($P_1,...,P_k$). The statistical entropy (H) of such a distribution is given by the following equation:

$$H(P_i) = -\sum_{i=1}^{k} P_i \cdot ld(P_i) \geq 0 \qquad \text{Eq. 1} \qquad \text{with} \qquad \sum_{i=1}^{k} P_i = 1 \qquad \text{Eq. 2}$$

P_i is the probability that event i (E_i) happens ($0 \leq P_i \leq 1$). The extrema of Equation 1 are found for definite distributions as shown in Figure 4. Any other existing distribution for the probabilities of the three events ($E_1,...,E_3$) as given in the simple example of Figure 4 is in the range between [0, Max].

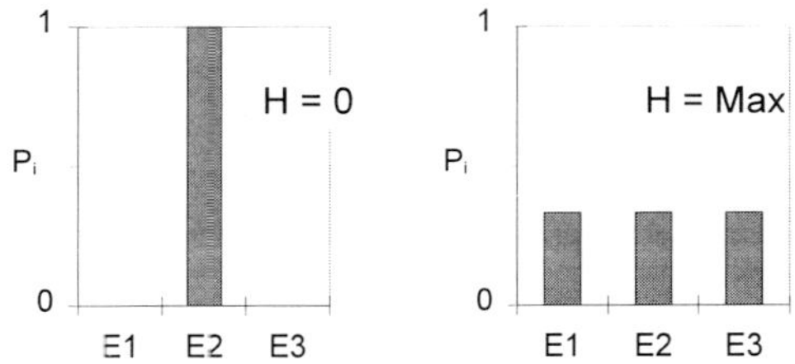

Figure 4 Distributions for extrema of H.

Generally, the statistical entropy measures the distribution pattern of a set of characteristics. A set of such characteristics for example are the concentrations of the system displayed in Figure 1. This goal orientated waste management system has a statistical entropy near zero. In order to apply the statistical entropy to the problem of substance balances Equation 1 has to be transformed. H now becomes a function of k mass-flows ($\underline{m}_i$) and k concentrations ($\underline{c}_{ij}$) of a substance j.

$$H(c_{ij}, m_i) = ld(X_j) - \frac{1}{X_j}\sum_{i=1}^{k} \underline{m}_i \cdot \underline{c}_{ij} \cdot ld(\underline{c}_{ij}) \qquad \text{Eq. 3}$$

with

$$X_j = \sum_{i=1}^{k} m_i \cdot c_{ij} \text{ ... the total substance flow through the investigated system}$$

and

$X_{ij} = m_i \cdot c_{ij}$ ... substance flow of output i

and

$$\underline{m}_i = \begin{cases} \dfrac{X_{ij}}{c_{j,geog.,g}} \cdot 100 \\ \dfrac{X_{ij}}{c_{j,geog.,a}} \cdot 100 \\ m_i \end{cases} \text{for} \begin{cases} i = 1,\dots,k_g \\ i = k_g + 1,\dots,k_g + k_a \\ i = k_g + k_a + 1,\dots,k \end{cases} \text{for} \left\{ \begin{matrix} \text{gaseous} \\ \text{aqueous} \\ \text{solid} \end{matrix} \right\} \text{emissions}$$

and

$$\underline{c}_{ij} = \begin{cases} c_{j,geog.,g} / 100 \\ c_{j,geog.,a} / 100 \\ c_{ij} \end{cases} \text{for} \begin{cases} i = 1,\dots,k_g \\ i = k_g + 1,\dots,k_g + k_a \\ i = k_g + k_a + 1,\dots,k \end{cases} \text{for} \left\{ \begin{matrix} \text{gaseous} \\ \text{aqueous} \\ \text{solid} \end{matrix} \right\} \text{emissions}$$

and

$k = k_g + k_a + k_s$

k_g...number of gaseous emissions
k_a...number of aqueous emissions
k_s...number of solid emissions
k....number of all emissions (= output flows) of the investigated system.

The concentrations with the index "geog." serve as reference values to describe the natural dilution of gaseous and aqueous emissions in the environmental compartments atmosphere and hydrosphere. The entire derivation of the entropy method for substance balances is described in detail in [1]. There the method was named MASTER which is short for "Method for the Analysis of the Substance concentrations of Treatment End-products and Recycling goods".

CALCULATIONS & RESULTS

In Figure 5 the statistical entropy H of the Cadmium flow of one ton MSW is shown for both scenarios (Step IVa and b). Additionally, three preceding steps are included for a better understanding of the generation and treatment of MSW and in order to relativize the impacts of the scenarios on the evaluation measure H. In Step I the different fractions of MSW (e.g. paper, plastics, metals) are still separated. The mean concentrations of Cadmium in these fractions are rather different (cf. Figure 6). Cadmium is mainly in the fractions plastics and metals. Step II represents the mixed fractions as they are found in a dustbin or landfill. As the fractions are mixed, information about the exact location of high or low concentrated goods has been lost. Only the mean concentration of the mixed fractions is known.

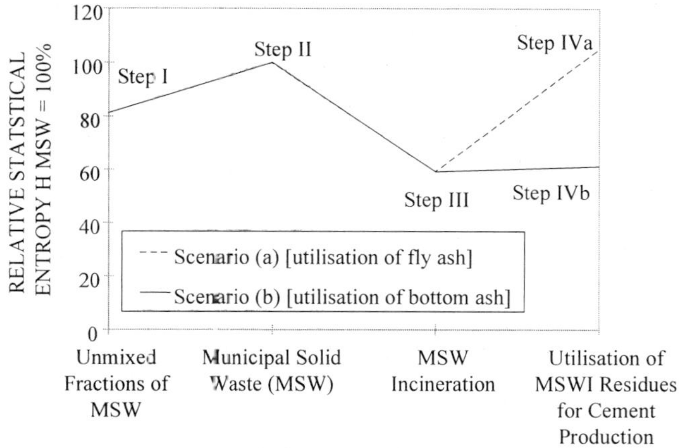

Figure 5 Relative Statistical Entropy for the distribution of Cadmium in different phases of the MSW life cycle

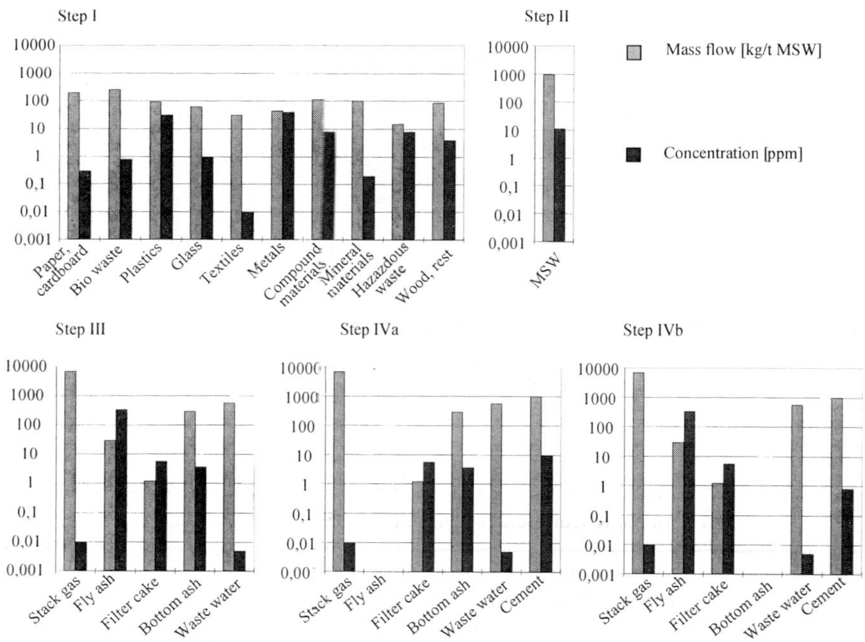

Figure 6 Distributions for Cadmium (mass-flows and concentrations) in different phases of the MSW life cycle

In Step III MSW is separated again into several product streams such as stack gas, wastewater and solid residues. Since the concentrations in these mass-flows are either very low (stack gas, wastewater) or rather high (fly ash) the statistical entropy is significantly decreased by the incineration step. The corresponding transfer coefficients (partitioning of a substance to outputs) for Cadmium are given in Table 2. The data are from a state-of-the-art incinerator equipped with a grate type furnace, a waste heat boiler, an electrostatic precipitator, two wet scrubbers, $DeNO_x$ plant as well as wastewater treatment [4]. Leaching of MSW as well as of incineration products in landfills and leaching of cement or concrete, respectively is not considered in the calculations. Leaching rates for landfilled MSW and bottom ash are that low, that they have no impact on the entropy during a time period of 100 years [1]. This means, that points II and III in Figure 5 would not make a relevant shift towards the X-axis when considering leaching processes. This is assumed for concrete as well (points IVa and b).

In Step IV either the total fly ash (scenario (a)) or the total bottom ash (scenario (b)) is utilised as a secondary material for cement production. The amounts for specific clinker, cement, and waste production (kg/capita and year) are estimated for average European Union conditions. A selection for some member states is given in Table 1. The figures about the rise of MSW are the result of a literature survey by de Fraja Frangipane et al. [6]. They are not consistent, as the authors explain that some countries summarise additionally bulky waste, commercial waste, sewage sludge, etc. under the term MSW. E.g. the actual rise of waste from private households which has to be treated (not recycled) in Austria is about 160 kg/capita and year [7] while de Fraja Frangipane et al. determined 320 kg/capita and year. However, in a first approach it can be assumed that the production figures for clinker, cement, and MSW are roughly equal. For the calculations a ratio between MSW and cement of 1:1 is taken. This means that in scenario (a) about 30 kg fly ash are mixed with clinker to produce 1 ton of cement. In scenario (b) one ton cement is generated with ca. 290 kg bottom ash. The results do not change significantly if the ratio between MSW and cement varies between 3:1 and 1:3. This range should cover the actual situation.

The difference between utilising fly ash or bottom ash in the case of Cadmium is evident (cf. Figure 5). Since the bottom ash is poor on Cadmium the mixing entropy with the low concentrated clinker is not relevant. On the other hand the utilisation of the high concentrated fly ash increases the entropy considerably. Hence, with regard to the management of Cadmium flows according to the above objectives there is a clear recommendation for utilising the bottom ash for cement manufacturing.

Table 1 Selected production and consumption figures for clinker, cement, and MSW in some European countries, [5], [6], figures rounded

COUNTRY	CLINKER PRODUCTION*	CEMENT PRODUCTION*	CEMENT CONSUMPTION**	WASTE PRODUCTION***
	kg/capita and year			
United Kingdom	200	210	220	560
Germany	340	400	420	440
France	260	310	320	470
Italy	460	590	590	350
Austria	360	470	630	320

* 1996, ** 1997, *** 1990ies

Table 2 Transfer coefficients for Cd, Pb, and Cu for a MSW incinerator, [4], figures rounded

	Cd	Pb	Cu
Stack gas	< 0.01	< 0.001	< 0.0001
Fly ash	0.90	0.22	0.043
Filter cake	< 0.001	< 0.001	< 0.001
Bottom ash	0.10	0.78	0.96
Waste water	< 0.001	< 0.0001	< 0.0001

In this paper Cadmium serves to asses the effect of volatile metals on the substance management. In the following, scenario (a) and (b) are shown for a less volatile (Pb) and a lithophile metal (Cu), cf. Figure 7. Contrary to Cadmium the entropy is more increased by the utilisation of the bottom ash than of the fly ash. This can be easily explained by considering the partitioning of these elements in the incineration process. Table 2 shows the transfer coefficients for Cadmium, Lead, and Copper in a state-of-the-art incinerator as described before [4].

All three elements are partitioned between fly ash and bottom ash since the flows via the remaining output goods are negligible (< 0.01). While for Cadmium the main flow is via the fly ash for Lead and even more for Copper it is via the bottom ash. The extreme differences in the partitioning between these metals prevent that one of the solid residues can be entirely utilised as a secondary resource without a significant increase in the entropy of the system.

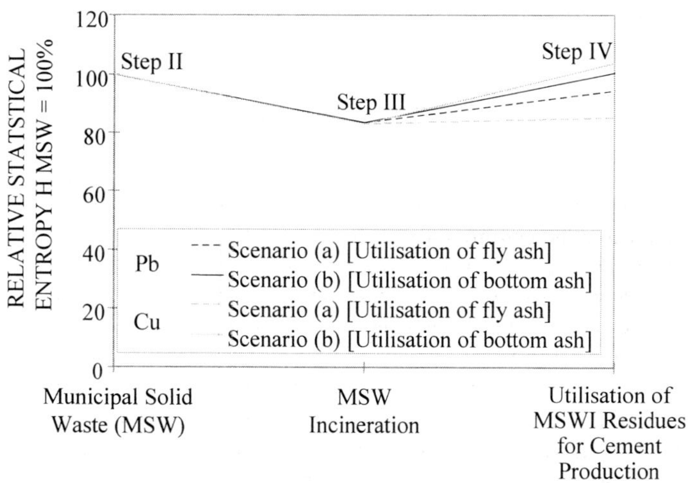

Figure 7 Relative Statistical Entropy for the distribution of Lead and Copper in different phases of the MSW life cycle

In order to avoid a significant re-increase in the entropy from Step III to IV there are several alternatives: (1) Only small amounts of the bottom ash are used for the cement production. The fly ash is not applied at all. (2) Bottom ash and/or fly ash are treated in further processes to reduce concentrations and substance flows via these residues. Such processes already exist in full or lab scale, respectively, and are matter of extensive research on university and industrial level.

With the introduced MASTER approach the necessary performance of these processes can be determined. E.g. the fly ash of an incineration plant can be treated in a chemical and thermal process in order to transfer metals into the filter cake. This has been realised in a plant in Austria [8]. If the partitioning of Cadmium is changed by the process like displayed in Figure 8, the utilisation of the fly ash would mean no significant increase in the statistical entropy compared to incineration without after-treatment. This theoretical distribution is almost achieved by the mentioned incinerator [8].

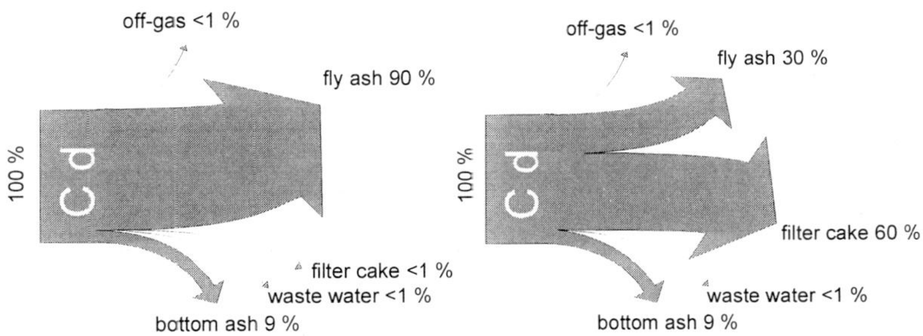

Figure 8 Required change in the partitioning of Cadmium through after-treatment of the fly ash to avoid an increase in the entropy when the fly ash is used for cement production

CONCLUSIONS

As MSW incinerators concentrate metals in their solid residues they are entropy reducing processes for relevant metals to waste management. Utilising incinerator residues for cement production re-increases the entropy for these metal distributions since they are diluted in the larger mass of the essentially lower concentrated clinker. Therefore neither fly ashes nor bottom ashes of MSW incinerators should come into the cement in large scales. This does not apply to residues which have been decontaminated from hazardous metals by after-treatment devices or further developed incineration processes. The MASTER approach can be used for the design of the required substance balances which have at least to be produced by these processes. It can be additionally applied to determine the amounts of the different residues which should be maximally used for manufacturing of cement in order to observe the objectives of sustainable waste or substance management, respectively.

REFERENCES

1. RECHBERGER H. Development of a method to evaluate substance balances in waste management. PhD-Thesis (in German), Institute for Water Quality and Waste Management, Vienna University of Technology, 1999

2. MORF L S. Development of an efficient method for routine-monitoring of substance flows through MSW incinerators, PhD-Thesis (in German), Institute for Water Quality and Waste Management, Vienna University of Technology, 1998

3. BACCINI P & BRUNNER P H. Metabolism of the Anthroposphere. Springer. New York, 1991

4. SCHACHERMAYER E, BAUER G, RITTER E & BRUNNER P H. A Material Balance of the Municipal Solid Waste Incineration Plant "Spittelau" via the Residues from Incineration. Federal Environment Agency, Vienna, Monographien/Band 56 (in German), 1995

5. STATISTICAL DEPARTMENT CEN BUREAU. 1999

6. DE FRAJA FRANGIPANE E, FERRARIO M & PASTORELLI G. Municipal solid waste management in European metropolitan areas: an integrated approach. ISWA 97 World Conference, Session Proceedings 1, p. 334-353, 1997

7. KRAMMER H J, PERZ K, DREIER P & DOMENIG M. Federal Waste Management Plan. Waste generated in Austria. Federal Environment Agency, Vienna, Monographien/Band 101 (in German), 1998

8. MORF L S, RITTER E & BRUNNER P H. Mass- and Substance Balance of the MSW Incinerator "Wels". Institute for Water Quality and Waste Management, Vienna University of Technology (in German), 1997

REUSE OF CEMENT-BASED S/S MSWI FLY ASH FOR LANDFILL DISPOSAL: MODELLING FOR LONG TERM ENVIRONMENTAL IMPACT PREDICTION

L Tiruta-Barna
A Imyim
R Barna
J Méhu
National Institute of Applied Sciences
France

ABSTRACT. The cement-based S/S MSWI fly ash was studied to investigate the mechanism of contaminant release under landfill disposal scenario. Waste behavioural parameters in scenario conditions have been studied according to the European standard ENV 12 920 "basic characterisation of leaching behaviour in specified conditions". The scenario definition is the disposal of S/S fly ash as a monolith washed off with rainwater. Samples containing 30% of fly ash, 23% of Blast Furnace Slag Cement, and 23% of Ordinary Portland Cement were prepared. The S/S waste was characterised by different leaching tests: Pore Water Test, influence of pH under stationary conditions and Monolithic Leaching Test. The results of the tests lead to identify the necessary input parameters of a new environmental behavioural model. The scenario simulation results obtained by modelling provide the profiles of pollutant release under the considered scenario. These profiles show the concentration of pollutants in leachate and their cumulative release amount over 100 years.

Keywords: Waste stabilisation, Leaching, Long-term prediction, Modelling, Diffusion

Dr L Tiruta-Barna, is Researcher/Lecturer at the Environmental Analysis and Industrial System Procedures Laboratory, National Institute of Applied Sciences of Lyon. Her research interests include waste assessment and environmental computation modelling.

Mr A Imyim, is a Research/Teaching staff at Department of Chemistry, Chulalongkorn University, THAILAND. He is actually a PhD student at National Institute of Applied Sciences of Lyon on the subject of waste treatment technology.

Dr R Barna, is Associated Professor at the Environmental Analysis and Industrial System Procedures Laboratory and Project Manager at Waste, Pollution and Environment Research Unit (Polden), National Institute of Applied Sciences of Lyon. He specialises in waste stabilisation/ solidification.

Mr J Méhu, is Director of Waste, Pollution and Environment Research Unit (Polden), National Institute of Applied Sciences of Lyon.

INTRODUCTION

Municipal wastes are increasingly produced and lead to an important environmental problem. These wastes must be treated by different procedures such as thermal treatment, biological treatment etc. The treatments can never totally eliminate the wastes, so it still remains secondary wastes called final waste. The Municipal Solid Waste Incineration (MSWI) is largely developed. This disposal route generates an important amount of MSWI fly ash. In France, according to the French regulation of final hazardous waste, it has to be treated by stabilisation/solidification (S/S) prior to landfilling. The cement S/S treatment process is largely used. This can reduce the leaching potential of contained pollutants in two ways [1]. First, the pollutant species are fixed in the cement structure that physically decreases their mobility. Second, they can react with another cementitious species during hydration reactions; this can chemically change the form of the contaminants and makes them less soluble and less mobile. This coupling of physical and chemical mechanisms has to be taken into account in the prediction of long term leaching behaviour.

One of the general problems of waste management is how to assess or provide a solution for waste disposal and/or its utilisation. A methodology is needed to determine the parameters of the leaching behaviour of waste under specified conditions. The methodology which is proposed as the European standard "basic characterisation of leaching behaviour in specified conditions" ENV 12 920 consists of several steps: definition of the problem and the solution sought, description of the scenario and the waste, determination of the influence of parameters on leaching behaviour, modelling of leaching behaviour, and behavioural model validation.

Since a great number of leaching tests exists, a selection of tests is required with scenario considerations. The leaching tests can be distinguished in two categories depending on the application context and on the scientific objective [2]. The first ones are not appropriate to understand leaching behaviour or for the prediction of release from S/S wastes. These tests, such as the compliance tests and in-situ tests, are applied for on-site verification and characterisation of unknown materials. The second ones lead to identify the parameters controlling the release mechanisms from the S/S wastes. The three most common are:

- Test of influence of pH under stationary conditions allowing to determine the acid neutralisation capacity (ANC) and the solubility pH dependence of considered species at equilibrium
- Monolithic Leaching Test (MLT) aiming at identifying the integration of physical and chemical parameters controlling the pollutant fluxes.
- Pore Water Test (PWT) aiming at determining the chemical composition of the pore solution

The release from monolithic S/S wastes is complex and time dependent. Such leaching tests have been applied to characterise the input parameters for the modelling step. Leaching models are useful to describe the release mechanism in order to predict the long-term behaviour in environment. Most of them are often based on the basic assumption that mass transfer of soluble contaminants during leaching can be described by Fick's law using an apparent diffusion mechanism [3]. A coupled dissolution/diffusion model [4 and 5] has been developed in this field. This more sophisticated model can better allow describing the leaching phenomena.

In this study, the purpose is to investigate the mechanism of contaminant release under landfill disposal scenario of the cement-based S/S MSWI fly ash by means of the application of the European standard ENV 12 920.

SCENARIO DEFINITION

The material is disposed as one monolith with a top cover letting a certain level of infiltration of rain. The scenario is schematically shown in Figure 1. The deposited S/S waste of 10-m height, 20-m width, and 50-m length is indirectly washed off by rainwater. The average annual rainfall in France is about 800 l/m^2/year. The monolith is covered by a permeable layer through which rainwater penetrates continuously with an approximate flow rate of 300 l/m^2/year. The chemical composition of rainwater is simplified that it doesn't contain any contaminant.

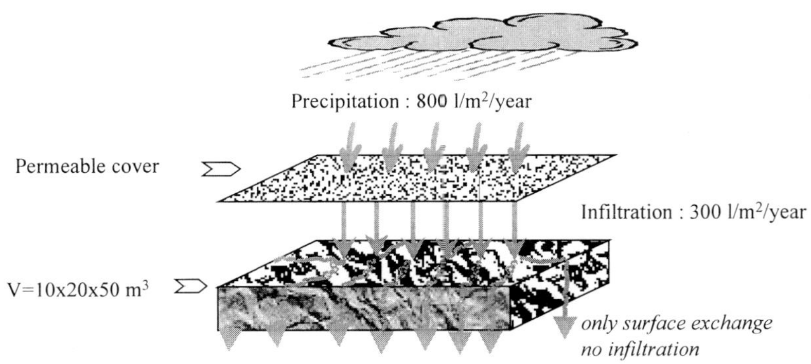

Figure 1 Schematic presentation of the landfill scenario

We assume that there is neither percolation trough the material nor modification of the physical properties and liquid/solid contact characteristics during the simulated time that will be 100 years. The principle mechanism of mass transfer toward the contact solution is only the leaching from the wet surface of the bloc. Only inorganic pollutants are considered, Pb in our case.

MATERIALS AND METHODS

Fly Ash and S/S Fly Ash

The total content of the MSWI fly ash and the S/S fly ash have been characterised by alkaline fusion and digestion in acid solution. The solution was analysed. The elementary chemical composition is presented in Table 1. The cement based S/S MSWI fly ash is composed of 30 % of fly ash, 23 % of Blast Furnace Slag Cement, 23 % of Ordinary Portland Cement, and 24 % of water. After 28 days of curing, its physical properties were determined and illustrated in Table 1.

The analyses were done by inductively coupled plasma-atomic emission spectroscopy (ICP-AES) for selected elements (Al, As, Ba, Bi, Ca, Cd, Cr, Cu, Fe, Ga, K, Mg, Mn, Na, Ni, Pb,

Sr, and Zn). Chloride ion concentration was analysed by ionic chromatography with conductivity detector.

Table 1 Metal contents and physical characteristics of the MSWI fly ash and the S/S fly ash

	FLY ASH	S/S FLY ASH (mg/kg)		FLY ASH	S/S FLY ASH (mg/kg)
Al	54711	38576	Mn	1052	828
As	174	98	Na	43974	15794
Ba	1000	515	Ni	103	42
Bi	81	22	Pb	5004	2552
Ca	166272	269454	Sr	13510	10540
Cd	270	104	Zn	18619	10540
Cr	566	210	Cl^-		34193
Cu	1299	675	*Physical characteristics*		
Fe	23610	13789	Density		2200 kg/m^3
Ga	720	401	Water Absorption Capacity(WAC)		29%
K	50612	20682	Humidity		14%
Mg	21251	15080	Porosity (calculated from WAC)		43%

Experimental Procedure of Leaching Tests

Three leaching tests have been performed as follows.

Pore Water Test (PW): The principle is to assess the solubility of constituents at equilibrium between fine fragmented product and demineralised water in closed vessel according to different liquid/solid ratios (L/S) at room temperature (23$\pm$1°C). The material was crushed to $\leq$ 315 μm and then put into contact with demineralised water with L/S of 10, 2, and 1 ml/g (dried material) in polyethylene vessels. The vessels were closed and agitated for 3 days by an end-over-end tumbler (10 rpm). The mixtures were filtered trough the 0.45 μm filter membrane under pressure of nitrogen gas to avoid the CO_2 uptake. The pH, conductivity, Redox potential were measured. Some of solutions are divided to determine the total solid dissolved (TDS). The solutions were analysed.

pH static Test: The test, which is conducted on a mixture of fragmented sample ($\leq$ 2 mm) and demineralised water with initial L/S 5 ml/g(dried material) at room temperature, is carried out at different controlled pH (4 -12) by addition of 1 M HNO_3 for 24 hours under magnetic stirring. The filtration, the parameter measurement, and the analysis are then performed by the same procedure as PW test.

Tank Leaching Test (NEN 7345 [6]): The test is conducted on monolithic cubic specimens (4x4x4 cm^3) with renewal of solution at 0.25, 1, 2, 4, 9, 16, 36 and 64 days. The samples were put in demineralised water acidified to pH 4$\pm$0.5 by 0.1 M HNO_3 at L/S 2.7 ml/g (dried material). Every change of solution, the filtration and the measurement of parameters must be done as well as the analysis of solutions.

The filtrates were analysed by ionic chromatography for chloride ion, then acidified to pH 2$\pm$0.5 with HNO_3 solution (65%) and analysed by ICP-AES for selected elements (Ca, Na, K, and Pb).

MODELLING OF LEACHING BEHAVIOUR

A behavioural model describing the constituent release mechanism from monolithic S/S wastes is considered in this step. The model [8] takes into account the couple of transport phenomena with major chemical reactions (dissolution, precipitation, and complexation). The chemical reactions are considered among 5 major elements (Ca, Na, K, Cl, and Pb). With taking into account of the activity of the species, an electroneutrality equation and a chemical equilibrium equation calculate concentrations of $Ca(OH)_2$ and/or CSH, Ca^{2+}, $Pb(OH)_2$, Pb^{2+}, $Pb(OH)_3^-$, Na^+, K^+, Cl^-, and also hydronium and hydroxide ions. One of the hypotheses of the model is that the dissolution of alkaline species controls mainly the leachant pH. The solubility of pollutants depends on the chemical context of the solution that is globally represented by the pH value.

The initial interest of the different leaching tests is to identify the parameters used as input values for the behavioural model. The necessary parameters for modelling are

- the initial concentration of each soluble species in the pore solution
- the initial concentration of each solid phase
- the diffusion coefficient of relevant mobile species obtained by adjustment of the simulation results to the experimental data of Tank Leach Test.

RESULTS AND DISCUSSION

Leach Test Results

The results of leaching tests presented below are pointed to interpret and determine the model parameters. The results of PW test are shown in Figure 2. The L/S ratios are plotted against the concentrations of elements (on principle axis) and the pH (on secondary axis). The pH of solution increases while the L/S ratios decrease, as well as the concentrations of K, Na, Cl and Pb. Contrarily, the Ca concentration decreases.

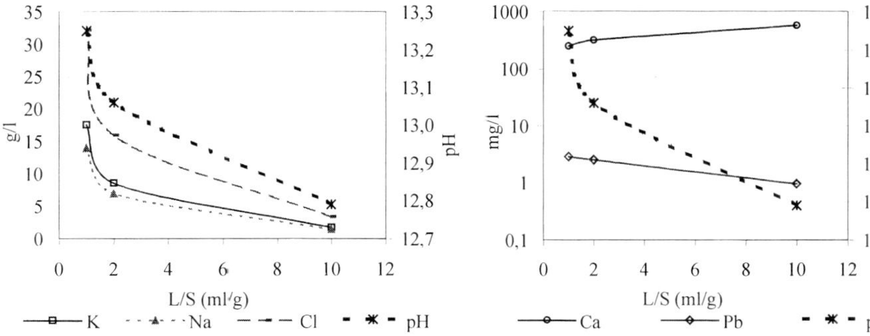

Figure 2 pH and concentration of relevant species in the pore solution

The pH of the solutions is mainly controlled by dissolution of alkaline species (K and Na). At high pH the solubility of calcium ion is limited by the solubility constant of Portlandite [7] or CSH [9]. The L/S ratio of the pore solution can be evaluated from the Water Absorption Capacity that is 29%. We can extrapolate the curves of the element concentrations toward the

L/S 0.29 ml/g. The initial concentrations of every species in pore solution are estimated and converted to mol/l; 2.08 M for Na, 1.58 M for K, 4.10^{-3} M for Ca, 1.34 M for Cl, and 1.10^{-5} M for Pb.

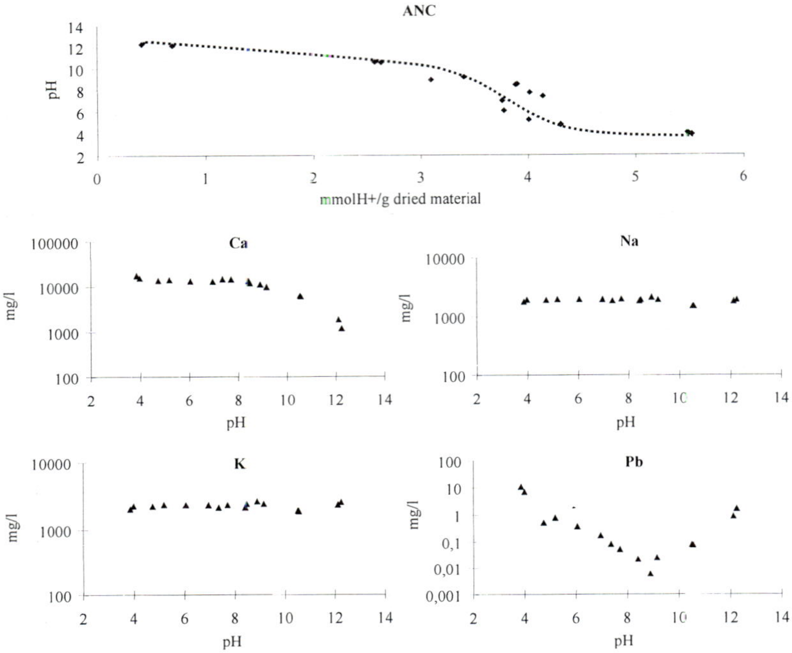

Figure 3 Concentration of elements in function of pH and ANC

The results of the pH static test (Figure 3) show the ANC of the material and the solubilisation of elements in function of pH. The pH non-dependent aspect of Na and K solubilisation has found. With regard to the graphic of Ca in Figure 3, the available amount of calcium can be estimated from the plateau of the curve between pH 4-8. We can identify the initial quantity of solid phases of calcium and afterward convert into pore concentration (input parameters of the model as presented above). Its value is 9.5 M. In the same way, the approximation of lead solid concentration in pore solution is deduced from the maximum extracted amount at the lowest pH . Its value is $1.1 \cdot 10^{-3}$ M.

The Pb concentration curve as a function of pH shows obviously the pH dependence. The chemical speciation of solid phases of lead in the cement matrix is unknown. We determine the Pb solubility in the pore solution by regression of the pH solubility curve in stead of the solubility constant of $Pb(OH)_2$. The equation (1) will next introduce in the model.

$$\log[Pb] = 0.1145pH^2 - 1.9654pH + 1.5882 \qquad (1)$$

The results of the Tank Leach Test are given in Table 2. The pH of the leachate remained virtually constant during the test, ranging from 11.95-12.16. It is controlled by leaching of soluble species from the cement matrix, mainly alkaline species, Portlandite, and CSH. The conductivity values lowered caused by the decrease of total ion concentrations in the leachate. The pH and calculated average flux of relevant species in *mmol/m²s* are plotted against mean square root of time (T_i in days) in Figure 4, represented by lozenge (•) and cross (x) symbols for the two series of tests.

Table 2 Physical parameters and concentration of elements (and their duplicates) in leachate of Tank Leach Test

RUNNING NO.	LEACH TIME (h)	pH	CONDUCTIVITY (mS 20°C)	REDOX POTENTIAL (mV)	TDS (g/l)	CONCENTRATION (mg/l)				
						Ca	K	Na	Pb	Cl
1	6	12.15	11.8	105	10.4	106	2023	1556	0.133	3955
1	24	12.17	9.2	114	7.5	109	1653	1311	0.121	3044
1	48	12.06	6.4	114	3.0	104	799	660	0.064	1569
1	96	12.11	5.4	129	3.5	109	630	508	0.043	1117
1	216	12.15	5.1	126	2.6	113	523	398	0.047	960
1	384	11.96	3.4	132	1.6	127	298	238	0.052	525
1	864	12.05	3.7	198	1.6	145	317	238	0.044	567
1	1536	11.95	2.8	221	1.1	141	218	175	0.016	349
2	6	12.16	10.1	101	7.5	103	1525	1162	0.125	3213
2	24	12.14	8.9	125	7.1	98	1393	1002	0.094	2601
2	48	12.06	6.0	125	2.7	98	736	600	0.060	1327
2	96	12.12	5.0	132	2.7	105	597	473	0.016	1047
2	216	12.16	5.2	131	2.6	106	544	396	0.023	980
2	384	11.98	3.8	143	1.8	119	359	264	bd*	733
2	864	12.11	4.5	184	2.1	133	450	313	0.040	789
2	1536	12.04	3.5	193	1.5	135	310	241	bd*	554

* *Below analytical detection limit*

Simulation Results

The parameters obtained from the leaching tests had been put into the model computing calculation program. The effective diffusion coefficients (D) of each species are evaluated by the best consistency of simulation results and experimental ones: $1.5 \cdot 10^{-10}$ m^2s^{-1} for Na^+, K^+, and Cl^-, $1 \cdot 10^{-10}$ m^2s^{-1} for Ca^{2+}, Pb^{2+}, and $Pb(OH)_3^-$. The experimental and simulation pH and constituent flux are illustrated in Figure 4.

We observe that the simulation values of the pH and the flux reproduce very well the experimental values. The simulated pH is a global parameter of the leaching system because it represents every physical and chemical equilibrium occurring during the leaching process. This shows the performance of the coupled physico-chemical interpretation method.

Scenario Simulation

The input parameters controlling the release were conserved and introduced in the specific computing calculation program for the scenario. The program needs also the scenario parameters as defined previously: the flow rates of effective rainwater (300 l/m^2/year), the chemical composition of rainwater, and the dimension of the monolithic bloc. The simulation has been done over 100 years. The model can provide the concentration and the cumulative

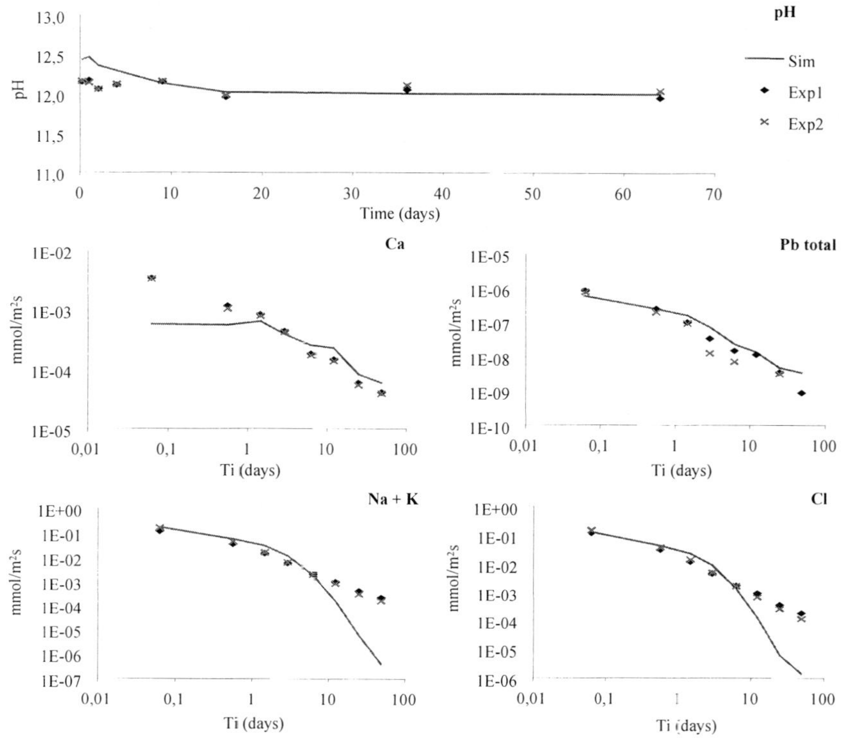

Figure 4 Experimental and simulation pH and constituent flux

released amount of each species in leachate. The prediction of pH and pollutant release (Pb) is proposed in Figure 5.

The leachate remains alkaline until 100 years, ranging from 13 to 12. This pH value is controlled by the dissolution of $Ca(OH)_2$ and CSH from the S/S waste. The cumulative extracted amount of Pb is 3×10^{-7} % of the total content. This proves obviously the performance of the S/S procedure in relation to the scenario conditions.

CONCLUSION

The integration of the interpretation of leaching test results and the modelling is a promising approach for long-term environmental impact prediction. The proposed complex leaching model provides a good simulation of the experimental results and forecast the leaching behaviour under the landfill scenario. However some sophisticated conditions (such as climatic conditions, biological reactions etc.) are not yet taken into account. The last step of the methodology, that is to say field & pilot confrontation, is lacking in this phase and constitutes the necessary next step of the study.

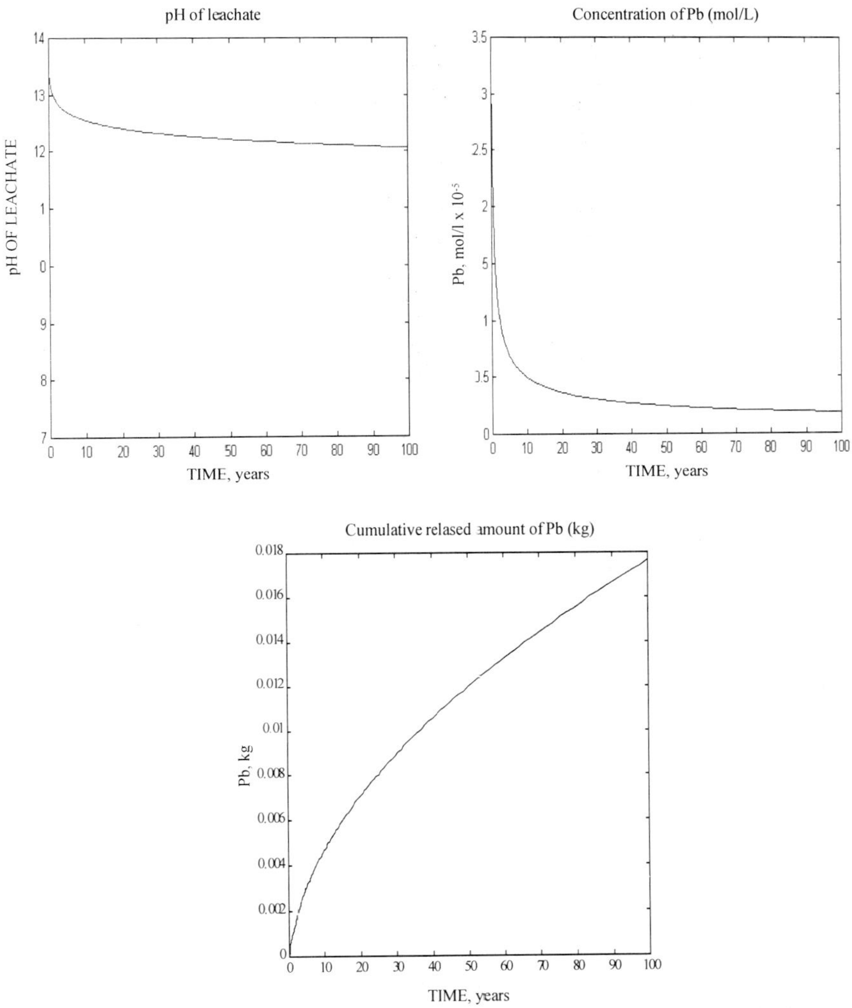

Figure 5 Scenario simulated concentration and cumulative release amount of Pb in the leachate

ACKNOWLEDGEMENTS

This study has been conducted in the framework of a European programme to support the Network of Harmonisation on Leaching/Extraction Tests (DGXII-S,M&T). The members of the co-ordination team are H.A. van der Sloot (ECN-NL-co-ordinator), L. Heasman (M.J. Carter Associates Ltd.-UK), G. Rauret (Universitat de Barcelona - S), I. Hohberg (Institut fur Bauforschung – G), O. Hjelmar (VKI – Denmark), M. van den Berg (NNI - The Netherlands), J.Turell (WRc-UK) J. Méhu (Polden-INSAVALOR – France), Ph. Quevauviller (EC DGXII , Standards, Measurement & Testing Programme).

In the project on Technical work in support of the Network Harmonization of Leaching/ Extraction Tests (SMT4-CT96-2066) the testing of the different materials (soil, contaminated soil, sediment, compost, sewage sludge, waste, stabilized waste, construction materials, aggregates, drinking water pipes and preserved wood) by the various selected leaching test methods has been completed. The data comparison, modelling and interpretation is in full progress (> 25000 data points). The information generated by this project will be available at the end of 1999.

REFERENCES

1. CONNER, J.R. Chemical fixation and solidification of hazardous wastes. New York: Van Nostrand Reinhold, 1990. 692 p.

2. VAN DER SLOOT, H.A., HEASMAN, L., QUEVAUVILLER, PH. Harmonisation of Leaching/ Extraction Tests. Amsterdam : Elsevier, 1997, Chapter 9-10, p 171-208.

3. MOSZKOWICZ, P., SANCHEZ, F., BARNA, R., MÉHU, J. Pollutants leaching behaviour from solidified wastes: a selection of adapted various models, Talanta, 1998, vol 46, p 375-383.

4. BATCHELOR, B. A numerical leaching model for solidified/stabilised wastes, Wat. Sci. Tech., 1992, vol 26, n° 1-2, p 107-115.

5. BARNA, R., SANCHEZ, F., MOSZKOWICZ, P., MÉHU, J. Leaching behaviour of pollutants in stabilised/solidified wastes, Journal of Hazardous Materials, 1997, vol 52, p.287-310.

6. NNI: NEN 7345, Leaching characteristics of building and solid waste materials – Leaching tests : Determination of the leaching behaviour of inorganic components from shaped building materials, monolithic and stabilised waste materials. The Netherlands.

7. DUCHESNE, J. AND REARDON, E.J. Measurement and prediction of Portlandite solubility interaction alkali solutions, Cement and Concrete Research, 1995, vol 25, n° 5, p. 1043-1053.

8. TIRUTA-BARNA, L., LACT, G., MOSZKOWICZ, P., BARNA, R. Emission des polluants dans differents scénarios de stockage et d'utilisation : Modèle de Transfert couplés avec réaction chimique pour la lixiviation de déchets stabilisés/solidifiés. In Wastes Stabilization & Environment 99 (Posters), Lyon, France, Société Alpine de Publications, 1999, p. 90-92.

9. FAUCON, P., ADENOT, F., JACQUINOT, J.F. ET AL. Long term behaviour of cement pastes used for nuclear wasted disposal: review of physico-chemical mechanisms of water degradation, Cement and Concrete Research, 1998, vol 28, n° 6, p. 847-857.

USE OF INCINERATOR FLY ASH AS A POZZOLANIC ACTIVATOR

J E Halliday

T D Dyer

R K Dhir

University of Dundee

United Kingdom

ABSTRACT. This paper investigates the use of municipal solid waste incinerator fly ash(IFA) as a cement replacement in ternary blends also containing Portland cement(PC) and pulverized-fuel ash (PFA). Mortar specimens where used to determine the compressive strength at 2,7 and 28 days and the porosity at 28 days of ternary blends, with PC levels between 100% and 50%. Hydration products were investigated, for blends of 65% PC, with the help of thermogravimetry, isothermal conduction calorimetry and x-ray diffraction. The compressive strength results show that optimum blends, for any given PC content, contain all three constituents, with a PFA:IFA ratio of 3:1 at all ages, and by 28 days these blends are much higher than PC/PFA only mix. The blends which display optimum compressive strength also possess the optimum porosity values for a given PC content. X-ray diffraction shows that Friedel's salts is an important component in the hydrated blends containing all three constituents and that inclusion of IFA accelerates the PFA reaction.

Keywords: Incinerator fly ash, Mortar, Compressive strength, Hydration products.

Miss J E Halliday, is a research student at the University of Dundee. She is currently completing her PhD on the subject of combustion waste materials as a cement replacement.

Dr T D Dyer, is a research / teaching fellow in the Concrete Technology Unit of the Department of Civil Engineering, University of Dundee. He is a materials scientist with interests in waste recycling and other aspects of sustainable construction.

Professor R K Dhir, is Director of the Concrete Technology Unit, University of Dundee. He specialises in binder technology, permeation, durability and protection of concrete. His interests also include the use of construction and industrial wastes in concrete to meet the challenges of sustainable construction. He has published and travelled widely and serves on many Technical Committees. He is a past chairman of the Concrete Society, Scotland.

INTRODUCTION

Every year more than 1 million tonnes of household solid waste is produced in the western world [1]. To avoid the health and the environmental problems that arise if left untreated, landfill disposal was once the most popular method of disposing waste. However, lack of available land, the implementation of landfill taxes and the increasing emphasis on ecological preservation have lead to other disposal solutions being sought.

One such solution is incineration, which is considered to be one of the better options, since it reduces the volume of incoming waste by 65-80% [2, 3]. Incineration of municipal waste produces residues, consisting of a combination of bottom ash (coarse) and fly ash (fine). Incineration of municipal solid waste usually yields fly ash in quantities of around 3-4% by weight of the original uncombusted material [4].

This paper deals with the use of incinerator fly ash (IFA) as a cement replacement, and the suitability of using it as a pozzolanic activator. The test program described below was carried out using binders containing ternary blends, i.e. PC / PFA / IFA. Portland cement (PC) was used as the driving force for other cementitious reactions and to give early strength, whilst pulverised-fuel ash (PFA) was used to harness the potential pozzolanic activation capabilities of the IFA. Figure 1 shows a ternary diagram indicating the composition of all the blends investigated, with percentages of each constituent also shown. The blends are numbered 11 to 26 (excluding 22 and 24).

MATERIALS

The incinerator fly ash used was derived from the combustion of municipal solid waste in a mass burn incinerator and the chemical composition is shown in Table 1. The chemical composition for both PC and PFA are also given in this table.

EXPERIMENTAL DETAILS

Compressive strength was determined at 2,7 and 28 days following BS EN 196 Part 1 [5], using 40x40x160mm mortar prisms. The three constituents(450g) were mixed together with 225g of tap water (0.5 w/c) and 1345g of CEN sand. The prisms were then filled in two layers, both compacted on a vibrating table. The specimens were then stored for 24 hours in a storage unit of 20 ^{o}C and a relative humidity of 95%. After storage, the prisms were demoulded and stored in water filled curing tanks (20±2 ^{o}C) until the desired time. Avery apparatus was used in obtaining the compressive strength with a loading rate of 2400 N/s. The mortar prisms were split in two, each placed in a 40x40mm bracket plate, and the average was obtained.

The porosity of each ternary blend mix was determined by submerging a 40x40x80mm mortar prism(from BS EN 196 Part1) [5] in tolulene under vacuum conditions for 2 hours in accordance with BS 1881 Part 124 [6]. These results were obtained after 28 days of water curing.

Table 1 Chemical compounds of constituents

OXIDE	CONCENTRATION (wt%)		
	PC	PFA	IFA
SiO_2	21.1	54.2	9.73
TiO_2	0.2	1.0	0.79
Al_2O_3	5.0	27.4	6.5
Fe_2O_3	2.7	5.4	0.8
MnO	0.1	0.1	0.03
MgO	1.6	2.1	1.0
CaO	64.9	4.3	30.25
Na_2O	0.3	1.4	2.9
K_2O	0.6	1.7	2.3
SO_3^{2-}	3.3	0.4	7.1
Cl-	0.03	0.01	29

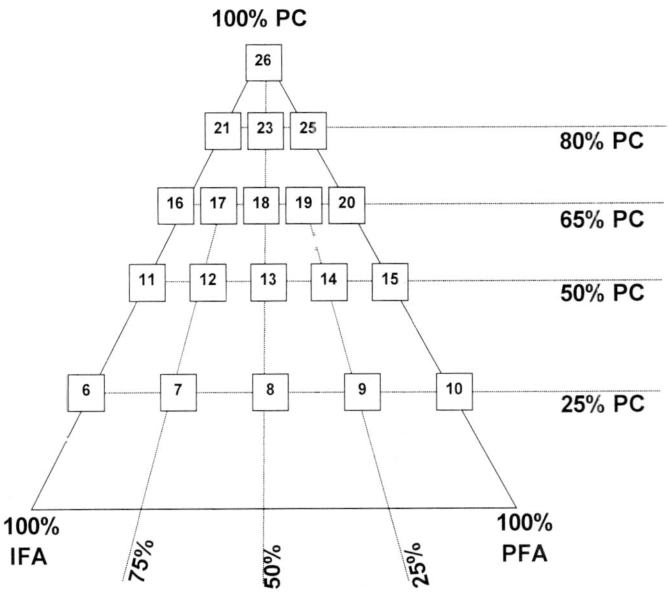

Figure 1 The blends investigated

Isothermal conduction calorimetry was carried out on paste samples to determine the effects of the inclusion of IFA in early hydration (the first 72 hours). The pastes were prepared with a water / solids ratio of 0.5 with distilled water using 30g of solid material. Heat evolution was monitored using a personal computer.

Pastes containing PC / PFA / IFA with 0.5 w/c (25g binder content and 12.5g of distilled water) were mixed and left to set in sealed cylindrical polyethlene containers for 24 hours, then opened and transferred to a water filled curing tank (20°C). The specimens were then removed at 2, 7 and 28 days and ground with acetone to stop further hydration. They were then left for 15 minutes, filtered and then dried in a vacuum for 2-8 hours at a temperature of 40oC. Once the specimens were dry, they were stored in a desiccator until analysis. The powdered samples were then used for various tests including x-ray diffraction (XRD) and thermogravimetry (TG).

XRD was carried out using a Philips diffractometer operating with a Cu-kα radiation source. The mineralogical composition of the hydrated pastes was estimated using Rietveld refinement techniques on the x-ray diffraction traces. The computer programs X-FIT and KOALARIET were used [7-12].

Thermogravimetry (TG) was also carried out, at 0, 2, 7 and 28 days, by heating 10mg of the powdered samples to 1000oC in a nitrogen atmosphere at a rate of 2oC/minute. Weight percentages of calcium-hydroxide (Ca(OH)2) and calcium-carbonate (CaCO3) of the residual ash were obtained.

An unknown quantity of amorphous material was present in all the samples as both CSH gel and the glassy phases of the slag. These unidentifiable phases present a problem in terms of quantification using Rietveld refinement. This problem may be overcome by the inclusion of a known quantity of an internal standard. However, since thermogravimetric analysis was also carried out on the pastes, portlandite contents were able to be determined accurately, thus allowing the quantities of this mineral present in the samples to be used as 'pre-existing' internal standards.

RESULTS AND DISCUSSION

Compressive Strength

The compressive strength results, at 2, 7 and 28 days, for ternary blends containing IFA are shown with the aid of ternary diagrams in Figure 2. The rate of strength development of blends containing PC/PFA/IFA are generally lower than those blends which only contain PC and PFA (which run along the right hand side of each ternary diagram). This can be seen at all ages with the exception of those mixes with an IFA/PFA ratio 1:3, i.e. mixes numbered 23, 19 and 14. The compressive strengths of these mixes are considerably higher than the mixes containing only PC and PFA at the same PC content (25,20 and 15 respectively). The optimum strength seems to occur at 80% PC with a 1:1 PFA/IFA ratio (which is greater than the 100% PC mix).

The rate of early strength development for mixes containing all constituents is slower than those mixes containing PC and PFA only. For example, mix 23 is lower than the PC mix at its early stages, but by 7 days it has gained strength and by 28 days is considerably higher than all the mixes.

Capillary Porosity

Capillary porosity results after 28 days of water curing are shown in Figure 3, and it can be seen that the mixes containing only PC and PFA have the lowest porosity values. However, higher strengths are obtained with blends containing IFA, which implies that the low porosities, which are presumably due to the excellent particle packing capabilities of PFA, are, in this case, less important than the nature of the hydration products formed.

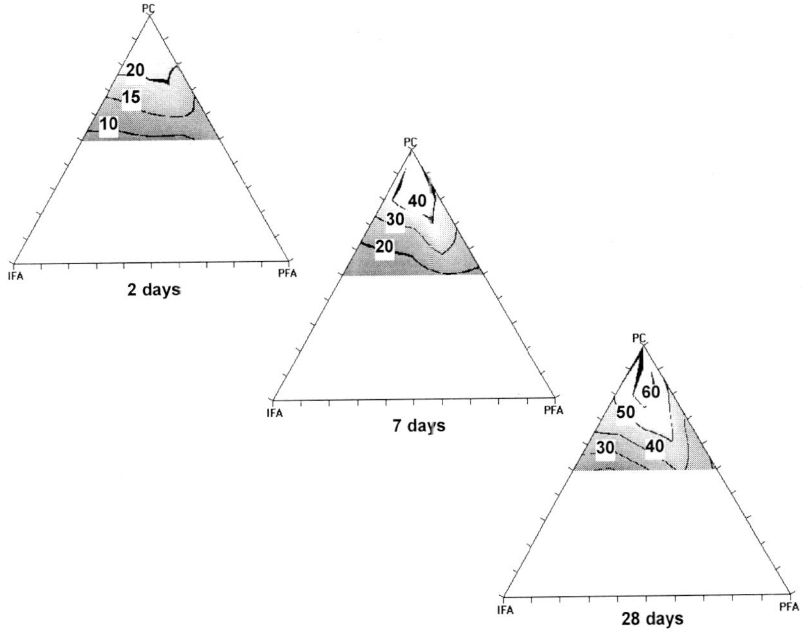

Figure 2 Ternary diagrams for compressive strength at 2,7 and 28 days (units = N/mm^2)

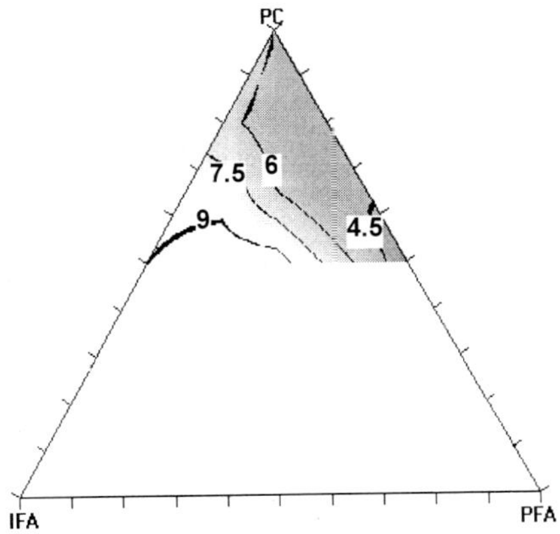

Figure 3 Ternary diagram for porosity at 28 days (units = %)

Isothermal Conduction Calorimetry

The three parameters that are used to determine the extent of acceleration or retardation of a particular blend are shown in Figure 4. The rate of heat evolution Q_{max}) ternary diagram shows that there is an increase in the value with mixes containing IFA. However, examination of the calorimeter traces shows that this increase is primarily due to the evolution of a significant quantity of heat immediately after mixing, presumably resulting from the wetting of the IFA.

The second ternary diagram shows the time at which the maximum rate of heat evolution occurs (t_{max}). Typically an increase in Q_{max} and a decrease in t_{max} indicates acceleration(and the opposite indicates retardation), but as it can be seen, t_{max} shows retardation at optimum blends, i.e. 23,19 and 14,. IFA contains components which both accelerate hydration (such as potassium chloride [13] and retard hydration (sulfate compounds [14]), and it is possible that the combination of such compounds would produce this effect.

Total heat generated by the test is shown in the third ternary diagram of Figure 4. This diagram shows that although there is little change in total heat between mixes containing IFA and those that do not, there is a slight increase with those containing PC / PFA / IFA.

Hydration Chemistry

The crystalline hydration products present in the ternary blends (with 65% PC only) are shown in Figure 5. The three graphs show the development of each product over a period of 2,7 and 28 days. The hydration products produced in ternary blends containing PC / PFA / IFA include ettringite ($3CaO.Al_2O_3.3CaSO_4.31H_2O$), which is produced during the hydration of the C_3A and C_4AF phases in the presence of gypsum. Ettringite also forms during the pozzolanic reaction of PFA. The growth of ettringite crystals is an expansive process which, when large quantities of the compound are formed, can cause cracking in concrete.

From Figure 5 it can been seen that at two and seven days, ettringite is present in all mixes, even those containing PC / PFA only. Mix 19 produces more ettringite than any other mix, but only in the early stages. By 28 days the PC / PFA only mix has run out of sulfate leading to a cessation in the production of ettringite, and the conversion of this compound to the AFm phase, calcium monosuphoaluminate hydrate ($3CaO.Al_2O_3.3CaSO_4.12H_2O$), often referred to as monosulfate.

Blends containing IFA do not display conversion of ettringite to monosulfate, since another AFm phase is favoured due to the presence of chloride ions [15]. Friedel's salt (calcium monochloraluminate hydrate) is produced when chloride ions are available in solution and once all available sulfate ions have been consumed, such as the case for blends containing IFA. For this particular set of mixes it can be seen that Friedel's salt is present to an increasing degree with an increase in initial chloride content (i.e. with an increase in IFA content). By 7 days the PC / IFA only mix (blend 16) has produced the maximum amount of Friedel's salt, whilst the remainder of the IFA containing mixes possess it in almost equal amounts.

When viewed as a whole, the production of crystalline hydration products (ettringite, Friedel's salt, monosulfate and portlandite) follows a relatively clear route. Initially the

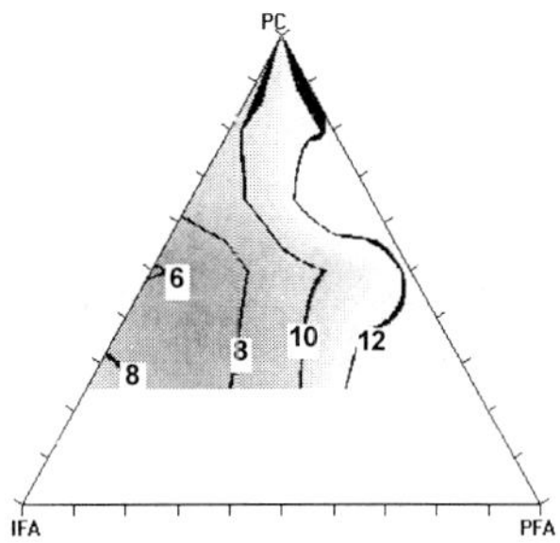

Q_{max} (units = W/kg)

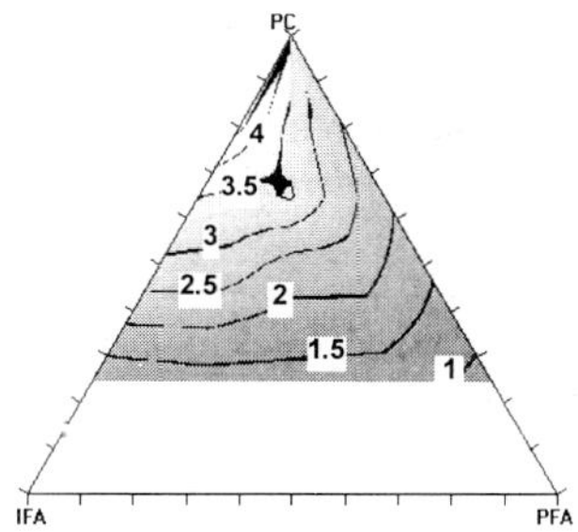

T_{max} (units = hours)

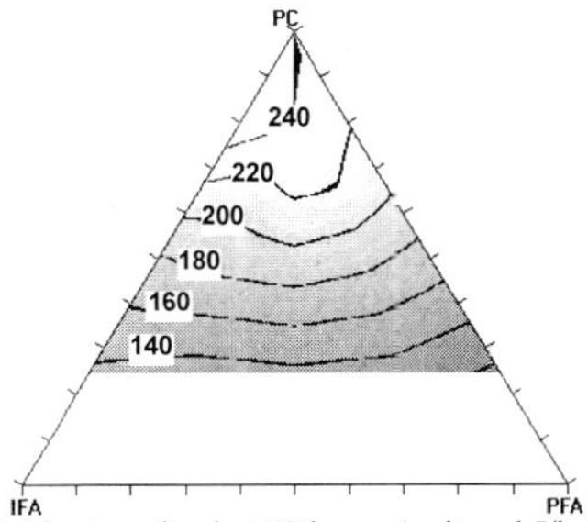

Total heat evolved at 72 hours (units = kJ/kg)

Figure 4 Ternary diagrams for isothermal conduction calorimetry parameters

ratio of 1:3. This corresponds with the high compressive strength observed in this blend largest quantity of crystalline hydrates is generated in blend 19, which possesses an IFA/PFA at all ages. Beyond 2 days, the quantity of hydrates in the blends containing higher quantities of IFA begins to increase, especially in blend 16. This increase is not reflected in terms of compressive strength development, which suggests that the increase in the quantity of these compounds in blends containing higher IFA levels is due to the leaching of soluble material, leaving behind the less soluble hydrates and probably creating porosity. In other words, the reason why blend 19 performs better than the other IFA-containing blends in terms of compressive strength development is that it is most appropriately proportioned for incorporating all the IFA components present within cement hydration products.

Calcium silicate hydrate (CSH) gel, which is a non-crystalline hydration product, cannot be measured using x-ray diffraction in this case, since it is impossible to differentiate between CSH gel and the amorphous components of both the IFA and PFA. However, an indication of the quantity present can be obtained from differentiated thermogravimetry traces, which are shown in Figure 6 for blends 16 to 20 at 2 and 28 days. A significant proportion of the dehydration of CSH gel occurs around 115°C, but overlaps the partial dehydration of ettringite. At 2 days the scale of this dehydration is largest in blend 19, which corresponds well with the compressive strength results. It should be noted that this blend contains the highest quantity of ettringite, and since it is extremely difficult to differentiate between the two features, it is impossible to determine precisely how much is actually CSH gel. By 28 days it would appear that leaching of material from the high IFA blends has caused the quantities of CSH in these blends to increase.

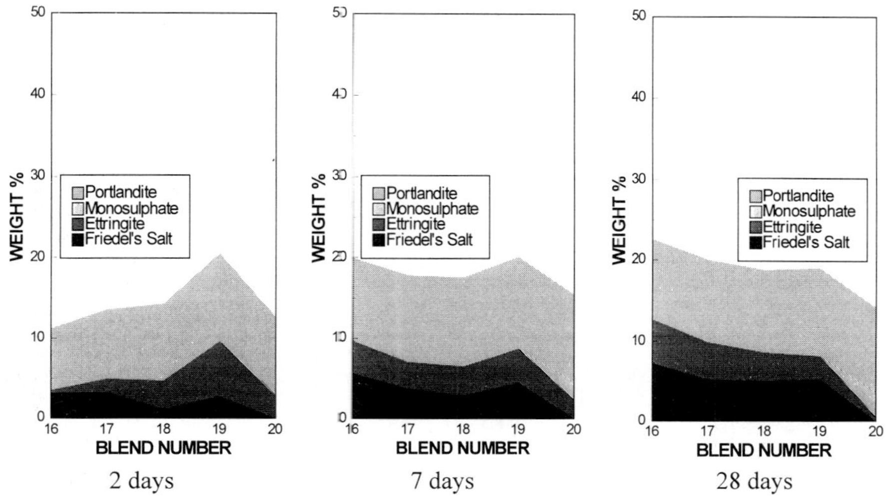

Figure 5 Crystalline hydration products at 2,7 and 28 days

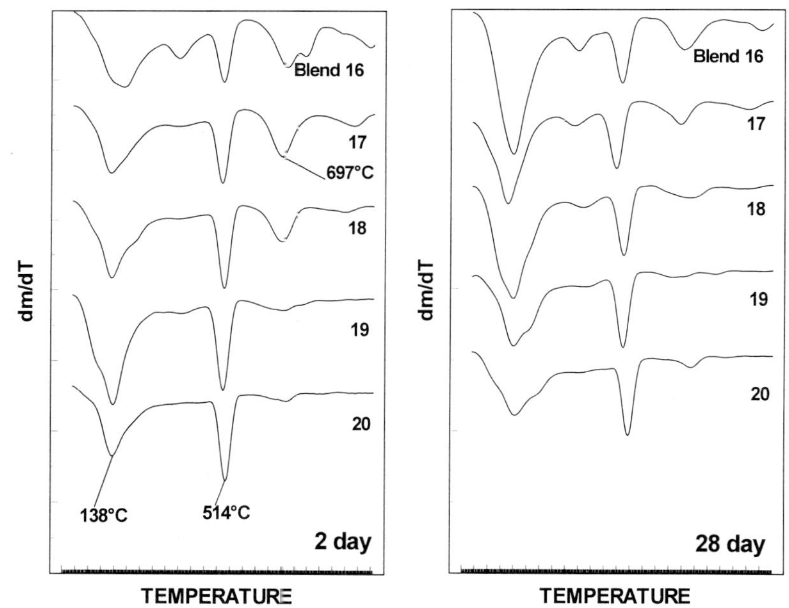

Figure 6 Differentiated thermogravimetry plots for blends 16 to 20 at 2 and 28 days

The quantities of anhydrous Portland cement phases in the blends are shown in Figure 7 at 0, 2, 7 and 28 days. It can be seen that at 2 days the quantity of compounds present has been greatly reduced for all mixes, with C_3A and C_3S reacting faster than the other phases, which is normal [16, 17]. In general there is little difference in the rate of reaction of PC phases for the different blends, although there is some indication of an acceleration of the reaction of the C_2S phase in blends containing IFA. Since the effect of IFA on the rate of reaction of the PC phases appears to be minor, it can be assumed that the improved compressive strength development observed in certain IFA blends is mainly due to enhanced interactions between IFA and PFA.

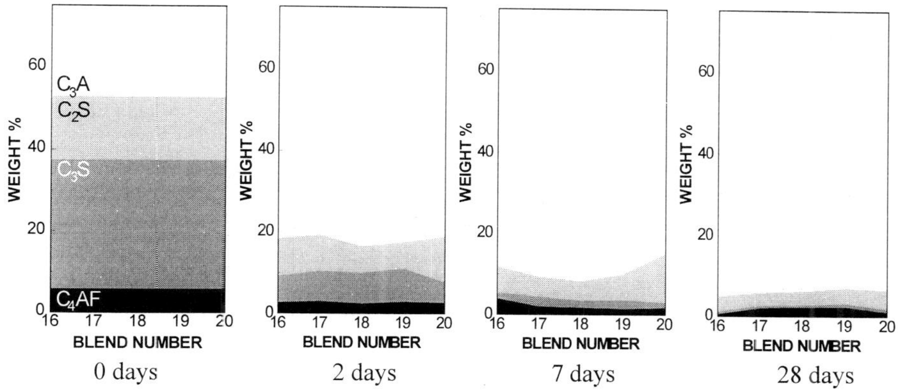

Figure 7 Cement compounds at 0,2,7 and 28 days

CONCLUSIONS

Incinerator fly ash (IFA) can be combined with Portland cement (PC) and Pulverized-fuel ash (PFA) to produce a binder which achieves higher compressive strengths than equivalent PC / PFA only mixes. This appears to be primarily due to the activation and acceleration of PFA reactions by IFA.
Ettringite and friedel's salt are key hydration products in hydrated IFA binders.

IFA usually contains environmentally sensitive trace metals in relatively enriched concentrations. Before fly ash can be considered as a recyclable material it will be essential to ensure long-term immobilisation of these species within the cementitious matrix. Leachate data from ash mixes investigated in this paper will be the subject of a subsequent paper.

IFA is unsuitable for concrete which contains reinforcement steel because of the high amounts of chlorides present, as this could cause corrosion problems in the lifetime of such concrete. However, other construction applications could tolerate these componenets, for example, the manufacture of paving slabs or other similar precast concrete products.

REFERENCES

1. KIELY, G., 'Environmental Engineering', 1st ed, McGraw-Hill, London, 1997, 979pp
2. TAY J.H. and CHEONG H.K., "Use of ash derived from refuse incineration as a partial replacement of cement," Cement and Concrete Composites, 1991, pp 171-175

3. ALI, M.T. and CHANG, W.F., "Strength properties of cement-stabilized municipal solid waste incinerator ash Masonry Units," ACI Materials Journal, Vol. 91, No. 3, 1993, pp 256-263

4. KIRBY, C.S., and RIMSTIDT, J.D., Mineralogy and surface properties of municipal solid waste ash, Environmental Science and Technology, Vol.27, 1993, pp652-660

5. BRITISH STANDARDS INSTITUTION, BS EN 196: Part 1, Methods of testing cement- Part 1: Determination of strength, 1995

6. BRITISH STANDARDS INSTITUTION, BS 1881: Part 124, Testing concrete. Methods for analysis of hardened concrete, 1988

7. BERGER, H., Study of the k-alpha emission-spectrum of copper, X-ray Spectrometry, Vol.15, 1986, pp241-243

8. CHEARY, R.W., and COELHO, A.A., A fundamental parameters approach of x-ray line-profile fitting, Journal of Applied Crystallography, Vol.25, 1992, pp109-121

9. CHEARY, R.W., and COELHO, A.A., Synthesising and fitting linear position-sensitive detector step-scanned line profiles, Journal of Applied Crystallography, Vol.27, 1994, pp673-681

10. CHEARY, R.W., and COELHO, A.A., Theoretical model for axial divergence with primary and secondary Soller slits in x-ray line-profile analysis, 1996, in preparation

11. MARQUARDT, D.W., Journal of the Society for Industrial and Applied Mathematics, Vol.11, 1963, pp431-331

12. NASH, J.C., Compact Numerical Methods for Computers, Adam Hilger, Bristol and New York, 1990

13. WILDING, C.R., WALTER, C.R., and Double D.D., A classification of inorganic and organic admixtures by conduction calorimetry, Cement and Concrete Research, Vol.14, 1984, pp185-194

14. KHALIL, S.M., WARD, M.A., Effect of sulfate content of cement upon heat evolution and slump loss of concretes high-range water reducers (superplasticizers), Magazine of Concrete Research, Vol.32, 1980, pp28-38

15. BIRNIN-YAURI, U.A., and GLASSER, F.P., Friedel's salt, $Ca_2Al(OH)_6(Cl,OH).2H_2O$: its solid solutions and their role in chloride binding, Cement and Concrete Research, Vol.28, 1998, pp1713-1723

16. HURLEY, F.R., and ANGSTADT, R.L., The early hydration of tricalcium silicate and □-dicalcium silicate in neat Portland cement pastes, Journal of Applied Chemistry, Vol.16, 1966, pp162-165

17. COLLEPARDI, M., MONOSI, S., MORICONI, G., and CORRADI, M., Tricalcium aluminoferrite hydration in the presence of lime and gypsum, Cement and Concrete Research, Vol.9, 1979, pp431-437

AN ELECTRICAL CONDUCTIVITY METHOD FOR MEASURING TRANSPORT PROPERTIES OF STABLIZED WASTES

A Caucheteux
E Garcia-Diaz
J M Siwak
Ecole des Mines de Douai
C Vernet
Laboratoire Central Lafarge
France

ABSTRACT. In this study, physical properties of municipal solid waste incineration fly ashes (mswifa), solidified with cementitious binder are evaluated. Compressive strengh and permeability are monitored. As for all cement based materials, the compressive strengh depends on the total porosity. The solidified waste presents several porous modes, measured by mercury intrusion porosimetry. Utilising a conductivity method and a conventional permeability method, we could identify the connected porous mode. In this way, the katz thompson theory, which relates permeability, conductivity and critical pore diameter, has been applied successfully. Contrary to compressive strengh and total porosity, there is no obvious relationship between compressive strengh and connected porous mode. Thus, in some cases, we obtained high porosity and small compressive strengh, but thanks to a small critical pore diameter, the permeability was good.

Keywords: Permeability, Conductivity, Mercury intrusion porosimetry, MSWIFA, Solidification / Stabilization, Compressive strengh, Critical pore diameter.

Antoine Caucheteux is in his first year of his doctoral studies. His thesis subject is about solidification / stabilisation of municipal solid waste incineration fly ashes with cementitious binder, and its long term efficiency.

Eric Garcia-Diaz is a research scientist teacher in the département Génie Civil of the Ecole des Mines de Douai. His research field concern chemical reaction in cement based materials.

Christian Vernet is a research senior manager in Lafarge company.

Jean Michel Siwak is responsible for the département Génie Civil of the Ecole des Mines de Douai.

INTRODUCTION

In spite of efforts made in wastes recycling, there still are some ultimate wastes. They are too polluting, so they have to be placed in dumping-ground, after being stabilized.

The most common used process to prevent pollutants to move throughout the environment, is solidification/stabilization with cementitious binder. The efficiency of the process is evaluated, in accordance with french law [1], according to a number of parameters, such as soluble content and compressive strengh.

These parameters do not permit to evaluate the long-term ability to immobilise the pollutant. A first approach to investigate this long-term ability consists in the evaluation of the transfer properties such as permeability. The usual methods are time consuming and difficult to perform. Thus we chose a new one which gives good results on concrete materials : electrical conductivity.

LITERATURE REVIEW

Solidification / Stabilization (s/s) [2]

We study two different Municipal Solid Waste Incineration Fly Ashes (MSWIFA), which come from two french municipal solid waste incinerator power plants. There are two processes to wash the flue gases which come from the incineration. The first one is a dry process : gases were cleaned over hydrated lime. The residue has a high content of lime ($Ca(OH)_2$) and CaClOH. There is a second process that is otherwise employed : the wet process. MSWIFA obtained by this way come from the electrofilter, and contain a lot of anhydrite ($CaSO_4$).

These residues are considered in France as ultimate residues. From this reason, they have to be solidified before their disposal in special landfills. The goal of the stabilization is to :

- improve physical properties of the waste,
- reduce the contact surface between solid and liquid phases,
- limit pollutant solubility and mobility,
- limit permeability of the final solidified waste.

One of the common used process is the solidification / stabilization with cementitious binder.

According to french law, the solidified waste ought to be massive (regarding to compressive strengh) and ought to present a soluble content lower than 10 % after 48 hours of leaching, with two replacements of water [NF X 31 - 210].

Conductivity and Permeability Measurements [3-4]

The permeability (k, m/s) of a material is define by Darcy's law :

$$J = -\frac{k}{\mu}\frac{\partial P}{\partial x} \tag{E1}$$

where J is the fluid velocity ($m^3.s^{-1}.m^{-2}$), μ is the fluid viscosity and P is the pressure.
For typical concrete, the average value of the permeability is of order 10^{-12} m/s [5].
Katz and Thompson (KT) [6] demonstrated in 1986, for rock formation, the following relationship with the conductivity :

$$k = c.d_c^{\ 2}.\frac{\sigma}{\sigma_0} \tag{E2}$$

where σ is the effective conductivity of the material, σ_0 is the conductivity of the bulk solution, c is a constant equal 1/226 and d_c is the critical pore diameter estimated by mercury intrusion porosimeter (MIP). It was defined as "the pore diameter above which a connected path can form from one side of the sample to the other" [7].

The Katz Thompson relationship has been tested successfully on cement based materials [7-8].

EXPERIMENTAL PLAN AND METHODS

We work on two different MSWIFA, one from each process. This leads to 6 different mixtures, three with each MSWIFA. From the dry process, the mixtures FD1, FD2, FD3, contain different amounts of cement (FD3 has the biggest rate, FD1 the lowest). FD2 contains more MSWIFA than the orthers. From the wet process, mixtures FW1, FW2 and FW3, have different water to solid (w/s) ratios (FW3 is the most wet).

The cylindrical samples of solidified wastes have been stored at 20°C under plastic film.

After different curing times, we measure :

- compressive strengh on 3 samples,
- porosity,
- transport properties.

We process MIP after a 50°C drying : it exhibits a porous distribution.

The total porosity we take into account is not the one determined by MIP. Indeed, at 50°C, not only the free water is dried, but some molecules which come from the hydration products. We carry out the drying of the grinded sample using acetone / ether. Next, we measure the absolute density of the powder using helium pycnometry. At last, knowing the humid density of the original sample, we calculate the porosity. This value seems to be closer to the real porosity.

Transport of a species through the sample occurs throughout the whole porosity. However, this is full of air, which does not let the current flow. Furthermore, to take into account the whole porosity in our measurements, it is necessary to carry out the water saturation of the sample. It is placed in humid atmosphere (a hermetic box with water, as shown in Figure 1), until a constant weight is reached.

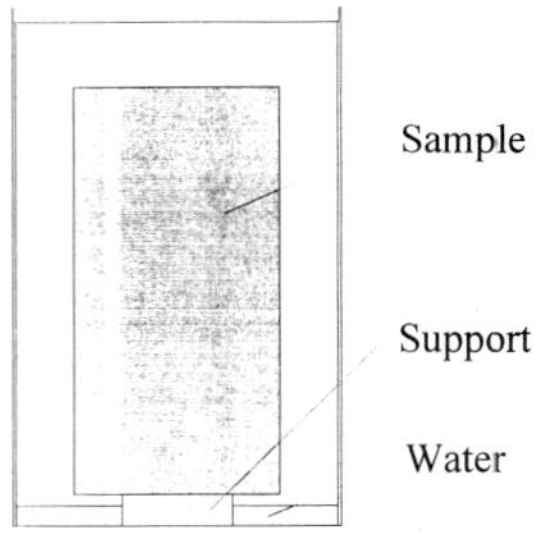

Figure 1 Sample saturation

Conductivity is measured by direct current. To take into account the electrode impedance, a multi-probe measurement is made. The sample is placed between two electrodes. The contact is achieved thanks to a conductive gel, as describe by Kyi and Batchelor [9] (figure 2).

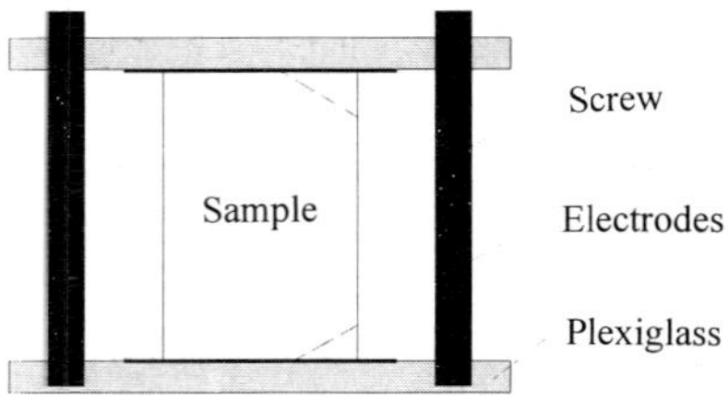

Figure 2 Schematic of the conductive cell used

The permeability is also evaluated by another methode : we measure water permeability with a triaxial cell (Figure 3) [10-11]. The sample is saturated in the same way as for the conductivity measurement. A pressure of 600 kPa is applied to the sample. The direction of flow is from the bottom to the top. We measure the water flow through the sample. Employing Darcy's law (E1), we can calculate the permeability.

This method also enables us to measure the pore fluid conductivity (σ_0). Indeed, the solution extracted in the first stage could be considered as the bulk solution.

RESULTS AND DISCUSSION

Compressive Strengh and Porosity

The results in the literature show, for concrete, a relationship between compressive strengh and porosity. In our case of solidified wastes, the relationship is not so well define (Figure 4). Indeed, for different mixtures, the materials are different and do not allow comparison with each others.

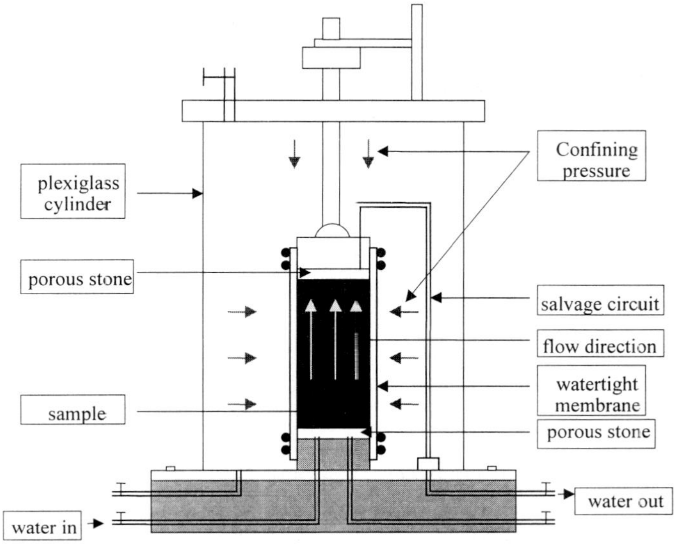

Figure 3 The permeability cell

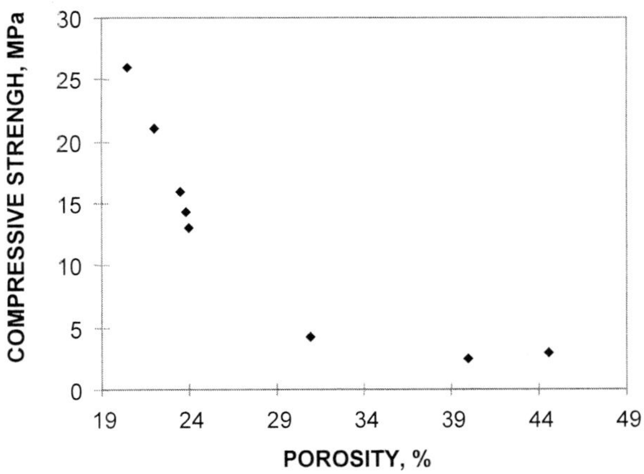

Figure 4 Correlation between compressive strengh and porosity

Usually, however, the lower the porosity is, the higher the compressive strengh is.

Evaluation of the Critical Pore Radius by MIP Measurement

The conductivity method permits determination of permeability (E2). With this aim, we have to dertermine the critical pore radius (r_c). The porous distribution of the material is determined by MIP. We obtained 3 principal sorts of diagrams.

The first one appears Figure 5. There are several porous mode :

- a first one has an average radius greater than 1 µm,
- a second one, wider, has smaller radius : between 0,01 and 0,3 µm.

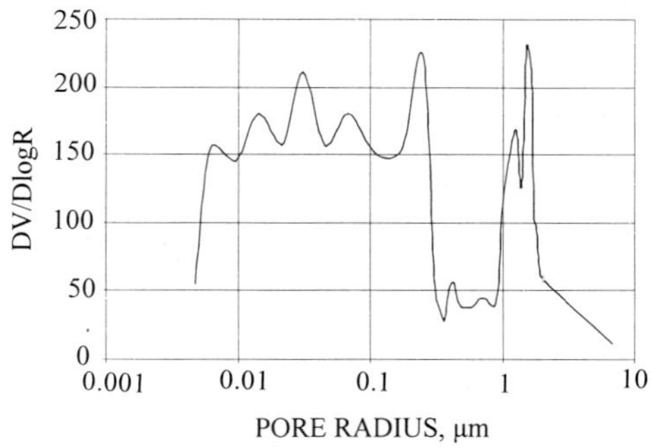

Figure 5 Determination of the critical pore radius from MIP curve : example 1

We have to choose the critical radius from these. Katz-Thompson describe the critical pore radius as the lergest radius which corresponds to the connected porosity. Thus we calculate the permeabillity (k_{KT}) with KT relationship, for different radii chosen from the MIP curve. We compared the results to the water flow permeability (k_{WF}) measurement (Table 1), for FD3 mixture.

Table 1 Influence of the critical pore diameter on the permeability by KT theory

r_c	K_{KT}	k_{WF}
1,52	1,4 10^{-8}	3 10^{-10}
0,25	3,9 10^{-10}	3 10^{-10}
0,006	2,2 10^{-13}	3 10^{-10}

The first porous mode (r_c about 1,5 µm) does not seem to be connected. The critical radius that give the best results is 0,25 µm, which corresponds to the right side of the second porous mode. This sort of result is given by FD1 and FD3 mixes.

For FD2, FW1 and FW3 mixes, the porous mode of 1 µm is far bigger than in the previous case (Figure 6). As we compare with the water flow method, we observe that it corresponds to the connected porosity. The critical radius is about 1 µm.

At last, for FW2, the chosen critical radius is about 0,1 µm (Figure 7).

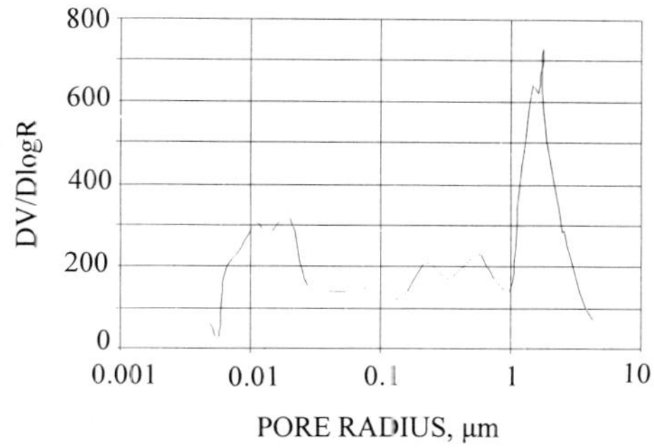

Figure 6 Determination of the critical pore radius from MIP curve : example 2

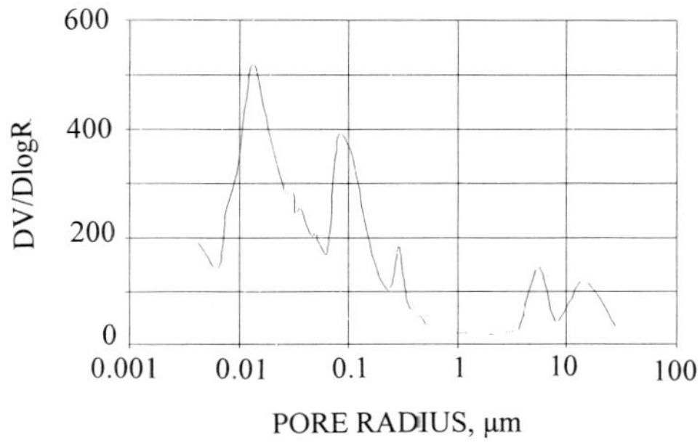

Figure 7 Determination of the critical pore radius from MIP curve : example 3

Permeability : Comparison with a Conventional Procedure

We carry out the measurement of permeability by the two methods. The first one is the KT method, using conductivity and MIP measurement. The second one is the water flow method. If we choose the appropriate critical radius (as described above), we have a good correlation between the two methods (Figure 8).

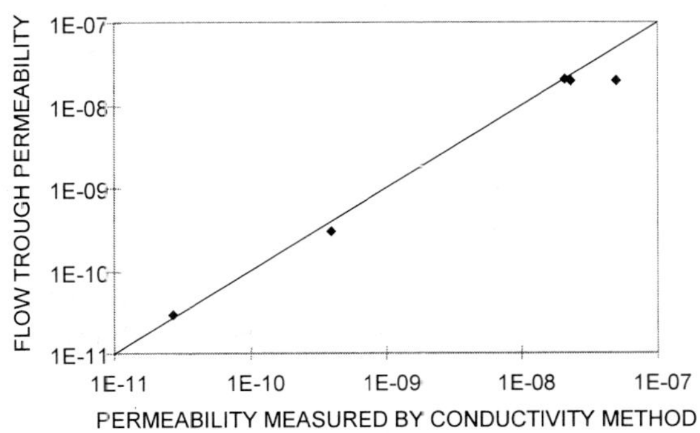

Figure 8 Comparison of permeability measured by conductivity method with permeability measured by water flow method

Compressive Strengh and Transport Properties

The evolution of the compressive strengh (CS) and the permeability (k) with the time have been observed for FD2 and FD3. For FD3, k is approximately constant with the time, as CS has nearly doubled (Table 2). FD2exhibits different behaviour. CS is constant, but k is one hundred times lower, between 28 and 90 days.

Table 2 Evolution of the compressive strengh and the permeability with the curing time

	CURING TIME (DAY)	COMPRESSIVE STRENGH (MPA)	PERMEABILITY (M/S)
FD3	28	16	$3{,}9\ 10^{-10}$
	90	20	$2{,}5\ 10^{-10}$
	360	26	$3{,}5\ 10^{-10}$
FD2	28	13	$5\ 10^{-8}$
	90	14	$4{,}5\ 10^{-10}$

Thus, we could have an increase in compressive strengh (*2) without any change in the permeability (FD3), or we could have a constant value for the compressive strengh with a reduction of the permeability (FD2).

Moreover, comparison of FD2 and FW2 at 28 days curing time gives very interresting results (Table 3).

Table 3 Comparison of FD2 with FW2

	FW2, 28 days	FD2, 28 days
compressive strengh	2,5 MPa	13 MPa
porosity	40 %	24 %
permeability (m/s)	$7{,}3\ 10^{-11}$	$5\ 10^{-8}$

Thus, with quite a high compressive strengh and small porosity (FD2), the sample has got very poor transport properties : $k = 5\ 10^{-8}$ m/s. But for FW2, the sample has a very low permeability ($7{,}3\ 10^{-11}$ m/s) with a very small compressive strengh (2,5 MPa) and large porosity (40 %). Indeed, even if there is a high porosity (which leads to small compressive strengh), a small connected porous mode guarantees the good transfer properties.

It follows that even if the compressive strengh appears to be a valid parameter to evaluate the massive aspect of the solidified waste, it is not a valid parameter to evaluate its tranfer properties.

CONCLUSION

If the measurement of compressive strengh and porosity, is to estimate the massive aspect of the solidified waste, it does not seem to be a valid parameter to determine the efficiency of a process in term of tranfer properties.

The evaluation of the permeability with conductivity measurement and Katz-Thompson relationship, give good results. This technique is less time consuming, easier to perform and more precise than classical methods.

Moreover, it allows determination of the critical pore diameter which seems to be an interesting gauge to evaluate the transfer properties.

However, transport properties are insufficient to provide conclusive evidence that the waste treatment is satisfactory long-term efficient. Indeed, all physico-chemical parameters, such as the efficiency of the chemical bond, are necessary to have the good idea of long-term efficiency of the process.

ACKNOWLEDGEMENTS

The authors would like to thank Christian Legendre, industrial realisation manager of the CIDEME company, for financial suport. CIDEME is a french firm specialised in waste management and concern itself with municipal solid waste collection and its subsequent incineration, recycling or stabilization.

REFERENCES

1. Journal Officiel de la République Française, « arrêté du 18 décembre 1992 relatif au stockage de certains déchets industriels spéciaux ultimes et stabilisés pour les installations nouvelles », Journal Officiel de la République Française, Paris 1992.
2. CAUCHETEUX, « stabilisation des REFIOM aux liants hydrauliques », rapport d'activité, Ecole des Mines de Douai, 1998, 48 p.
3. J.D. SHANE, J.H. HWANG, D. SOHN, T.O. MASON, H.M. JENNINGS, E.J. GARBOCZI, « Recent developments in the measurement of transport properties in cement based materials », Mechanism of Chemical Degradation of Cement Based Systems, proceding of the materials research society's symposium on mecanisms of chemical degradatyion of cement-based systems, november 27-30, 1995, Boston, MA, E & FN Spon, London, England, K.L. Scrivener , J.F. Young Editors, pp 413-423, 1997
4. E.J. GARBOCZI, « permeability, diffusivity, and microstructural parameters : a critical review », Cement and Concrete Research, 1990, vol 20, pp 591-601.
5. N. S. MARTYS, « Survey of concrete transport properties and their measurement », NISTIR 5592, 1995, february, 46 p.

6. A.J. KATZ AND A.H. THOMPSON, « a quantitative prediction of permeability in porous rock, Physical Review B, 1986, vol 34, december, pp 8179-8181.

7. A.S. EL-DIEB, R.D. HOOTON, « evaluation of the Katz-thompson model for estimating the water permeability of cement based materials from mercury intrusion porosimetry data », Cement and Concrete Research, 1994, vol 24 n°3, pp 443-455.

8. B.J. CHRISTENSEN, R.T. COVERDALE, R.A. OLSON, S.J. FORD, E.J. GARBOCZI, H.M. JENNINGS, T.O. MASON, « Impedance spectroscopy of hydrating cement based materials : measurement, interpretation, and application », Journal of the American Ceramic Society, 1994, vol 77, pp 2789-2804.

9. AYE AYE KYI, BILL BATCHELOR, « an electrical conductivity method for measuring the effects of additives on effective diffusivities in portland cement pastes », Cement and Concrete Research, 1994, vol 24 n°4, pp 752-764.

10. P-A. ANDREANI, « Rapport de stage de DEA Instrumentation et Analyses Avancées : Essais de percolixiviation sur déchets stabiliséss » UST Lille, 1998, 27p.

11. E. LE YAOUANQ, « Rapport de stage de DEA Instrumentation et Analyses Avancées Essais de percolixiviation sur déchets stabilisés » UST Lille, 1999, 30p.

CLOSING KEYNOTE PAPER

REUSE OF INCINERATOR ASH – CURRENT AND FUTURE TRENDS

K-Y Show

J-H Tay

H-K Cheong

Nanyang Technological University

Singapore

ABSTRACT. This paper presents a brief review of the development in the reuse of incinerator ash in civil engineering. Based on reported data and findings, several engineering and environmental factors to be considered in selecting, designing, and implementing incinerator ash utilization are pointed out. A review of current studies of incinerator ash reuse for various applications, such as road aggregates, fill material, admixture in soil stabilization, masonry bricks, cement replacement material and artificial reef, are also presented. Properties of the products derived from the ash, and their applications or potential use are examined. The issue of legitimate concerns of reusing the ash is addressed, and is categorized as engineering, environmental and economic concerns. The problems encountered or envisaged in reusing the ash, the quality and performance of the products derived from the ash, and the costs and environmental acceptability are discussed.

Keywords: Incinerator ash, Reuse, Civil engineering, Applications, Environment, Legitimate concerns.

Dr Kuan-Yeow Show is an Assistant Professor in the Division of Environmental & Water Resources Engineering, School of Civil & Structural Engineering, Nanyang Technological University, Singapore. His research interests include concrete materials and waste management, with special emphasis on wastes recycling & reuse.

Dr Joo-Hwa Tay is a Professor and Head of the Division of Environmental & Water Resources Engineering, and the Director of the Environmental Engineering Research Centre, School of Civil & Structural Engineering, Nanyang Technological University, Singapore. His research interests focus on the hazardous and industrial waste management, waste minimization, recycling and reuse.

Dr Hee-Kiat Cheong is a Professor and Dean of the School of Civil & Structural Engineering, Nanyang Technological University, Singapore. His research interests include use of wastes in cement/concrete, behaviour of concrete under multi-axial stress conditions, structural performance of concrete structures, behaviour of upgraded RC beams, and blast effects in soil and on piles

INTRODUCTION

With increasing environmental pressures, escalating energy and prices of raw materials, and higher waste disposal costs, it is critical that industries develop comprehensive waste management strategies for both current and future residual-generation streams. While such strategies are multi-faceted and obviously involve liquid and gaseous emissions, and energy accounting, the major concern here is municipal solid waste (MSW) incinerator ash utilization in civil engineering. Two types of ash are generated during incineration. Bottom ash is the relatively coarse unburned or partly combusted residue retained on furnace grates following combustion and solids that pass through the grates (siftings). Fly ash is normally referred to as entrained particulates that are trapped, and residues generated, by acid gas scrubbers and subsequently removed by fabric filters and/or electrostatic precipitators. Fly ash typically amounts to about 5 to 15 percent of the total ash (1).

The amount of MSW ash generated in the United States is about 2.8 to 5.5 million tons per year (2). This may increase two to five times depending on how many planned facilities are built. While the available quantities are large, much of the potential of utilizing MSW incinerator ash remains to be exploited. Although research and demonstration projects have indicated that the ash can be beneficially utilized, utilization of MSW residues is not commonly practised in most parts of the world. Some countries, including Denmark, the Netherlands, France, Germany, Switzerland, and Japan are further advanced in ash utilization and in the establishment of a systematic process for evaluating and selecting disposal and utilization options. This paper presents a brief review of civil engineering applications for incinerator ash. The major technical, economic and environmental concerns of incinerator ash utilization are discussed. The issue of legitimate concerns categorized as engineering, environmental and economic concerns of reusing the ash are addressed.

DEVELOPMENT OF INCINERATOR ASH UTILIZATION

Ash utilization is particularly attractive since it couples resource conservation with attenuation of disposal problems. Although extensive studies have indicated that the ash can be beneficially utilized, utilization of MSW residues is not routinely practised or mandated in most countries. In some European countries, 50% or more of the ash is utilized, while less than 5% of the ash is being utilized in the U.S. The following summarizes ash utilization in member countries of the International Ash Working Group (IAWG):

United States

Primary applications under consideration in the U.S. include the use of bottom or fly ash as a substitute in Portland cement, as bituminous pavement, as sub-base in roads, as fill material under contained conditions such as covered embankments, and as daily cover in landfills (3). Recently there has been increased interest in marine applications, such as inshore erosion control and artificial reefs construction.

Germany

Approximately 50% of bottom ash is utilized primarily as road bases and sound barriers on autobahns. There are also pilot-scale test being conducted to evaluate fly ash as grout in coal mines. Another consideration involves the use of fly ash to form alinite cement.

The Netherlands

The goal in The Netherlands is to utilize at least 80% of the residues. Currently 40% of the fly ash produced is used as a fine aggregate in asphalt. More than 2 million tons of bottom ash (approximately 60%) have been used as road base, embankments, aggregate in concrete, and aggregate in asphaltic concrete. To reach this goal, however, The Netherlands has established a policy and regulatory framework which establishes standard leaching tests, composition requirements, evaluation protocols for building materials, and certification of residues for utilization.

Denmark

Utilization of bottom ash was initiated in Denmark in 1974. Approximately 72% of the bottom ash, including grate siftings and heat recovery ash, is used as sub-base at parking lots, bicycle paths, and paved and unpaved roads. Requirements for ash composition for utilization are pH > 9.0 in a 1% slurry, alkalinity greater than 1.5 eq/kg, Pb content below 3,000 mg/kg, and Hg content less than 0.5 mg/kg, as determined by HNO_3 digestion. In addition, there are restrictions which include a 20 m minimum distance to water wells, 1-2 m maximum thickness under a pavement, and a maximum of 200 m^3 and 0.30 m maximum thickness for unpaved applications.

Sweden

Sweden has adopted a policy that utilization should improve the general environmental conditions and create a smaller environmental impact than disposal. Regulations for MSW ash utilization are currently being developed.

Based on technical data and findings by the IAWG (4), the following factors should be considered in selecting, designing, and implementing incinerator ash utilization:

- Bottom ash can be effectively utilized as an aggregate substitute in several applications if appropriate engineering (structural) criteria and environmental performance guidelines are met. Large-scale ash-utilization projects processed as structural soil in Europe include its use in land reclamation from the ocean, wind barriers, sound barriers, road sub-bases, and parking lot and bike path bases. Demonstrations in the U.S. have used bottom ash as aggregate substitutes in bituminous pavements, sometimes at high substitution rates. In other demonstrations, bottom ash has been used as an aggregate substitute in Portland cement applications for marine reefs, shore protection devices, and service buildings.
- The high alkalinity caused by the addition of cement in some applications influences the release of some metals from the ash in a negative manner. This must be considered in the design.

- Use of bottom ash as structural fill in large applications such as embankments may lead to high salt loads and relatively high leachability of salts. Pretreatment to remove soluble salts may be necessary in some applications.
- Bottom ash should be stockpiled prior to utilization for an estimated one to three months with adequate moisture, so swelling, hydration, carbonation, and oxidation reactions can occur. This benefits structural durability and chemical stability. Appropriate methods should be used to control and manage any leachate and/or runoff during stockpiling.
- The design, construction, and implementation of any ash utilization project should be based on a complete understanding and knowledge of the ash characteristics, the ash product behaviour in the situation, environmental conditions, physical requirements, regulatory requirements, and similar factors. This will necessitate appropriate testing and evaluation protocols, including tests to determine elemental concentration, availability of trace metals and total soluble salts for release, estimates of contaminant release over time and under field conditions, physical durability, and similar factors. This cannot be done by relying on a single test such as the Toxicity Characteristics Leaching Procedure (TCLP).
- The acceptability of a product for utilization should be based on the amount of anticipated contaminant release (assuming that physical and other specifications are met) over the expected product lifetime. This should take into consideration both contaminant release potential and release rates. The acceptable level of contaminant release should be defined for each utilization – this can be based on release rates, the potential of similar products from natural materials, and environmental considerations.
- The time frame of intended use should be considered and an analysis conducted on the ultimate fate of the material following completion of the specified use period. Many of the utilization scenarios for bottom ash have release rates or release potentials similar to natural materials.
- Granular material (e.g., roadbase) will behave differently compared to monolithic applications and require different testing protocols (e.g., compacted granular leach test). Monolithic ash products should be evaluated using appropriate monolithic leach test (e.g., tank leaching) and also for durability.
- Appropriate quality control and quality assurance procedures should be developed and implemented to maintain ash product quality.

There are a number of other factors to consider, such as distance to sensitive natural areas, local requirements, monitoring during the product use and similar factors.

INCINERATOR ASH TREATMENT AND REUSE

Once incinerator ash is collected, other management options besides landfilling exists. Ferrous and non-ferrous metals in the ash are recovered at some facilities by using screens, magnets, and other mechanical processes. Incinerator ash can be treated chemically or thermally to decrease the likelihood of leaching. Untreated ash can be stabilized or solidified and then used in different applications.

Treatment Options

For utilization purposes, there are several options available for treating MSW combustion residues. The treatment options include:

Compaction

The physical properties of the residues are such that compaction at optimum moisture content can be effective in significantly reducing the permeability of the residues for utilization options. This reduces infiltration of water and, therefore, the potential for release of contaminants.

Classification

Screening or separation of the ash into oversized and under-sized fractions can be beneficial and is, in some cases, necessary for utilization purposes. Screening coarser material from bottom ash improves the characteristics as a coarse aggregate. From a physical perspective, fly ash can be screened to meet specifications for fine cement aggregate. The size distribution of the fly ash may also make it suitable for use as a soil aggregate for paving applications.

Solidification

Ash can be solidified by adding Portland cement or lime and dampening the mixture so that a concrete-like product forms. In theory, this immobilizes metals and inhibits leaching, allowing the blocks to be used for different civil engineering applications.

Vitrification and other thermal processes

Vitrification results in the melting of the ash with glass forming additives to incorporate the contaminants into an alumina-silicate matrix. The process can substantially reduce the volume of materials by 60% or more and usually results in a product more resistant to leaching suitable for landfilling or reuse as aggregate. Heating of the residues can also cause sintering, mineral respeciation or melting to "fuse" the material into a slag.

Beneficial Ash Reuse

There has been significant interest in evaluation of ash reuse alternatives including a number of research and demonstration projects. A brief review of some of the applications is discussed.

Road aggregates

Reusing ash as a road aggregate was suggested decades ago. Bottom ash is used as road aggregate in several European countries (5, 6, 7, 8), usually under some kind of guidelines. Denmark's rules for reusing ash residues in road construction, for example, require that the ash contain less than 25 percent fly ash and that reuse does not occur less than 20 metres from drinking water wells (8). Over two dozen U.S. companies have expressed interest in using solidified ash in road beds or concrete construction (9).

The Franklin Institute Research Laboratory in Philadelphia developed a process of densifying (fusing) the unfused incinerator residue and converting it into useful construction materials (10). The incinerator residue was ground with a hammer mill, preheated at 688°C, and then melted and fused at 1093°C to form a column of semi-molten product which was later cooled and crushed to the required aggregate sizes. The preliminary laboratory evaluation indicated that the fused incinerator residue was not suitable for hot-mix asphalt mixtures because it did not comply with the specified gradation, and the mixtures did not meet design criteria for stability, air voids, or voids in the mineral aggregate. The incinerator residue is not suitable for use in hot-mix asphalt mixtures unless it is processed further. Subsequent improvements in the process, however, produced a high quality aggregate material which performed well in a hot-mix asphalt surface course installation. The following recommendations have been made from the study: (a) incinerator residues should be well-burned out with loss-on ignition of less than 10%, (b) hot-mix asphalt mixtures for base courses containing 50% natural aggregate and 50% incinerator residue hold the most promise, and (c) 2% lime should be added to minimize stripping problems.

Incinerator ash was also analyzed for potential application as an aggregate for asphalt concrete (11). The physical and chemical properties of the 'washed' bottom ash passing the 4.75 mm sieve indicated potential utilization in bituminous surfacing. Marshall specimens prepared with incinerator residue-asphalt mixtures produced Marshall stability and flow values acceptable both to the Public Works Department of Singapore and ASSHTO criteria. Adding granite aggregate as coarse materials into the residue-asphalt mixtures helped to improve Marshall stability and reduce air voids. A reduction in asphalt content for optimum coating from 11.5% to 12% by weight and a marked improvement in the physical properties of this mixture were also observed when 1% hydrated lime was added. From the results, potential applications are possible in areas where high quality aggregate is not required, such as residential areas or secondary roads limited to light commercial vehicles.

Little research has been conducted, however, on the long term fate of metals in road aggregate and on concentrations of leachate compared with other sources such as surface runoff of oils, greases, and lead. Some initial research had been conducted in the U.S. (12). A street was paved with an asphaltic concrete aggregate containing up to 15% incinerator ash treated with industrial reagents. The leaching characteristics of metals from the treated incinerator ash, asphaltic aggregates and the pavement were examined. The results are promising, but because test data were limited to about 1 year, additional research is needed.

Fill material and admixture in soft soil stabilization

Potential use of incinerator residues in geotechnical applications was looked into (13). Both the washed bottom ash and fly ash were analyzed for use as replacement materials for conventional fill materials, and as an admixture in the stabilization of soft soil. The lower density of the incinerator residues is advantageous for construction of embankments and fills on soft, compressible soils in place of conventional fill materials, as the smaller loads imposed on the soft soils from using the lighter incinerator residues would result in smaller ground settlements. The relatively high strength parameters of the residues indicate that the materials would be adequate for bearing capacity and stability considerations. The high permeabilities of the incinerator residues, which are of the same order of magnitude as sand, are also advantageous as consolidation is sped up so that a major portion of the foundation settlement would occur during construction.

The soil stabilization tests that were carried out by mixing fly ash with the soft marine clay showed promising results. After curing, the treated clay samples showed improved shear strengths, a reduction in plasticity, lower compression indexes, higher coefficients of consolidation and higher permeabilities in comparison to the untreated clay samples. The higher shear strengths indicate that the fly ash can be used in soil stabilization in place of lime or cement. Further gains in strength would be expected if the fly ash is used in combination with lime or cement. The lower compression indices would result in smaller settlements. In addition, the higher coefficients of consolidation and permeability in relation to the untreated clay are advantageous as they imply faster rates of foundation settlement.

Masonry bricks

A research was carried out to determine the strength and durability of bricks made with municipal solid waste incinerator fine ash, using cement and limestone screenings (14). The ash was obtained from a waste-to-energy facility as a combined ash. Different cement contents ranging from 4% to 10% were used. The results confirmed that bricks with high compressive strength and modulus of rupture can be produced under static compaction pressure of 41.4 and 82.7 MPa using nominal amounts of Portland cement. The soaked compressive strength and modulus of rupture were also improved by adding small amounts of cement. A 28-day air-dry compressive strength as high as 51.7 MPa was obtained at 40% ash and 10% cement content. Modulus of rupture above 5.17 MPa was also achieved in this investigation. In addition, the bricks showed excellent resistance to abrasion and a lower water absorption.

Ash-based concrete

A study was carried out to examine the compressive strength properties of MSW fly ash-based concrete from refuse-derived fuel (RDF) and mass-burn plants (15). Mass-burn fly ash-based concrete required more water than RDF fly ash-based concrete to maintain the same workability. The setting times of the concrete were increased by cement replacement of the ashes. The concrete achieved a higher strength with up to 30% replacement of cement with RDF fly ash. The final strengths of up to 34.5 MPa were reported to exceed those of the control concrete. However, the rate of strength gain was slower at early ages, and the strength decreased with an increased amount of MSW fly ash.

Artificial reefs construction

Research studies had been conducted in examining the feasibility of using stabilized incinerator ash, in the form of blocks manufactured from crushed combined ash and Portland cement, for artificial reef construction (16, 17, 18, 19). Blocks were placed in two structures in the marine waters of Long Island Sound. After 1 year of submersion, the compressive strength of the ash blocks was unchanged and was well above the standard for marine disposal of stabilized products. In comparison, the strength of cement blocks declined by almost 30% in the same period. Tests were conducted on the submerged blocks to see whether they retained metallic components. Of 13 metals tested, the only significant differences after 380 days of submersion were an increase in magnesium and a decrease in potassium. Metals such as lead, chromium, copper and cadmium were retained within the block. The retention was attributed to the high alkalinity of the ash, the Portland cement additive, and alkalinity of the seawater – all of these favour formation of less soluble metal compounds and adsorption of metal ions (17).

CONCERNS OF INCINERATOR ASH REUSE

Over the past several years, there has been significant controversy concerning the proper management of the residues from combusting municipal solid waste (MSW) and their regulatory classification as hazardous or non-hazardous waste. This controversy and other factors such as a lack of guidance, heavy metal content and leaching of these metals are most often cited as the reasons the material should be managed as a hazardous waste. If not managed properly, contamination of groundwater by leaching of soluble salts from the ash may also be a concern. Legislation has been enacted in recent years to either mandate the use of the ash or to examine the feasibility of such usage. This has raised the following legitimate concerns:

Engineering Concerns

The following concerns must be addressed from the engineering viewpoint:

Properties of products incorporating incinerator ash

Since the incinerator ash will replace or modify the properties of the products or structures, it will affect the engineering properties (such as strength and durability) following the ash incorporation. Therefore, the materials containing the ash must be re-evaluated thoroughly and carefully both in the laboratory and the field. The long term physical integrity and behaviour of stabilized ash products need to be assessed.

Impact of production

Incorporation of the ash may require a change in the material production equipment and/or processes and, therefore, the production is likely to be affected.

Variability in quality

It is important to characterize the properties of incinerator ash and evaluate their suitability before utilizing them in civil engineering applications. A broad range of organic and inorganic compounds is found in incinerator ash. The ash, however, is difficult to be characterized in terms of its composition, because it reflects: (1) different types and sizes of facilities; (2) unknown sample sizes at each facility; (3) inter-laboratory variation in testing procedures (even using the same test); and (4) variable inputs into the ash itself (e.g., heterogeneous MSW). The major engineering constraints in having the ash widely accepted for use in civil engineering are related to their inherent variability and the need to meet applicable agency standards and specifications. The incinerator ash may vary in quality (chemical or physical properties) and, thus affect the engineering properties of the products.

Environmental Concerns

The following concerns must be dealt with from environmental and health viewpoints:

Leaching

Components of incinerator ash may be susceptible to leaching and thus cause storm water and/or ground water contamination.

Handling and processing procedure

Handling and processing procedures are important but these procedures are not fully developed for incinerator ash. The ash could be hazardous and could pose health related risks to workers in handling the material.

Long term effects

The long term effects of incorporating incinerator ash is not fully known and need to be assessed. Long term environmental testing needs to be conducted before full-scale ash reuse is allowed to protect human health and environment.

Economic Concerns

From an economic viewpoint, the following concerns must also be addressed:

Price

There are no standard specifications being laid to measure the quality of incinerator ash. Requiring incinerator ash to meet the strict specifications of its competing materials, such as Portland cement, may create the perception that there is inconsistent ash quality and thus quantity, and that prices for "spec" ash are higher than the cost of competing materials. In reality and if allowed or mandated by environmental authorities, there is a vast supply of incinerator ash available at reasonable or subsidized rates from the utilities. However, mandated use of incinerator ash is likely to raise the price of the products. For example, virgin aggregates are generally available locally at lower costs compared to the ash which must be hauled.

Quantity

The quantity available at one location must be large enough to justify the development of handling, processing and transportation systems. The transportation distances involved must be reasonable in terms of competition with conventional materials.

Life cycle costs

Life cycle costs need to be evaluated for the materials containing incinerator ash. It is quite possible that the ash may not affect the initial engineering properties of the products but may reduce its service life.

Disposal costs

If the materials containing incinerator ash cannot be recycled, then its disposal costs need to be determined.

Salvage value

Salvage values of the products containing incinerator ash are not known.

Lack of incentive

At the present time there is a lack of monetary incentives to use incinerator ash in civil engineering.

In summary, there are two general themes that are prominent when viewing the concerns discussed above. The first is that incinerator ash is perceived as being non-competitive with natural materials, most likely because of the perception that there is inconsistent ash quality and quantity. The second theme is that there is a 'knowledge deficiency' associated with ash reuse, such as a lack of data on environmental and health effects, lack of physical/engineering data, and a lack of familiarity with potential ash applications. The lack of standard specifications for using incinerator ash and the lack of regulations and guidelines from environmental authorities addressing beneficial use are most likely directly related to the general 'knowledge deficiency' theme that has been identified. Overcoming this constraint through education could lead to the development of regulations by environmental authorities that specifically address and allow the use of incinerator ash in civil engineering applications. The educational efforts should be directed at legislators, regulators, designers, and contractors. Once the regulatory issue is resolved, development of specifications that readily allow, or do not discriminate against, the use of incinerator ash could then follow.

CONCLUDING REMARKS

Reuse of incinerator ash in civil engineering applications should be encouraged. Civil engineering structures containing incinerator ash are likely to perform as well or better than conventional structures. It should also be environmentally safe both for the initial construction and its service life. However, it is necessary to address the engineering, environmental and economic concerns mentioned in this paper before large scale use of the ash is considered. The use of incinerator ash should only be mandated after regulations are developed to address these questions. More research should be initiated to provide in-depth studies exploring distinct, identifiable advantages if incinerator ash is to be effectively used in civil engineering applications.

REFERENCES

1. HAHN, J . L. AND SUSSMAN, D. B. Municipal waste combustion ash: testing methods, constituents and potential risks, Resource Recovery 2(5):16-18, December 1988.

2. GOVERNMENTAL ADVISORY ASSOCIATES, INC., Resource recovery yearbook, New York, 1986.

3. CARTON, C.W. Municipal solid waste combustion ash: state-of-the-knowledge, Municipal Solid Wastes- Problems and Solutions. Ed. Landreth, R. E. L. and Rebers, P. A. CRC Press, Lewis Publishers, Boca Raton, 1997, 276.

4. THE INTERNATIONAL ASH WORKING GROUP (IAWG). An international perspective on municipal waste incineration residue characterization and management, International Conference on Environmental Implications of Construction Materials and Technology Developments, Maastricht, The Netherlands, 1-3 June 1994.

5. GORTZ, W. Problems in evaluating the risk of water contamination by solids from incinerator residues, translation, Gewaesserschutz, Wasser 99:144-169, 1987.

6. HARTLEN, J. Incinerator ash utilization in some countries in Europe, Paper presented at 1988 Conference on Ash Utilization, Philadelphia, 1988.

7. MODIG, S. Swedish view of the ash issue, Paper presented at An Environmentally Acceptable solution to the Solid Waste Problem, Symposium sponsored by Swedish Trade Council and National Resources Recovery Association, Washington, D C., March 23, 1988.

8. CENTER FOR RISK MANAGEMENT. Resource for the future, Management of Municipal Waste Combustor Ash, Washington, D. C., 1989.

9. ENVIRONMENT REPORTER. EPA to develop ash management strategy, present it for ASB review within two months, Environmental Reporter, 18(53), 2548-2549, April 29, 1988.

10. MILLER, R. H. AND COLLINS, R. J. Waste materials as potential replacements for highway aggregates, NCHRP Report 166, Transportation Research Board, Philadelphia, 1976.

11. LUM, K. M. AND TAY, J. H. A study on the utilization of incinerator residue for asphalt concrete, Utilization of Waste Materials in Civil Engineering Construction. Ed. Inyang, H. I. and Bergeson, K. L.., ASCE, New York, 1992.

12. WASTE MANAGEMENT ENERGY SYSTEMS, INC. McKay Bay refuse-to-energy facility, Tampa, Florida, McKaynite Asphalt Test Report, Tampa, Florida, December 1987.

13. TAY, J. H. AND GOH, A. T. C. Engineering properties of incinerator residue, Journal of Environmental Engineering, American Society of Civil Engineers, 1997, 117(2), 224-235.

14. MUJAHID, T. A. AND WEN, F.C. Strength properties of cement-stabilized municipal solid waste incinerator ash masonry bricks, ACI Materials Journal, 1993, 91(3), 256-263.

15. HAMERNIK, J. D. AND FRANTZ, G. C. Compressive strength of municipal solid waste fly ash concrete. ACI Materials Journal, 88(5), 1991, 508-517.

16. ROETHEL, F. J. Ash disposal solution is 2000 year old, Waste Age, February 1987, 66-69.

17. ROETHEL, F. J. AND BRESLIN, V. T. Unique method of ash disposal can benefit marine life, Solid Waste & Power, 2(5), October 1988, 42-48.

18. ROETHEL, F. J. AND BRESLIN, V. T. Stabilized incinerator ash tested in construction of artificial reef, Waste Management Research Report, 1(2), Spring 1989, 3-8.

19. ROETHEL, F. J., SCHAEPERKOETTER, V. GREGG, R. AND PARK, K. The fixation of incineration residues, Final Report, prepared for New York State Legislative Commission on the Water Resource Needs of Long Island, 1986.

INDEX OF AUTHORS

SUBJECT INDEX

This index has been compiled from the keywords assigned to the papers, edited and extended as appropriate. The page references are to the first page of the relevant paper.